HANDBOOK OF DREDGING ENGINEERING

HANDBOOK OF DREDGING ENGINEERING

John B. Herbich, Ph.D., P.E.

W. H. Bauer Professor of Dredging Engineering
Ocean Engineering Program
Department of Civil Engineering
Texas A&M University

McGRAW-HILL, INC.

New York St. Louis San Francisco Auckland Bogotá
Caracas Lisbon London Madrid Mexico Milan
Montreal New Delhi Paris San Juan São Paulo
Singapore Sydney Tokyo Toronto

Library of Congress Cataloging-in-Publication Data

Handbook of dredging engineering / [edited by] John B. Herbich.
     p.   cm.
   Includes bibliographical references and index.
   ISBN 0-07-028360-5
   1. Dredging. I. Herbich, John B.
TC187.H33   1992   91-39282
627'.73—dc20   CIP

1 2 3 4 5 6 7 8 9 0 DOC/DOC 9 8 7 6 5 4 3 2

ISBN   0-07-028360-5

The sponsoring editor for this book was Joel Stein, the editing supervisor was Nancy Young, and the production supervisor was Pamela A. Pelton. This book was set in Times Roman by McGraw-Hill's Professional Book Group composition unit.

Printed and bound by R. R. Donnelley & Sons Company.

To my wife, Margaret Pauline,
and children,
Ann, Barbara, Gregory, and Patricia,
who have been a source of encouragement
and inspiration during the preparation
of this book.

CONTENTS

CONTRIBUTORS

Jones, R. Anne   *Vice President, G. Fred Lee & Associates, El Macero, CA (CHAP. 9, "Water Quality Aspects of Dredging and Dredged Sediment Disposal")*

Landin, Mary C.   *Research Biologist, U.S. Army Engineer Waterways Experiment Station, Vicksburg, MS (CHAP. 9, "Need, Construction, and Management of Dredged Material Islands for Wildlife")*

Lee, G. Fred   *President, G. Fred Lee & Associates, El Macero, CA (CHAP. 9, "Water Quality Aspects of Dredging and Dredged Sediment Disposal")*

Palermo, Michael R.   *Research Civil Engineer, Environmental Engineering Division, Environmental Laboratory, U.S. Army Engineer Waterways Experiment Station, Vicksburg, MS (CHAP. 8, "Long-Term Storage Capacity of Confined Disposal Facilities")*

Pankow, Virginia R.   *Hydraulics Laboratory, U.S. Army Engineers Water Resources Support Center, Fort Belvoir, VA (CHAP. 10, "Laboratory Evaluation of Production Meter Components")*

Sanderson, W. H.   *Dredging Consultant, The Sand Hen Corporation, Wilmington, NC (CHAP. 9, "Dredging Contracts")*

Schiller, R. E., Jr.   *Professor Emeritus of Ocean and Civil Engineering, Texas A & M University, College Station, TX, life member, A.S.C.E., N.S.P.E., and A.S.E.E. (CHAP. 6, "Sediment Transport in Pipes")*

Venezian, Giulio   *Associate Professor, Physics Department, Southeast Missouri State University, Cape Girardeau, MO (CHAP. 7, "Effect of a Ladder Pump on the Cavitation Characteristics of a Cutterhead Dredge")*

Weeks, Colin G.   *President, Hydrographic Associates, Inc. (CHAP. 10, "Automation, Survey, and Profitable Dredging")*

Wright, Thomas D.   *Environmental Laboratory, U.S. Army Engineer Waterways Experiment Station, Vicksburg, MS (CHAP. 9, "Evaluation of Dredged Material for Open-Water Disposal: Numerical Criteria or Effects-Based?")*

FOREWORD

Domestic and international maritime trade are dependent upon suitable access to the seas through navigation channels which must be improved and maintained by dredging, often on an annual basis. Waterways are vital to the flow of commerce because they offer the most economical and energy efficient transportation mode for bulk cargoes, such as coal, grain, petroleum products, chemicals, iron ore, and steel. The navigation channels of the United States exceed 25,000 linear miles and serve over 400 ports, which handle in excess of 2 billion tons of commerce each year. The waterways serve 130 of the nation's 150 largest cities.

In spite of its importance to our national economy and well-being, dredging is probably the least understood element of the construction industry. This is not surprising, particularly in the navigation field, because the material is excavated from the bottoms of waterways, and the operations are often located in open-water areas inaccessible to the public. Another factor contributing to the limited national and international awareness of the importance of dredging technology is the small volume of literature which was available on the subject for many years. Fortunately, the number of publications has increased many fold during the past 25 years, including a wide variety of papers related to the protection and enhancement of the environment.

In my view, the significant increase in the number of publications is due to the creation of the nonprofit, nonpolitical World Organization of Dredging Associations in 1967, the evolving public concern for the protection and enhancement of the environment, and the annual series of Texas A & M Dredging Seminars arranged and conducted by Dr. John Herbich during the past 25 years.

The *Handbook of Dredging Engineering* is another major contribution to the engineering and scientific literature on dredging technology by Dr. John B. Herbich. I am quite sure the Handbook will prove to be of significant and long-term value to the Texas A & M curriculum, which far exceeds that of any university in the United States on this subject, and it will also be used extensively by the international management and engineering officials engaged in all aspects of dredging technology. Dr. Herbich has the unique distinction of being a major contributor to the advancement of the profession in both the applied and theoretical fields.

William R. Murden, NAE, P.E.
President, Murden Marine Ltd.

PREFACE

This book has gradually evolved from courses developed at Lehigh University and Texas A & M University. The courses taught were at a senior and graduate level. At the 1968 World Dredging Conference, John Huston said, "On premise that a profession is known by its literature, dredging might well be eliminated. Its literature is almost nil." However, tremendous progress has been made in the last 24 years. This is substantiated by the fact that it has been necessary to update many chapters of the book since the 1975 edition, *Coastal & Deep Ocean Dredging* was published. In addition, several new chapters have been added that deal with the disposal of dredged material and the environmental aspects of dredging.

Since very few books on dredging exist in English, one of the main purposes for writing this book was to collect all the available relevant information under one cover for easy reference. The book was written for the use of graduate students in coastal engineering, for practicing coastal and ocean engineers, and, above all for the engineering and management personnel of dredging companies and manufacturers. It was also written for the personnel of state and federal regulating agencies involved in establishing and enforcing environmental standards. Better understanding of dredging methods and processes will assist in establishing and enforcing realistic standards for the protection of the environment without excessively increasing the cost of dredging.

Some dredging methods and practices are still in the "art" rather than "science" category of engineering. It is hoped this book will contribute toward dredging's becoming more of a science and less of an art.

ACKNOWLEDGMENTS

Dr. John B. Herbich's interest in coastal engineering was initiated at the Technical University of Delft and followed by research and studies at the Saint Anthony Falls Hydraulic Laboratory, University of Minnesota. While on the faculty of Fritz Engineering Laboratory, Lehigh University, Dr. Herbich conducted research on dredge pumps, dredging systems and dredging technology. Portions of Chap. 3 are the results of that work.

In 1967, Herbich joined Texas A & M University to head the coastal, hydraulic and ocean engineering program as a professor of civil and ocean engineering.

In 1968, a Center for Dredging Studies was established at the university for the purpose of conducting basic and applied research and for dissemination of information.

The Center, supported by the dredging industry, affords involvement with "real world" problems confronted by dredging operators and manufacturers. Additional research in dredging technology has been conducted in recent years, resulting in better understanding of dredge pump cavitation and seaworthiness of a cutterhead dredge.

Since 1968, professional development short courses have been conducted on an annual basis at Texas A & M University's Center for Dredging Studies. The 21st Annual Short Course on Dredging Engineering was held in January 1992. Many of the short course lecturers contributed to this Handbook. The following is a list of contributors to this Handbook: Dr. Robert E. Schiller, Jr., Dr. Giulio Venezian, Dr. Michael R. Palermo, Dr. G. Fred Lee, Dr. R. Anne Jones, Dr. Thomas D. Wright, Dr. Mary C. Landin, Mr. Colin G. Weeks, Ms. Virginia R. Pankow, and Mr. W. H. Sanderson. Their contributions are most appreciated.

John B. Herbich

HANDBOOK OF
DREDGING
ENGINEERING

CHAPTER 1
INTRODUCTION

Dredging is an ancient art but a relatively new science. Although work on primitive dredging can be traced back for several thousand years, it is only relatively recent that the art has been transformed into a science covering the design of dredges and dredging techniques.

Dredging may be defined as raising material from the bottom of a water-covered area to the surface and pumping it over some distance. This, however, covers a wide range of activities[1] from deepening of drainage canals to a very complex technique of dredging for marine minerals offshore.

The art of dredging began along the Nile, Euphrates, Tigris, and Indus Rivers many thousands of years ago as described by Gower.[2] There are many references in history of canal dredging in Sumeria and Egypt about 4000 B.C. Dredging of canals of Babylon and between the Euphrates and the Tigris were under the direction of Nebuchadnezzar about 600 B.C.[3] The first canal between the Nile and the Red Sea was started by Nikau II about 600 B.C. and completed under Darius I about 500 B.C. These early forms of dredging were carried out by primitive methods with spades and baskets. The Roman infantry, slaves, and prisoners of war were often employed in large-scale excavation works.

Agitation dredging was also used in early times. Tree trunks weighted by stones were dragged behind a boat on the Indus River to stir the mud into suspension. River current was employed to carry the suspended material downstream. This method could not be used today because of the environmental consideration, but then, when mechanical power was not available, the agitation dredging method was efficient.

The scraper dredge, also using the agitation principle and relying on the ebb current to carry the sediment out to sea, was first used in Zeeland in 1435 A.D. This dredge, called the *Krabbelaar,* is shown in Fig. 1.1.[4]

In the Netherlands a new basic dredging tool was developed during the Middle Ages. This tool, known as the "bag and spoon," was a development of spade and a basket. It was an efficient tool operated by two men, one holding the spoon and another pulling it by a rope attached to iron banding.

The need to dredge harbors and ship channels quickly developed in England in the second part of the sixteenth century. The first known offshore mining operation was off the Essex coast in England when bisulphide of iron was dredged from the sea bed.[5]

The "mud mill" was developed toward the end of the sixteenth century in Delft, Holland.[6,7] The mill, activated by a revolving chain, scooped up the mud onto a chute. Figures 1.2 and 1.3 present scale models of the mud mill.

FIGURE 1.1   The *Krabbelaar* scraper. (*Courtesy, Ports and Dredging*)

A grab dredge, forerunner of a clam shell, was developed in the sixteenth century both in Italy and Holland.[4,8]

Development of a steam engine by James Watt in the eighteenth century finally gave the long-needed energy to propel ships and dredges, and the development of a centrifugal pump by LeDemour in 1732 gave birth to modern dredges.

Bazin presented the idea of a suction dredge at the Paris Exposition in 1867. His design, which incorporated a rotating harrow under the bow of the ship and suction pipes under the stern, was applied in the dredging of the Suez Canal.[6,9]

Lebby[10] conceived the first hydraulic hopper dredge *General Moultrie,* which operated in the United States in 1855. The 365-ton dredge had the following features:

1. Wooden null, 150 ft (45.7 m) long, 10 ft 3 in (3.1 m) deep with a beam of 26 ft 8 in (8.1 m)

2. Steam engine, maximum steam pressure 60 psi (413,655 N/m^2), maximum speed 50 rpm

3. Four-bladed propeller, 9 ft (2.7 m) diameter, 18 ft (5.5 m) pitch

4. Centrifugal pump, probably with a 6-ft (1.8-m) diameter impeller rotating on a vertical axis

5. 19-in suction pipe

FIGURE 1.2 The *Mud Mill*. (*Courtesy, Ports and Dredging*)

FIGURE 1.3 The *Mud Mill*. (*Courtesy, Ports and Dredging*)

Average production for this dredge in 1857 was about 328 yd³ (251 m³) per working day. There are many types of equipment used in dredging that can be classified into two broad categories: mechanically or hydraulically operating. The greatest improvements during the last 30 years have been in the dredges operating on the hydraulic principle.[11,12] Not only are the dredges more efficient, which translates into a lower unit cost of dredging, but the modern dredges are fully instrumented and partially or fully automated. It is anticipated that the twenty-first century will bring further advances in instrumentation, automation, and positioning and better accuracy of dredging.

Modern hydraulic dredges have compact power plants, efficiently designed equipment, such as pumping and swell compensating devices, and electronic equipment for automatic controls, measurement, and navigation. Even these, however, are often not quite adequate since today's requirements include dredging, deepening, and maintaining navigation channels in the sea to depths exceeding 88 ft (27 m), working offshore in deep water on mining projects, on land and in the water for reclamation, and along coasts for beach nourishment. Designs have been developed for offshore nuclear plants, airports, artificial islands, and deep-water ports.

Dredging developments, which continued at a fairly constant pace over the years, were accelerated by three major outside influences in the second half of this century:

1. The disastrous floods of 1953 caused by a North Sea storm led the Netherlands to undertake a massive sea defense project (Delta Project).

2. The low cost of oil in the 1950s led to an increase in ship sizes in the 1960s and 1970s, resulting in a major increase in the depth requirements at various ports.

3. Containerization made many established ports obsolete because of insufficient space and created a need for new or enlarged ports, resulting in a continuous interaction between the designers of harbors and the developers of dredging equipment.[11,13]

In addition, factors such as dredging in deep water, ocean mining, and environmental constraints have forced the development of new techniques. This chapter briefly describes the problems involved and the new techniques and equipment developed to cope with the requirements in the field of dredging.

DREDGING PROBLEMS

The nature of problems involved in the development of dredging equipment and techniques is largely dependent on the type of work to be accomplished and could be broadly categorized as follows:[1]

Type of work	Problems
1. Dredging of navigational channels in the open sea under various conditions such as waves, tides, currents, winds, etc., and in depths of the order of 82 to 98 ft (30 to 45 m).	Design of equipment to work under these conditions.
2. Dredging in the nearshore region for rehabilitation of eroded beaches with sand found in the vicinity.	Design of equipment to work in the surf zone, dredging and transport.[14]

Type of work	Problems
3. Dredging for the recovery of material from seafloor mining in nearshore and offshore regions.	Design of equipment for very deep areas. The dredging operation must be complemented by an upgrading process aimed at retaining valuables such as gold, tin, diamonds, etc., and disregarding the other matrix material.
4. Cleanup of pollution as a result of dredging of contaminated material and disposal of same.	Design of suitable dredging equipment that conforms to the environmental constraints to minimize pollution level.
5. Dredging in hard clayey and rocky material in deep areas.	Designs for hard-materials excavation.
6. Land-based dredging such as dredging in open-cast coal and ore mines, marshy areas, etc.	Design of suitable equipment for work on land.
7. Transportation of equipment to dredging sites.	Portable dredges.
8. Creation of artificial islands.	Design of dredging and transport equipment to handle very large quantities in a very short time.

Literature on dredging engineering and applications is somewhat limited, especially if one considers that about $1 billion is spent annually on maintaining and deepening navigation channels. Four books were published since 1970 by Huston (1970),[15] Herbich (1975) (this handbook is an expanded and updated version of that book),[12,16] Bray (1979),[17] and Turner (1984).[18] There are several magazines dealing with dredging: *World Dredging and Construction, International Dredging Review, Dock and Harbour Authority, Terra et Aqua, Dredging and Port Construction, Ports and Dredging,* and *Port Construction and Ocean Technology.*

WODCON Association has sponsored WODCON Conferences for many years and published proceedings of these conferences (1967–1980).

World Organization of Dredging Associations has been sponsoring World Dredging Congresses.

The American Society of Civil Engineers (ASCE) has sponsored two specialty conferences:

1. Dredging and Its Environmental Effects, 1976[19]
2. Dredging '84, 1984[20]

The British Hydromechanics Research Association (BHRA) has been sponsoring "International Symposia on Dredging Technology." BHRA also provides an abstracting service.

The Center for Dredging Studies at Texas A&M University (TAMU) has been sponsoring annual dredging seminars and dredging engineering short courses since 1967. The Center has published the *Bibliography on Dredging* and is providing an abstracting service.

Western Dredging Association (WEDA), Central Dredging Association

(CEDA), and Eastern Dredging Association (EEDA) are members of the World Organization of Dredging Associations and organize meetings on a regular basis.

REFERENCES

1. Donkers, J. M., "Dredging at Sea," *Ports and Dredging,* no. 73, 1973.
2. Gower, G. L., "A History of Dredging," *Dredging Symposium, Proc.,* Institution of Civil Engineering, England, 1968.
3. Camp, L. Sprague De., *The Ancient Engineers,* Souvenir Press, London, 1963.
4. *Ports and Dredging,* vol. 1 et seq., IHC Holland, Rotterdam.
5. Dickin, E. P., *A History of Brightlingsea,* Wiles and Son, Colchester, England, 1939.
6. Doorman, G., "Dredging, History of Technology," vol. 4, ch. 21, Clarendon Press, Oxford, England, 1958.
7. Van Veen, J., *Dredge, Drain, Reclaim. The Art of a Nation,* Martinus Nijhoff, The Hague, the Netherlands, 1962.
8. MacCurdy, E., *The Notebooks of Leonardo da Vinci,* Alden Press, Oxford, England, 1938.
9. Prelini, C., *Dredges and Dredging,* Crosby Lockwood and Sons, London, England, 1912.
10. Scheffauer, F. C., "The Hopper Dredge," U.S. Government Printing Office, Washington, DC, 1954.
11. De Koning, J., "Customer's Requirements, Dredging Operations and Equipment Development," *Proc., World Dredging Conference,* WODCON VIII, pp. 89–113, 1973.
12. Herbich, J. B., *Coastal and Deep Ocean Dredging,* Gulf Publishing Company, Houston, TX, 622 pp., 1975.
13. Vroege, C. J., and Leeuw, R. D., "Developments in the Design of Dredging Equipment," *Proc., World Dredging Conference,* WODCON VIII, pp. 115–130, 1978.
14. Richardson, T. W., "Beach Nourishment Techniques," Technical Report H-78-13, Report 1, U.S. Army Engineers Waterways Experiment Station, Vicksburg, MS, 1976.
15. Huston, J., *Hydraulic Dredging,* Cornell Maritime Press, Centreville, MD, 332 pp., 1970.
16. Herbich, J. B. (ed.), *Handbook on Coastal and Ocean Engineering,* vol. III, Gulf Publishing Company, Houston, TX, 1992.
17. Bray, R. N., *Dredging—A Handbook for Engineers,* Edward Arnold Publishers, London, 1979.
18. Turner, T. M., *Fundamentals of Hydraulic Dredging,* Cornell Maritime Press, Centreville, MD, 215 pp., 1984.
19. "Dredging and Its Environmental Effects," *Dredging '84,* ASCE, Clearwater, FL, Nov. 1984.
20. "Dredging," *XIIth World Dredging Congress,* Western Dredging Association, Orlando, FL, 1084 pp., May 1989.

CHAPTER 2
BASIC FLUID MECHANICS

Some knowledge of basic fluid mechanics is of great importance to anyone involved in hydraulic dredging operations. This is in spite of the fact that mechanics of flow in hydraulic pumps are very complex. The 90° turns, reduction, and increase in cross-sectional areas, centrifugal action, etc., all add complications to theoretical evaluation of the flow inside the pump. Stepanoff[1] warns that simple fluid mechanics relationships established for ideal flows may give erroneous answers to problems involving dredge pumps. However, there are other areas of dredging such as flow in pipes, head losses in elbows and valves, and cavitation where fluid mechanics principles are relatively simple and may be directly applied to the problem on hand.

There are three areas of interest in fluid flow:

1. Fluid statics is concerned with fluids at rest. The application of fluid statics in dredging will pertain to settling tanks and dredged material ponds, manometry, etc.
2. Fluid kinematics is concerned with moving liquids where consideration is given only to velocity distribution but not to forces acting or energy.
3. Fluid dynamics is also concerned with moving liquids where pressures, forces, and energy are considered.

The terms *hydromechanics* and *hydrodynamics* are also used. The former is synonymous with fluid mechanics, but the latter usually refers to theoretical fluid mechanics or mechanics of ideal flow. The ideal flow is defined as flow of ideal, nonviscous fluid as opposed to flow of real, viscous fluid. The nonviscous fluid is also referred to as inviscid fluid. Since dredging deals only with real fluids, only mechanics of viscous fluids are normally considered.

DEFINITIONS

A fluid is a substance which can deform continuously under shear stress of any magnitude. Shear stress is equal to a shear force component tangent to a surface divided by the area of the surface.

One of the distinctions between a fluid and a solid is the spacing between the molecules. In a fluid the spacing is much longer than in a solid where the attraction between the molecules permits the solid to retain its shape. In a fluid the tangential stresses are proportional to the velocity of deformation and are reduced to zero as the velocity becomes zero.

Fluid properties may be defined in terms of basic dimensions of mechanics: force F, mass M, length L, and time T. The second law of Newton states:

$$\text{force} = (\text{mass})(\text{acceleration}) \tag{2.1}$$

or

$$F = \frac{ML}{T^2} \tag{2.2}$$

Mass density ρ is the mass of fluid contained in a unit of volume. Specific weight γ is the weight of fluid contained in a unit of volume. The units of mass density in the United States have dimensions of slugs per cubic foot and units of specific weight are expressed as pounds (weight) per cubic foot.

Thus the dimensions of Eq. (2.1) are

$$1 \text{ lb} = (1 \text{ slug})(1 \text{ ft/s}^2)$$

Since weight W is related to its mass, from Eq. (2.1)

$$W = Mg \tag{2.3}$$

where g = acceleration due to gravity (ft/s^2).

Similarly the specific weight may be expressed in terms of mass density and gravitational acceleration, or

$$\gamma = \rho g \tag{2.4}$$

Dimensions of ρ may be found from Eq. (2.4), that is,

$$\rho = \frac{\gamma}{g} \tag{2.5}$$

or

$$[\rho] = \frac{[\gamma]}{[g]} = \frac{\text{lb/ft}^3}{\text{ft/s}^2} = \frac{\text{lb s}^2}{\text{ft}^4} \tag{2.6}$$

The specific weight of liquids may be calculated from an expression relating it to the specific gravity ($S.G.$) of the liquid and specific gravity of water (or ratio of its density to density of water):

$$\gamma_{\text{substance}} = (S.G._{\text{substance}})(\gamma_{\text{water}}) \tag{2.7}$$

The specific weight of perfect gases may be calculated from the equation of state

$$\gamma = \frac{p}{RT} \tag{2.8}$$

where p = absolute pressure in pounds per square foot (kilograms per square meter

R = gas constant in foot pounds per pound degrees Rankine

T = temperature in degrees Rankine (°F + 460)

All real fluids are viscous and produce friction when in motion. Viscosity is a property of the fluid which is a measure of its resistance to shear or angular deformation. Further discussion of viscosity is in Chap. 3.

Tables in Appendix I summarize the more important properties of liquids and gases which may be encountered in dredging operations.

Fluid Statics

The study of fluid friction shows that shear forces can exist only where there is relative motion between fluid particles. Since there is no relative motion between fluid particles in fluid statics, there can be no shear forces and only body forces and pressure forces need to be considered. A body force depends on the mass or volume of the body, and the most important body force is weight.

Figure 2.1 examines a cube of fluid at rest, with respect to the axis system and completely immersed in fluid.

The only forces acting on this free body are the pressure forces on each face and the weight of the cube which acts at the center of the cube. Take the pressure of the center of the cube equal to p. The pressure at the top of the cube is the pressure at the center, plus some infinitesimal increment of pressure. The change of p with respect to a change in z is called the partial derivative of p with respect to z. (It represents the change of p in the z-direction.) When we multiply the rate of change of p in the z-direction by the distance moved in the z-direction, we obtain a close approximation to the change of p.

Thus the pressure on the top of the element is

$$p + \frac{\partial p}{\partial z}\frac{dz}{2} \tag{2.9}$$

Similarly, the pressure at the bottom of the cube is

$$p - \frac{\partial p}{\partial z}\frac{dz}{2} \tag{2.10}$$

and pressures on the other faces are determined using a similar procedure.

Summing the forces on the cube in the z-direction, we obtain

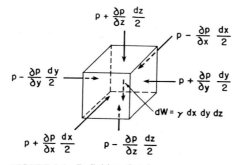

FIGURE 2.1 Definition sketch.

$$\left(p - \frac{\partial p}{\partial z}\frac{dz}{2}\right) dx\, dy - \left(p + \frac{\partial p}{\partial z}\frac{dz}{2}\right) dx\, dy - dW = 0 \tag{2.11}$$

or

$$-\frac{\partial p}{\partial z}\, dx\, dy\, dz = dW = \gamma\, dx\, dy\, dz$$

which further reduces to

$$\frac{\partial p}{\partial z} = -\gamma \tag{2.12}$$

Summing the forces in the x- and y-directions, we obtain

$$\frac{\partial p}{\partial x} = 0 \tag{2.13}$$

and

$$\frac{\partial p}{\partial y} = 0 \tag{2.14}$$

Equations (2.12) and (2.13) indicate that no change of pressure exists with a change of x or y; thus p is only a function of z and independent of x and y.
Therefore, we can write

$$\frac{dp}{dz} = -\gamma \quad \text{or} \quad \int dp = -\int \gamma dz \tag{2.15}$$

Equation (2.15) is the fundamental equation of fluid statics. Since we are dealing here with fluids of constant density, Eq. (2.15) may be written as

$$\int dp = -\gamma \int dz$$

and integrated:

$$p_2 - p_1 = \gamma(z_1 - z_2) \tag{2.16}$$

Equation (2.16) is the basic equation of incompressible fluid statics.

In many practical problems p_1 may be taken as atmospheric pressure (or pressure at the free surface) and h as the depth below the free surface ($y_1 - y_2$). Total or absolute pressure (p_g) or (p_{abs}) is equal to atmospheric and gauge pressure (p_g), or

$$p_{abs} = p_{atm} + p_g \tag{2.17}$$

Thus, we can express any liquid pressure in terms of an equivalent column of liquid, that is,

$$h = \frac{p}{\gamma} \tag{2.18}$$

The relationship between pressures is shown schematically in Fig. 2.2.

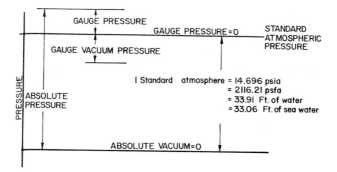

FIGURE 2.2 Pressure relationship.

MANOMETERS

The most important application of equations of fluid statics is in the area of manometry. Manometers are commonly used to measure pressures in pipelines at the suction or on the discharge side of the pump.

The basic types of manometers are:

1. Piezometer
2. Open-end manometer
3. Differential manometer
   a. U-tube
   b. Inverted U-tube
4. Inclined manometer

Applications of Basic Equations

Consider an open-end manometer, shown in Fig. 2.3, in which all distances and densities are known and pressure p_b is to be found. Over horizontal planes within continuous columns of the same liquid pressures are equal, therefore,

$$p_1 = p_2 \tag{2.19}$$

and from Eq. (2.16)

$$p_1 = p_b + \gamma y \tag{2.20}$$

and

$$p_2 = 0 + \gamma_1 h \tag{2.21}$$

Equating Eqs. (2.20) and (2.21) we obtain

$$p_b = \gamma_1 h - \gamma y \tag{2.22}$$

Note that p_b is gauge pressure, because the atmospheric pressure (p_{atm}) was taken as 0 in Eq. (2.21).

In the case of an open-end manometer, the height of the column of water, h_{H_2O}, is balanced by the column of mercury of height h_{Hg} in the other tube, or

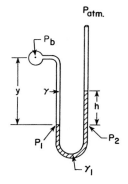

FIGURE 2.3 Open-end manometer.

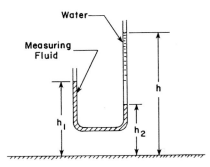

FIGURE 2.4 Simple U-tube manometer.

$$h_{H_2O} = S.G._{Hg}(h_{Hg})$$

and

$$S.G._{Hg} = \text{specific gravity of mercury} = \frac{h_{H_2O}}{h_{Hg}} \qquad (2.23)$$

Any fluid which is insoluble in water may be used as a gauge fluid. Mercury is suitable for measuring fairly high pressure heads. By placing several U-tubes using mercury in series, it is possible to measure large pressures without using inconveniently high columns. For measuring small differences of pressure, fluids having specific gravities nearer that of water are more suitable because the difference in head is multiplied. By diluting carbon tetrachloride with gasoline, the specific gravity of the mixture may be reduced to any desired value between 0.75 and 1.58. A mixture having a specific gravity of about 1.25 is often used in work with the pitot tube. The value of the gauge coefficient $(S.G. - 1)$ is then about 0.25, which means that the gauge reads approximately 4 times the difference in head. Greater multiplications can be achieved by making the specific gravity of the mixture nearer to unity but without a corresponding gain in accuracy because the meniscus becomes unstable. Compressed air is often used as a gauge fluid in an inverted gauge. The specific gravity of the air is assumed to be zero, hence the gauge coefficient $(1 - S.G.)$ is unity.

The size of the glass tubes affects the height of the fluid in the gauge. Mercury is depressed and most other gauge fluids are raised by this effect. For water in a glass tube, the rise from capillary action is about $0.046/d$, in which d is the internal diameter of the tube in inches. If both legs of the U-tube were exactly the same diameter, the effect of capillary action in the two legs would cancel, but there is almost always some difference in diameter along the length of glass tubes and between two tubes. However, for tubes having an internal diameter greater than ⅜ in, the effect of capillary action is negligible.

Other types of manometers are shown in Figs. 2.4 through 2.6. Figure 2.4 represents a simple U-tube manometer, Fig. 2.5 a differential manometer with gauge liquid heavier than water, and Fig. 2.6 an inverted U-tube manometer. Figure 2.7 shows an open-end manometer.

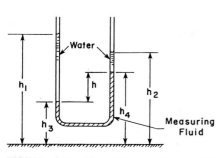

FIGURE 2.5 Differential manometer with gauge liquid heavier than water.

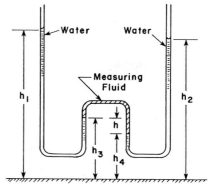

FIGURE 2.6 Inverted U-tbue manometer with gauge liquid lighter than water.

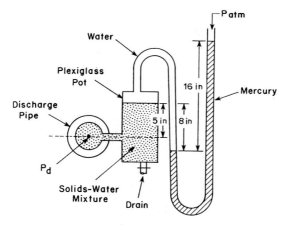

FIGURE 2.7 Open-end manometer.

BUOYANCY AND FLOTATION

The Archimedes' principle states that a floating body displaces its own weight of the liquid in which it floats. It may also be said that a body immersed in a fluid is buoyed up by a force equal to the weight of the fluid displaced.[2]

Consider the rectangular pontoon floating in water shown in Fig. 2.8. The upward buoyant force F_B is the product of the pressure on the bottom of the pontoon and the bottom area, that is,

$$F_B = pB(a)b$$
$$= \gamma h(a)b$$
$$= \gamma V_D \qquad (2.24)$$

where V_D = volume of fluid displaced. The buoyant force acts at the center of

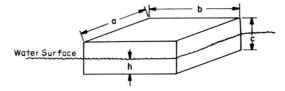

FIGURE 2.8   Pontoon floating in water.

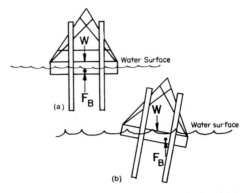

FIGURE 2.9   Stability of a cutterhead dredge (*a*) in still water and (*b*) under action of waves.

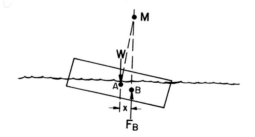

FIGURE 2.10   Metacentric height.

gravity of the displaced fluid.

The stability of a floating pontoon will depend on the relative location of the buoyant force and the weight of the body. Examine the cutterhead dredge shown in Fig. 2.9. The weight of the dredge acts through the center of gravity. In still water the weight force and the buoyant force counteract each other, resulting in a stable situation. Under action of waves the dredge may heave to the right causing a moment between the two forces as the forces move out of vertical alignment. Since the forces are unbalanced, a righting moment will permit the dredge to return to its normal floating position. Consider Fig. 2.10 showing location of the metacenter (point M) which may be found by drawing lines from points A and B. Distance MA is known as the metacentric height, which is an indication of the stability of the dredge. The greater the metacentric height the greater capability of the dredge to return to normal floating position.

DYNAMICS OF FLOW

Flow of fluid may be either steady or unsteady. The flow may be considered steady if none of the variables changes with time, and conversely the flow is unsteady if one of the variables affecting the flow changes with time.

In steady flow, tangents drawn at any point in the direction of velocity will form streamlines, as shown in Fig. 2.11. Such a streamline pattern visually shows regions of high velocity (or low pressure) and of low velocity (or high pressure). In dealing with bounded systems streamlines form a streamtube as shown in Fig. 2.12.

Both velocity and acceleration are vector quantities, possessing both magnitude and direction, while mass density and discharge are scalar quantities since they have only magnitude. The velocity V along a streamline is a function of both the distance along the streamline s and time t, or

$$V = f(s, t) \tag{2.25}$$

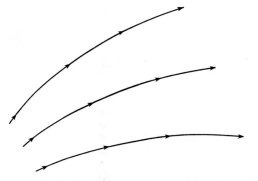

FIGURE 2.11   Streamline pattern.

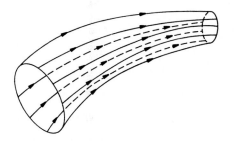

FIGURE 2.12   A streamtube.

If velocity of a particle is taken as an increment of distance ds over an increment of time dt, then

$$V = \frac{ds}{dt} \tag{2.26}$$

Tangential acceleration then is the first derivative of the velocity with respect to time:

$$a_s = \frac{dV}{dt} = \left(\frac{d}{dt}\right)\left(\frac{ds}{dt}\right) = V\frac{dV}{ds} \tag{2.27}$$

and normal acceleration is

$$a_r = -\frac{V^2}{r} \tag{2.28}$$

where r is the radius of curvature.

Acceleration in the s-direction is composed of two components, the convective acceleration and the local acceleration.

$$a_s = V \underbrace{\frac{\partial V}{\partial s}}_{\text{convective}} + \underbrace{\frac{\partial V}{\partial t}}_{\text{local}} \tag{2.29}$$

If local acceleration is equal to zero, the flow is steady, which is the assumption made in most practical dredging problems. However, it should be realized that most of the flows of solid-liquid mixture in dredging processes are turbulent and consequently not steady. In most of the dredging engineering problems a one-dimensional flow is assumed in the zones where streamlines are essentially parallel. The average velocity across a section is computed by dividing the discharge by the cross-sectional area.

Principle of Continuity

Application of the principle of conservation of mass to a steady flow in a streamtube produces what is known as a continuity equation.

Consider a two-dimensional flow in Fig. 2.13. Assume that the mass of fluid entering section 1 in time dt is the same as the mass of fluid leaving section 2 in the same time dt.

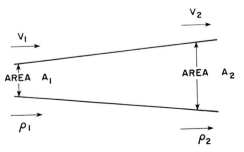

FIGURE 2.13   Flow in a streamtube.

$$\text{Mass flow rate} = \frac{\rho_1 A_1 \, dx_1}{dt} = \frac{\rho_2 A_2 \, dx_2}{dt} \qquad (2.30)$$

Assuming a one-dimensional flow principle, the mean velocities are equal to dx/dt, thus

$$V_1 = \frac{dx_1}{dt} \quad \text{and} \quad V_2 = \frac{dx_2}{dt} \quad \text{and} \quad \rho_1 A_1 V_1 = \rho_2 A_2 V_2 \quad (2.31)$$

Example 2.1

Thirty cubic feet of water flows through a pipe reducer. Using Fig. 2.14 calculate the mean velocities in two sections of the pipe and the discharge in gallons per minute:

1. Velocity in 24-in pipe = Q/A = $30/[\pi/4 \, (24/12)^2]$ = 9.55 fps (2.9 m/s)
2. Velocity in 20-in pipe = $30/[\pi/4 \, (20/12)^2]$ = 13.76 fps (4.2 m/s)
3. Discharge = 30 cfs = 30 (62.4) 60 lb/min
   = 112,320 lb/min
   = (112,320)/(7.48) gpm
   = 15,016.0 gpm (947.2 liters/s)

FIGURE 2.14 Flow in a reducer.

Euler Equation of Motion

Consider the small streamtube in Fig. 2.15. The forces acting in the direction of

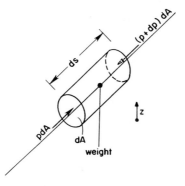

FIGURE 2.15 Flow in a streamtube.

flow on the ends of a streamtube are pressures at each end and a component of the weight of the streamtube.

Pressure forces are

$$pdA - (p + dp)(dA) = -dp\,dA \tag{2.32}$$

The component of weight in the direction along the streamline is

$$-\gamma ds\,dA\,\frac{dz}{ds} \tag{2.33}$$

Applying Newton's Second Law of Motion

$$dF = (dM)(a) \tag{2.34}$$

where a = acceleration.

$$-dp\,dA - \gamma ds\,dA\left(\frac{dz}{ds}\right) = (\rho ds\,dA)(a) \tag{2.35}$$

or

$$-\frac{1}{\rho}\frac{dp}{ds} - \frac{\gamma}{\rho}\frac{dz}{ds} = a \tag{2.36}$$

and

$$a = V\frac{\partial V}{\partial s} + \frac{\partial V}{\partial t} \tag{2.37}$$

therefore

$$\frac{1}{\rho}\frac{dp}{ds} + g\frac{dz}{ds} + V\frac{\partial V}{\partial s} + \frac{\partial V}{\partial t} = 0 \tag{2.38}$$

For steady flow $\partial V/\partial t = 0$, and

$$\frac{1}{\rho}\frac{dp}{ds} + g\frac{dz}{ds} + V\frac{dV}{ds} = 0 \tag{2.39}$$

or

$$\frac{dp}{\rho} + g\,dz + V\,dV = 0 \tag{2.40}$$

or

$$\frac{dp}{\gamma} + dz + \frac{dV^2}{2g} = 0 \tag{2.41}$$

For incompressible one-dimensional flow, by integration

$$\underbrace{\frac{p}{\gamma}}_{\substack{\text{pressure}\\\text{head}}} + \underbrace{z}_{\substack{\text{elevation}\\\text{head}}} + \underbrace{\frac{V^2}{2g}}_{\substack{\text{velocity}\\\text{head}}} = \underbrace{H}_{\substack{\text{total}\\\text{head}}} \tag{2.42}$$

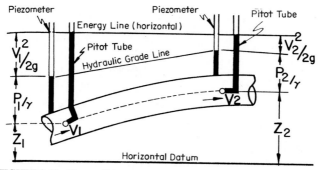

FIGURE 2.16 Energy lines.

Equation (2.42) is very useful as it relates p, V, and z. The relationship between various heads is presented graphically in Fig. 2.16.

The Impulse-Momentum Equation

The impulse-momentum equation is concerned only with external forces and provides useful results without requiring detailed processes within the fluid. The impulse-momentum equation of mechanics states that the product of the force and the increment of time over which it acts (i.e., impulse of the force) are equal to the resulting change in the product of the mass of the body on which the force acts and the velocity of the body (i.e., change in the momentum of the body). Both impulse and momentum are vector quantities. In the x-direction the equation is

$$F_x \, dt = d(MV)_x \qquad (2.43)$$

Let dF_x represent the differential force acting upon an incremental length of a stream filament

$$dF_x \, dt = d(\rho \, ds \, dA \, V)_x \qquad (2.44)$$

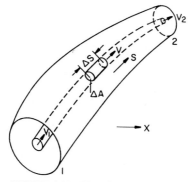

FIGURE 2.17 Flow in a streamtube.

Figure 2.17 shows that for steady flow at ρ = constant

$$d(\rho \ ds \ dA \ V)_x$$
$$= (\rho \ ds \ dA)dV_x$$
$$= \rho \ ds \ dA \ dV_x$$
$$= \rho \ ds \ dA \ \frac{\partial V_x}{\partial s} \ ds$$
$$= \rho \ ds \ dA \ \frac{\partial V_x}{\partial s} \ V \ dt$$
$$= \rho \ \frac{\partial V_x}{\partial s} \ ds \ dQ \ dt \qquad (2.45)$$

or per unit time

$$dF_x = \rho \ \frac{\partial V_x}{\partial s} \ ds \ dQ \qquad (2.46)$$

Since dQ = constant along the stream filament

$$\Delta F_x = \rho[(V_x)_2 - (V_x)_1]dQ \qquad (2.47)$$

by integration between sections 1 and 2 or

$$\Sigma F_x = \rho \int [(V_x)_2 - (V_x)_1]dQ \qquad (2.48)$$

or in general form

$$\Sigma F_x = \rho Q(V_{x2} - V_{x1}) \qquad (2.49)$$

Since the velocity will generally vary across the two end sections of the zone in question, the integration must be done either graphically or analytically. Similarly,

$$\Sigma F_y = \rho Q(V_{y2} - V_{y1}) \qquad (2.50)$$

Using Fig. 2.18 it can be shown that

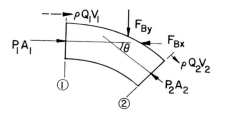

FIGURE 2.18   Forces on an elbow.

$$\Sigma F_x = p_1 A_1 - p_2 A_2 \cos \theta - F_{Bx} = \rho Q(V_{x2} - V_{x1}) \qquad (2.51)$$

Example 2.2

Figure 2.19 shows that

$$F_{By} = F_{By}$$

Force in y-direction is equal to zero since the jet is split in two directions.

$$- F_{Bx} = p_1 A_1 = \rho Q(0 - V_{x1}) \qquad (2.52)$$

Example 2.3

From Fig. 2.20 it is seen that

$$\Sigma F_y = \rho Q(V_{y2} - 0) \qquad (2.53)$$

Example 2.4

Using Fig. 2.21 find the force exerted on a fixed curved blade.

$$\begin{aligned}
Q &= 0.5 \text{ cfs } (0.014 \text{ m 3/s)water} \qquad (2.54)\\
\Sigma F_x &= \rho Q(V_{x2} - V_{x1})\\
\Sigma F_x &= -F_{Bx} = \rho Q(V_{x2} - V_{x1})\\
&= \rho Q(V_{x2} \cos \theta - V_{x1})\\
-F_{Bx} &= (1.94)0.5[50 \cos \theta - 100]\\
-F_{Bx} &= 1.94(0.5)[50 (- 0.707) - 100] = -62.71 \text{ lb } (28.44 \text{ kg})
\end{aligned}$$

That is, the force of the blade against the water is 62.71 lb (28.44 kg) (to the left) and 62.71 lb (28.44 kg) (to the right).

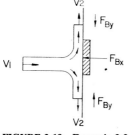

FIGURE 2.19 Example 2.2.

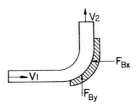
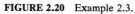

FIGURE 2.20 Example 2.3.

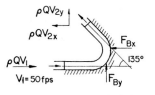

FIGURE 2.21 Example 2.4.

$$\Sigma F_y = F_{By} = 1.94(0.5)[50 (0.707) - 0] = 34.29 \text{ lb } (15.55 \text{kg}) \text{upward}$$

Force of the water against the blade is equal and opposite 34.29 lb (15.55 kg) downward.

REFERENCES

1. Stepanoff, A. J., *Centrifugal and Axial Flow Pumps,* John Wiley, New York, 1962.
2. Vennard, J. K., *Elementary Fluid Mechanics,* 4th ed., John Wiley, New York, 1961.

CHAPTER 3
DREDGE PUMPS

THEORY AND APPLICATION OF CENTRIFUGAL PUMPS

A centrifugal pump may be called a rotodynamic pressure generator. Rotating vanes create a forced vortex as the low-pressure liquid is drawn from a "suction" line near the axis, and centrifugal force carries it outward into a high-pressure region. Mechanical energy supplied to the wheel is converted partly into kinetic and partly into pressure energy, and some of the energy is lost (mostly through heat dissipation). The potential energy is immediately available, but the kinetic energy of liquid with water velocity at 30 to 200 ft/s (9.1 to 61.0 m/s) must be gradually converted to extra pressure energy by reduction of the discharge velocity to 4 to 20 ft/s (1.2 to 6.1 m/s) after leaving the impeller. This conversion is incomplete and some of the kinetic energy is also "lost."

Impeller Types

The following applies to impeller types:

1. They can be open or closed. The open type consists of vanes set in a ring; it has lower efficiency, no axial thrust, and low probability of clogging. The closed type consists of vanes set between discs. Early dredge pumps were the open type; modern dredge pumps are almost exclusively the closed type (Fig. 3.1).

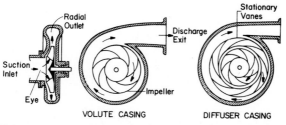

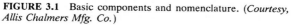

FIGURE 3.1 Basic components and nomenclature. (*Courtesy, Allis Chalmers Mfg. Co.*)

3.1

2. They can be single or double suction. In a single suction type water enters from one side, while in a double suction type the water enters from both sides. Dredge pumps are generally single suction.

3. The ratios of outside radius to inside radius decrease with increase in specific speed. The radial-type impellers have the highest ratio, and for axial-type impellers the ratio is equal to 1 (Fig. 3.2). The majority of dredge pumps are radial flow types.

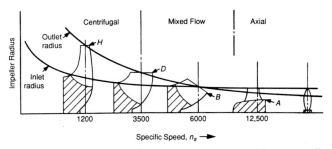

FIGURE 3.2 Pump evolution by specific speed. (*Courtesy, Allis Chalmers Mfg. Co.*)

Casing Types

1. *Constant velocity volute:* The casing surrounding the impeller has an increasing cross-sectional area so that velocity of discharge at all sections is constant at about one-half of the impeller peripheral velocity. A diffuser cone can reduce this high velocity to the pressure head. Diffuser conversion efficiency may be up to 90 percent (Fig. 3.1).

2. *Variable velocity volute:* The passage area increases at a greater rate than necessary for constant velocity, thus reducing liquid velocity along the volume. The resultant large pump and head conversion is not efficient. Efficiency may be as low as 40 percent.

3. *Diffusion vane casing:* Fixed guide vanes surround the impeller and provide diverging channels to reduce discharge velocity. These gain 10 to 20 percent in conversion, yielding total efficiencies up to 70 percent (Fig. 3.1).

Modern dredge pumps have constant velocity volutes. Optimum efficiency of water pumps as a function of dimensional specific speed is shown in Fig. 3.3. The specific speed is defined as

$$N_S = \frac{N \sqrt{Q}}{H^{3/4}} \tag{3.1}$$

where N = rotational speed of the pump in revolutions per minute (rpm)
Q = rate of flow in gallons per minute (gpm)
H = total head developed in feet of water

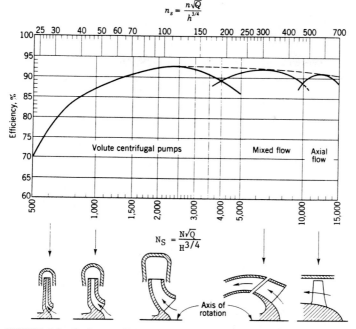

$$n_s = \frac{n\sqrt{Q}}{h^{3/4}}$$

$$N_S = \frac{N\sqrt{Q}}{H^{3/4}}$$

FIGURE 3.3 Optimum efficiency of water pumps as a function of specific speed.

Theory of Centrifugal Pumps

An idealized one-dimensional approach is used here, employing mean velocities at cross sections of the impeller channels and neglecting friction and eddy losses and assuming that the fluid enters and leaves the impeller vanes tangential to the vane tips.

The velocities described in Table 3.1 are shown graphically in Fig. 3.4.

TABLE 3.1 Symbol Convention

	Entrance	Exit	Components	
			Tangential	Radial (meridional)
Impeller, peripheral velocity	u_1	u_2	u_1	u_2
Water, relative velocity				
(to impeller vane)	w_1	w_2	w_{u1}	w_{u2}
absolute velocity	C_1	C_2	C_{u1}	C_{u2}

Theoretical Input Head. The Euler head (H_e) is computed from the change in angular momentum (or "moment of momentum") of the water. Shaft torque is the rate of change of angular momentum and power is given by (torque) × (angular velocity) = $QH_e\gamma$. Thus,

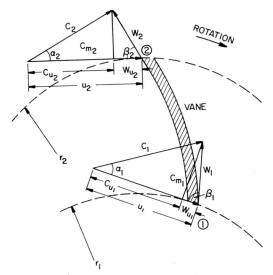

FIGURE 3.4 Theoretical velocity triangles.

Torque $T = \dfrac{d}{dt}(\text{mass} \times C_u r)$

$$= \frac{\text{mass}}{\text{sec}} \times \text{change in } C_u r \qquad (3.2a)$$

$$= Q\rho(C_{u2}r_2 - C_{u1}r_1)$$

$$\text{Power} = T\omega = Q\rho(C_{u2}\omega r_2 - C_{u1}\omega r_1) \qquad (3.2b)$$

$$= Q\rho(u_2 C_{u2} - u_1 C_{u1})$$

$$= Q\gamma H_e$$

$$\therefore H_e = \frac{u_2 C_{u2} - u_1 C_{u1}}{g} \text{ ft} \cdot \text{lb/lb, or ft (m} \cdot \text{kg/kg, or m)} \qquad (3.3)$$

which is called the Euler's head.

Figure 3.4 illustrates the entrance and the exit velocity triangles.

Flow Through the Impeller.   The impeller does not impart sufficient power (input head) to produce the Euler head because:

1. Pressure distribution is higher on the front face of the vane, as shown in Fig. 3.5*a*. Velocity distribution is thereby affected in the opposite way, as in Fig. 3.5*b*.

2. Velocity is distorted because of the liquid's change in direction of motion on entering the impeller.

3. Relative circulation, or eddy in impeller passage, caused by inertia of fluid elements, sets up a circulatory motion. This results in an absolute discharge angle α_2

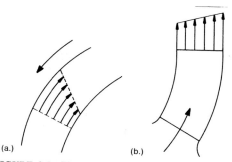

FIGURE 3.5 Flow through impeller. (*a*) Pressure distribution; (*b*) velocity distribution.

less than that necessary for a smooth exit along the vane. At the inlet, circulation gives the effect of positive prerotation increasing β_1 to β_2 (Fig. 3.6).

4. Pressure difference between the front and back of the vane drops to zero at the exit tip. Velocity of the vane is active in driving water.

As a result, the actual velocity triangles differ from the theoretical velocity triangles, as shown in Fig. 3.6.

Euler ideal head-discharge relationship plots as a straight line on arithmetic graph paper, as shown in Fig. 3.7. Because of leakage, disk friction, and mechanical losses, the actual head developed is lower than the ideal Euler head. A comparison between the ideal Euler head and the actual curves is shown schematically in Fig. 3.7.

The distribution of power in a pump operating at variable head and constant speed is shown in Fig. 3.8.

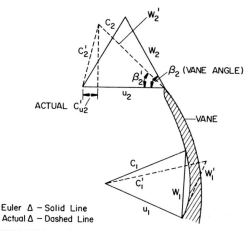

Euler Δ – Solid Line
Actual Δ – Dashed Line

FIGURE 3.6 Comparison of actual and Euler velocity triangles.

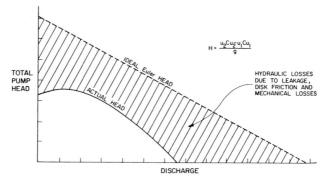

FIGURE 3.7   Comparison of ideal and actual head curves.

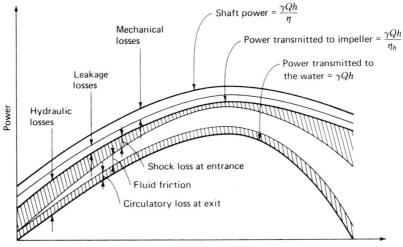

FIGURE 3.8   Disposition of power in a pump operating at variable head and constant speed.

ELEMENTS OF DREDGE PUMP DESIGN

General

In the design of a centrifugal dredge pump, it is necessary to consider more factors than those customarily considered in the design of a centrifugal water pump. The nature of the dredging operation is such that sufficient clearances must be provided through the pump so that occasional gravel, rocks, and debris may pass through the pump without jamming.[1] This requirement means that there is a practical limit to the number of vanes which the impeller may contain and also that clearances in parts of the pump must be made in excess of those which highest

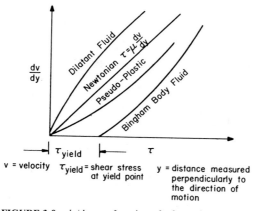

FIGURE 3.9 dv/dy as a function of τ for a given concentration and temperature.

performance would dictate. An example of the above is the clearance at the cutwater between the impeller and the volute.

The increased wear and abrasion caused by particles in suspension require the use of special materials. Provision must also be made for easy access to parts of the pump for maintenance and renewal of components.

The type of material pumped has been shown to have a great effect on the performance of a centrifugal pump. The dredge pump may have to handle a mud-like fluid (silt-clay-water mixture) common to dredging operations in the vicinity of New York Harbor, Chesapeake Bay, Houston Ship Channel, or coral sand off the coast of Florida.

Besides its variation in specific gravity, which is greater than that of water, depending on the weight of solids in suspension, a mixture of silt, clay, and water has a highly variable viscosity. In addition, the mixture behaves like a fluid with suspended solids at lower concentrations up to about 1200 grams per liter, and the solids settle readily. At high concentrations the mixture is more homogeneous, and the solids do not settle for a considerable length of time.

The viscosity of the material depends on the concentration, temperature, past history, and rate of shearing stress after an initial yield value has been reached. Figure 3.9 shows the variation of viscosity with rate of shearing stress for a certain concentration. As shown in this figure, different types of fluids may have different relations between shear stress and rate of strain. If the shear stress τ is directly proportional to the rate of strain (dv/dy) and if $\tau = 0$, when $(dv/dy) = 0$, such fluid is called *newtonian fluid,* and the constant of proportionality μ is called the dynamic viscosity, or viscosity. The newtonian relation between shear stress and the rate of strain may be shown as

$$\tau = \mu \left(\frac{dv}{dy}\right) \tag{3.4}$$

Fluids such as solid-water mixtures, paints, plastics, etc., have a variable proportionality between stress and strain. Such fluids are called *nonnewtonian* and the science of rheology concerns itself with nonnewtonian fluids. There are different types of nonnewtonian fluids: (1) dilatant fluid, (2) pseudo-plastic fluid, and (3) Bingham Body fluid, etc. The relationships for different types of fluids are shown in Fig. 3.9.

The general equation relating shear stress and the rate of strain may be written as

$$\tau = \eta \frac{dv}{dy} + \kappa \tag{3.5}$$

where $\eta = f(dv/dy)$
$\kappa = \text{constant} = \tau_{\text{yield}} = \tau_y$

The equation for dilatant fluid is then

$$\tau = \eta_1 \frac{dv}{dy} \tag{3.6}$$

when η_1 is increasing with increasing rate of strain.

The expression for a pseudo-plastic fluid may be given as

$$\tau = \eta_2 \frac{dv}{dy} \tag{3.7}$$

where η_2 is decreasing with increasing rate of strain. Note that for a Newtonian fluid $\kappa = 0$ and $\eta = \mu = \text{constant}$.

The behavior of Bingham Body fluid, which is characteristic of a silt-clay-water mixture, is more complicated, as shown in Fig. 3.9. The material exhibits both plastic and viscous properties. It is like an ideal plastic in that it flows when a given yield shearing stress τ_y is reached. It is unlike a plastic in that once flow starts, it is retarded by an increased resistance to shear. If the shearing stress τ is less than τ_y, the material will deform elastically but will not flow. It flows when τ exceeds τ_y. It should be noted that the diagram in Fig. 3.9 holds for one concentration only; a similar diagram with different values is required for each concentration.

The equation for a Bingham Body fluid may be expressed as

$$\tau = \eta_3 \frac{dv}{dy} + \tau_c \tag{3.8}$$

where η_3 may ($\eta_3 \neq \text{constant}$) or may not change with an increasing rate of strain ($\eta_3 = \text{constant}$), and $\tau_c = \kappa$. The relationship between shear stress and time is shown in Fig. 3.10.

The effects of viscosity are important considerations in any pump analysis or design.[2] The distribution of velocity of flow through any passage is dependent, to a large extent, on viscosity; an increase in viscosity will produce a corresponding reduction in the effectiveness of the area available. Loss of head through the passage also increases with increased viscosity. The power absorbed by the pump increases with an increase in viscosity because of the greater resistance of the fluid to rotation of the impeller.

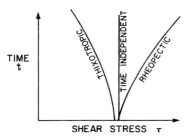

FIGURE 3.10 Shear stress as a function of time.

There is considerably more difficulty in working with a material which has a viscosity dependent on a number of variables than there is in working with a fluid such as water which possesses a viscosity which is dependent on temperature only. Thus the type of application and the material to be handled impose difficulties on design and analysis which must be kept in mind.

Suction Inlet

The suction inlet design is of great importance, particularly if the pump is to be cavitation-free. Some of the geometric variables are the suction pipe diameter, curvature of inlet, eye diameter, and leading edge of the vane.[3]

Ideal conditions for approach exist when a sufficient length of straight pipe without valves or other disturbances precedes the entrance to the pump. Well upstream, a normal velocity distribution can be expected, but as the distance to the pump decreases, a phenomenon known as prerotation may develop.

Prerotation is the condition that exists in the approach pipe to a centrifugal pump when the flow is moving both axially toward the pump and rotating about the longitudinal axis of the pipe. Peck[4] has carried out investigations on prerotation and has obtained the following information from Pitot tube traverses at the suction flange of a pump handling water. At shut-off head conditions, a forced vortex was created in the suction pipe. The total head and pressure head curves were obtained, and the differences gave the velocity head from which the curves of axial flow and circumferential flow components were plotted. Curves of this type are shown in Fig. 3.11. This indicates an axial flow away from the pump through a narrow annular ring extending from the pipe wall. For small discharges the forced vortex was confined to a somewhat larger annular ring than previously, and while there was still a small component of flow away from the pump, axial flow toward the pump occurred in the center portion of the pipe. The extent of the prerotation, with respect to distance from the pump, was found to decrease as the flow increased. The swirl gradually disappeared because of friction between the pipe wall and the water.

Stepanoff[5] explains prerotation as follows: Referring to Fig. 3.12, at section 1, sufficiently distant from the pump pressure p_1 is uniform across the suction, and a normal pipe velocity distribution prevails. At section 2, near the pump the pressure p_2, as measured at the pipe wall, is higher than at section 1. This has been found to be true experimentally when prerotation is present.[4] Since the energy gradient must decrease from section 1 to section 2 if there is to be flow, the higher pressure at 2, at the pipe wall, can come about only at the expense of the energy level of the center portion of fluid. A paraboloid of pressure distribution is de-

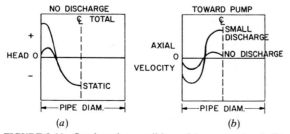

FIGURE 3.11 Suction pipe conditions, (*a*) pressure head, (*b*) velocity. (*Peck, 1951*)

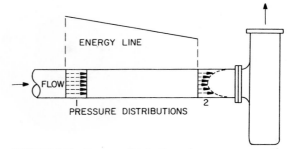

FIGURE 3.12 Pressure distributions along pump approach.
(*Stepanoff, 1957*)

veloped at section 2 with the pressure at the periphery higher and the pressure at the center lower than the pressure existing at section 1. Pipe wall pressure taps at a location such as section 2 will indicate too high a mean pressure and thus often introduce error into experiments. The absolute velocities at the periphery of section 2 are higher than those in the middle as a result of the addition of a tangential component caused by the rotation of the stream.

Prerotation is caused by the tendency of the fluid to follow a path of least resistance on its way to enter the impeller channels. Prerotation is most evident at flows less than the design capacity and practically disappears at that point. The angle at which the fluid enters the impeller channel is influenced by prerotation. Since the geometry of a given impeller is fixed, the entrance angle for the impeller vanes is calculated for the design capacity with no prerotation considered. It is customary to increase the design flow by a small percentage when calculating the entrance angle, since this allows for the unavoidable leakage losses. It is evident that at flows other than design, prerotation will come into being, and the effect will be detrimental to pump performance. When a pump must be operated at some flow other than the flow at maximum efficiency, there is one solution for overcoming the undesirable effect of prerotation. This remedy consists of placing guide vanes in the approach to the pump at such an angle as to make the fluid conform to the entrance angle of the impeller. Care must be taken to see that the guide vanes are so constructed as to prevent separation, cavitation, or high losses. The nature of dredge pump application makes this method impractical.

Model dredge pump studies by Herbich[6] indicated that no prerotation existed at design rates of flow and at design speed; however, prerotation existed at low and at high rates of flow.

Inducing Section. The purpose of the inducing section of the pump is to accept the incoming fluid from the suction pipe in the correct manner and to turn it so that its relative velocity is along the axis of the impeller channels.[7] This means that the flow must be turned through 90° for the case of the radial impeller. This turning of the fluid causes losses and disordered flow. The transition must be made as gradually as possible, with no irregularities, so that the flow is delivered to the impeller channels as ideally as possible. Low suction velocities and generously proportioned suction sections are recommended for pumps handling high-viscosity liquids.[8] The factor of manufacturing economy enters here to set a practical length which may be used for this turning transition. According to NACA investigations,[9] the inducing section has a marked effect on performance. An inducing section of comparatively large axial length gave very good results since it provided a much gentler curve for the transition from tangential to axial flow.

Flow at Impeller Entrance. It is important to limit the relative velocity at the inlet of the pump since a high relative velocity could lead to cavitation. Assuming a uniform axial component of velocity at the eye, or inducing section, the critical region is located at the eye tip. This is where the impeller speed is highest for the inducing section, and hence the relative entrance velocity is also great here. Therefore, flow conditions should be checked at this point. The combination of variables determining relative velocity at this point may be arranged to result in a minimum value. A high relative velocity can result from either of the two following situations:

1. A large eye area, resulting in a small axial entrance velocity, combined with a high impeller velocity
2. A small eye area, resulting in a high axial entrance velocity, combined with a low impeller speed

Using the following equation, a formula for minimum relative entrance velocity may be developed combining u and v:

$$w_1 = \sqrt{v_1^2 + u_1^2} \quad \text{or } w_1 = \left[\left(\frac{4Q}{\pi d_1^2}\right)^2 + \left(\frac{\pi d_1 N}{60}\right)^2\right]^{1/2} \quad (3.9)$$

where w_1 = relative velocity at entrance in feet per second (m/s)
$u_1 = \pi N d_1/60$ = tangential velocity at the eye tip in feet per second (m/s)
N = impeller speed in revolutions per minute
$v_1 = 4Q/\pi d_1^2$ = radial velocity in feet per second of through flow at entrance, assuming no prerotation
d_1 = diameter of eye in feet (m)
Q = flow in cubic feet per second (m/s)

The effect of any one variable of w_1 may be seen by holding all other variables constant. For instance, with Q and N fixed, the variation in w_1 with d_1 can be found. The minimum value of d_1 can be obtained by differentiation or trial and error. Figure 3.13 shows the variation for this case, and it will be noted that for a certain value of d_1 there is a minimum value of w_1 which will tend to prevent cavitation.

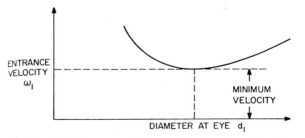

FIGURE 3.13 Entrance velocity as a function of eye diameter.

Impeller

The variables affecting the impeller performance are the inlet angle, vane shape, vane area, and exit angle.

Flow in the Impeller Channel. Because of viscosity, turbulence, and separation, the velocities in an actual centrifugal pump impeller are seldom uniform over a given section. The number of turns in the impeller channel approach and the impeller profile increases the velocity distortion. In radial flow and mixed flow impellers, the fluid must make nearly a full 90° turn before it is acted upon by the vanes.

In an established flow, whether rotating or straight as in open channel flow, a body must tend to move faster than the established velocity of flow in order to exert any force on the fluid flowing in the same direction. Thus, an impeller vane must tend to move faster than the fluid in order to transmit energy from the impeller vane to the fluid. This means that the pressure on the leading face of the vane should be higher than the pressure on the trailing face and that there will be a higher velocity there producing a velocity variation across the channel. The result of this is that relative to the vane the fluid leaves the vane tangentially only at the high-pressure, or leading, side of the trailing edge. The fluid has a circumferential component relative to the vane across the channel from one leading face to the trailing face of the adjacent vane, with the result that the fluid is discharged from the impeller at a mean angle relative to the impeller which is less than the vane angle. The absolute discharge velocity is less than that assumed by using the vane angle itself—this deviation is called the *slip* of the impeller.

Figure 3.14 illustrates the difference between the case of slip and no slip. It is important to realize that changing the vane discharge angle from β_2 to β_2' will only mean that the fluid will again lag behind the vane and discharge at some smaller angle β_2'' less than β_2'. The net result of the nonuniform velocity and the slip is to reduce the theoretical head based on the simple ideal velocity diagram. It may be shown that the head produced by a varying velocity distribution is less than that produced by a uniform velocity, given the same rate of flow.[5]

The occurrence of slip in an actual pump has been studied by Peck.[4] Analysis of data from Pitot tube transverses indicated a difference between the mean absolute velocity discharge angle and the calculated angle, assuming that the relative velocity leaves the impeller parallel to the actual vane angle.

Actual observations of flow in impeller channels have been made using a pump constructed of a transparent material.[10] The flow was seen to be far from ideal at most capacities except those at the design rate of flow where it was often surprisingly good. Fisher's investigations into flow in impeller channels revealed a large area of separation from the trailing face of the vane. This dead water region formed a considerable portion of the passage area at low flow rates. Reverse flow in these areas was also present in some instances. Similar work by Binder and Knapp[11] showed almost uniform absolute discharge velocity across the width of the impeller passage at normal capacity, with small gradients existing at other capacities.

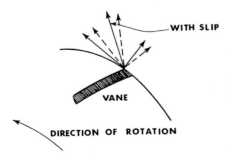

FIGURE 3.14 Change in discharge velocity caused by slip.

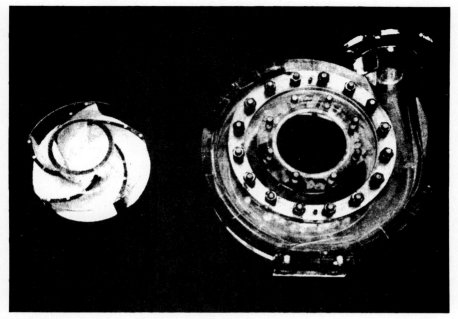

FIGURE 3.15   Plexiglas volute casing and impeller.

Herbich[12] employed high-speed photography to observe and analyze particle motion through the impeller and in the volute casing of a model dredge pump. The model dredge pump was of the volute type, a 1:8 scale model of the pumps installed on the U.S. Army Engineers dredge *Essayons*. The model pump had a 4½-in suction pipe diameter, 4-in discharge pipe diameter, and 10½-in impeller diameter. The volute casing and suction side head as well as the suction side shroud on the impeller were made of transparent plexiglas (Fig. 3.15).

In this study particles ⅛ in in diameter with a specific gravity of 1.19 were introduced into the suction line of the pump and allowed to pass through the pump. The particles were photographed with a high-speed motion picture camera at a speed of 6000 to 8000 frames per second, and the usable portions of 100-ft rolls of film were taken at an elapsed time of about 0.7 s.

The exit angle of the particles was found by tracing the path of about 30 beads per film as projected onto a piece of paper and measuring the included angle between particle path and the tangent to the edge of the impeller. An example of these traces is shown in Fig. 3.16. In Fig. 3.17 the exit angles were plotted versus pump speed and compared to theoretical curves for water. As can be seen from the figure, the exit angles of the beads are in all cases greater than those theoretically determined for water. This must be caused by radial velocities higher than theoretical or tangential velocities lower than theoretical or, as will be seen to be the case, a combination of both. Also, from this figure it can be seen that the variation in exit angle with pump speed follows a trend similar to that theoretically determined for water.

The vane exit angle was also studied for a number of flow rates and a number of pump speeds.

Sample plots showing the variation of particle exit angle are shown in Fig. 3.18

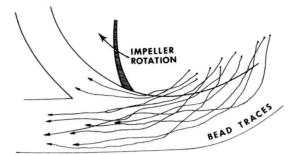

FIGURE 3.16 Bead paths through volute.

for a relatively low discharge (one-fifth of the design discharge) and the design speed of the pump (1440 rpm). In Fig. 3.19 plots are shown for the design speed and are 40 percent higher than the design discharge.

The exit angles of the particles relative to the vane were also found in this study by direct plots of single particles and the trailing vane for alternate frames. Analyses of high-speed movies indicate that the particle exit angle is greater, on the average, than the vane angle. Very few particles left the impeller at an angle less than the vane angle. Sample plots are shown in Fig. 3.20 for 60 percent design discharge and a design speed and in Fig. 3.21 for the design conditions.

The vane angle is defined as the angle that the high-pressure side of the vane makes with a tangent to the edge of the impeller at the tip of the vane. This study was made to check experimental velocity triangles and also to note the tendency of the particles to scour the high-pressure side of the vane.

Because the movement of the bead relative to the vane could not be traced directly from the films, a frame by frame plot was made holding the vane in a fixed position on paper and plotting the movement of a single bead. The analysis was completed only for a few films since it was very time-consuming. The films used were 1440 rpm with 600, 800, 1000, and 1200 gpm. About six plots were made per film.

These plots indicated an angle that is greater, on the average, than the vane angle. Since the vane angle is determined from the theoretical velocity triangle to coincide with the direction of relative velocity of flow, this discrepancy indicates again a larger radial velocity for the particles than that determined theoretically for water. The plots also showed an increase in turbulence and separation within the vane for the smaller discharges.

The previous investigations have indicated a radial particle velocity that is higher than and a tangential velocity that is lower than those that would be expected from theoretical considerations. In this investigation these radial and tangential velocities are measured directly, and the reasons for the values thus determined are discussed.[12]

The study is similar to that for the determination of exit angles except that the bead was plotted frame by frame rather than tracing a moving particle. The bead location was plotted relative to the casing, and the location of the following vane was marked as the bead left the impeller shroud. This is to determine the bead location between vanes.

Since these plots were made at the same scale as the actual model pump, the distance between bead locations gives their actual distance traveled per frame.

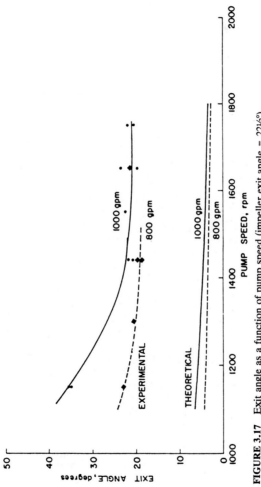

FIGURE 3.17 Exit angle as a function of pump speed (impeller exit angle = 22½°).

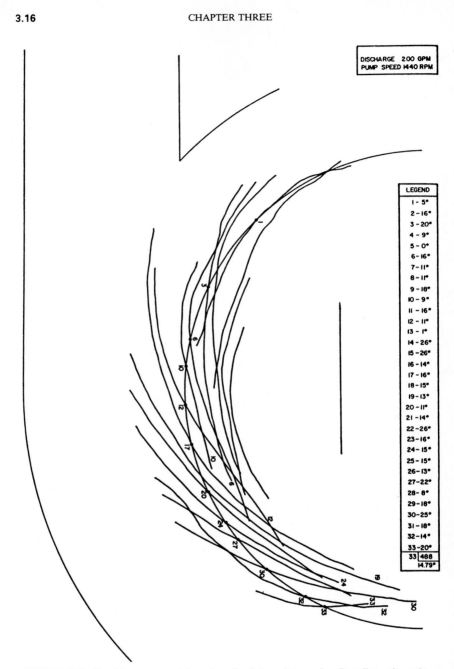

FIGURE 3.18 Particle movement from impeller into volute casing (impeller exit angle = 22½°, discharge = 200 gpm).

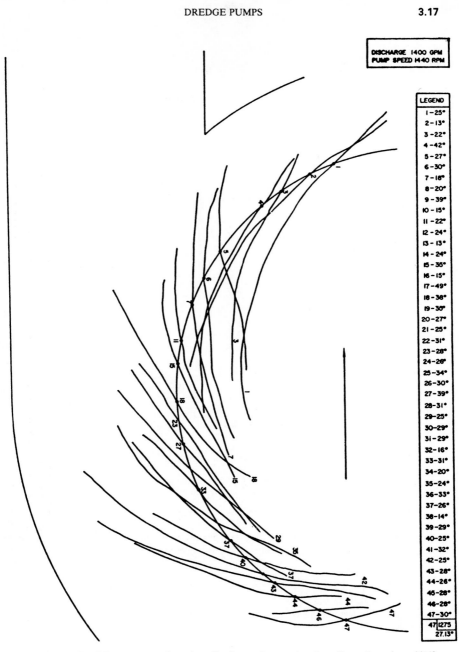

FIGURE 3.19 Particle movement from impeller into volute casing (impeller exit angle = 22½°, design discharge = 1400 gpm).

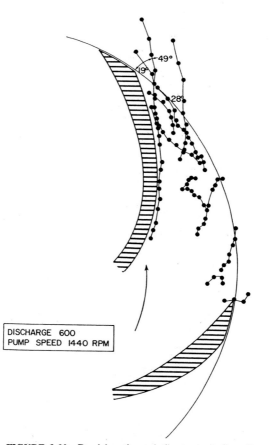

FIGURE 3.20 Bead location relative to vane (impeller exit angle = 22½°, discharge = 600 gpm, pump speed = 1440 rpm).

The absolute velocity of the particle can be determined by counting the number of frames per pump revolution and using the following relation:

$$c = \frac{1}{60} s\, FN \qquad (3.10)$$

where c = absolute velocity, ft/s (m/s)
  s = distance traveled by particle, foot per frame (meter per frame)
  F = frames per pump revolution
  N = pump speed, rpm

By measuring the exit angle of the bead, radial and tangential velocities can be calculated as components of the absolute velocity. These studies were made for 20 beads from each of two films—1440 rpm and 1000 gpm (63 l/s) and 1650 rpm and 1000 gpm. Figures 3.22 and 3.23 show the variation of radial and absolute velocities across the vane opening for 1440 rpm and 1000 gpm (63 l/s). The sus-

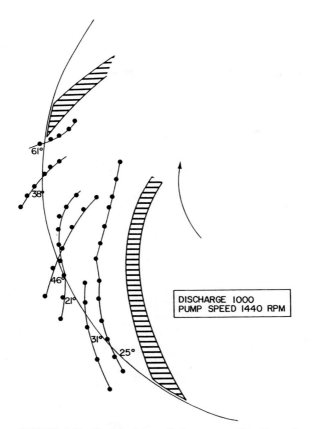

FIGURE 3.21 Bead location relative to vane (impeller exit angle = 22½°, discharge = 1000 gpm, pump speed = 1440 rpm).

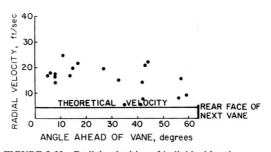

FIGURE 3.22 Radial velocities of individual beads.

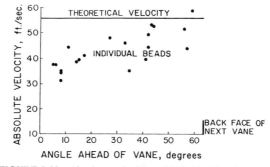

FIGURE 3.23 Absolute velocities of individual beads.

picion is then confirmed that radial velocities are high and tangential velocities are low compared to theory. The variation in exit angle across the vane opening is similar to that for radial velocity.

The velocities of the transporting liquid were calculated at various locations in and near the pump in search of an explanation for the high radial velocities observed. The results for 1440 rpm and 1000 gpm (63 l/s) are as follows:

Suction line	axial velocity = 19.3 ft/s (5.9 m/s)
Suction ring of impeller	axial velocity = 6.92 ft/s (1.9 m/s)
Discharge ring of impeller	radial velocity = 4.21 ft/s (1.3 m/s)

Before entering the impeller, the bead is seen to have a high axial velocity and, disregarding prerotation, no tangential component of velocity. The radial and tangential velocities at the discharge ring of the impeller are given in Table 3.2.

In other words, since the solid particles are not controlled by liquid continuity (AV = constant), they tend to retain their original velocity components because of their inertia, being changed only by frictional interaction with the transporting liquid.

These measured velocities have an important effect upon the relative exit angle of the particles, and the effect depends upon the vane angle of the pump. The theoretical and experimental velocity triangles are shown in Fig. 3.24 for two impellers. One impeller had a 22½° vane angle (Fig. 3.24a), and another had a 35° vane angle (Fig. 3.24b).

In Fig. 3.24a, the experimental velocity triangles result in a relative angle greater than the vane angle. This indicates that the particles in suspension tend to avoid contact with the vane and thus minimize the possibility of scour. On the other hand, in Fig. 3.24b, the relative angle is less than the vane angle, indicating contact between the vane and the particles and resulting in scour.

TABLE 3.2 Comparison of Theoretical and Measured Velocities

Velocity	Theoretical	Measured
Radial	4.21 ft/s (1.3 m/s)	15.6 ft/s (4.8 m/s)
Tangential	55.80 ft/s (17.0 m/s)	39.6 ft/s (5.2 m/s)

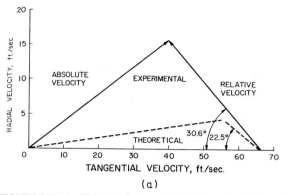

FIGURE 3.24(a) Exit velocity triangles for impeller with 22½° vane angle.

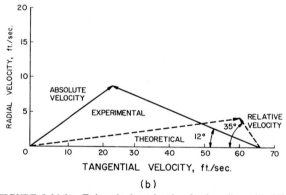

FIGURE 3.24(b) Exit velocity triangles for impeller with 35° vane angle.

The Number and Shape of the Impeller Vanes. Theoretically, an infinite number of vanes is required to produce the head indicated by the ideal velocity triangle. In an actual pump, the head and efficiency increases with the number of vanes[13] until the additional losses produced by the larger number of vanes reach some point where the efficiency is a maximum. The available flow area is also reduced by the finite thickness of the vanes. This is especially critical at the inlet where space is limited and the problem of cavitation may occur. Friction losses in a duct are minimal for the largest hydraulic radius. For a quadrangular passage, this is best suited by a square cross section. For a typical five-vane dredge pump impeller, the vane spacing to passage ratio may be about 1:7 at the radius of the impeller, corresponding to mid-distance along the vane. With the addition of another value, making a total of six, this ratio would be 1:4. Of course, it must be kept in mind that the flow conditions in a duct are not strictly the same as those occurring in the impeller channels as some of the investigations mentioned earlier have shown.

An incompressible nonviscous fluid with radial and rotational motion will ide-ally follow a spiral, possibly logarithmic, path in which the streamlines (imagi-nary lines, every point on which is tangent to the velocity vector at that point) at any point have a constant inclination with the tangent to the radius at that point.[7] Since this has a certain rational basis, and since a logarithmic spiral is geometri-cally simple, pump vanes are often designed in this manner. An involute curve may also be used in place of a spiral, although there seems to be no theoretical basis for employing such a curve.

Manufacturing cost and simplicity are also factors to be considered. Some-times pump vanes are curves which are a portion of a simple circular arc, al-though it is known that this does not give the best results. The circular arc does not give as satisfactory a flow path as the logarithmic spiral. Since the many cur-rent dredge pump impellers have vanes defined by a circular arc, or by several circular arcs, more sophisticated vane shapes should be employed as discussed in the impeller design section.

The vane tips used in many dredge impellers are blunt, as shown in Fig. 3.25. This will tend to cause disturbances in the volute. This effect may be partially or entirely eliminated by tapering the vanes as shown.[4] A number of different vane shapes and profiles have been tried by researchers in order to improve pump ef-ficiency. One of these trials consisted of utilizing only the active flow part of the impeller passage. Therefore, the areas of dead water mentioned previously were removed by shaping the vanes to occupy this area. "Club-headed" vanes, the width of which was greatly increased at the outlet to reduce the water passage to that occupied by the useful flow, did not produce the improvements which might be expected in efficiency and power at small flows.[14]

The discharge angle is one of the most important aspects of impeller design. It had been explained earlier that a fluid does not leave the vane tangent to the sur-face of the vane. Stepanoff[5] suggests 22½° as the best vane exit angle for water. This angle is the acute angle formed by a line tangent to the vane at its end and a line tangent to the impeller periphery, at the point. Angles from 17½° to 27½° are the usual limits for centrifugal water pump vane exit angles. For pumping viscous liquids, an increase in the discharge angle up to 60° is suggested. The discharge angle has an effect on the head-capacity characteristic curve of the pump. Studies were made on a water pump, keeping every variable except vane discharge angle and vane profile constant. Vane profile had to change to result in the various dis-charge angles tested. Moderate values of the angle, 20° to 30°, gave relatively flat head curves with good efficiencies. As the angle was increased, the head pro-duced increased a little, and efficiency fell off a little. The head curve tended to be more rounded with some maximum value. This would not be desirable if pumps of this type were to operate in parallel. The discharge angle of a typical dredge may be 35°. Tests on the model have shown the head capacity curves to

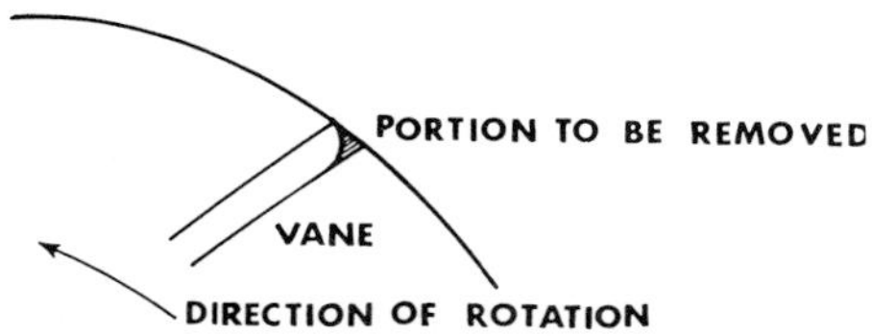

FIGURE 3.25   Rounding off of trailing edge of blunt vane tips.

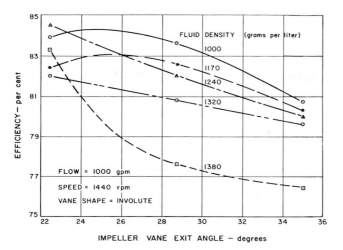

FIGURE 3.26   Effect of impeller vane exit angle on pump efficiency.

be rising to a maximum value and then falling off. A slight decrease in angle, say to 30°, may tend to produce a flatter curve and still not affect the efficiency.

Studies[6] on the model *Essayons* pump indicated that the maximum efficiencies were obtained with vane exit angles between 22½° and 28°45′ for fluid densities of 1000 and 1170 grams per liter (or specific gravity of 1.00 and 1.17) and close to 22½° (the lowest vane exit angle tested) for densities between 1240 and 1380 grams per liter (Fig. 3.26). These comparisons were made at best efficiency points for the pump tested.

Another factor related to the impeller which influences the efficiency of a pump is the clearance between the impeller and the cutwater or volute tongue. An excessive gap here leads to reduction in efficiency.[6] The gap on many dredge pumps is quite large, and if practical considerations would allow, this distance should be decreased with a resulting increase in efficiency. However, the clearance should be at least twice that recommended for a water pump, since friction effects will be more pronounced with a liquid of high viscosity.

Recirculation of particles with the volute casing was also studied on the model *Essayons* pump. Recirculation was measured by direct counts of beads passing back through the cutwater opening as compared to those leaving the pump through the discharge line. These counts were made by stopping the film at some arbitrary frame and counting all the beads within the two boundaries shown in Fig. 3.27. The length of the boundary is directly proportional to the particle velocity in the opening as determined by observation. This was repeated until a total of 100 to 400 beads had been counted for each film. In this way, the flow rate in each opening was directly proportional to the number of beads counted.

In summary, on the basis of the factors discussed, the following changes in the design of the hypothetical dredge pump impeller are recommended.

1. Change the profile of the vanes from a circular arc to a logarithmic spiral or an involute curve.

2. Increase the number of vanes from five to six if silt, clay, sand, gravel, and

water mixtures are pumped and no large particles are expected. Presence of large objects may limit the number of vanes to four.

3. Decrease the vane exit angle to a value of about 22½°.
4. Round off the blunt trailing edges of the vanes with a small radius fillet (Fig. 3.25).
5. Decrease the impeller-volute tongue clearance to a value at least twice that recommended for centrifugal water pumps, if practical considerations will allow it.

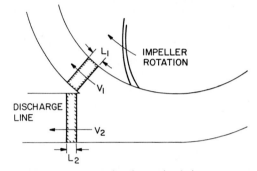

FIGURE 3.27 Boundaries for recirculation measurements.

MODEL TESTING OF PUMPS

Models are used to study flow phenomena which cannot be solved by analytical methods or by available prototype experimental results. The aerospace engineer obtains the design information from model tests in wind tunnels; the hydraulic engineer studies the models of hydraulic structures and rivers; the coastal engineer evaluates the models of tidal estuaries and seashore; the mechanical engineer tests models of turbines, blowers, compressors, and pumps to predict performance of full-size (prototype) machines; the naval architect tests ship models in towing tanks; the chemical engineer obtains design information from model tests, etc.

There are many types of similitude, all of which should be obtained if complete similarity is to exist between fluid flow phenomena. However, in some cases effective and useful similarity might be obtained without satisfying the complete similitude. In certain cases of models of rivers, harbors, and estuaries, departure from geometric similarity is permitted, resulting in distorted models.

The application of a special similarity law depends upon the forces determining the particular flow. Frequently, the flow depends not on one ratio of forces only but on two or possibly three ratios.

Dimensional Analysis—Buckingham π-Theorem

Modern fluid mechanics is based on a combination of physical analysis and experimental observations. The general objective is to provide dependable practical results and a thorough understanding of fundamental flow features. The dimensional analysis in conjunction with dynamic similarity has proved very useful in the organization, correlation, and interpretation of experimental data.

The basis of the development rests on the concept of dimensions of a physical quantity. The term *physical quantity* represents any property of a body and may be illustrated by length, weight, viscosity, mass, temperature, volume, etc. Each of these has both magnitude and character. These properties may be measured in terms of a given physical quantity or by combinations of fundamental physical quantities (i.e., length L, mass M, and time T). Thus the dimensional formula for velocity is L/T. Each fundamental dimension, however, may be divided into various units of measurement. Suppose a distance is equal to 24,000. This does not have a meaning unless we specify the dimension. To a sailor the speed of wind is meaningless unless it is expressed on the scale of Beaufort (i.e., number 10—a very strong gale).

A distance may be measured in any one number of units, all having the same dimension—length. So let us choose the length as our basic dimension; then the velocity is the rate of change of length in time. However, we have to adopt the standard time and standard length so that others know what we mean. Similarly, the acceleration is the rate of change of velocity L/T^2 and so on.

If a quantity (such as a pressure gradient or a hydraulic slope) is a ratio, and all of the fundamental dimensions disappear, the quantity is said to be dimensionless. It will then be the same numerically, regardless of what consistent system of units is used.

The F-L-T system, usually referred to as the engineer's system, involves the dimensions of force, length, and time. The M-L-T system, involving mass, length, and time, is called the absolute or physicist's system. They differ only in the distinction between mass and force.

In the engineer's system, F-L-T, the unit mass is defined as that which acquires unit acceleration when acted on by a unit force. This unit of force depends upon gravitational acceleration (g) and varies with location. American units are the pounds for force, the foot for length, and the second for time. The unit of mass is the slug.

The entire problem of data organization involves that of obtaining a convenient set of coordinates. In some problems it may be possible to discover convenient coordinates very easily; in other problems, π-theorem, first devised by Buckingham (1915)[15], is generally used.

A general function of n variables or physical quantities may be represented by

$$f(Q_1,Q_2,Q_3,\ldots,Q_n) = 0 \qquad (3.11)$$

The theorem states that if these n variables can be given in terms of m dimensional units, the general equation may then be expressed as a function of $n - m$ dimensionless π-terms, and that each term will have $(m + 1)$ variables, of which only one need be changed from term to term. Hence,

$$F_1(\pi_1,\pi_2,\ldots,\pi_{n-m}) = 0 \qquad (3.12)$$

The criterion that this equation shall be dimensionally homogeneous is that each term, when expressed in dimensionless units of M-L-T or F-L-T, must contain identical powers of each of the respective dimensions.[16]

The variables that can influence fluid motion are (1) geometry of the process, that is, a series of linear dimensions that defines the boundaries, such as a, b, c, d; (2) flow of the process, certain kinematic and dynamic quantities, such as mean velocity, a pressure increment or gradient; and (3) fluid in the process, the physical properties of the fluid, such as density, specific weight, viscosity, surface tension, elastic modulus.

Thus, the general equation of flow is

$$f_2(a, b, c, d, V, \Delta p, \rho, \gamma, \mu, \sigma, E) = 0 \qquad (3.13)$$

where V = velocity, Δp = pressure change, ρ = density, γ = specific weight, μ = dynamic viscosity, σ = surface tension, and E = modulus of elasticity.

Each term can be expressed in three-dimensional units. There are 11 terms, therefore $11 - 3 = 8$ dimensionless π-terms in the function.

$$f_3(\pi_1, \pi_2, \ldots, \pi_8) = 0 \qquad (3.14)$$

In general, it is expedient to use a length, velocity, and density, with the remaining eight terms appearing singly in each group, with a negative exponent. (A positive exponent can be used also.)

$$
\begin{aligned}
\pi_1 &= a^{x_1} V^{y_1} \rho^{z_1} b^{-1} \\
\pi_2 &= a^{x_2} V^{y_2} \rho^{z_2} c^{-1} \\
\pi_3 &= a^{x_3} V^{y_3} \rho^{z_3} d^{-1} \\
\pi_4 &= a^{x_4} V^{y_4} \rho^{z_4} \Delta p^{-1} \\
\pi_5 &= a^{x_5} V^{y_5} \rho^{z_5} v^{-1} \\
\pi_6 &= a^{x_6} V^{y_6} \rho^{z_6} \mu^{-1} \\
\pi_7 &= a^{x_7} V^{y_7} \rho^{z_7} \sigma^{-1} \\
\pi_8 &= a^{x_8} V^{y_8} \rho^{z_8} E^{-1}
\end{aligned}
\qquad (3.15)
$$

Let the quantities in each term be expressed in the dimensional units and equate the sum of the exponents to zero. There will be three simultaneous linear equations to find three dimensionless unknowns.

$$\pi_1 = L^{x_1}\left(\frac{L}{T}\right)^{y_1}\left(\frac{M}{L^3}\right)^{z_1} L^{-1}$$

$$\text{For } L: \quad x_1 + y_1 - 3z_1 - 1 = 0$$
$$\text{For } T: \quad -y_1 = 0 \qquad (3.16)$$
$$\text{For } M: \quad z_1 = 0$$
$$\therefore x_1 = 1 \; y_1 = 0 \; z_1 = 0$$

$$\pi_1 = \frac{a}{b}$$

similarly

$$\pi_2 = \frac{a}{c}$$

$$\pi_3 = \frac{a}{d}$$

Also

$$\pi_4 = \frac{\rho V^2}{\Delta p} = N_E = \text{Euler number}$$

$$\pi_5 = \frac{\rho V^2}{\gamma a} = N_F = \text{Froude number}$$

$$\pi_6 = \frac{\rho V a}{\mu} = N_R = \text{Reynolds number}$$

$$\pi_7 = \frac{\rho V^2 a}{\sigma} = N_W = \text{Weber number}$$

$$\pi_8 = \frac{\rho V^2}{E} = N_C = \text{Cauchy number}$$

The equation is then:

$$f_3\left(\frac{a}{b}, \frac{a}{c}, \frac{a}{d}, \frac{\rho V^2}{\Delta p}, \frac{\rho V^2}{\gamma a}, \frac{\rho V a}{\mu}, \frac{\rho V^2 a}{\sigma}, \frac{\rho V^2}{E}\right) = 0 \qquad (3.17)$$

If an expression for velocity is sought, one of the terms which includes velocity may be taken out of the functional expression, that is,

$$\frac{\rho V^2}{\Delta p} = f_4\left(\frac{a}{b}, \frac{a}{c}, \frac{a}{d}, \frac{\rho V^2}{\gamma a}, \frac{\rho V a}{\mu}, \frac{\rho V^2 a}{\sigma}, \frac{\rho V^2}{E}\right) \qquad (3.18)$$

or

$$V^2 = \frac{\Delta p}{\rho} f_5\left(\frac{a}{b}, \frac{a}{c}, \frac{a}{d}, N_F, N_R, N_W, N_E\right) \qquad (3.19)$$

Example

If in a fluid drag problem the variables are pressure p, length L, mass density ρ, dynamic viscosity μ, and velocity V, then

$$f(L, V, p, \rho, \mu) = 0 \qquad (3.20)$$

or substituting dimensions,

$$f\left(L, \frac{L}{T}, \frac{M}{LT^2}, \frac{M}{L^3}, \frac{M}{LT}\right) = 0 \qquad (3.21)$$

Since there are five variables, or $n = 5$, and since all three fundamental dimensions—L, M, T—are involved, $m = 3$. Hence, the number of dimensionless π-terms is $n - m = 2$, each with possibly $m + 1 = 4$ variables in it.

Select m (i.e., 3) of the variables as repeating, making sure that all dimensions L, M, and T are included in the three selected and preferably selecting one variable from each of the groups of linear variables, kinematic and dynamic variables, and fluid properties.

$$f = (\underbrace{L}_{\substack{\text{linear} \\ \text{variables}}}, \underbrace{V, p}_{\substack{\text{kinematic} \\ \text{and dynamic} \\ \text{variables}}}, \underbrace{\rho, \mu}_{\substack{\text{fluid} \\ \text{properties}}}) = 0 \qquad (3.22)$$

Select L, V, and ρ.

It should be noted that another selection of repeating variables would yield a correct result dimensionally but possibly a less useful form of dimensionless numbers.

$$\pi_1 = L^{x_1} V^{y_1} \rho^{z_1} p^{-1} \tag{3.23}$$

or

$$\pi_1 = [L]^{x_1} \left[\frac{L}{T}\right]^{y_1} \left[\frac{M}{L^3}\right]^{z_1} \left[\frac{M}{LT^2}\right]^{-1} = M^0 L^0 T^0 \tag{3.24}$$

Exponents of M: $\quad\quad z_1 - 1 = 0$

Exponents of L: $x_1 + y_1 - 3z_1 - 1 = 0$

Exponents of T: $\quad -y_1 \quad\quad + 2 = 0$

or

$$x_1 = 2, y_1 = 2, z_1 = 1$$

$$\pi_1 = \frac{L^2 V^2 \rho}{p} \tag{3.25}$$

$$\pi_2 = L^{x_2} V^{y_2} \rho^{z_2} \mu^{-1} \tag{3.26}$$

or

$$\pi_2 = [L]^{x_1} \left[\frac{L}{T}\right]^{y_2} \left[\frac{M}{L^3}\right]^{z_2} \left[\frac{M}{LT}\right]^{-1} = M^0 L^0 T^0 \tag{3.27}$$

Exponents of M: $\quad\quad z_2 - 1 = 0$

Exponents of $\dot{L}$: $x_2 + y_2 - 3z + 1 = 0$

Exponents of T: $\quad -y_2 \quad\quad + 1 = 0$

or

$$x_2 = 1, y_2 = 1, z_2 = 1$$

$$\pi_2 = \frac{LV\rho}{\mu} \tag{3.28}$$

$$f_1(\pi_1, \pi_2) = 0$$

$$\pi_1 = f_2(\pi_2)$$

or

$$\frac{1}{\pi_1} = f_3(\pi_2)$$

$$\frac{p}{L^2 V^2 \rho} = f_3 \frac{LV\rho}{\mu} = f_3(N_R) \tag{3.29}$$

or

$$p = L^2 V^2 \rho f_3(N_R) \tag{3.30}$$

This equation may be compared with the normal drag equation:

$$p = \tfrac{1}{2} C_D A V^2 \rho$$

thus

$$C_D = f_4(N_R) \tag{3.31}$$

as may be expected.

Choice of other repeating variables yields a different result which is less useful but which can be rearranged to yield the same result as before.

Let us choose L, p, and μ as repeating variables:

$$\pi_1 = L^{x_1} p^{y_1} \mu^{z_1} \rho^{-1} \qquad = \frac{\mu^2}{p\rho}, \tag{3.32}$$

$$\pi_2 = L^{x_2} p^{y_2} \mu^{z_2} V^{-1} \qquad = \frac{p}{\mu V L} \tag{3.33}$$

$$\pi_2 = f_1(\pi_1) \qquad \text{or} \qquad \frac{p}{\mu V L} = f_1 \frac{\mu^2}{p\rho} \tag{3.34}$$

This does not represent a satisfactory formula for pressure p since p occurs on both sides of the formula. However the π-terms may be combined to form new and more suitable ones.

Hence, since $\pi_2 = f_1(\pi_1)$, it can be written

$$\pi_1 \pi_2^2 = f_2 \frac{1}{\pi_1 \pi_2} \tag{3.35}$$

or

$$\pi_3 = f_2(\pi_4)$$

where

$$\pi_3 = \pi_1 \pi_2^2 = \left(\frac{\mu^2}{p\rho}\right) \left(\frac{p}{\mu V L}\right)^2 = \frac{p}{\rho V^2 L^2} \tag{3.36}$$

and

$$\pi_4 = \frac{1}{\pi_1 \pi_2} = \left(\frac{p\rho}{\mu^2}\right) \left(\frac{\mu V L}{p}\right) = \frac{V L \rho}{\mu} \tag{3.37}$$

that is,

$$\frac{p}{\rho V^2 L^2} = f_2 \frac{V L \rho}{\mu} \tag{3.38}$$

which was the same result as obtained with L, V, and ρ as repeating variables.

Buckingham's π-theorem of developing dimensionless numbers is but one of several methods. As early as 1899 Rayleigh[17] applied dimensional analysis to the problem of temperature effect on viscosity. Although Rayleigh's method is different from the π-theorem, it produces the same results.

A convenient method of developing dimensionless numbers is described by Langhaar.[18] This method employs matrix algebra and gives the most complete and useful results.

Similitude Relationships

Many textbooks and handbooks describe the necessity for geometric, kinematic, and dynamic similarity and discuss the physical meaning of various dimensionless ratios such as Reynolds, Froude, Weber, Mach, etc.[17,18,19]

Similarity Conditions Developed from Navier-Stokes Equations. The Navier-Stokes equations of motion may be written in Cartesian coordinates as in Eq. (3.39).[20]

$$\frac{\partial u}{\partial t} + u\frac{\partial u}{\partial x} + v\frac{\partial u}{\partial y} + w\frac{\partial u}{\partial z} = -g\frac{\partial h}{\partial x} - \frac{1}{\rho}\frac{\partial p}{\partial x} + \frac{\mu}{\rho}\left(\frac{\partial^2 u}{\partial x^2} + \frac{\partial^2 u}{\partial y^2} + \frac{\partial^2 u}{\partial z^2}\right)$$

$$\frac{\partial v}{\partial t} + u\frac{\partial v}{\partial x} + v\frac{\partial v}{\partial y} + w\frac{\partial v}{\partial z} = -g\frac{\partial h}{\partial y} - \frac{1}{\rho}\frac{\partial p}{\partial y} + \frac{\mu}{\rho}\left(\frac{\partial^2 v}{\partial x^2} + \frac{\partial^2 v}{\partial y^2} + \frac{\partial^2 v}{\partial z^2}\right) \qquad (3.39)$$

$$\frac{\partial w}{\partial t} + u\frac{\partial w}{\partial x} + v\frac{\partial w}{\partial y} + w\frac{\partial w}{\partial z} = -g\frac{\partial h}{\partial z} - \frac{1}{\rho}\frac{\partial p}{\partial z} + \frac{\mu}{\rho}\left(\frac{\partial^2 w}{\partial x^2} + \frac{\partial^2 w}{\partial y^2} + \frac{\partial^2 w}{\partial z^2}\right)$$

It will be noted that for isothermal flows and incompressible flows in a field of gravity, there are four flow terms which appear in the equation of motion. These are three velocity components, u_1, v_1, and w_1 and pressure p.

The Navier-Stokes equations, in addition to the incompressible continuity equation,

$$\frac{\partial u}{\partial x} + \frac{\partial v}{\partial y} + \frac{\partial w}{\partial z} = 0$$

permit a solution of a fluid flow problem if the boundary conditions are known.

Equation (3.39) may be written in terms of dimensionless quantities defined as follows:

$$x_1 = \frac{x}{L} \qquad \text{where } L = \text{a characteristic length} \qquad (3.40)$$

$$y_1 = \frac{y}{L}$$

$$z_1 = \frac{z}{L}$$

$$h_1 = \frac{h}{L}$$

$$u_1 = \frac{u}{V} \qquad \text{where } V = \text{a characteristic velocity}$$

$$v_1 = \frac{v}{V}$$

$$w_1 = \frac{w}{V}$$

$$t_1 = \frac{t}{T} = \frac{t}{L/V}$$

$$p_1 = \frac{p}{\rho V^2}$$

ρ = constant
μ = constant
g = constant

Substituting the above dimensionless equations in Eq. (3.40) results in Eq. (3.41):

$$\left(\frac{V^2}{L}\right)\frac{\partial u_1}{\partial t_1} + \left(\frac{V^2}{L}\right)u_1\frac{\partial u_1}{\partial x_1} + \left(\frac{V^2}{L}\right)v_1\frac{\partial u_1}{\partial y_1} + \left(\frac{V^2}{L}\right)w_1\frac{\partial u_1}{\partial z_1}$$

$$= -(g)\frac{\partial h_1}{\partial x_1} - \left(\frac{V^2}{L}\right)\frac{\partial p_1}{\partial x_1} + \frac{\mu V}{\rho L^2}\left(\frac{\partial^2 u_1}{\partial x_1^2} + \frac{\partial^2 u_1}{\partial y_1^2} + \frac{\partial^2 u_1}{\partial z_1^2}\right) \qquad (3.41)$$

Equation (3.41) may be made dimensionless by dividing by V^2/L, as shown in Eq. (3.42):

$$\frac{\partial u_1}{\partial t_1} + u_1\frac{\partial u_1}{\partial x_1} + v_1\frac{\partial u_1}{\partial y_1} + w_1\frac{\partial u_1}{\partial z_1}$$

$$= -\left(\frac{gL}{V^2}\right)\frac{\partial h_1}{\partial x_1} - \frac{\partial p_1}{\partial x_1} + \left(\frac{\mu}{\rho VL}\right)\left(\frac{\partial^2 u_1}{\partial x_1^2} + \frac{\partial^2 u_1}{\partial y_1^2} + \frac{\partial^2 u_1}{\partial z_1^2}\right) \qquad (3.42)$$

Note that all of the quantities are dimensionless, and two of the groups, (gL/V^2) and $(\mu/\rho\,VL)$, have a particular physical significance.

Group (gL/V^2) may be formed by dividing the inertia force by the gravity force. If the gravity force $F_g = mg = \rho L^3 g$ and the inertia force $F_i = ma = \rho V^2 L^2$, then

$$\frac{F_i}{F_g} = \frac{\rho V^2 L^2}{\rho L^3 g} = \frac{V^2}{Lg} = \pi_5 \qquad (3.43)$$

One form of the dimensionless number which involves gravity is called the Froude number (N_F), and

$$N_F = \sqrt{\pi_5} = \frac{V}{\sqrt{gL}} \qquad (3.44)$$

Group $(\mu/\rho VL)$ may be formed by dividing the inertia force by viscous force. If the viscous force $F_v = \mu VL$, then

$$\frac{F_i}{F_v} = \frac{\rho V^2 L^2}{\mu VL} = \frac{\rho VL}{\mu} = \pi_6 \qquad (3.45)$$

This dimensionless number is called the Reynolds number (N_R) and

$$N_R = \pi_6$$

Equation (3.42) can be written as

$$\frac{\partial u_1}{\partial t_1} + u_1\frac{\partial u_1}{\partial x_1} + v_1\frac{\partial u_1}{\partial y_1} + w_1\frac{\partial u_1}{\partial z_1} = -\frac{1}{N_F^2}\frac{\partial h_1}{\partial x_1} - \frac{\partial p_1}{\partial x_1}$$

$$+ \frac{1}{N_R}\left(\frac{\partial^2 u_1}{\partial x_1^2} + \frac{\partial^2 u_1}{\partial y_1^2} + \frac{\partial^2 u_1}{\partial z_1^2}\right) \tag{3.46}$$

The same groups may be obtained from the y and z components of Eq. (3.40).

Similitude Ratios.   Table 3.3 summarizes scale ratios for the laws of Froude, Reynolds, Weber, and Cauchy. The scale ratios may be derived for other laws whenever needed. A sample derivation for time ratio based on Froude law is presented below.

From the second law of Newton, the force ratio is given by

$$F_r = M_r\frac{L_r}{(T_r)^2} \tag{3.47}$$

where M = mass
   L = linear dimension
   T = time dimension
   subscript r = ratio

or

$$F_r = \rho_r\,(L_r)^3\frac{L_r}{(T_r)^2} = \rho_r\frac{(L_r)^4}{(T_r)^2} \tag{3.48}$$

Also, it may be shown that the force ratio based on gravity effects $F_r = M_r g_r$ or $F_r = \rho_r L_r^3 g_r$, simplifying $F_r = \gamma_r L_r^3$ and equating Eqs. (3.47) and (3.48)

$$\rho_r\frac{(L_r)^4}{(T_r)^2} = \gamma_r\, L_r^3$$

or

$$T_r = \sqrt{\frac{L_r}{g_r}} \tag{3.49}$$

This represents a time ratio based on the Froude Law. Equation (3.49) may be rewritten in terms of values for model and prototype:

$$\frac{T_m}{T_p} = \frac{\sqrt{L_m/g_m}}{\sqrt{L_p/g_p}} \tag{3.50}$$

TABLE 3.3 Similitude Ratios[16]

Symbols	Characteristics	Dimensions	Scale Ratios for the Laws of:			
			Froude	Reynolds	Weber	Cauchy
			Geometric similarity			
L	Length	L	L_r	L_r	L_r	L_r
A	Area	L^2	$(L_r)^2$	$(L_r)^2$	$(L_r)^2$	$(L_r)^2$
V	Volume	L^3	$(L)^3$	$(L)^3$	$(L)^3$	$(L_r)^3$
			Kinematic similarity			
T	Time	T	$(\sqrt{L\rho/\gamma})_r$	$(L^2\rho/\mu)_r$	$(\sqrt{L^3\rho/\sigma})_r$	$(L\sqrt{\rho/E})_r$
V	Velocity	L/T	$(\sqrt{L\gamma/\rho})_r$	$(\mu/L\rho)_r$	$(\sqrt{\sigma/L\rho})_r$	$(\sqrt{E/\rho})_r$
a	Acceleration	L/T^2	$(\gamma/\rho)_r$	$(\mu^2/L^3\rho^2)_r$	$(\sigma/L^2\rho)_r$	$(E/L\rho)_r$
Q	Discharge	L^3/T	$(L^{5/2}\sqrt{\gamma/\rho})_r$	$(L\mu/\rho)_r$	$(L^{3/2}\sqrt{\sigma/\rho})_r$	$(L^2\sqrt{E/\rho})_r$
v	Kinematic viscosity	L^2/T	$(L^{3/2}\sqrt{\gamma/\rho})_r$	$(\mu/\rho)_r$	$(\sqrt{L\sigma/\rho})_r$	$(L\sqrt{E/\rho})_r$
			Dynamic properties			
M	Mass	M	$(L^3\rho)_r$	$(L^3\rho)_r$	$(L^3\rho)_r$	$(L^3\rho)_r$
F	Force	ML/T^2	$(L^3\gamma)_r$	$(\mu^2/\rho)_r$	$(L\sigma)_r$	$(L^2E)_r$
ρ	Density	M/L^3	ρ_r	ρ_r	ρ_r	ρ_r
γ	Specific gravity	M/L^2T^2	γ_r	$(\mu^2/L^3\rho)_r$	$(\sigma/L^2)_r$	$(E/L)_r$
μ	Dynamic viscosity	M/LT	$(L^{3/2}\sqrt{\rho\gamma})_r$	μ_r	$(\sqrt{\rho L\sigma})_r$	$(L\sqrt{E\rho})_r$
σ	Surface tension	M/T^2	$(L^2\gamma)_r$	$(\mu^2/L\rho)_r$	σ_r	$(LE)_r$
p	Pressure intensity	M/LT^2	$(L\gamma)_r$	$(\mu^2/L^2\rho)_r$	$(\sigma/L)_r$	$(E)_r$
I	Impulse and momentum	ML/T	$(L^{7/2}\sqrt{\rho\gamma})_r$	$(L^2\mu)_r$	$(L^{5/2}\sqrt{\rho\sigma})_r$	$(L^3\sqrt{E\rho})_r$
E	Energy and work	ML^2/T^2	$(L^4\gamma)_r$	$(L\mu^2/\rho)_r$	$(L^2\sigma)_r$	$(L^3E)_r$
P	Power	ML^2/T^3	$(L^{7/2}\gamma^{3/2}/\rho^{1/2})_r$	$(\mu^3/L\rho^2)_r$	$(\sigma^{3/2}\sqrt{L/\rho})_r$	$(L^2E^{3/2}/\rho^{1/2})_r$

Note: When g is the same for model and prototype, $(\gamma/\rho)_r = 1$.

The discharge ratio based on the Froude Law may be obtained as:

$$Q_r = A_rV_r = L_r^2V_r \qquad (3.51)$$

and

$$V_r = \sqrt{L_r\frac{\gamma_r}{\rho_r}} \qquad (3.52)$$

Hence,

$$Q_r = L_r^2\sqrt{L_r\frac{\gamma_r}{\rho_r}} = (L_r)^{5/2}\sqrt{g_r} \qquad (3.53)$$

Similitude Ratios for Hydraulic Machinery.   Similitude requirements for hydraulic
machines follow those developed for close-conduit flow. Similarity of the flow in
hydraulic machines requires a constant ratio between the fluid velocities and the
peripheral velocities of the runner at all geometrically similar points of the machines compared. This condition may be expressed by the following ratio:

$$\frac{V}{u} = \text{constant} \tag{3.54}$$

where V = fluid velocity
$\quad\; u$ = peripheral velocity

For pumps and turbines:

$$\frac{Q/D^2}{\omega D} = \frac{Q}{\omega D^3} = \text{constant} \tag{3.55}$$

Dimensional Analysis of a Pump.   Three important quantities in connection with
the design or selection of a pump are: the head produced, H; the power input
required, P; and the efficiency, η.[21] For a machine of given design, each of these
is a function of the following independent variables: $\rho, \omega, D, Q,$ and μ. Here D
represents the impeller diameter, chosen as a convenient length characteristic of
the machine. The head H depends also on g, since H represents shaft work per
unit weight of fluid. The product gH, however, or the shaft work per unit mass of
liquid, is independent of g because the flow through the machine is totally enclosed and the fluid assumed incompressible. Accordingly, for a given design:

$$gH = f_1(\rho, \omega, D, Q, \mu) \tag{3.56}$$
$$P = f_2(\rho, \omega, D, Q, \mu) \tag{3.57}$$
$$\eta = f_3(\rho, \omega, D, Q, \mu) \tag{3.58}$$

Applying the π-theorem and letting $\rho, \omega,$ and D be the repeating variables, we
obtain,

$$\pi_1 = \frac{gH}{D^2\omega^2} \qquad \pi_2 = \frac{Q}{\omega D^3} \qquad \pi_3 = \frac{\rho\omega D^2}{\mu}$$

and

$$\frac{gH}{\omega^2 D^2} = f_4\left(\frac{Q}{\omega D^3}, \frac{\rho\omega D^2}{\mu}\right) \tag{3.59}$$

similarly

$$\frac{P}{\rho\omega^3 D^5} = f_5\left(\frac{Q}{\omega D^3}, \frac{\rho\omega D^2}{\mu}\right) \tag{3.60}$$

$$\eta = f_6\left(\frac{Q}{\omega D^3}, \frac{\rho\omega D^2}{\mu}\right) \tag{3.61}$$

If the effect of viscosity is neglected:

$$\frac{gH}{\omega^2 D^2} = f_4 \frac{Q}{\omega D^3} \tag{3.62}$$

$$\frac{P}{\rho \omega^3 D^5} = f_5 \frac{Q}{\omega D^3} \tag{3.63}$$

$$\eta = f_6 \frac{Q}{\omega D^3} \tag{3.64}$$

Equations (3.62) to (3.64) are useful in indicating the effect of changes in size or speed on the performance at maximum efficiency η_{max}. If $\eta = \eta_{max}$, Eq. (3.64) shows that $Q/\omega D^3$ has a definite constant value. Therefore,

$$\frac{Q}{\omega D^3} = C_1 \tag{3.65}$$

$$\frac{gH}{\omega^2 D^2} = C_2 \tag{3.66}$$

$$\frac{P}{\rho \omega^3 D^5} = C_3 \tag{3.67}$$

In other words, Q is proportional to ωD^3, H is proportional to $\omega^2 D^2$, and P is proportional to $\rho \omega^3 D^5$.

The concept of specific speed is applied in the selection of a pump to satisfy known operating conditions. The discharge Q, the head H, and the speed are usually required to have certain values in any proposed installation. A dimensionless product of these quantities that has a definite value for a machine of given design operating at maximum efficiency is obtained from Eqs. (3.65) and (3.66).

$$\frac{\omega Q^{1/2}}{(gH)^{3/4}} = \frac{C_1^{1/2}}{C_2^{3/4}} = \text{constant} \tag{3.68}$$

For pumps, the speed N is usually expressed in revolutions per minute (rpm), and the discharge in gallons per minute (gpm). The head H is expressed in feet. From the above relation, we can define:

$$N_s = \frac{N (Q)^{1/2}}{H^{3/4}} \tag{3.69}$$

where N_s, the specific speed, has a definite value for a pump of given design operating at maximum efficiency. For single-stage centrifugal pumps the values of N_s range between 500 and 5000, while for single-stage mixed flow pumps N_s lies between 3700 and 10,000 and for single-stage axial flow pumps it is between 10,000 and 15,000.

The specific speed for pumps may also be expressed in a dimensionless form as follows:

$$n_s = \frac{\omega Q^{1/2}}{(gH)^{3/4}} \tag{3.70}$$

where ω is in radians per second
Q is in cubic feet per second (cubic meters per second)
g is in feet per second squared (meters per second squared)
H is in feet (meters)

Note. The metric system may also be used, and the dimensionless number will have the same numerical value as in the foot-pound-second system.

If we assume that the only forces acting on the fluid are inertia forces, it is possible to establish a definite relation between the forces and the velocities under similar flow conditions. Using the impulse-momentum principle, it can be shown that

$$\frac{p}{\gamma} = H = C\frac{V^2}{g} \tag{3.71}$$

where C = constant

or by applying Eq. (3.71) specifically to the total head H under which the machine is operating, it is possible to obtain the following relationships:

$$H = C\frac{V^2}{2g} \quad \text{or} \quad H = C\frac{u^2}{2g} \tag{3.72}$$

where u = peripheral velocity of the runner or

$$\frac{2gH}{u^2} = \text{constant} \tag{3.73}$$

The dynamic relations [Eqs. (3.71), (3.72), and (3.73)], together with kinematic conditions of similarity Eq. (3.47), are the basis of all fundamental similarity relations for the flow in turbomachinery.

The hydraulic power P may be obtained from

$$P = \gamma Q H \tag{3.74}$$

where Q = discharge.
According to Eq. (3.75),

$$\frac{Q_m}{Q_p} = \frac{\omega_m D_m^3}{\omega_p D_p^3} \tag{3.75}$$

Also from dimensional or other considerations,

$$\frac{gH}{\omega^2 D^2} = \text{constant} \tag{3.76}$$

Therefore,

$$\frac{H_m}{H_p} = \frac{\omega_m^2 D_m^2 g_p}{\omega_p^2 D_p^2 g_m} \tag{3.77}$$

if

$$g_p = g_m \qquad (3.78a)$$

therefore,

$$\frac{H_m}{H_p} = \frac{\omega_m^2 D_m^2}{\omega_p^2 D_p^2} \qquad (3.78b)$$

Hence,

$$\frac{P_m}{P_p} = \frac{Q_m H_m}{Q_p H_p} = \frac{\omega_m^3 D_m^5}{\omega_p^3 D_p^5} \qquad (\text{if } \gamma_p = \gamma_m) \qquad (3.79)$$

A corresponding relationship for shaft torque T may be derived from the fact that

$$P \propto M\omega \qquad (3.80)$$

$$\frac{T_m}{T_p} = \frac{\omega_m^2 D_m^5}{\omega_p^2 D_p^5} \qquad (3.81)$$

The relationships may be summarized as follows:

1. For geometrically similar pumps operating at constant speed, or ω = constant

$$\frac{H_m}{H_p} = \frac{\omega_m^2 D_m^2}{\omega_p^2 D_p^2} \qquad \text{or} \qquad H_r = \omega_r^2 D_r^2 \qquad (3.82)$$

Since

$$\omega_r^2 = \text{constant}$$
$$H \propto D^2$$

Similarly

$$Q \propto D^3$$
$$P \propto D^5$$
$$T \propto D^5$$

2. For a given pump operating at variable speed, where D = constant

$$\frac{Q_m}{Q_p} = \frac{\omega_m D_m^3}{\omega_p D_p^3} \qquad \text{or} \qquad Q_r = \omega_r D_r^3 \qquad (3.83)$$

Since

$$D = \text{constant}$$
$$Q \propto \omega$$

Similarly

$$H \propto \omega^2$$
$$T \propto \omega^2$$
$$P \propto \omega^3$$

3. In addition, it can be stated that the specific speed N_s describes a combination of operating conditions that permits similar flow conditions in geometrically similar machines.

PUMP CHARACTERISTICS

Dimensional Presentation

Presentation of dredge pump characteristics essentially follows the methods used by the water pump industry. The types of graphs, curves, etc., evolved gradually over the years and everyone dealing with manufacturing, sales, or use of pumps became familiar with typical curves, such as

1. Total head H as a function of discharge Q
2. Brake horsepower (bhp) as a function of discharge Q
3. Efficiency η as a function of discharge Q
4. Net positive suction head (NPSH) as a function of discharge Q

Head-Discharge Curves.   The head in this case is usually the total head and not the suction head or the discharge head. The total head is given by the following equation:

$$H = H_d - H_s \tag{3.84}$$

where H_d = discharge head in units of energy per pound of fluid or ft · lb/lb = ft, or m kg/kg = m.

$$H_s = \text{suction head}$$

since

$$H_d = \frac{P_d}{\gamma} + \frac{V_d^2}{2g} + z_d \tag{3.85}$$

where P_d = pressure in discharge pipe
  V_d = mean velocity in discharge pipe
  γ = specific weight of fluid
  z_d = elevation of centerline of discharge pipe with respect to centerline of the pump

and

$$H_s = \frac{P_s}{\gamma} + \frac{V_s^2}{2g} + z_s \tag{3.86}$$

where P_s = pressure in suction pipe
  V_s = mean velocity in suction pipe
  z_s = elevation head with respect to centerline of the pump

Substituting in Eq. (3.84),

$$H = \left(\frac{P_d}{\gamma} + \frac{V_d^2}{2g} + z_d\right) - \left(\frac{P_s}{\gamma} + \frac{V_s^2}{2g} + z_s\right) \tag{3.87}$$

$$H = \left(\frac{P_d}{\gamma} - \frac{P_s}{\gamma}\right) + \left(\frac{V_d^2}{2g} - \frac{V_s^2}{2g}\right) + (z_d - z_s) \tag{3.88}$$

or

H = difference in pressure + difference in velocity + difference in
    heads between dis-   heads between dis-   elevation
    charge and suction   charge and suction

or

$$H = \frac{\Delta P_{d-s}}{\gamma} + \frac{\Delta V_{d-s}^2}{2g} + \Delta z_{d-s} \tag{3.89}$$

since

$$\frac{P_d}{\gamma} = H_d \tag{3.90}$$

where

$$H_d = \text{discharge head}$$

and

$$\frac{P_s}{\gamma} = H_s$$

where

$$H_s = \text{suction head} \tag{3.91}$$

$$H = H_d - H_s + \frac{V_d^2 - V_s^2}{2g} + z_d - z_s \tag{3.92}$$

In water pumps the suction and velocity pipe diameters are usually equal and $(V_d^2 - V_s^2)/2g = 0$, but in dredge pumps the suction pipe is usually larger than the discharge pipe and the difference in velocity heads may be large and should not be overlooked.

The head as given in manufacturers' specifications is usually expressed in feet of water. This is erroneous and in the case of water-solids mixtures the total head should be expressed in feet of actual mixture flowing.

The discharge is usually the actual measured discharge (volume divided by time). A typical head-discharge plot is shown in Fig. 3.28.

Brake Horsepower-Discharge Curves. The power input to the pump used to be measured by a friction "brake" device, so the power input is referred to as brake horsepower. A more correct expression is the shaft horsepower. One horsepower is equal to 550 ft · lb/s. A typical brake horsepower-discharge curve is shown in Fig. 3.29.

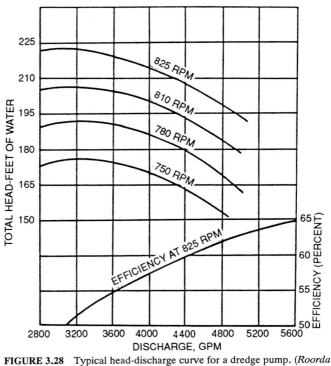

FIGURE 3.28 Typical head-discharge curve for a dredge pump. (*Roorda and Vertregt, 1963*)

Efficiency-Discharge Curves. The overall efficiency of dredge pumps is usually defined as

$$\eta = \frac{\text{water horsepower}}{\text{brake horsepower}} = \frac{\text{whp}}{\text{bhp}} = \frac{\gamma HQ}{550 \text{ bhp}} \tag{3.93}$$

Net Positive Suction Head-Discharge. NPSH at some location is usually defined as the total absolute head minus the vapor pressure head, or

$$\text{NPSH} = \frac{P_a}{\gamma} + \frac{P_s}{\gamma} + \frac{V_s^2}{2g} - \frac{P_v}{\gamma} \tag{3.94}$$

$$= \text{total absolute head} - \text{vapor pressure head}$$

where P_a = local barometric pressure
P_v = vapor pressure of liquid

NPSH should also be expressed in feet of liquid pumped. Typical NPSH discharge curves are shown in Fig. 3.30.

Examples. Examples of dimensional plots given by the manufacturers or produced elsewhere are presented in Figs. 3.29 through 3.34.

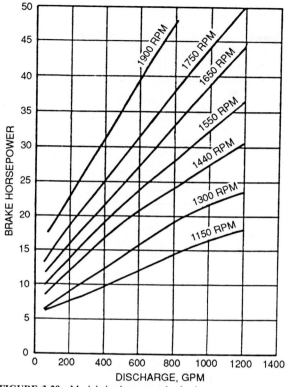

FIGURE 3.29 Model dredge pump brake horsepower require-ments as a function of discharge. Fluid mixture specific gravity = 1.17. (*Herbich and Vallentine, 1961*)

Figure 3.30 shows the total head discharge curve for a number of impellers of different sizes, presumably in the same volute casing. The total head discharge curve is given for 24- to 30-in impellers. The lines of equal horsepower are also shown from 60 to 350 hp. This is presented instead of the usual brake horsepower discharge curves and one can determine from the graph the total head for a given motor. For example, if a 225-hp (168-kW) motor is available, a pump with a 26-in (66-cm) impeller will produce a total head of 80 ft (24.4 m) of water and will pump 7500 gpm (473 l/s) at about 70 percent total efficiency; or a pump with a 28-in (71-cm) impeller will pump 5050 gpm (318 l/s) developing a 124-ft (38-m) total head at 73 percent efficiency. The NPSH value in the first case will be about 28 ft (8.5 m) and in the second case about 12 ft (3.6 m).

The constant efficiency lines shown in Fig. 3.30 are sometimes referred to as "efficiency hill," since, if one is familiar with topographic maps, the efficiencies remind one of a hill.

Another dimensional plot is reproduced here from *Effect of Impeller Design Changes on Characteristics of a Model Dredge Pump*[1] in Fig. 3.31. This plot presents results for a model pump handling different concentrations of silt-clay-water mixtures. Total head in this case is expressed in pounds per square inch

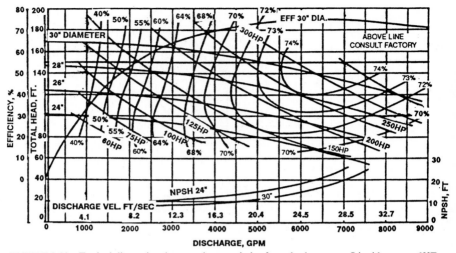

FIGURE 3.30 Typical dimensional pump characteristics for a dredge pump. Liquid: water, 60°F, pump size, 10 CK, speed, 700 rpm. (*Courtesy, Morris Machine Works*)

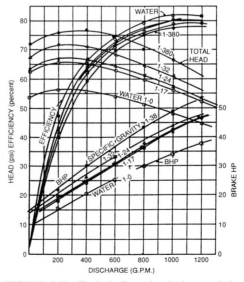

FIGURE 3.31 Typical dimensional characteristic curves for one speed (1750 rpm) and a range of specific gravities (1.0–1.38). (*Herbich and Vallentine, 1961*)

and is plotted as a function of discharge. Similarly the efficiency and the brake horsepower are plotted as a function of discharge. The pertinent information regarding the model dredge pump is given in Table 3.4.

TABLE 3.4 Information Pertinent to Model Dredge Pump

Impeller diameter	10½ in (26.7 cm)
Entrance vane angle	45°
Exit vane angle	22½°
Number of vanes	5
Width	2⅝ in (6.7 cm)
Suction pipe diameter	4½ in (11.4 cm)
Discharge pipe diameter	4 in (10.2 cm)
Liquid mixture	Silt-clay-water

As an example let us consider operation of the pump at maximum efficiency for fluid mixture of specific gravity equal to 1.38. The brake horsepower required is of the order of 52.5 hp, the total head developed is equal to 62.5 psi (430.9 kN/m²) for a total discharge of 1150 gpm (72.5 l/s) and efficiency of 78.2 percent.

De Groot[22] brought up an interesting diagram on restrictions in pump working range (Fig. 3.32). If power available to the pump is limited, the total head developed by the pump at any given discharge is determined by the pump geometric characteristics, such as impeller diameter, the number of impeller vanes, and the

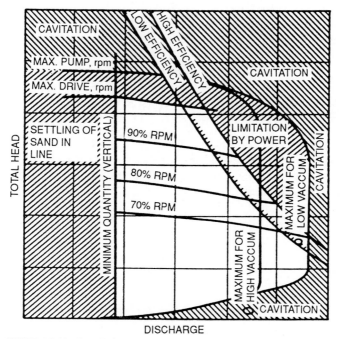

FIGURE 3.32 Restrictions in pump working range. (*De Groot, 1963*)

shape and width of impeller, and by the dynamic characteristic, such as speed, and by the cavitation characteristics of the pump, which depend to a large extent on suction pressure and on inlet and impeller geometry. The cavitation limits dredge pump operation at high heads and/or higher discharges. The working range of the pumps is also restricted by the speed and torque of the motor as well as the efficiency of the pump. The efficiency η may be denoted as the ratio of water horsepower (whp) developed to the brake horsepower (bhp) or

$$\eta = \frac{\text{whp}}{\text{bhp}} = \frac{\gamma QH}{\text{bhp}} \tag{3.95}$$

where γ = specific weight of mixture. Thus, the higher the efficiency, the higher the value of the product of discharge and head. This is indicated in Fig. 3.32, where the unshaded area represents the working range of the dredge pump.

Figure 3.33 is taken from *Floating Dredges*[23] and gives characteristics for a dredge pump installed on a trailing suction dredge.

An interesting comparison of water pump and dredge pump characteristics is given in Fig. 3.34.[24] In this semidimensionless plot the dimensionless pressure

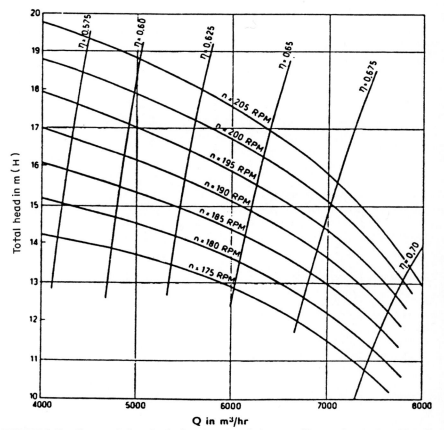

FIGURE 3.33 Characteristics of a dredge pump installed on a trailing suction dredge. (*Roorda and Vertregt, 1963*)

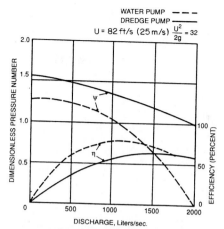

FIGURE 3.34 Comparison of water pump and dredge pump performance. (*Marnitz et al., 1963*)

number Ψ and efficiency η are given as a function of discharge Q. The dimensionless pressure number is defined as

$$\psi = \frac{H}{u^2/2g} = \frac{2gH}{u^2} \tag{3.96}$$

where

$$u = \text{rotational velocity in ft/s (m/s)} = \pi D n$$

where D = diameter
n = revolutions per second

The dimensional-type plotting is inefficient in itself, requiring a separate curve for each pump speed, impeller diameter, specific gravity of mixture, etc.

Everyone involved in dredging technology is urged to adopt the dimensionless parameters and dimensionless plotting described in the section on dimensionless presentation.

Semidimensionless Presentation

In this type of presentation one of the axes (the abscissa or the ordinate) is dimensionless while the other is dimensional. An example of this type of plot is shown in Fig. 3.34 and has certain advantages over the strictly dimensional presentation. However, a more useful presentation is a fully dimensionless plot or the one in which both the abscissa and the ordinate are dimensionless.

Dimensionless Presentation

The dimensionless expressions, parameters, or numbers have been discussed in a previous section. The parameters which are particularly useful in presenting dredge pump characteristics are:

$$\frac{gH}{\omega^2 D^2} \tag{3.97}$$

$$\frac{Q}{\omega D^3} \tag{3.98}$$

$$\frac{P}{\rho \omega^3 D^5} \tag{3.99}$$

where g = acceleration due to gravity in ft/s^2 (m/s^2)
H = total head in ft · lb/lb (m kg/kg)
ω = pump speed in radians per second
D = impeller diameter in ft (m)
Q = discharge in ft^3/s (m^3/s)
P = input power in ft · lb/s (Nm/s)
ρ = fluid density in slugs per cubic foot (kg/m^3)

The dimensionless characteristic curves thus become $(gH)/(\omega^2 D^2)$ as a function of $Q/(\omega D^3)$ or dimensionless head as a function of dimensionless discharge. It is comparable to head as a function of discharge curves. This type of plot is shown in Fig. 3.35.

In the same figure, efficiency, which is also a dimensionless parameter, is plotted as a function of dimensionless discharge. $P/(\rho \omega^3 D^5)$ as a function of $Q/(\omega D^3)$, or the dimensionless power as a function of dimensionless discharge, is shown in Fig. 3.36.

It should be noted that for any given pump there will only be one curve for several sizes of impeller diameters, for all speeds and for all fluid densities with comparable rheological characteristics. Some discrepancies may be observed if one were to plot on the same graph the results of tests for silt-clay-water mixtures of Bingham Body type and for coarse gravel mixture. Even then, in many cases the differences will be small from a practical point of view in field applications where the type of mixture and concentration may vary from hour to hour.

Complete characteristics of a given pump may thus be presented on one page, as shown in Fig. 3.37 where the total head, brake horsepower, and efficiency are all plotted as a function of discharge, all in dimensionless terms.[25,26]

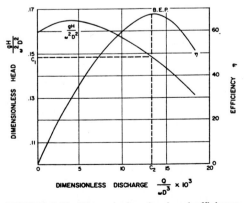

FIGURE 3.35 Dimensionless head and efficiency as a function of dimensionless discharge for a dredge pump. (*Herbich, 1975*)

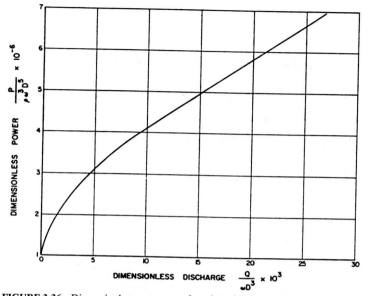

FIGURE 3.36 Dimensionless power as a function of dimensionless discharge for a dredge pump. (*Herbich, 1975*)

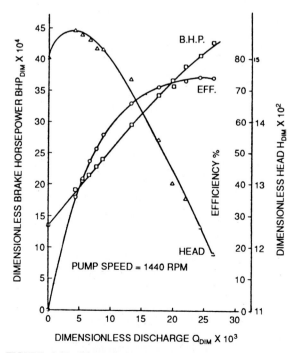

FIGURE 3.37 Dimensionless dredge pump characteristics. (*Herbich, 1975*)

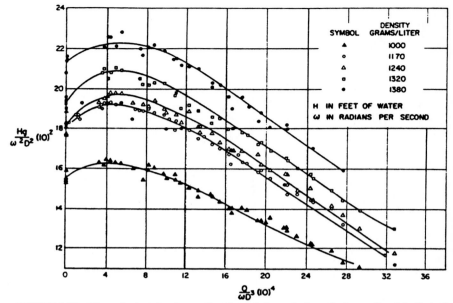

FIGURE 3.38 Dimensionless head as a function of dimensionless discharge (head in feet of water). (*Herbich, 1975*)

An example of dimensionless head (with head *H* expressed in feet of water) plotted as a function of dimensionless discharge is shown in Fig. 3.38. This is done purposely to stress the fact that all units must be compatible and that the head must be expressed in terms of feet of liquid mixture. In this diagram there

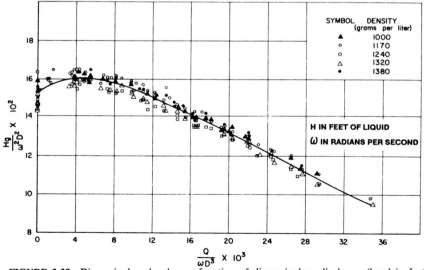

FIGURE 3.39 Dimensionless head as a function of dimensionless discharge (head in feet of liquid). (*Herbich, 1975*)

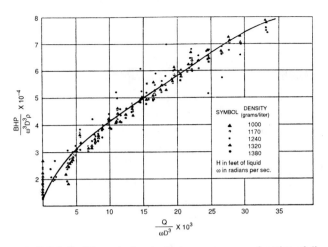

FIGURE 3.40 Dimensionless brake horsepower as a function of dimensionless discharge. (*Herbich, 1975*)

are five curves for each liquid density, since the head is expressed in feet of water rather than in feet of liquid.

However, if the head is computed in terms of liquid, only one curve results for all liquid densities, as shown in Fig. 3.39.

Dimensionless brake horsepower as a function of dimensionless discharge for a range of fluid densities is shown in Fig. 3.40. Although some experimental error exists, there is no doubt that a single curve can represent the relationship between brake horsepower and discharge for practical purposes.[27]

The actual values of head, discharge, or efficiency may be fairly quickly computed for any speed of rotation and fluid density. Some familiarization with dimensionless numbers is necessary but may be quickly acquired.

Example

Look up the value of $(gH)/(\omega^2 D^2)$, $Q/(\omega D^3)$, and $bhp/(\rho\omega^3 D^5)$ at the best efficiency point or any other point, that is,

$$\frac{gH}{\omega^2 D^2} = C_1 \tag{3.100}$$

$$\frac{Q}{\omega D^3} = C_2 \tag{3.101}$$

$$\frac{bhp}{\rho\omega^3 D^5} = C_3 \tag{3.102}$$

If the desired pump speed $= \omega_1$, pump diameter $= D_1$, and fluid density $= \rho_1$, then

$$H_1 = \frac{C_1(\omega_1)^2 (D_1)^2}{g} \tag{3.103}$$

$$Q_1 = C_2\omega_1(D_1)^3(D_1)^5 \tag{3.104}$$

$$\text{bhp} = C_3(\rho_1)(\omega_1)^3(D_1)^5 \tag{3.105}$$

There are many other dimensionless parameters which, it is hoped, will be adopted by everyone in the hydromachinery industry. Some of the other dimensionless parameters which are useful are as follows:

$$\frac{\text{NPSH}}{V_s^2/2g} \tag{3.106}$$

where NPSH = net positive suction head in feet of liquid mixture
V_s = suction pipe velocity

and

$$\frac{S}{V_s^2/2g} \tag{3.107}$$

where S = suction specific speed.

The cavitation parameters are fully discussed in the next section.

Another important parameter is the specific speed which can be made dimensionless in the following form:

$$n_s = \frac{\omega\sqrt{Q}}{(gH)^{3/4}} \tag{3.108}$$

where ω = radians per second
Q = cubic feet per second (cubic meters per second)
H = feet of liquid mixture (meters of liquid mixture)

EFFECT OF SOLID-WATER MIXTURES ON PUMP PERFORMANCE

The energy required to move solids in suspension is derived from the fluid, which acquires its energy from the impeller. The next effect of the presence of solids is an increase in the average specific gravity of the mixture. It has been shown that for a homogeneous mixture pipe friction loss is the same as for clear water if expressed in meters of mixture. Stepanoff[28] indicates that the head in meters of mixture at the best efficiency point should be the same as that for clear water, reduced only by the additional hydraulic losses caused by the presence of solids in the pump passages. Thus the efficiency for pumping solid-water mixtures will be reduced in the ratio of head reduction below that on clear water, or

$$\frac{H_m}{H} = \frac{\eta_m}{\eta} \tag{3.109}$$

where H_m = head in meters of mixture
η_m = efficiency when pumping solid-water mixture
H = head of clear water in meters
η = efficiency when pumping clear water

For homogeneous mixtures at a fixed capacity (BEP), the power will increase in the ratio of specific gravity (S.G.) increase, or

$$\text{bhp}_{\text{mixture}} = (\text{bhp}) \, \text{S.G.}_{\text{mixture}} \tag{3.110}$$

or

$$\text{bhp}_{\text{mixture}} = \frac{Q H_m S_m}{\eta_m} \quad \text{for mixture} \tag{3.111}$$

or

$$\text{bhp}_{\text{water}} = \frac{Q H}{\eta} \quad \text{for water} \tag{3.112}$$

Hence,

$$\frac{\text{bhp}_{\text{mixture}}}{(\text{bhp}) \, \text{S.G.}_{\text{mixture}}} = \frac{H_{\text{mixture}}}{H_{\text{water}}} \frac{\eta_{\text{water}}}{\eta_{\text{mixture}}} \tag{3.113}$$

Model dredge pump performance studies were conducted on a 1:8 model of a modified dredge pump installed on a hopper dredge *Essayons*. Information regarding the model pump is given in Table 3.5, and dimensionless pump characteristics are shown in Fig. 3.41.

TABLE 3.5 Information Pertinent to Model Dredge Pump

Impeller diameter	10.5 in (267 mm)
Entrance vane angle	45°
Exit vane angle	22.5°
Number of vanes	5
Width	2.6 in (66.7 mm)
Suction pipe diameter	5 in (114.3 mm)
Discharge pipe diameter	4 in (101.6 mm)
Liquid mixture	Silt-clay-water

Figure 3.42 presents the characteristics of a 1:8 model dredge pump. The solids consisted of 55 percent silt, 14 percent sand, 11 percent clay, and 20 percent colloid; 99.5 percent of the mixture was finer than 0.155 mm and 20 percent was finer than 0.001 mm. The pump was tested at four speeds and the performance follows the affinity laws within the accuracy of the test. At and near the best efficiency point (BEP), the discharge pressure varies directly as the specific gravity of the mixture. The head reduction is proportional to the efficiency drop.

Figure 3.43 shows the head in feet of mixture plotted as a function of capacity in gallons per minute (l/s) at one speed (1750 rpm) for the range of specific gravities of mixture from 1.0 (water) to 1.38. At specific gravity higher than 1.2, the mixture displays properties of a Bingham Body. Although the viscosity and thus the Reynolds number varies slightly, for practical dredging purposes the Reynolds number effects may be ignored.

Figure 3.44 shows three sets of performance curves of a 76-mm pump at 1450 rpm for fly ash (-200 mesh, specific gravity = 20.4), for sand (0.5–1.0 mm, specific gravity = 2.63), and for gravel (6–10 mm, specific gravity = 2.63). The results confirm that for a given rate of flow (at BEP) the brake horsepower varies

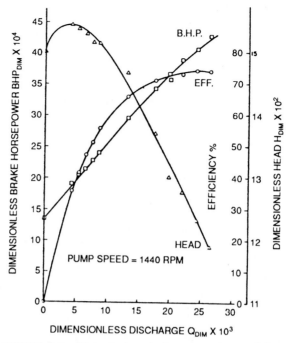

FIGURE 3.41 Dimensionless dredge pump characteristics. (*Herbich, 1975*)

directly as specific gravity. The head reduction varies directly as the efficiency drop. The discharge pressure increases directly as the specific gravity. Note that the efficiency reduction is greater for larger sizes of sediment.

Figure 3.45 shows a plot of pipe friction per 100-m length of 2-in (50.8-mm) pipe as a function of velocity at different consistencies (concentrations) by weight. For concentrations up to 21.1 percent, the mixtures display the properties of newtonian fluids; at higher concentrations the mixtures become Bingham Body fluids.

Figure 3.46 shows the ratios of $\eta_{\text{mixture}}/\eta_{\text{water}}$ as a function of the average grain size for a range of concentrations by volume from 6 to 36 percent. The data are taken from several published results as shown in Table 3.6. There are sufficient data to indicate that the points tend to align along a straight line on semilog scale for a given concentration.

Summary

1. For a given rate of flow (such as at BEP), the head in meters of mixture produced by the impeller is the same as that for clear water reduced only by the additional hydraulic losses caused by the presence of solids in the pump passages.

2. For pump discharge, pressure obtained with water is increased in proportion

to the specific gravity of the mixture, reduced only in the ratio of pump efficiencies.

3. The brake horsepower of the pump increases in the same ratio when the effects of increased apparent viscosity do not appear.

4. For homogeneous mixtures, the pipe friction head loss expressed in meters of mixture is essentially the same as for clear water.

5. The pump best efficiency discharge remains essentially the same as for clear water.

6. The efficiency of the pump handling solid-water mixtures is reduced in the same ratio as the head reduction below that obtained on clear water, or

$$\frac{H_{\text{mixture}}}{H} = \frac{\eta_{\text{mixture}}}{\eta}$$

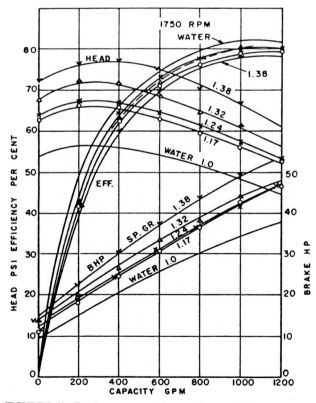

FIGURE 3.42 Dredge pump model test. (*Stepanoff, 1965*)

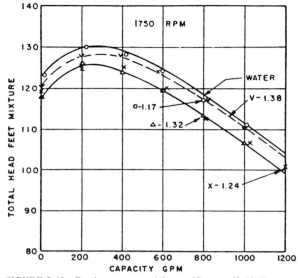

FIGURE 3.43   Dredge pump model test. (*Stepanoff, 1965*)

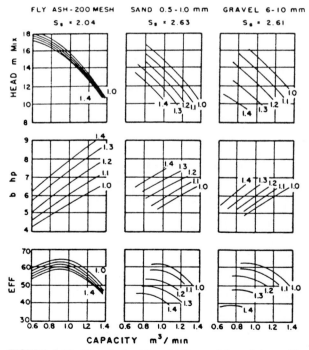

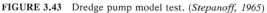

FIGURE 3.44   Three-inch pump, 1450-rpm 1.0 to 1.4 specific gravity. (*Hasegawa et al., 1958*)

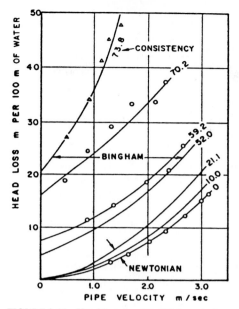
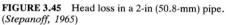

FIGURE 3.45 Head loss in a 2-in (50.8-mm) pipe. (*Stepanoff, 1965*)

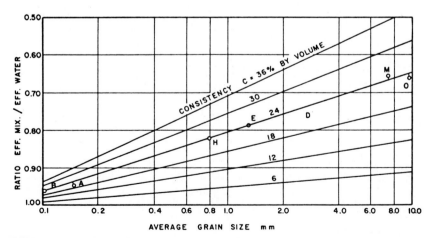

FIGURE 3.46 Ratio of efficiency mixture to efficiency water versus grain size. (*Stepanoff, 1965*)

TABLE 3.6 Legend for Figure 3.46

Symbol	Solids	Size, grain (mm)	Source	Remarks
A	Clay-silt	<0.15	Herbich et al.	$H_m/H = e_m/e$
B	Phosphate matrix		Antunes	$H_m/H = e_m/e$
C	Fly ash	<0.074	Hasegawa et al.	$H_m/H = e_m/e$
D	Sand	2.5	Terada	$H_m/H = e_m/e$
E	Sand	1.3	Terada	$H_m/H = e_m/e$
F	Sand	0.5	Terada	$H_m/H = e_m/e$
G	Gypsum		Antunes	$H_m/H = e_m/e$
H	Sand	0.8	Fairbanks	$H_m/H = e_m/e$
J	Sand	0.5–1.0	Hasegawa et al.[29]	$H_m/H = e_m/e$
K	Gravel	6–10	Hasegawa et al.	$H_m/H = e_m/e$
L	Sand	<0.5	O'Brien et al.	$H_m/H = e_m/e$
M	Coal	0–0.5	Sasaki et al.	$H_m/H > e_m/e$
N	Coal	5–10	Sasaki et al.	$H_m/H > e_m/e$
O	Coal	10–15	Sasaki et al.	$H_m/H > e_m/e$
P	Different	Mixture	Hotta	$H_m/H = e_m/e$

Source: Stepanoff, 1965.

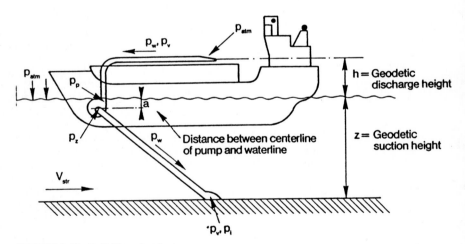

FIGURE 3.47 Definition sketch.

Other Effects

The other factors affecting the performance of a hopper dredge (Fig. 3.47) can be shown in a series of three-dimensional graphs.[30] Figure 3.48 presents the effect of depth of pump below the waterline. Figure 3.49 shows the effect of dredging depth, Fig. 3.50 presents the effect of suction pipe diameter, and Fig. 3.51 shows the effect of soil type.

Graphical representation of the conditions in a proposed pumping system is quite useful in the system analysis.

System friction curve: Friction-head loss in a pumping system is a function of pipe size, length, minor losses, mixture flow rate, and specific gravity of the mixture. A plot of head as a function of flow rate is known as the *system friction curve* (Fig. 3.52). (Note that the curve passes through the origin since there is no flow in the system at zero head.)

System-head curve: The system-head curve is obtained by combining the system friction curve with a head-rate of flow curve for a given pump. The pump should be selected to operate where the best efficiency point meets the system friction curve (Fig. 3.53). Variable speed drives shift the *H-Q* curve up or down and slide the operating point along the system curve.[31]

Types of system-head curves: When selecting a pump for a dredging system, it may be necessary to use a range of points on the system-head curves since the length of discharge pipeline will generally vary on a given project and the lift may also vary. If there is no lift and all losses are caused by friction, the system-head curve is quite steep (Fig. 3.54). If lift is significant and the discharge line is very short, the system-head curve is quite flat, as shown in Fig. 3.55.[32]

Cutterhead Pipeline Dredge

In a typical cutterhead dredge operation with a long discharge pipeline friction, losses are significant and lift is generally low, resulting in a steep system curve (Fig. 3.56).

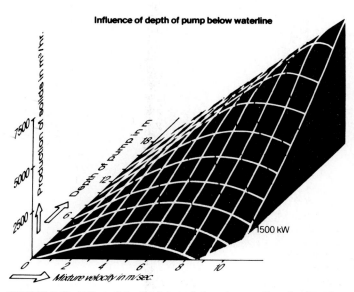

FIGURE 3.48 Effect of depth of pump below the waterline. Suction pipe diameter = 0.96 ft (800 mm), ρ_s = 1900 kg/m³, and dredging depth = 65.6 ft (20 m). *U* = 60 is a specific diameter ≈ 0.165 mm. (*Ports & Dredging, IHC Holland, 1986*)

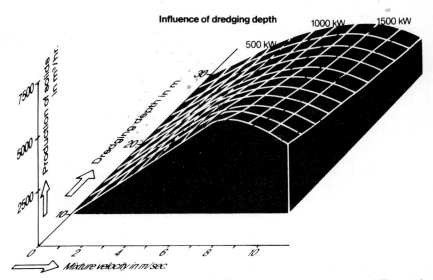

FIGURE 3.49 Effect of dredging depth. Distance between pump centerline and waterline = 29.5 ft (9 m), suction pipe diameter = 0.96 ft (800 mm), and ρ_s = 1900 kg/m³. U = 60 is a specific diameter ≈ 0.165 mm. (*Ports & Dredging, IHC Holland, 1986*)

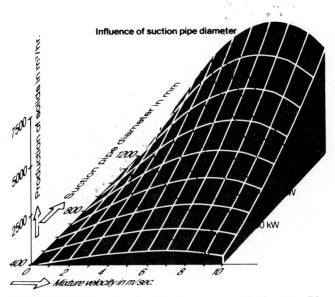

FIGURE 3.50 Effect of suction pipe diameter on production rates. Distance between pump centerline and water line = 29.5 ft (9 m), dredging depth = 65.6 ft (20 m), and ρ_s = 1900 kg/m³. U = 60 is a specific diameter ≈ 0.165 mm. (*Ports & Dredging, IHC Holland, 1986*)

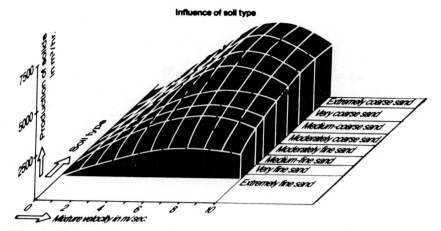

FIGURE 3.51 Effect of soil type on production rates. Suction pipe diameter = 0.96 ft (800 mm), distance between pump centerline and waterline = 29.5 ft (9 m), and dredging depth = 65.6 ft (20 m). (*Ports & Dredging, IHC Holland, 1986*)

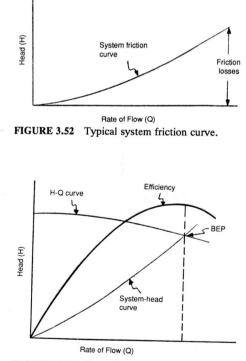

FIGURE 3.52 Typical system friction curve.

FIGURE 3.53 Typical centrifugal pump characteristics.

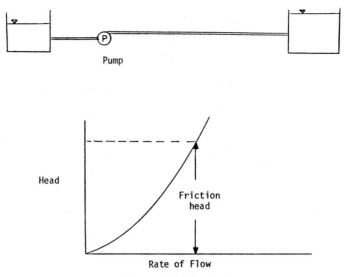

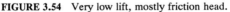

FIGURE 3.54   Very low lift, mostly friction head.

Variable Speed Pump

The variable speed affects the operating point along the system curve, as shown in Fig. 3.57. For example, a higher pump speed will increase both the head developed by the pump and the rate of flow. Thus, the same pump can pump a higher rate of flow for a greater distance at a higher pump speed, while a lower pump speed will produce lower head and a lower rate of flow.

Rate of Flow as a Function of System-Head Curves

Flat system curves (i.e., in a hopper dredge) produce greater capacity (rate of flow) loss than steeper system curves, as shown in Fig. 3.58.

Pumps in Parallel or Series Operations

Pumps can be operated in series or in parallel depending on the requirements. Figure 3.59 shows the head-rate of flow curves for two unlike pumps in parallel operation. Figure 3.60 shows a cutterhead dredge operating with a long discharge line necessitating the installation of a booster pump. It will be noted that the head developed is almost doubled in such a case while the rate of flow remains the same. It should also be noted that the booster pump will generally have the same diameter on the suction and discharge sides of the pump. The booster pump is located so that there is a low positive pressure on the suction side of the pump. The discharge pressure of the booster pump will depend on the length of the discharge line.

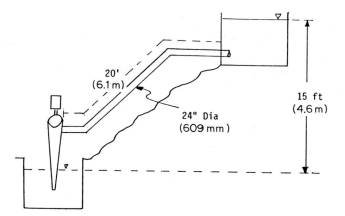

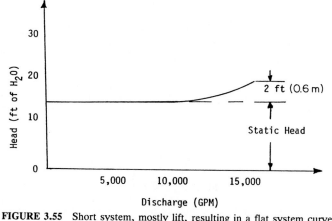

FIGURE 3.55 Short system, mostly lift, resulting in a flat system curve. (*Hicks, 1957*)

CAVITATION OF DREDGE PUMPS

Introduction

The word *cavitation* originated about 1904. The problem of cavitation, however, was first recognized as an engineering problem around 1900, when marine engineers noticed loss of efficiency and damage to propellers. The emphasis on general cavitation problems in structures did not start until 1940, when the hydraulic engineers began designing higher-head hydraulic structures with the resulting increase in velocities.

The *Reader's Digest Great Encyclopedic Dictionary* defines *cavity* as "a hollow or sunken space; hole" and *cavitation* as "the formation and collapse of

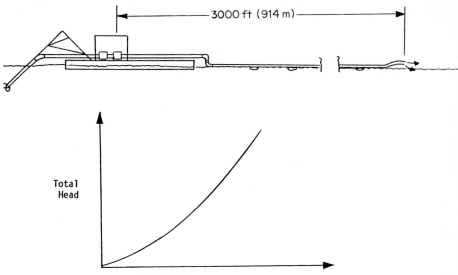

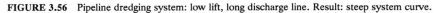

FIGURE 3.56 Pipeline dredging system: low lift, long discharge line. Result: steep system curve.

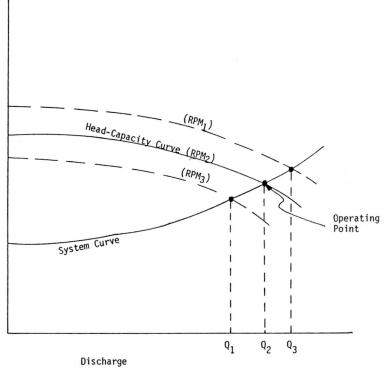

FIGURE 3.57 Variable speed pump.

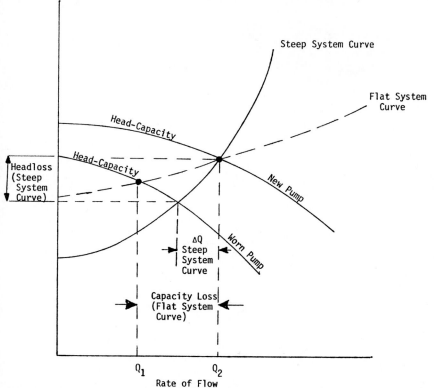

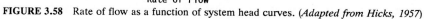

FIGURE 3.58 Rate of flow as a function of system head curves. (*Adapted from Hicks, 1957*)

low-pressure vapor cavities in a flowing liquid, often resulting in serious damage to pumps, ship propellers, etc." It was originally thought that cavitation was caused by electrochemical action (or corrosion), chemical action, or high tension in the water. But more recent studies have shown that these theories were not entirely correct and that the damage to surfaces is essentially caused by mechanical action. Experiments[1] indicated pressures on the order of 200,000 psi (1.38

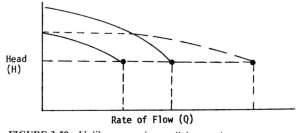

FIGURE 3.59 Unlike pumps in parallel operation.

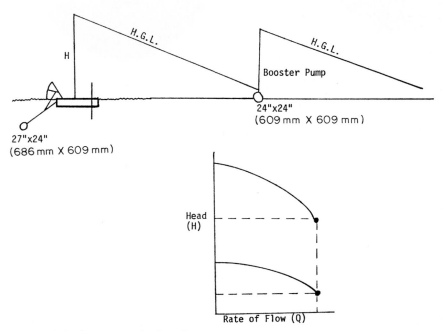

FIGURE 3.60 Pumps operating in series.

GN/m^2) based on the analysis of strain waves in a photoelastic specimen exposed to cavitation. Pressures of this intensity appear to coincide with those estimated from measurements of damage caused by cavitation.

The Physical Process

When liquid pressure is reduced at constant temperature by either static or dynamic means, vapor* filled cavities will form. The cavities will also form when the temperature of liquid is increased to boiling at constant pressure. Cavitation occurs only in liquids and never in a gas or solid.

Since liquid cannot expand or support tension stress, the cavities form when the absolute pressure falls below or close to the vapor pressure of liquid. See Fig. 3.61. These cavities form about any minute particles of foreign matter which are in suspension. In sea water, which is super-saturated and contains a large number of undissolved nuclei, the bursts of cavitation can occur at pressures above vapor pressure.

Cavitation bubbles which form in an area of low pressure may grow rapidly, then flow into an area of higher pressure and collapse, or implode, inside a turbomachine or a hydraulic structure. The process from cavitation inception to

---

*What is vapor? Vapor, such as steam or ammonia, differs from a gas by being readily condensable to a liquid. All liquids tend to vaporize, which they do by projecting molecules into the space above their surfaces. Vapor pressure increases with temperature.

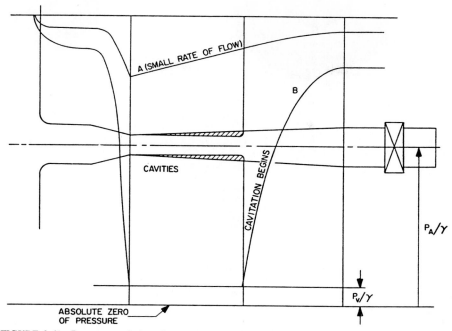

FIGURE 3.61 Pressure variations in a convergent/divergent passage with and without cavitation.

implosion may take only a few thousandths of a second. The bubbles collapse, or implode, causing pressure forces which may severely damage the metal surface of a hydromachine or the concrete surface of a hydraulic structure.

Occurrence of Cavitation. In addition to a localized pressure reduction, which causes formation of a cavitation bubble at that point in the fluid next to the boundary, the elevation also has its effect on cavitation. For example, (1) cavitation appears more readily at the top of a conduit than at the bottom, (2) turbines located above tailwater are more subject to cavitation, and (3) propellers of surface vessels are more susceptible to cavitation than propellers of submerged submarines.

High-velocity flow along a boundary will decrease pressure according to the Bernoulli's equation, and if the pressure is reduced to the vapor pressure at the corresponding temperature, cavities will occur. The regions of high velocity (or low pressure) will form at constricted passages.

Flow curvature, a third and possibly the most important factor in occurrence of cavitation, is caused or may be accompanied by eddies, vortices, or separation of flow. See Fig. 3.62. The curvature flow also causes an increase in velocity and a decrease of pressure. Flow curvatures occur either on a surface of conduit; at a pump entrance; on the blades of a propeller, turbine, or pump; or at abrupt changes of direction, such as a sharp corner. Localized regions of high velocity and low pressure may also exist at the centers of vortices and eddies and may produce a rather unpredictable type of cavitation.

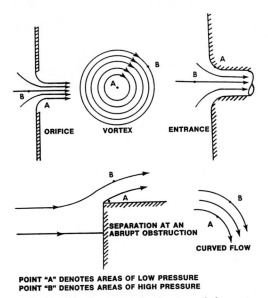

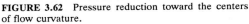

POINT "A" DENOTES AREAS OF LOW PRESSURE
POINT "B" DENOTES AREAS OF HIGH PRESSURE

FIGURE 3.62 Pressure reduction toward the centers of flow curvature.

Nature of Cavitation. The cavitation process consists of four parts: (1) the formation of the cavity, (2) its travel, (3) its final implosion, or collapse, and (4) the consequences of its collapse.

1. Cavities are formed in regions of low pressure, with these pressures being larger than the vapor pressure of the liquid by the partial pressure of the air in the cavity. The frequency of cavity formation is directly proportional to the velocity of flow and inversely proportional to the length of the cavity. As the cavity increases in size, the liquid vaporizes and dissolved gases come out of the solution, but this process is quite unpredictable.

2. From the point of formation, the cavity moves to the point of collapse. Bubbles which are attached to the boundary have been observed to move downstream with approximately half the speed of the liquid.

3. The cavity is destroyed when it moves downstream into the zone of higher pressure and collapses. Tremendous pressures are developed by a cavity collapsing in this zone. Lord Rayleigh computed the pressure within a spherical cavity to be 68 tons per square foot and 756 tons per square foot when the diameter had been reduced to 1/20 and 1/100, respectively, of its original diameter.[33]

   The cavities are usually of irregular shape and contain small droplets of liquid mixed with the bubbles. As the cavity collapses, the liquid rushes into the void and the droplets may be shot at the solid surface. The speed of the droplets is sufficient to produce deformation and eventual destruction of the metal surface. The mass of liquid rushing into the void produces a great local pressure which is transmitted radially outward with the speed of sound, followed by a negative pressure wave which may lead to one or more repetitions of the vaporization-condensation cycle. Boundary materials are thus subjected to rapidly repeated stress reversals and may eventually fail by fatigue.

Because of turbulent flow, the bubbles may follow a random path before collapse; therefore, some of the bubbles will collapse within the free flow and out of contact with the boundary. The resulting collapse will create compression waves sent out in all directions, which may endanger a structure by setting up forced vibrations.

Increased velocity produces no further pressure reduction but does elongate the cavity. Bubble sizes will increase until a stable vapor pocket is formed which is similar in shape to the zone of separation next to an unstreamlined boundary.

4. Some of the effects from cavitation are: (a) pitting of solid boundaries, (b) reduction in efficiency, and (c) vibrations in structures and machines. When a cavity collapses next to a solid surface or a droplet of liquid strikes the surface at high speed, the action is similar to striking the surface with a ball-peen hammer. The pitting of materials is primarily a fatiguing action in which the surface skin of the boundary is continuously hammered by millions of tiny bubbles until it cracks or chips off.

Remedies for Pitting. The prevention of cavitation can be accomplished only through cooperation between the designer and the laboratory. The designer should make every effort to streamline a structure with easy curves to prevent flow curvatures, vortices, eddies, separation, or high local velocities.

In completed structures which are pitted and cannot be altered, various methods could be used to prevent further cavitation. (1) Eroded parts can be replaced by a tougher, more resistant material. (2) In hydraulic structures, air can be admitted to regions of low pressure. The high air content cushions the collapse of the cavity and reduces its destructive effects. This eliminates true cavitation, however, because there is no longer a vacuum.

Cavitation in Dredge Pumps. Although much theoretical and experimental work has been done on cavitation, the mechanism of cavitation has not been definitely established on a quantitative basis in fluid mechanics.

It is difficult to calculate exactly the local pressures for some complicated flow at a certain point inside a dredge pump. It is frequently necessary to investigate experimentally the performance of the model dredge pump to determine the upper limits below which cavitation effects would not occur. The effects of cavitation on a dredge pump's performance have, in many cases, been experimentally established. Various parameters, such as specific speed and dimensionless ratios, which include vapor pressure, head, net positive suction heads, have been used to describe the effects of cavitation.

Cavitation in dredge pumps can cause noisy operation, reduction of effectiveness of the pump, mechanical vibration, and with time actual damage to metal passageways. The intensity effects on dredge pump functions are schematically depicted in Fig. 3.63.

A large zone of cavitation will obstruct the flow of the water-solids mixture and reduce the pump's performance. This will result in a failure of the pump to maintain the relations between head and rate of flow, or power and rate of flow which exist without cavitation. See Fig. 3.64.

Intensity of Cavitation. There are a number of terms which have been developed to describe the intensity or stages of cavitation. Some of the terms are discussed in *Cavitation,* by Knapp, Dailey, and Hammitt,[34] and a complete listing is given below:

1. *Threshold* conditions—a boundary between detectable cavitation and no cavitation. The cavities may form and may disappear.

2. *Incipient* cavitation—usually describes the stage of the cavitation process in which intermittent vapor bubbles are formed. There is usually a limited zone over which cavitation occurs.

3. *Desinent* cavitation—describes the disappearance of cavitation conditions.

4. *Developed* cavitation—describes continuous cavitation under fairly constant pressure, velocity, and temperature conditions.

5. *Industrial* cavitation—a term first used by Herbich[3] in dredge pump applications. The reduction of total head by the occurrence of cavitation is used as a criterion to describe this stage.

6. *Supercavitation* condition—a term used to describe a condition where a vapor cavity covers a large part or encompasses the whole body.

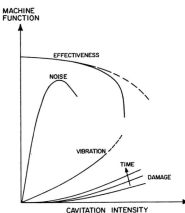

FIGURE 3.63   Schematic of intensity effects on machine function. (*Knapp et al., 1970*)

Knapp et al.[34] also suggest the following classification of cavitation according to the principal physical characteristics:

1. *Traveling* cavitation

2. *Fixed* cavitation

3. *Vortex* cavitation

4. *Vibratory* cavitation

It should be noted that there is no uniformity between various authors regarding the adjectives used with the word *cavitation*.

Knapp et al. describe *traveling cavitation* as composed of "individual transient cavities or bubbles which form in the liquid and move with the liquid as they expand, shrink and then collapse." See Fig. 3.65. *Fixed cavitation* is described as a cavitation "in which the liquid flow detaches from the rigid boundary of an im-

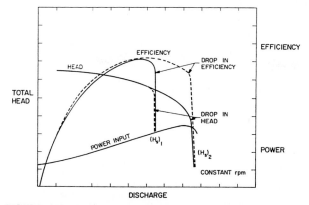

FIGURE 3.64 Performance with variable suction head: abrupt onset of cavitation.

mersed body or a flow passage to form a pocket or cavity attached to the boundary." See Fig. 3.66. *Vortex cavitation* may be described as formation of cavities at tips of propeller blades. *Vibratory cavitation* is caused by a continuous series of high-frequency, high-amplitude pressure pulsations. High-velocity flow is absent in this type of cavitation; in fact vibratory cavitation may occur in stationary liquid.

Cavitation Parameters. One of the more important parameters in dealing with cavitation is the NPSH produced by a pump for any given discharge. NPSH is

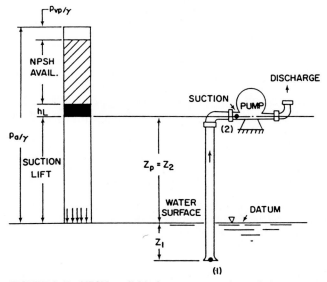

FIGURE 3.65 NPSH available for water pump located above water level.

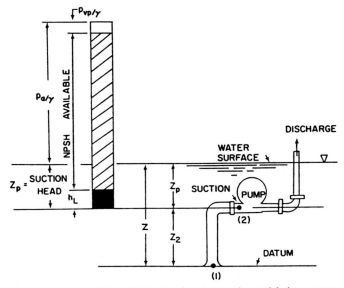

FIGURE 3.66 NPSH available for water pump located below water surface.

defined as the total head available to the pump above the vapor pressure and is computed as:

$$\text{NPSH} = H_{\text{atm}} + \frac{P_s}{\gamma} + \frac{V_s^2}{2g} - h_v = \frac{P_a}{\gamma} + \frac{P_s}{\gamma} + \frac{V_s^2}{2g} - \frac{P_v}{\gamma} \qquad (3.114)$$

where H_{atm} = atmospheric head in feet of liquid
   P_a = local barometric pressure
   P_s = pressure measured at the pump inlet in pounds per square foot (kg/cm^2)
   γ = specific weight of liquid in pounds per cubic foot (kg/m^3)
   V_s = velocity in the suction pipe at the pump inlet in feet per second (m/s)
   g = 32.2 ft/s^2 (9.81 m/s^2)
   h_v = head of liquid in feet (m) corresponding to the vapor pressure of the liquid
   P_v = vapor pressure of liquid in pounds per square foot (N/m^2)

 Values of the minimum NPSH for incipient cavitation can be determined by tests for different rates of discharge; these minimum values for incipient cavitation can be plotted as a function of discharge or specific speed to form a smooth curve. Values of the minimum NPSH for incipient cavitation, referred to as the critical NPSH, can then be determined for any rate of discharge by interpolation from the graph. If a centrifugal pump is operated below this critical NPSH, cavitation will develop and the total head will be reduced by an amount ΔH and the efficiency decreased.

Usually, in order to ensure the safe operation of a pump, it is operated at a value of NPSH, referred to as the plant NPSH or $NPSH_p$, which is higher than the critical NPSH, $(NPSH_c)$, by the amount $\Delta NPSH$. In equation form, the plant NPSH is equal to the sum of the critical NPSH and $\Delta NPSH$ or

$$NPSH_p = NPSH_c + \Delta NPSH \qquad (3.115)$$

Dividing this equation by the total pump head H and calling the resulting ratio sigma σ we have

$$\sigma_P = \sigma_c + \frac{\Delta NPSH}{H} \qquad (3.116)$$

Until recently, it was generally accepted that for all similar pumps operating at corresponding points on the Q-H curve the value of plant sigma σ_p was constant, regardless of head. This statement was frequently referred to as *Thoma's Law*. Recently, however, many cases have been discovered which deviate considerably from this rule. Laboratory tests prove that when the ratio $\Delta H/H$, where ΔH is the amount of head lost through cavitation effects, is held constant, the critical sigma will also be constant.

When two similar pumps are operated at different heads at their best efficiency point so that ΔH in each case is equal to a constant value, Tenot's equation holds:

$$\frac{\sigma_{p1} - \sigma_c}{\sigma_{p2} - \sigma_c} = \frac{H_2}{H_1} \qquad (3.117)$$

where the subscripts 1 and 2 refer to the first and second pumps, respectively, and σ_c is equal for both pumps if they are similar. Rearranging:

$$(\sigma_{p1} - \sigma_c)H_1 = (\sigma_{p2} - \sigma_c)H_2 \qquad (3.118)$$

or

$$(NPSH)_{p1} - (NPSH)_c = (NPSH)_{p2} - (NPSH)_c$$
$$\Delta(NPSH)_1 = \Delta(NPSH)_2$$

thus proving that both points have the same margin of safety against cavitation.

If the two similar pumps are operated at the same head, as is usual in model-prototype testing, this equation reduces to Thoma's Law:

$$\sigma_{p1} = \sigma_{p2} \qquad (3.119)$$

Thus it is seen from the above that when σ_c and $(NPSH)_{p1}$ or $(NPSH)_{p2}$ are determined for any head by experiment, the plant NPSH for any other head can be directly calculated.

Liquids Other than Cold Water. For liquids other than cold water, the same theories can essentially be used as above, but a "corrected" value of the cold water NPSH, which can be obtained from plots of empirical and thermodynamical data, must be used. Thus, if the cavitation characteristics of any pump are known for cold water, these characteristics can be obtained for the same pumps handling oil, butane, etc.

Cavitation of Dredge Pumps. The material which passes through a dredge pump is essentially a mixture of water and suspended particles. As such it is separate from the two classifications found above. It is neither pure water nor a homogeneous liquid entirely different from water, such as butane, oil, etc., whose thermodynamic properties are known at all times.

Available NPSH. Available NPSH indicates the amount of energy available (represented as NPSH) to prevent the pump from cavitating. If the available NPSH is less than that required by the pump, the pump will cavitate; if the available NPSH is greater than that required by the pump, the pump will not cavitate. Thus, it is important to be able to determine the available NPSH for any given installation. This can be done by applying the hydrodynamic principles to the system.

A number of examples presented by Basco[37] are reproduced here to indicate the methods used in determining the available NPSH. The first case presents a pump above the water surface pumping water (Fig. 3.65). Let us write the energy equation between the entrance to the suction line (point 1) and the pump inlet (point 2):

$$\frac{P_1}{\gamma} + \frac{V_1^2}{2g} + Z_1 = \frac{P_2}{\gamma} + \frac{V_2^2}{2g} + Z_2 + h_l \qquad (3.120)$$

where P/γ = the pressure head in foot-pounds per pound of water (m-kg/kg of water)

$V^2/2g$ = velocity head in foot-pounds per pound of water (m-kg/kg of water)

Z = the elevation head in foot-pounds per pound of water (m-kg/kg of water)

h_l = the head loss between points 1 and 2 in foot-pounds per pound of water (m-kg/kg of water)

The pressure head P_1/γ may be expressed in absolute terms as follows:

$$\frac{P_1}{\gamma} = \left(\frac{P_a}{\gamma} - \frac{P_v}{\gamma}\right) + Z_1 \qquad (3.121)$$

If Z_1 is taken at the water surface and if entrance velocity head is neglected, the equation becomes

$$\frac{P_a}{\gamma} - \frac{P_v}{\gamma} + Z_1 + \frac{V_1^2}{2g} - Z_1 = \frac{P_2}{\gamma} + \frac{V_2^2}{2g} + Z_2 + h_l \qquad (3.122)$$

or, the NPSH available is equal to

$$\frac{P_2}{\gamma} + \frac{V_2^2}{2g} = \frac{P_a}{\gamma} - \frac{P_v}{\gamma} - Z_2 - h_l \qquad (3.123)$$

If $Z_p = Z_2$, then

$$\text{Available NPSH} = \frac{P_a}{\gamma} - \frac{P_v}{\gamma} - Z_p - h_l \qquad (3.124)$$

If the atmospheric pressure is 33.9 ft · lb/lb of water at sea level, and if the vapor pressure is about 1 ft · lb/lb of water for temperatures below 80°F, and the losses are, say, 5 ft · lb/lb of water and the elevation of the pump is 15 ft, the

Available NPSH = 33.9 − 1 − 15 − 5 = 12.9 ft (3.93 m)

If the required NPSH for a given pump is 15 ft (4.6 m), the pump will cavitate; if it is 12.0 ft (3.6 m), the pump will not cavitate.

In the second case the pump is located below the water surface (Fig. 3.66). The energy equation may again be written and it will be similar to Eq. (3.121).

The pressure head in absolute terms may be written as

$$\frac{P_1}{\gamma} = \frac{P_a}{\gamma} - \frac{P_v}{\gamma} + Z \qquad (3.125)$$

or the available NPSH is equal to

$$\frac{P_2}{\gamma} + \frac{V_2^2}{2g} = \frac{P_a}{\gamma} - \frac{P_v}{\gamma} + Z - Z_2 - h_l \qquad (3.126)$$

Since $Z_p = Z - Z_2$, the available NPSH is equal to

$$\frac{P_a}{\gamma} - \frac{P_v}{\gamma} + Z_p - h_l \qquad (3.127)$$

If the conditions are again the same as in the first case, and if $Z = 10$ ft (3.0 m),

$$\text{Available NPSH} = 33.9 - 1 + 10 - 5 = 37.9 \text{ ft (11.55 m)} \qquad (3.128)$$

This indicates that the dredging depth can be considerably increased by locating the pump below the water surface.

The third case applies to a dredging installation. The pump is located below the water surface and the suction line is either equipped with a drag or a cutterhead (Fig. 3.67).

When dealing with mixtures of water and solids, the pressure heads, NPSH, etc., must be expressed in foot-pounds per pound of liquid flowing (not feet of water). Thus, the heads must be expressed in feet of mixture (containing water and solids), and not in feet of water.

The energy equation is now

$$\frac{P_1}{\gamma_m} + \frac{V_1^2}{2g} + \frac{Z_1}{\text{S.G.}_m} = \frac{P_2}{\gamma_m} + \frac{V_2^2}{2g} + \frac{Z_2}{\text{S.G.}_m} + h_l \qquad (3.129)$$

where $\text{S.G.}_m = \gamma_m/\gamma$ = specific weight of mixture/specific weight of water. Note that all parts of the equation are now expressed in feet (or meters) of mixture.

The pressure head in absolute terms is now

$$\frac{P_1}{\gamma_m} = \frac{P_a}{\gamma_m} - \frac{P_v}{\gamma_m} + \frac{d}{\text{S.G.}_m} \qquad (3.130)$$

where d is the digging depth, or difference between the water surface and the entrance to the suction line.

The available NPSH is now equal to

$$\frac{P_2}{\gamma_m} + \frac{V_2^2}{2g} = \frac{P_a}{\gamma_m} - \frac{P_v}{\gamma_m} + \frac{d}{\text{S.G.}_m} - Z_2 - h_l \qquad (3.131)$$

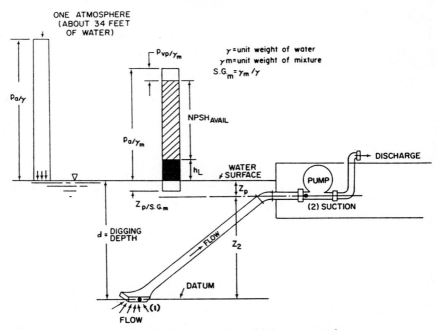

FIGURE 3.67   NPSH available for water pump located below water surface.

Since

$$\frac{Z_p}{S.G._m} = \frac{d - Z_2}{S.G._m} \qquad (3.132)$$

$$\text{Available NPSH} = \frac{P_a}{\gamma_m} - \frac{P_v}{\gamma_m} + \frac{d}{S.G._m} - Z_2 - h_l \qquad (3.133)$$

It should be noted again that the NPSH is now expressed in feet of mixture.

Example 3.1. Consider a dredge pump located 10 ft below the water surface. The digging depth is 50 ft, and the anticipated specific gravity of mixture is 1.20. The estimated total head loss is 10.5 ft of mixture, and the vapor pressure for a given location is 0.36 lb/in² at 70°F. The atmospheric pressure is assumed to be 14.4 lb/in².

The solution is:

$$S.G._m = \frac{\gamma_m}{\gamma}$$

or

$$\gamma_m = \gamma(S.G._m) = 1.2(62.4) = 74.8 \text{ lb/ft}^3 \ (11.751 \text{ kN/m}^3)$$

$$\frac{P_a}{\gamma_m} = \frac{14.4(144)}{74.8} = 27.7 \text{ ft (8.4 m) of mixture}$$

$$\frac{P_v}{\gamma_m} = \frac{0.36(144)}{74.8} = 0.70 \text{ ft (0.2 m) of mixture}$$

$$\text{Available NPSH} = \left(\frac{P_a}{\gamma_m}\right) - \left(\frac{P_v}{\gamma_m}\right) + \left(\frac{d}{\text{S.G.}_m}\right) - Z_2 - h_l$$

$$= (27.7) - (0.70) + \left(\frac{50}{1.2}\right) - 40 - 10.5$$

$$= 18.17 \text{ ft (5.5 m) of mixture}$$

As mentioned above, in cavitation testing of dredge pumps it is convenient to measure the NPSH at the suction side of the pump:

$$\text{NPSH} = H_a + \frac{P_s}{\gamma} + \frac{V_s^2}{2g} - h_v \tag{3.134}$$

where H_a = atmospheric pressure in feet of liquid = P_a/γ_m
 P_a = atmospheric pressure in pounds per square inch (N/cm^2)
 P_s = suction pressure
 γ = specific weight of fluid
 P_s/γ = suction pressure head in feet of liquid (m of liquid)
 $V_s^2/2g$ = velocity head in feet of liquid (m of liquid)
 h_v = vapor pressure in feet of liquid = P_v/γ_m
 P_v = vapor pressure in pounds per square inch (N/cm^2)

In water pumps the units in feet of liquid are equivalent to units of feet of water; however, in dredge pumps the units of NPSH are in feet of liquid (or feet of mixture of solids-water being pumped). In this case NPSH' will indicate that the units employed are feet of liquid (or meters of liquid if metric system is used).

A dimensionless index is usually computed for comparing cavitation performance of dredge pumps. This index, called *sigma* (σ), is a ratio of NPSH to the total head developed by the pump H_t

$$\sigma = \frac{\text{NPSH}}{H_t} = \frac{\text{NPSH}'}{(H_t)'} \tag{3.135}$$

where H_t' = total head developed in feet of liquid.

Another useful expression employed in evaluating cavitation conditions is the suction specific speed S, defined as

$$S = \frac{N\sqrt{Q}}{\text{NPSH}^{3/4}} \tag{3.136}$$

where N = speed in revolutions per minute
 Q = discharge in gallons per minute (liters/minute)

Note that this expression is not dimensionless. A dimensionless expression for suction specific speed is

$$S_{\text{dim}} = \frac{\omega\sqrt{q}}{[g(\text{NPSH}')]^{3/4}}$$
(3.137)

where ω = speed in radians per second
      q = discharge in cubic feet per second (m³/s)
      g = acceleration due to gravity in feet per second squared (m/s²)

A variety of plots may be used to indicate the effects of cavitation. The following include some of the possible plots which can be employed:

1. Dimensional plots
   a. H_t^1 as a function of h_s^1 where h_s^1 = suction head in feet (or meters) of liquid
   b. H_t^1 as a function of NPSH^1
   c. NPSH^1 as a function of Q
2. Semidimensional plots
   a. H_{dim} as a function of h_s^1 where $H_{\text{dim}} = [g(H_t)^1]/\omega^2D^2$
   b. H_{dim} as a function of NPSH^1
   c. NPSH^1 as a function of Q_{dim} where $Q_{\text{dim}} = Q/\omega D^3$
   d. σ as a function of N_s where $N_s = (N\sqrt{Q})/(H_t)^{3/4}$. *Note: N_s is dimensional*
3. Dimensionless plots
   a. H_{dim} as a function of $h_s^1/(V_s^2/2g)$ where V_s = suction velocity in fps
   b. H_{dim} as a function of $\text{NPSH}^1/(V_s^2/2g)$
   c. $\text{NPSH}^1/(V_s^2/2g)$ as a function of $Q/\omega D^3$
   d. σ as a function of n_s and S_{dim} where $n_s = (\omega\sqrt{q})/[g(H_t^1)]^{3/4}$

The dimensional plots, although in common use, are inefficient in themselves, requiring separate curves for each pump speed, for each specific gravity or liquid density of mixture, etc. The dredge pump industry is urged to adopt the dimensionless parameters.[25] Examples of various plots are given in later chapters.

Effects of solid-water mixtures were studied experimentally by Mariani and Herbich[36] and by Herbich and Cooper.[35]

Experimental Facilities

The experimental facility used a 1:8 model of the prototype pump used on the Corps of Engineers hopper dredge *Essayons*.[6]

The test loop for pumping the river silt-clay-water mixture consisted of a tank and a dredge pump operated by a calibrated direct current 40-hp (29.8-kW) motor. The motor was specially designed to provide a wide speed range with accurate regulation. Total flow was measured by a magnetic flowmeter and continuously recorded on a circular recorder. All pressure measurements were by means of manometers to ensure accuracy, and pump speed was measured by means of a tachometer generator and calibrated hand tachometer.[27]

The experimental facility used in the second study consists of three major components:

1. Dredge pump and drive unit
2. Vacuum tank
3. Instrumentation

The facility can accommodate a dredge pump up to 8-in (203-mm) suction by 8-in discharge size, requiring up to 300 hp (223 kW) drive unit. The motor is a 300-hp (223-kW), 440-V, 1785-rpm electric induction motor. Speed variation is accomplished with a variable speed eddy coupling, permitting pump operation between 50 and 1750 rpm.[35]

The vacuum tank is 11 ft (3.4 m) in diameter and 15 ft (4.5 m) high with a maximum volume of 800 ft^3 (22.7 m^3). It was designed to withstand a vacuum pressure equal to 29 in (73.7 cm) of mercury. The pressure inside the tank is regulated with a rotary piston vacuum pump. The instrumentation includes (1) suction pressure manometer, (2) discharge pressure gauge, (3) 6-in magnetic flow meter, (4) nuclear density meter, (5) torque sensor to measure shaft input horsepower, cross-point display and (6) electronic counter. Temperature of the fluid mixture is measured with a thermometer. A transparent plexiglas suction side head and a transparent plexiglas suction pipe were used for high-speed photographic observation of cavitation when pumping clear water. Figure 3.68 is a schematic diagram of the experimental facility, and Fig. 3.69 is a view of the test facility.

Experimental Studies. Three pumps were used in experimental studies:

Pump A
  Impeller diameter: 10½ in (267 mm)
  Number of vanes: 5
  Suction pipe: 4½ in (114 mm)
  Discharge pipe: 4 in (102 mm)
Pump B
  Impeller diameter: 14 in (356 mm)
  Number of vanes: 5
  Suction pipe: 6 in (152 mm)
  Discharge pipe: 6 in (152 mm)
Pump C
  Impeller diameter: 24 in (610 mm)
  Number of vanes: 4
  Suction pipe: 8 in (203 mm)
  Discharge pipe: 6 in (152 mm)

Silt-clay-water mixtures were employed in pump A, fine sand in pump B, and medium sand in pump C. The pumps were not geometrically similar so no direct comparisons could be made between pump cavitation characteristics determined. However, since the main purpose of the investigation was to determine the effects of materials on pump cavitation, the studies on three different pumps did not present any particular problems.

Silt-Clay-Water Mixtures. The characteristics of the silt-clay material were:

Colloids	20 percent
Clay	11 percent
Silt	55 percent
Sand	14 percent

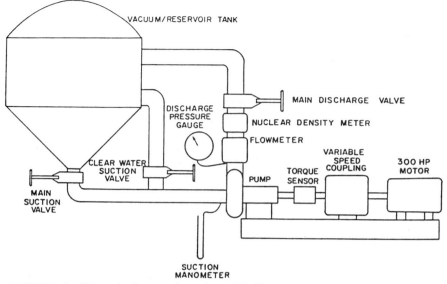

FIGURE 3.68   Schematic diagram of experimental facility.

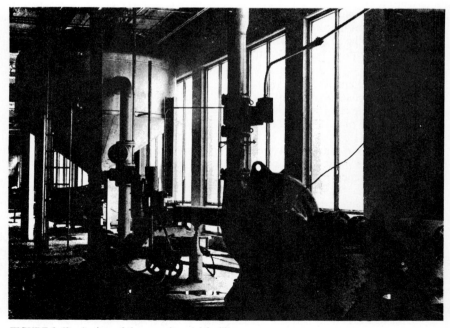

FIGURE 3.69   A view of the experimental facility.

The experiments were conducted with mixtures having densities of 1170 grams per liter and 1320 grams per liter. This corresponds to specific gravities of 1.17 and 1.32, respectively, and covers the range normally encountered in dredging operations for this type of material.[36]

Some of the results of this study are presented herein. For example, Fig. 3.70 is a plot of cavitation index as a function of dimensional specific speed N_s. The data are shown for three different pump impellers. It will be noted that there are three separate sets of curves, each for a different specific gravity. This might indicate that pumping a heavier liquid would cause cavitation to occur at a lower specific speed, if the specific speed is defined as $(N\sqrt{Q})/(H_t)^{3/4}$ with H_t in feet (or meters) of water. However, if the specific speed is defined as $N_s^1 = (N\sqrt{Q})/(H_t^1)^{3/4}$ where H_t^1 is in feet (or meters) of liquid mixture, the resulting curves for each liquid density will fall on one line as shown in Fig. 3.71. Similar results can be expected if the cavitation

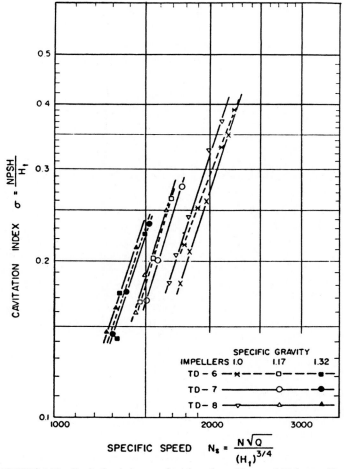

FIGURE 3.70 Cavitation index as a function of specific speed N_s (pump A).

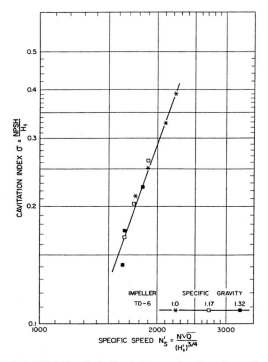

FIGURE 3.71 Cavitation index as a function of specific speed N_s' (pump A).

index σ is plotted against the dimensionless specific speed $n_s = (\omega\sqrt{q})/(gH_t^1)^{3/4}$.

This would indicate that the specific gravity of the mixture has little, if any, effect on cavitation characteristics. Such a plot as Fig. 3.71 or a plot of σ versus N_s may then be used to find cavitation conditions for a dredge pump handling a silt-clay-water mixture of any specific gravity. Since silt-clay-water mixtures form nonnewtonian fluids, the effect of "apparent" viscosity was also investigated.[25] Figure 3.72 presents the flow curves for the silt-clay-water mixture employed. Figure 3.73 is a plot of specific speeds as a function of the Reynolds number. The Reynolds number is defined as

$$N_R = \frac{V_s D_s}{\nu^1}$$

where V_s = velocity in the suction pipe
D_s = diameter of suction pipe
ν^1 = apparent kinematic viscosity

It will be noted that if the specific speed is determined in terms of feet of liquid mixture, the lines are essentially horizontal. Thus, it may be tentatively concluded that the viscosity of the mixture has very little effect, if any, on the cavitation index σ.

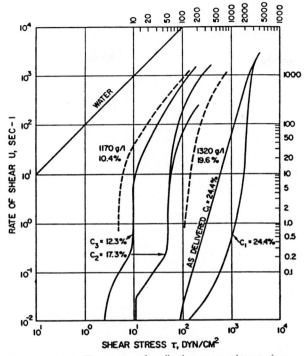

FIGURE 3.72 Flow curves for silt-clay-water mixtures (pump A). (*Mariani and Herbich, 1966*)

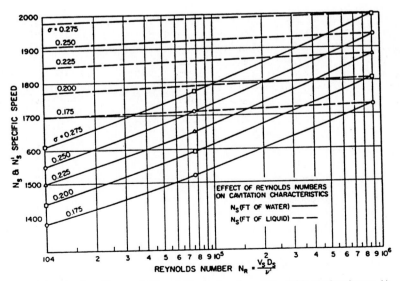

FIGURE 3.73 Specific speeds N_s and N'_s as a function of Reynolds number (pump A). (*Mariani and Herbich, 1966*)

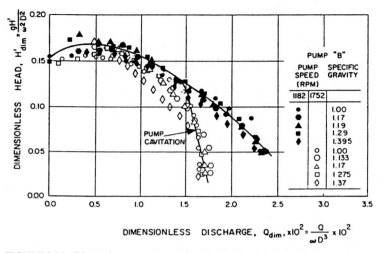

FIGURE 3.74 Dimensionless head as a function of dimensionless discharge (pump B). (*Herbich and Cooper, 1971*)

Fine Sand-Water Mixtures. Pump B was employed in the study and uniformly graded sand was used. Sand-water mixture specific gravities ranged from 1.00 (clear water) to 1.43. The importance of using dimensionless-type plots was stressed in an earlier discussion.[3] Two characteristic pump curves are shown, one for dimensionless total head plotted as a function of dimensionless discharge (Fig. 3.74) and another for dimensionless power plotted as a function of dimensionless discharge (Fig. 3.75).

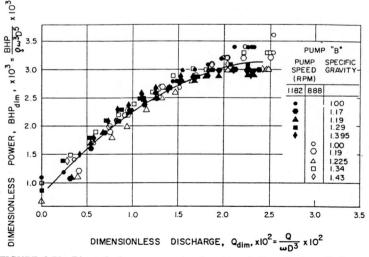

FIGURE 3.75 Dimensionless power as a function of dimensionless discharge (pump B).

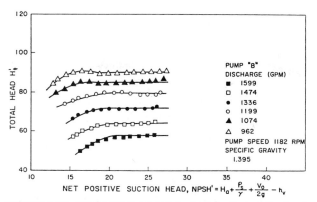

FIGURE 3.76 Total head as a function of net positive suction head. Pump speed = 1182 rpm, specific gravity of mixture = 1.395 (pump B).

For a very wide range of mixture's specific gravities and pump speeds all data fall approximately on one curve, except when the pump is cavitating. It may be concluded here that dredge pump tests with water may be used to predict pump performance with sand-water mixtures.

A convenient method of presenting cavitation data is to plot the total head of liquid (H_t) as a function of NPSH[1] in feet of liquid. This is a dimensional plot which may be used to determine critical values of NPSH (Fig. 3.76). It will be noted that the head-NPSH curve does not fall off as rapidly as does the head-discharge or efficiency-discharge curve, and that the critical values of NPSH may be defined differently. One proposed definition is that the point of "industrial" cavitation is reached when the total head developed falls 3 percent or is reduced to 97 percent of that developed for noncavitating conditions. The exact reduction percentage can be preset as may be desirable for any given pump and/or project.[35]

A more general plot may be prepared combining all pump speeds and liquid densities by making both the total head and the NPSH dimensionless. A sample plot of this nature is presented for pump C (Fig. 3.77).

Medium Sand-Water Mixture. Sand passing mesh 30 and retained on mesh 40 was used in this part of the study. Its median size was 0.50 mm.

Several significant plots are shown herein, the first being a dimensionless head plotted as a function of NPSH (Fig. 3.77). This is a semidimensionless plot; the industrial cavitation conditions for any head reduction percentage may be determined from this plot. Figure 3.78 is completely dimensionless; although some scatter exists, a single curve, or a narrow band may be drawn through the data. It will be noted that in this case the data for all speeds form one narrow band rather than separate curves for different speeds, as shown in Fig. 3.77.

By far, the most significant plots are those shown in Figs. 3.79 and 3.80. The cavitation index σ was plotted against the dimensional specific speed N_s in Fig. 3.79. The constant values of suction specific speed S are also shown on this graph. A completely dimensionless plot is shown in Fig. 3.80 where the cavitation index σ was plotted against the dimensionless specific speed n_s. Constant lines of dimensionless suction specific speed S are also indicated on the graph. Plots of this type provide complete information regarding the cavitation characteristics

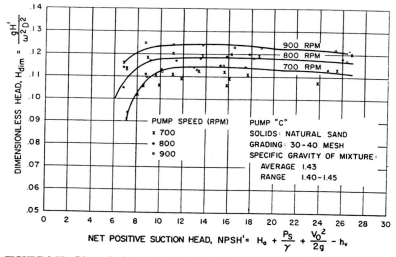

FIGURE 3.77   Dimensionless head as a function of NPSH (pump C).

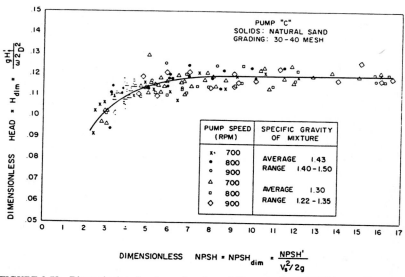

FIGURE 3.78   Dimensionless head as a function of dimensionless NPSH (pump C).

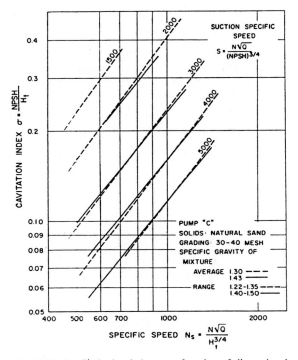

FIGURE 3.79 Cavitation index as a function of dimensional specific speed (pump C).

of a series of geometrically similar dredge pumps. It may be tentatively concluded that the specific gravity has virtually no effect on the cavitation index σ for medium sand-water mixtures.

Conclusions

1. Basco[37,38] put forward a hypothesis that since cavitation inception and intensity levels are related to when and how much liquid in the water-solids mixture vaporizes, the NPSH values expressed in feet of liquid will not depend on the density, or size of solid particles. It appears that the specific gravity of the mixture and size of particles have no effect on industrial cavitation characteristics of dredge pumps.
2. It will be noted that the experimental studies conducted with silt-clay-water, fine sand-water, and medium sand-water mixtures were essentially concerned with homogeneous mixtures having solid particles uniformly dispersed in slurry flows. There is a possibility that findings of future studies with relatively large stones in water may be different from those stated above. One of the objectives of the future investigation is to determine if and when the effect of solid size would affect the cavitation requirements of dredge pumps.
3. Additional, more detailed findings are as follows:

a. The specific gravity of silt-clay-water mixtures has little, if any, effect on cavitation characteristics.
b. The Reynolds number based on the apparent viscosity of the silt-clay-water mixture has very little effect, if any, on the cavitation index.
c. Dimensionless presentations of pump and cavitation characteristics are not only the most efficient way of handling a large amount of data but also are most useful in relating to characteristics of other sizes of pumps, speeds, and specific gravities of solid-water mixtures and are extremely valuable in checking experimental results.[1,3,6]
d. The specific gravity has virtually no effect on the cavitation index σ for medium sand-water mixtures.

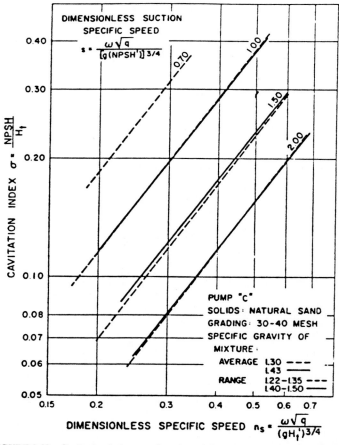

FIGURE 3.80 Cavitation index as a function of dimensionless specific speed (pump C).

Example 3.2.   Compute the following:

1. $NPSH'$ and NPSH
2. Allowable suction head in feet of liquid mixture and in feet of liquid.

Given

Discharge 200 ft^3/s (5.66 m^3/s)
Dredge pump speed 500 rpm
Total head 150 ft (45.7 m) of water
Suction pipe velocity 15 ft/s (4.6 m/s)
Specific gravity of mixture 1.2
Material: sand (grading 30–40 mesh)
Dredge pump to operate at 95 percent head (industrial cavitation)
Assume p_a = 14.7 psi (10.1 N/cm^2), temperature of mixture 90°F
Impeller diameter = 3.5 ft (1.07 m)

The solution is found by computing the following:

$$\frac{gH'}{\omega^2 D^2} = \frac{32.2(150/1.2)}{[500(0.1047)]^2 \, (3.5)^2} = \frac{32.2(125)}{(52.3)^2(12.26)}$$

$$= \frac{32.2(125)}{2740(12.26)} = \frac{4020}{33,600} = 0.1196$$

Dimensionless head for "industrial" cavitation is 0.1196(0.95) = 0.1138

$$\frac{NPSH'}{V_s^2/2g} = 4.95 \qquad \text{(from Fig. 3.78)}$$

$$NPSH' = 4.95 \, \frac{V_s^2}{2g}$$

$$V_s = 15 \text{ ft/s (4.6 m/s)} \qquad \frac{V_s^2}{2g} = \frac{15^2}{64.4} = \frac{225}{64.4} = 3.5$$

$$NPSH' = 4.95(3.5) = 17.3 \text{ ft (5.3 m) of mixture}$$

$$NPSH = 17.3(1.2) = 20.8 \text{ ft (6.3 m) of water}$$

$$NPSH' = \frac{P_a}{\gamma} + \frac{P_s}{\gamma} + \frac{V_s^2}{2g} - h_v$$

$$\frac{P_s}{\gamma} = h_s = NPSH' - \frac{P_a}{\gamma} - \frac{V_s^2}{2g} + h_v$$

$$= 17.3 - \frac{33.9}{12} - 3.5 + \frac{1.61}{1.2}$$

$$= 13.1 \text{ ft (4.0 m) of mixture} = -15.7 \text{ ft (4.8 m) of water}$$

GAS EFFECT AND REMOVAL

Introduction

The modern sea-going hopper dredge and the cutterhead dredge are the result of progressive development in the United States and abroad during the past century. They have found increasing importance in improving harbors, seaways, and channels along the coast. The hopper dredges are of hydraulic suction type, equipped with special machinery which enables them to dredge the material from the ocean bed or channel bottom, discharge it into hoppers, transport it, and discharge it at disposal sites. The cutterhead dredges usually discharge through a pipeline and some are already being used in an offshore environment.

The heart of the dredge is the pump. It is of centrifugal radial type and must be designed to withstand heavy wear and abrasion. While in operation, it may encounter a variety of mixtures made up of liquids, solids, and gases.

If materials containing a considerable amount of gas are encountered, the gas is drawn into the suction pipe, causing an appreciable decrease in vacuum and volume of solids discharged. In some cases a complete stoppage of the pump may occur. Occasionally, the combustible gases will ignite when a match is tossed into the hoppers.[39]

In many areas of estuaries, bays, harbors, and tidal rivers, particularly near the industrial or municipal centers, one can detect the odor of hydrogen sulfide or methane. The hydrogen sulfide, because of its high solubility property, is usually present in the dissolved form. Estuary or ship channel beds often contain trapped or dissolved gases including hydrogen sulfide, methane, and nitrogen. Any disturbance of the silt-clay bed releases bubbles of gas which travel to the surface.

Literature review indicates that in the last 40 years there were numerous investigations on the flow of gas-liquid mixtures in pipes.[44] The natural gas-petroleum industry investigated such flows in connection with the possible transportation of gas-liquid petroleum mixtures. However, almost invariably these studies have dealt only with the mechanics of flow within the pipe itself. Little has been written on how these gas-liquid mixtures affect the performance of the pumps they must go through. The presence of air in a suction system may cause problems and the effect of air on a pump's performance should be known to the dredge pump operators.

Dredge pumps may encounter mixtures consisting of widely varying proportions of solids, liquids, and gases. No particular difficulty is experienced when liquid-solid mixtures are pumped, except that the pump may choke if the density of the material in the suction line is too high. In normal dredging operations, the specific gravity of the mixture pumped is about 1.2, although laboratory experiments indicate that the dredge pump would operate satisfactorily for a silt-clay mixture of specific gravity of up to about 1.45, which corresponds to the consistency of a thick catsup. For sand-water mixtures, a pump will operate satisfactorily up to a specific gravity of about 1.55. A choking condition is alleviated by either lifting the drag head out of the mud or by admitting water to the suction line.

On the other hand, when material containing considerable amounts of entrapped or dissolved gas is encountered, the gas which enters the suction line adversely affects dredging performance. The gas tends to collect in the pump and can severely reduce the flow of water and solids or even cause the pump to lose its prime. The latter condition is very serious. A hopper dredge will miss part of a pass while the pump is being reprimed. Sudden stoppage of pumping on a pipe-line dredge can result in water hammer in the disposal line with possible danger to the pipeline or to the pump casing. In recent years, the difference between actual choking and stoppage of the pump because of excessive gas has been recognized and the need for gas removal from the suction line has become apparent.

The gases are products of the decomposition of organic matter in the bottom material. They are dissolved in the water forming a part of the in situ material and if the water is saturated, bubbles form throughout the material. Since mud usually has high viscosity such bubbles may remain in the mixture for years. Analysis of samples taken in American ports indicates[39] a gas mixture with a range of methane (CH_4) from 0 to 85 percent, nitrogen (N_2) from 0.2 to 98.8 percent, hydrogen (H_2) from 0.6 to 29.8 percent, oxygen (O_2) from 0.4 to 12.1 percent, and carbon dioxide (CO_2) from 1.0 to 14.6 percent. Methane gas is flammable and the need to remove it from the suction line is also important for fire safety. The actual results of tests taken by the U.S. Army Engineers are shown in Table 3.7.

The most soluble gas encountered is carbon dioxide, followed by methane. So it appears that the most soluble mixture which can be encountered by dredges consists of 85 percent methane and 15 percent carbon dioxide.

Effect of Air Content on Dredge Pump Performance

Laboratory studies were conducted in 1962 to 1967 on the effect of air content on a dredge pump's performance.[40] These studies are briefly summarized below.

Test Facility. The test facility consisted of a storage tank, suction pipe, discharge pipe, discharge tank, and return pipe all connected in a continuous recirculating flow loop. External to this flow system was the pump motor and an air compressor.

The model dredge pump assembly was geometrically similar to the prototype installation on the U.S. Army Engineers dredge *Essayons*.[3] The main exception was that the draghead itself was not reproduced in the model. Air was used in the model instead of gas because of safety requirements and because the solubility of

TABLE 3.7 Gas Samples Analyzed

No.	Date	Dredge	CO_2	O_2	H_2	CH_4	N_2	Other
1. Raritan River	3–21–47	Atlantic	2.1	7.4	.6	.0	89.8	0.1
2. Raritan River	3–31–47	Atlantic	2.8	3.0	18.7	8.0	67.5	.0
3. Bay Ridge Ch. NY	3–31–47	Atlantic	14.6	2.0	29.8	1.6	52.0	.0
4. Del. R. Cherry I. Range	10–23–47	Delaware	10.7	9.9	1.1	.6	77.7	.0
5. Del. R. Cherry I. Range	10–23–47	Delaware	7.1	9.3	1.1	1.4	81.1	.0
6. Schuylkill River	12– 3–47	Delaware	5.4	2.1	3.9	6.0	82.2	.4
7. Del. R. Mifflin Range	12–10–47	Delaware	9.4	9.8	3.2	46.8	30.8	.0
8. Del. R. Marcus Hook Range	12–10–47	New Orleans	4.6	.4	9.5	85.3	.2	.0
9. Edgewater Ch. NY	3–11–49	Goethals	3.0	7.3	6.2	44.4	37.5	1.6
10. Del. R., Liston Range	3–28–50	Goethals	1.0	12.1	1.0	.4	85.5	.0

Source: U.S. Government Printing Office. *The Hopper Dredge—Its History, Development and Operations.*

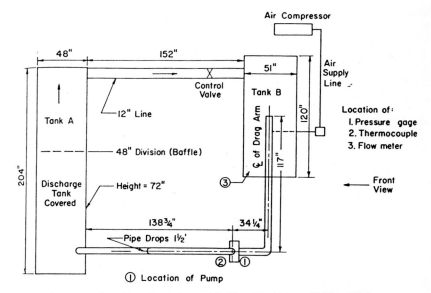

FIGURE 3.81   Plan view of experimental facility. (*Herbich and Miller, 1970*)

air closely approximates that of methane, the gas most frequently encountered in the field (Figs. 3.81 and 3.82).

The pump selected was a 1:8 model of the *Essayons* pump. The centrifugal pump was of a radial type designed for handling solid-water mixtures. The impeller was modified as a result of another study to improve performance of dredge pumps,[1,3] and it had the following characteristics: diameter 10 1/2 in (267 mm), inlet angle 45°, exit angle 22 1/2°, involute vane shape, and five vanes. Both the impeller suction side and the suction head of the pump were made of transparent plexiglas to permit observations by means of high-speed movies.

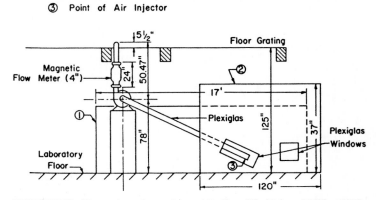

FIGURE 3.82   Front view of experimental facility. (*Herbich and Miller, 1970*)

The 4.5-in (114-mm) diameter suction pipe was also made of plexiglas. The air was provided by a rotary compressor driven by a 7.5-hp (5.6-kW) ac electric motor. The compressed air was filtered and cooled before injection into the suction line. Air injection was accomplished by a manifold with 16 flexible hoses connected to 1/16-in (0.16-cm) diameter holes drilled through the end flange of the suction pipe.

The air flow was measured with a rotameter calibrated to read standard cubic feet of air per minute at 25 psia (172.375 kN/m²), 70°F. The air temperature at the flow meter was measured with a calibrated resistance wire temperature gauge. The air volumes were corrected to the standard conditions given above.

The pump discharge was measured with a magnetic flow meter. Under test conditions, the discharge was a mixture of air and water. This meter read the volume flow rate of the total air-water mixture. The meter was installed vertically to ensure that the pipe was flowing full at all times. The readout from the meter was by means of a Dynalog recorder.

The pump speed was measured with a speed indicator and monitored frequently with a stroboscopic tachometer. The suction and discharge heads were measured with manometers.

Characteristics of Pump without Air Injection. Pump characteristics were first obtained for zero air-flow content for subsequent comparisons with various air-water mixtures.[3] A dimensionless plot of pump characteristics is given in Fig. 3.83.

Test Variables. The main variables in the study were the fluid rate of flow, air injection rate, and pump speed.

Test Procedures. In general, steady-state flow was established in the test facility for the desired flow rate, pump speed, and draghead depth. After all readings of discharge speed, suction discharge pressures, and power input were taken, the desired air content was injected into the suction line and new readings were recorded.

The rates of air flow were then increased in steps until the model pump collapsed.

Results of Study
Model-Prototype Relationships. The results presented herein are based on tests performed on a 1:8 scale model pump.[40]

The use of models to study prototype pumps requires geometric similitude as well as geometrically similar vector diagrams of the velocity entering and leaving the moving parts. Viscous effects may be neglected since it is generally impossible to satisfy all of the geometric similarity requirements and have equal Reynolds numbers in the model and in the prototype.[41]

The most important pump action is the dynamic transfer of energy from the rotating impeller to the moving fluid. If two pumps are geometrically similar and have similar vector diagrams, they are homologous. Homologous pumps also have similar streamlines, and for practical purposes, dynamic similitude exists. It is assumed that the scale model pump used for these experiments can accurately predict the performance characteristics of the prototype pump. Although the pump is a model, the total head and the velocity are not scaled down. These quantities remain the same in both the model and the prototype.

The prediction of prototype performance from tests of a homologous model requires the use of certain general laws of similarity. These similarity laws define a group of dimensionless terms which in turn can be used to predict the performance of the prototype pump.

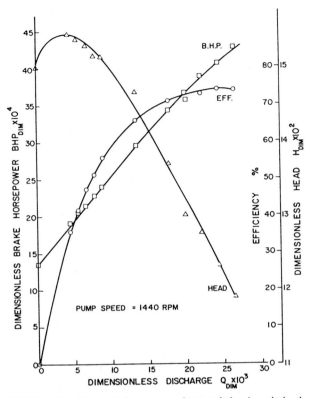

FIGURE 3.83 Dimensionless pump characteristics (no air in the fluid). (*Herbich and Miller, 1970*)

Dimensionless Terms

1. The dimensionless head H_{dim} may be defined as

$$H_{\text{dim}} = \frac{gH}{\omega^2 D^2} \tag{3.138}$$

where g = acceleration caused by gravity
   H = total head
   ω = pump speed in radians per second
   D = impeller diameter

When the values of the dimensionless head in the model are equated to those in the prototype, and if the model studies are conducted at the same head as the prototype head, the speed in the prototype ω is equal to one-eighth of the model speed, ω_m, or

$$\omega_p = \frac{\omega_m}{8} \qquad (3.139)$$

2. The dimensionless discharge Q_{dim} is defined as

$$Q_{\text{dim}} = \frac{Q}{\omega D^3} \qquad (3.140)$$

When the values of model and prototype dimensionless discharges are equated and the model and prototype heads are equal, the following relationship results:

$$Q_p = 64 \, Q_m \qquad (3.141)$$

or the prototype discharge is equal to 64 times the model discharge.

3. Dimensionless power (P_{dim})

$$P_{\text{dim}} = \frac{\text{bhp}}{\rho \omega^3 D^5} \qquad (3.142)$$

where   bhp = brake horsepower
         ρ = fluid density

The power relationship is thus

$$\text{bhp}_p = 64 \, \text{bhp}_m \qquad (3.143)$$

or the prototype brake horsepower is equal to 64 times the model power.

Effect of Air Injection on Water Discharge

There was little reduction in discharge if the amount of injected air was less than 2 percent of the water discharge. This was true regardless of the initial water discharge (Fig. 3.84).

The water discharge remained fairly constant until a certain air input was reached. At this point the discharge fell off rapidly. Right at this drop-off point, the conditions were very unstable. A loud vibrating noise was set up by the pump and the discharge began to oscillate. For a given discharge and orifice the drop-off point always occurred at the same air input.

The reduction in water discharge with increasing air flow is shown in Figs. 3.84 and 3.85. These curves indicate that after the initial drop-off the discharge began to stabilize up to the collapse point of the pump.

The actual amount of air injection which caused the pump to collapse was difficult to define exactly. However, the transparent suction pipe allowed visual observation of the flow. As the air injection was increased, the flow gradually progressed into a slug flow in which the air was no longer all entrained in the water. A slug of water with entrained air was usually followed by a slug of air which completely filled the suction pipe. Pump collapse followed shortly after this condition was reached.

The water discharge-standard air flow curves in Fig. 3.85 were plotted primarily to record the collapse point. The points plotted directly on the abscissa are the collapse points where water discharge was zero. Putting a vertical tangent at

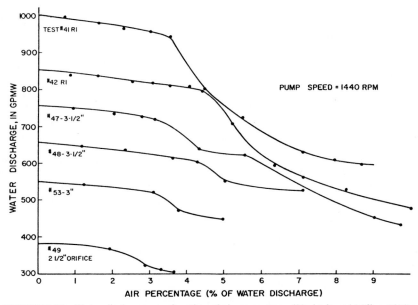

FIGURE 3.84   Water discharge as a function of air percentage. (*Herbich and Miller, 1970*)

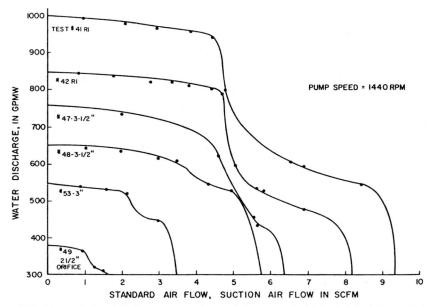

FIGURE 3.85   Water discharge as a function of standard air flow. (*Herbich and Miller, 1970*)

these points distorts the lower part of the curve. However, no intermediate points could be obtained because of the instability of the collapsing pump. The lower part of these curves should, therefore, be considered purely qualitative.

Effect of Air Injection on Total Head Developed

The effect of the injected air on the total dynamic head of the pump is shown in dimensionless form in Fig. 3.86. The head was expressed in feet of water in this plot rather than in feet of mixture. The percentage of air was based on discharge air content. This plot represents the results of a series of seven tests. It was not possible to set the air injection rates at equal air flows during the different tests because of the varying air temperatures and pressures for each test. The curves in Fig. 3.86 are the result of an interpolation procedure to obtain the head at whole number increments in air flow. The interpolation was made from individual test plots based on the exact air percentage for each test point.

The head was little affected by the low air injection rates, but an air flow of 3 to 4 percent caused an abrupt drop in the total head. The spread of the data points for the 3 to 5 percent range of air flows indicated instability of the system around the drop-off point. The fairly consistent data points for 7 to 10 percent air flows indicate the relative stability prior to collapse of the pump.

When the total head is expressed in feet of air-water mixture and the air flow is calculated for pump suction conditions, the percentage of air is increased.[42] Figure 3.87 presents the data plotted in this form. The main difference between Figs. 3.86 and 3.87 is the increase in percentage of air content.

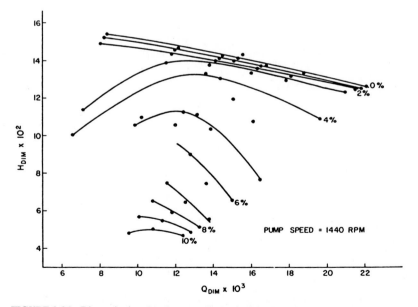

FIGURE 3.86 Dimensionless head versus dimensionless discharge as a function of discharge air content. (*Herbich and Miller, 1970*)

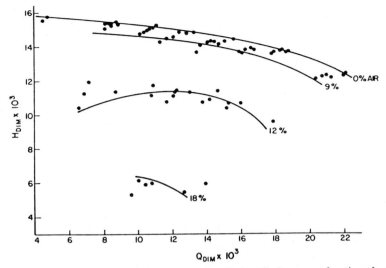

FIGURE 3.87   Dimensionless head versus dimensionless discharge as a function of suction air percent. (*Herbich and Miller, 1970*)

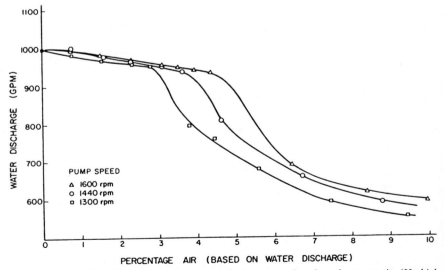

FIGURE 3.88   Effect of pump speed on water discharge as a function of percent air. (*Herbich and Miller, 1970*)

Effect of Time of Air Injection on Total Head.   The analysis of some early test data indicated a hysteresis effect in the total head when the air flow was reduced from high to low values. The total head near the drop-off point did not return to its original value but remained at a slightly depressed level. This could have been due simply to the instability of the pump or possibly some air was collecting in the pump casing or at the eye of the impeller.

Two tests were conducted to determine if the pump characteristics were affected by the length of time of air injection. In the first test, the discharge was set at 1000 gpm (63.1 l/s) and an air flow of 2.4 scfm (1.13 standard l/s) (1.9 percent) was set. The system was run continuously for over 2 hours at these conditions. Head and power data were recorded periodically. In the second test the initial charge was again 1000 gpm (63.1 l/s), but the air flow was increased every 15 min in 1-scfm (0.47-standard l/s) increments [near drop-off the increment was reduced to 0.5 scfm (0.24 standard l/s)].

The results of both tests indicated that the pump performance was independent of the time interval over which the air was injected. There were no measurable changes in either the head or the brake horsepower as the time of air injection increased.

It was concluded that there was no cumulative effect of air in the pump and that the inconsistencies in head data near the drop-off point were the result of pump instability at that particular air content.

Effect of Pump Speed on Injected Air-Pump Discharge Relationship.   The design speed for the pump used in this study was 1440 rpm. Almost all the tests were run at this speed. However, several tests were run at pump speeds greater and less than the design speed to determine what effect the speed had on the air-discharge relationship. The results of these tests are shown in Fig. 3.88.

The discharge began to fall off at 3 percent air at a speed of 1300 rpm. However, at 1600 rpm it required 4.5 percent air to decrease significantly. If the air percentage was less than 3, the speed had very little effect on the discharge. At higher air flows, the range in discharge was only about 50 gpm (3.15 l/s) for the three pump speeds.

Conclusions

1. Low rates of air injection have only a small effect on pump performance. Even at low discharge rates, the pump can operate effectively with a 3 percent air flow.
2. The pump discharge decreases as the amount of air injected increases. This reduction in discharge is gradual up to about 2 percent air flow (based on discharge air content); then the discharge drops off rapidly.
3. Excessive amounts of air will cause a complete collapse of the pump. At an initial model discharge of 1000 gpm (63.1 l/s) [64,000 gpm (4037 l/s) in the prototype] the pump will collapse at an air flow of about 10 percent (based on discharge air content).
4. The length of time that air is injected has no effect on pump performance.
5. The useful pumping range is extended from an air flow of 4.5 percent if the model pump speed is increased from 1300 to 1600 rpm (prototype pump speed of 163 to 200 rpm).

GAS REMOVAL SYSTEMS

The Corps of Engineers (CE) developed the first gas removal system[39] which was successful in removing gas from the suction pipe of the dredge *Atlantic* in 1946. This system consisted of a simple steam jet ejector which had its suction side connected to the top of the suction side of the dredge pump (near the pump). The ejector discharged directly overboard and its selection was essentially made by a trial and error procedure.[39] By experimenting with several ejectors, two were found to be approximately equally efficient. One of the ejectors had a capacity of 240 ft³/min (6.79 m³/min) at 20 in (50.8 cm) of vacuum. Based on these experiments, a method for determining the required gas discharge was established.

$$Q = 0.0383 \ Vd^2 \qquad (3.144)$$

where Q = cubic feet per minute of gas to be removed from the suction pipe (l/s)
V = velocity in suction pipe in feet per second (m/s)
d = diameter of suction pipe in inches (mm)

The CE's experience indicates that this equation gave sufficient capacities for ejectors on dredges *Atlantic, Essayons,* and *Goethals,* although none of these vessels continuously pumped material of high gas content except during limited test periods. Usually when high gas contents are encountered, the suction line velocities are reduced from somewhere around 16 ft/s (4.9 m/s) to below 10 ft/s (3.05 m/s). The density of pumped mixtures may vary from 1.10 to 1.45 and the suction head from 12 to 29 in (30.5–73 cm) of mercury. Since such a large variation in densities and pressures may be encountered, it is recommended by the CE that the velocity of 20 ft/s (6.1 m/s) be used in Eq. (3.144) to obtain satisfactory estimates for sizing the gas removal equipment.

Hoffman[43] was issued a number of patents on the subject of practical systems for handling gases in the suction pipe of a hydraulic dredge. These patents contain basic claims covering methods for preventing reduction of dredge pump efficiency due to gases released from dredged material. Hoffman received an award for the gas removal concept from the U.S. Army in their "Suggestion Awards Program" and he subsequently gave the CE the right to use the system.

In the early 1950s more sophisticated gas removal systems were installed on the U.S. Army Engineers' dredges *Goethals* and *Essayons* using Nash Hytor pumps in connection with Leveltrol controls. To protect against the problem caused by excessive quantities of gas in the dredge pump, gas removal systems have been installed on many government and private hopper dredges in the field. However, the effectiveness of these gas removal systems has not been evaluated scientifically. Prototype evaluation of gas removal is not readily accomplished because of several factors such as: (1) uncertain location of gaseous sediments, (2) complex three-phase flow system, and (3) large quantities to be measured. Consequently, a model study sponsored by the U.S. Army was undertaken by Herbich, Adams, Isaacs, Shindala, and Gupta to determine in the laboratory the performance of a typical gas removal system.[40,42,44,45,46]

The common type of gas removal system now in use on CE dredges consists of a gas accumulator and a source of vacuum. The accumulator is a vertical cylinder placed on top of the suction pipe at either the highest point on the line or at a location near the dredge pump. The cylinder is approximately the same diameter as the suction pipe and has a height equal to two or three

pipe diameters (Fig. 3.89). The purpose of the accumulator is to collect
the gas and to separate it from the
water-solid mixture being dredged.
Once the gas enters the accumulator,
it is removed through the top of the
accumulator by the vacuum system.
The vacuum is produced either by a
rotary liquid piston type of vacuum
pump or by a steam or water-driven
ejector. The vacuum system should

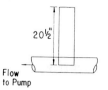

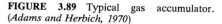

FIGURE 3.89 Typical gas accumulator. *(Adams and Herbich, 1970)*

be capable of handling the maximum or design gas flow at a vacuum pressure
approximately the same as dredge pump suction pressure.

Model Test Facility and Procedures

The laboratory dredge pump was a one-eighth scale model of the dredge pump
installed in the hopper dredge *Essayons*.[42] The basic test facility was essentially
the same as the one used for determination of effect of air content on dredge
pump performance with appropriate modification for study of gas removal systems (Fig. 3.90). Pertinent dimensions of model and prototype systems are given
in Table 3.8.

The discharge ratio Q_p/Q_m shown in Table 3.8 was based on Reynolds number
pump similarity requirements and modern pump testing procedures. This reflected the initial intention to determine the gas removal system's effectiveness
by observing their influence on dredge pump performance. However, if gas removal is considered directly, it is reasonable to treat the problem as one of two-
phase flow. In this type of problem, which involves the buoyant force on the gas
bubbles as the primary cause of motion of the gas relative to the water in the
suction line, a different similitude relationship should be used. A form of Froude
number may be used in this case:

$$\text{Froude number} = N_F = \frac{V}{[(\gamma_f - \gamma_g/\gamma_f)L]^{0.05}} \tag{3.145}$$

where V = mean pipe velocity in feet per second (m/s)
γ_f = unit weight of liquid in pounds per cubic foot (N/m³)
γ_g = unit weight of gas in pounds per cubic foot (N/m³)
L = some characteristic length in feet (m)

The bubble diameter should probably be used as characteristic length in this
case; however, it would be quite difficult to determine bubble diameters either in
the laboratory or in the field. Consequently, the suction pipe diameter may be
used since it represents the distance between the flow boundaries and is closely
related to the dimension of the accumulator. Using this Froude number, the velocity ratio, V_p/V_m = 8, and the discharge ratio, Q_p/Q_m = 181, or the prototype
discharge of 64,000 gpm (4037 l/s) would correspond to 350 gpm (22.1 l/s) in the
model. Another similitude relationship may be based on the time that gas bubbles
pass the accumulator opening. Since the pump similarity in modern pump testing
requires that the model velocities be equal to prototype velocities, and the length

(a)

(b)

FIGURE 3.90 Original accumulator. (a) Suction pipe in foreground; (b) accumulator between 90° elbow and pump. (*Adams and Herbich, 1970*)

TABLE 3.8 Dimensions of Model and Prototype Systems

Quantity	Model	Prototype
Impeller	10.50 in (26.7 cm)	84.0 in (213 cm)
Suction pipe diameter	4.50 in (11.4 cm)	36.0 in (91 cm)
Discharge pipe diameter	4.00 in (10.2 cm)	32.0 in (81 cm)
Pump speed	1440 rpm (1440 rpm)	180.0 rpm (180 rpm)
Discharge	2.21 ft^3/s (1.04 l/s)	141.4 ft^3/s (62.6 l/s)
Discharge	1000.0 gpm (63.1 l/s)	64,000.0 gpm (4037 l/s)
Total head	81.00 ft H_2O (24.7 m H_2O)	81.0 ft^2 H_2O (24.7 m H_2O)
Accumulator size	4.50 in^2 (29.0 cm^2)	36.0 in^2 (232.3 cm^2)
Accumulator height	1.70 ft (0.52 m)	6.5 ft (2.0 m)
Gas capacity	25.00 ft^3/min (11.8 l/s)	1000.0 ft^3/m (472 l/s)

of accumulator opening in the model is only one-eighth of the length in the prototype, the gas bubbles do not have a similar chance to enter the accumulator. If equal time is used as a criterion, the velocity ratio (V_p/V_m = 8) and discharge ratio (Q_p/Q_m = 512) would correspond to the model discharge of 125 gpm (7.9 l/s). In view of an unusually wide range of scaling values as based on different considerations, a large range of discharges was employed in the study.

The vacuum system used in the experiments was the Ingersoll-Rand V244 single-stage reciprocating vacuum pump which has an air capacity of 25 ft^3/min (0.708 m^3/min) at 20 in (50.8 cm) of mercury vacuum. The scrubber tank and filter were provided to protect the vacuum pump from water carryover. A water ejector was also used for some tests. The ejector selected was Penberthy Model 190A, 4-in (10.2-cm) size, capable of handling the following air volumes with a water supply of 80 gpm (5.05 l/s) at 40 psi (275.8 kN/m^2) pressure:

1. 14.7 scfm (24.97 m^3/s) at 5 in (12.7 cm) of mercury vacuum
2. 8.2 scfm (13.93 m^3/s) at 10 in (25.4 cm) of mercury vacuum

The discharge from the ejector was measured with a magnetic flowmeter.

Gas was provided by an Allis-Chalmers rotary compressor Model 6CCA, rated at 45 ft^3/min at 30 psi (206.85 kN/m^2) gauge pressure. After passing through an after-cooler and an oil trap, the air was injected into the suction line through a series of fine orifices surrounding the inlet. This was later changed to a system that allowed steady or slug injection through a 2-in (5.1-cm) plug valve and the injection tube was placed in the center of the suction pipe inlet. Measurements were made of quantities related to pump performance, gas removal, and gas injection. Power input to the dredge pump was determined from voltage and current readings at the dc motor and the manufacturer's efficiency curve for the motor. Pump speed was measured with a Hasler tachometer or a Strobotac. Pump suction and discharge pressures were indicated on mercury manometers. The total flow through the pump was measured by a magnetic flowmeter in the discharge line. For continuous air injection, rotameters and a pressure gauge were used to measure the injection rate. For slug injection, an orifice with a gauge and differential pressure transducers of the strain gauge type were used to record the injected air quantity. The gas removal from the accumulator was determined by a similar orifice and transducer device for both types of injection. All transducer outputs were recorded on a six-channel Brush Recorder.

Performance with Original Accumulator. The effects of continuous gas injection without removal may be summarized by three conditions. First, gas flows less than 5 percent at pump discharge or 9 percent at pump suction cause a drop in water discharge of less than 10 percent and do not affect the pump head. Increased gas injection causes a sudden drop in both head and water discharge, producing unstable performance and possibly pump oscillation. This may cause the pump to collapse. For higher initial flow rates the pump performance stabilizes at a low level until gas flows of about 10 percent at discharge or 20 percent at suction conditions cause collapse.

The first tests with the gas removal system installed were conducted with no vacuum applied to the accumulator. The pump performance with and without gas injection was not affected by the geometric change due to the presence of the original accumulator in the suction line.

Next, a series of tests was run with the vacuum applied to the accumulator as the primary variable. For accumulator vacuum lower than pump suction, air was drawn into the suction line through the relief valve which controls the vacuum level in the gas removal system. This condition resulted in more air passing through the dredge pump and thus caused a further decrease in dredging performance. If the accumulator vacuum is higher than the pump suction, water as well as gas is removed from the accumulator. The scrubber tank alone served merely to collect a volume of water before the gas removal system had to be shut down to prevent water damage to the dry-type vacuum pump. A 34-ft-high hose loop was used to simulate a riser loop used on prototype dredges. Vacuum levels equal to or slightly higher than the pump suction produced liquid levels within the model accumulator or the hose loop. In no case was removal of useful quantities of gas achieved. The air bubbles were easily seen in the transparent suction pipe. However, the air streamed past the bottom of the accumulator except at vacuums high enough to draw water out of the accumulator. In the case of high accumulator vacuum, no effect on dredging performance was noted during the short run times available without endangering the vacuum pump.[42]

A Fisher Level-Trol was used in an attempt to maintain the liquid level in the accumulator with higher vacuum applied. The response time of the Level-Trol and diaphragm-operated relief valve was long and the water level oscillated from below the lower control point to above the top of the accumulator. Improvement in gas removal capacity was achieved only at high air flows. The air injection rate that caused collapse was increased about 10 percent with the Level-Trol in operation.

Performance with Modified Accumulator. It was obvious from the studies with the original accumulator that the entrance to the accumulator and its height must be increased.[42,44] Because of the fluid column in the accumulator, the vacuum to maintain the liquid level within the height of the accumulator is higher than pump suction by the hydrostatic head. Thus, the gas must enter the accumulator under no driving force other than buoyancy. Elongation of the opening in the top of the suction pipe is a possible way to provide more time for gas to enter the accumulator. The modified accumulator with a sloping entry section on the upstream side was designed to approximately double the distance in which gas could rise above the top of the suction pipe into the accumulator.

The modified accumulator is depicted in Figs. 3.91 and 3.92. Gas removal was observed in all tests with either continuous or slug injection of gas. The effectiveness is clearly dependent on the initial (no gas flow) conditions. Air flow to the dredge pump, initial water discharge, and injected air flow are the features that best describe the gas removal system performance.

The air flow rates are a direct measure of the gas removal system behavior since the air flow to the pump is the difference between the injection and removal air flow rates. The water discharges are needed to evaluate the effect of the gas removal system on dredging performance.

Figures 3.93 and 3.94 show the relation between air percentage at pump suction and total air injection rate divided by the initial water discharge. Air percentage at pump suction is given by the volume flow rate of air at

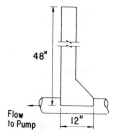

FIGURE 3.91 Modified accumulator. (*Adams and Herbich, 1970*)

pump suction conditions divided by the water discharge at that air flow rate. The initial water discharge is used as reference for the injected air flow rate at standard conditions so that the abscissa varies only with the air injection rate.

Figure 3.93 presents experimental results for an initial flow rate of 400 gpm (25.2 l/s). This rate is slightly higher than that indicated by Froude scaling ratios. Using no gas removal system, but using the original accumulator, the triangular data points were obtained. The pump collapsed when the air flow rate increased above 8.5 percent of the water flow rate. The gas removal system with the improved accumulator was used for the tests that produced the line with circular data points. Two things may be noted on this figure. One is the extension of dredging to an air content of over 25 percent and another is an indication of the gas being removed. For example, at an injected air ratio of 4 percent the air content has been reduced from 7.5 to 3.7 percent.

Figure 3.94 is a similar plot for an initial water discharge of 1000 gpm (63 l/s). This is the scaled flow rate for pump testing scale ratios. The results are similar to those for an initial water discharge of 400 gpm (25.2 l/s) but do differ in some respects. The air percentage at collapse is lower for the improved accumulator, but the injected air ratio is considerably higher. The influence of the improved accumulator is shown by the continuation of dredging at injected air ratios above 6 percent. Using the same injected air ratio of 4 percent, the air content at pump suction conditions is reduced from 11.5 to 7 percent.

The effect of gas flow to the dredge pump is shown in Figs. 3.95 and 3.96. For an initial water discharge of 400 gpm (25.2 l/s) the water flow rate is affected similarly in both systems for air contents under 5 percent. For higher air contents the water flow rate is decreased more for the system using the modified accumulator. However, at an air content of 9 percent at pump suction the original system has reached pump collapse with a total flow rate equal to zero, while the modified system is still operating, although at 67 percent of the initial water discharge. Using this value of air content, Fig. 3.93 indicates an air injection ratio of 8 percent, which is about twice the injection rate that caused collapse in the original system.

For the 1000-gpm (63-l/s) starting point there are several differences that are noticeable in Fig. 3.96. First, the original system appears to yield higher water flow rates for air contents above 3 percent. The low water flow rates show the influence of the geometry of the entry to the modified accumulator. The sudden drop in flow rate and pump head is accompanied by a change in pump suction pressure and a drop in liquid level in the accumulator. With the modified accumulator installed the larger opening in the top of the suction pipe causes this drop to occur at lower air contents. This drop is very abrupt and unsteady. The grad-

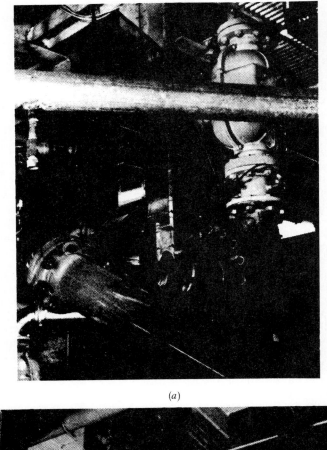

(a)

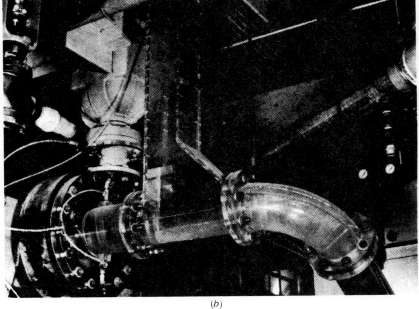

(b)

FIGURE 3.92 Modified accumulator. (a) Suction pipe in foreground; (b) accumulator between 90° elbow and pump. (*Adams and Herbich, 1970*)

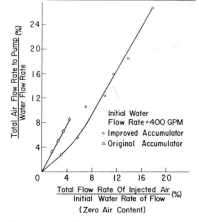

AIR PERCENT VS. INJECTED AIR RATIO

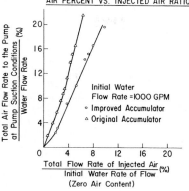

AIR PERCENT VS. INJECTED AIR RATIO

FIGURE 3.93 Air percent as a function of injected air ratio for original and modified accumulators. (Initial water flow ratio = 400 gpm). (*Adams and Herbich, 1970*)

FIGURE 3.94 Air percent as a function of injected air ratio for original and modified accumulators. (Initial water flow ratio = 1000 gpm). (*Adams and Herbich, 1970*)

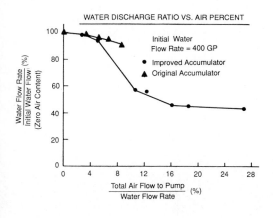

WATER DISCHARGE RATIO VS. AIR PERCENT

FIGURE 3.95 Water discharge ratio as a function of air percent for original and improved accumulator. (Initial water flow = 400 gpm). (*Adams and Herbich, 1970*)

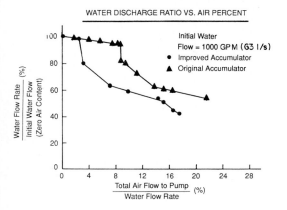

WATER DISCHARGE RATIO VS. AIR PERCENT

FIGURE 3.96 Water discharge ratio as a function of air percentage for original and improved accumulator (Initial water flow = 1000 gpm). (*Adams and Herbich, 1970*)

ual drop in performance that is associated with the air percentage increasing above this "break point" value is relatively stable until collapse occurs.

Now consider Figs. 3.94 and 3.96 together. At an air content of 10 percent in the original system the injected gas ratio is 3.6 percent. This corresponds to an air content of 6 percent with the modified system. Thus, the comparable water discharge ratios are 77 and 68 percent for the original and modified systems. The comparison favors the original system with an ineffective gas removal system for injection ratios less than 3 percent. This injection ratio results in an air content of 8 percent and a water discharge ratio of 94 percent. An injection ratio of 5 percent produces the same water discharge ratio of 60 percent of the initial discharge in both systems. For higher injection rates the modified system with effective gas removal results in improved water flow rates.

Gas Removal by Vacuum Pump

The continuously steady-air injection tests were conducted with essentially constant liquid level in the accumulator. The water level was kept in the middle portion of the accumulator. This was done to provide good flow conditions in the suction line and to prevent water from getting into the vacuum pump.

The results of these tests are shown in Fig. 3.97. The percentage of injected gas which is removed is plotted as a function of injected air flow-initial water discharge ratio for a range of water discharges. For the lower water flow rates the percentage

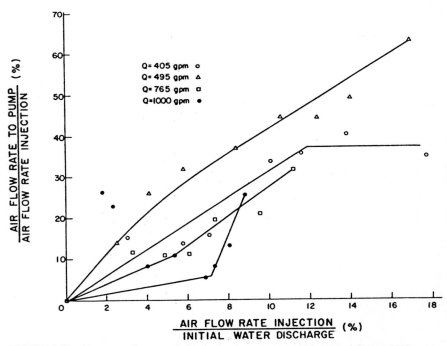

FIGURE 3.97   Air removed versus air injected: vacuum pump. (*Adams and Gupta, 1970*)

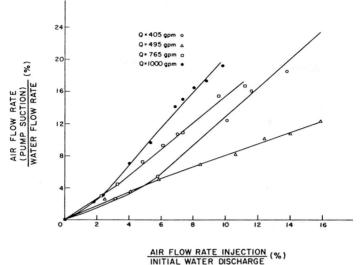

FIGURE 3.98 Air percent at pump suction versus air injected: vacuum pump. (*Adams and Gupta, 1970*)

of gas removed augments with increasing air injection, while at the higher water flow rates a little air is removed until the injected air-water ratio reaches about 6 percent. The pump collapse occurs at about 17 percent of the injected air-initial water discharge ratio for lower rates of flow and at about 10 percent for higher rates of flow.

Figure 3.98 shows a relative air flow-water flow ratio as a function of injected air flow-initial water discharge ratio.

Figure 3.99 shows the effect of air flow rate on pump discharge expressed in dimensionless form. There is a considerable drop in dimensionless discharge for higher rates of water flow for a low air flow-water flow ratio and a gradual decrease in discharge with increasing air flow-water flow ratios.

Example 3.3

Given

1. Initial water discharge = 400 gpm (25.2 l/s)
2. Air flow in suction line = 4.0 ft³/min (1.9 l/s)
3. Pump speed = 1440 rpm
4. Impeller diameter = 10.5 in (266 mm)

Compute

1. Dimensionless discharge Q_{dim}
2. Discharge Q

Solution

$$\frac{\text{Air flow rate}}{\text{Initial water discharge}} = \frac{4.0 \ (7.48) \ \text{gpm}}{400 \ \text{gpm}} = 0.0748 = 7.48\%$$

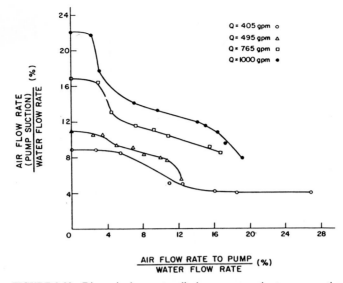

FIGURE 3.99 Dimensionless water discharge versus air at pump suction: vacuum pump. (*Adams and Gupta, 1970*)

From Fig. 3.97 this corresponds to 23 percent gas removal. From Fig. 3.98 the air flow rate at pump suction-water flow rate ratio is 8.2 percent. For the 8.2 percent ratio the dimensionless discharge can be obtained from Fig. 3.94 and is equal to 6.6 percent

$$Q_{\text{dim}} = \frac{Q}{\omega D^3}$$

or

$$Q = Q_{\text{dim}}\omega D^3$$

$$= 6.6(10)^{-3}(1440)\frac{(2\pi)}{60}\frac{(10.5)^3}{12^3}$$

$$= 0.68 \text{ ft}^3/\text{s}$$

$$= 0.68(7.48)\ 60 \text{ gpm}$$

$$= 304 \text{ gpm } (19.2 \text{ l/s})$$

Gas Removal by Ejector

Similar studies of gas removal were conducted with an ejector instead of a vacuum pump. Figures 3.100 through 3.102 present the same type of information for ejector tests as Figs. 3.97 through 3.99 for vacuum pump tests. Thus Fig. 3.100 shows the air flow to pump-injected air flow rate as a function of injected air flow-initial water discharge ratio, and Fig. 3.101 shows the relative air flow-water flow ratio as a function of injected air flow-initial water discharge ratio. Figure 3.102 presents the effect of air flow rate on pump dimensionless discharge.

Several photographs are reproduced to visually indicate the behavior of gas in the accumulator for steady flow conditions. Thus Figs. 3.103 through 3.105 show

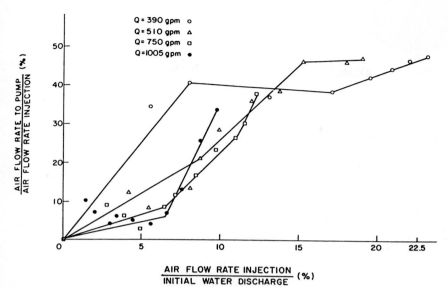

FIGURE 3.100   Air removed versus air injected: ejector. (*Adams and Gupta, 1970*)

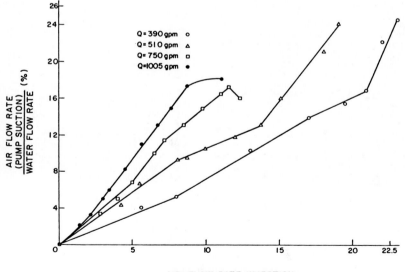

FIGURE 3.101   Air percent at pump suction versus air injected: ejector. (*Adams and Gupta, 1970*)

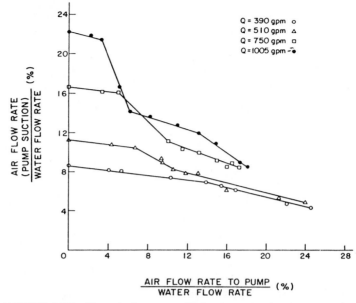

FIGURE 3.102 Dimensionless water discharge versus air at pump suction: ejector. (*Adams and Gupta, 1970*)

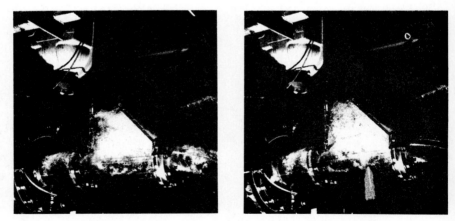

FIGURE 3.103 (*a*) Left; (*b*) right. Steady gas removal. (*Adams and Gupta, 1970*)

the conditions at different initial flow rates and two water levels. Figure 3.103*a* is at 8 in and Fig. 3.104 is at 40 in. The percentage of gas removal corresponding to these flow rates and water levels is shown in Table 3.9.

It can be seen that the gas is concentrated in the upper part of the suction line (Figs. 3.103 and 3.104*a*), and there are large bubbles visible in the lower part of the accumulator. However, with an increasing rate of water discharge it is ob-

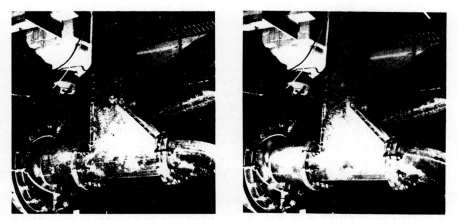

FIGURE 3.104 (*a*) Left and (*b*) right. Steady gas removal. (*Adams and Gupta, 1970*)

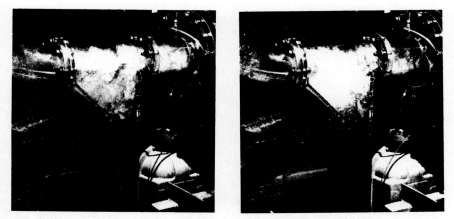

FIGURE 3.105 (*a*) Left; (*b*) right. Steady gas removal. (*Adams and Gupta, 1970*)

TABLE 3.9 Gas Removal Rates

Figure no.	Percentage
Figure 3.103*a*	30
Figure 3.103*b*	40
Figure 3.104*a*	40
Figure 3.104*b*	20
Figure 3.105*a*	6
Figure 3.105*b*	4

served that there is more gas in the suction line and the accumulator contains smaller bubbles (Fig. 3.104*b* and Fig. 3.105).

Unsteady Gas Flow

Since the gas flow in actual dredging operations will most likely be unsteady, Adams and Gupta[46] experimented with the effect of discrete slugs of gas on the effectiveness of the gas removal system. In some tests the drop in dredge pump total discharge was about 5 percent and a drop in total head of 25 percent for discharge of 400 gpm (25.2 l/s) and 33 percent for flow of 750 gpm (47.25 l/s). Pump collapse normally did not occur because of the slow response of the hydraulic system associated with a dredge pump installation.

Figures 3.106 through 3.110 show a slug of air passing the accumulator opening. The slug had a volume of approximately 1 ft^3 and was released over a period of about 1 s.

Comparison between Vacuum Pump and Ejector

There are several operational differences between a vacuum pump and an ejector. Most vacuum pumps are damaged by water but an ejector is unaffected by liquid-gas mixtures. The vacuum pump can be controlled by an air inlet valve but the controls of an ejector are more complicated and involve a bypass valve or a discharge valve and a regulation of pump speed.

The vacuum pump was found to be most effective with liquid level control in the middle portion of the accumulator. The ejector is most effective with the liquid level in the upper portion of the accumulator.

One of the ways of comparing the effectiveness of the two systems is the effect on the dimensionless discharge. Figure 3.111 indicates that there is not much difference between the two methods of evacuating air up to 7 percent air-water flow ratio, but that at higher ratios the ejector system is superior. If one compares

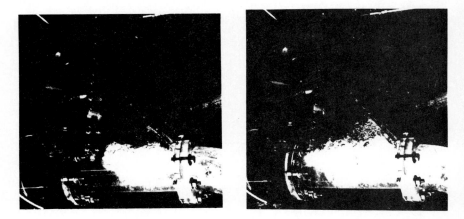

FIGURE 3.106 (*a*) Left; (*b*) right. Gas slug just entering the open base of the accumulator. (*Adams and Gupta, 1970*)

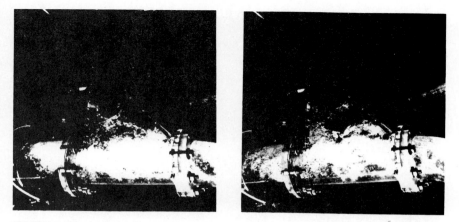

FIGURE 3.107 (*a*) Left; (*b*) right. Gas trapped by impinging on vertical side of accumulator. (*Adams and Gupta, 1970*)

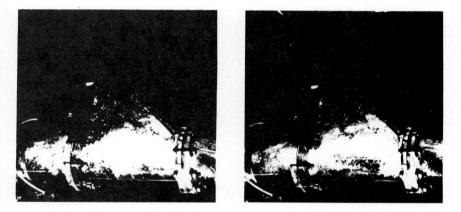

FIGURE 3.108 (*a*) Left; (*b*) right. Formation of eddies containing gas bubbles. (*Adams and Gupta, 1970*)

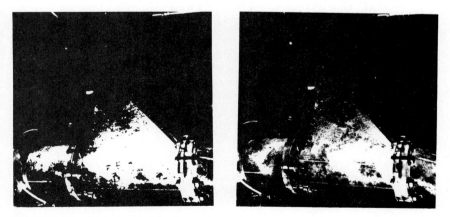

FIGURE 3.109 (*a*) Left; (*b*) right. Large slugs rising in the accumulator. (*Adams and Gupta, 1970*)

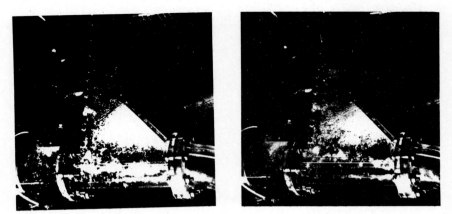

FIGURE 3.110 (*a*) Left; (*b*) right. Slug ended and entrapped air being removed. (*Adams and Gupta, 1970*)

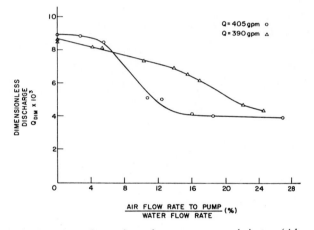

FIGURE 3.111 Comparison of vacuum pump and ejector. (*Adams and Gupta, 1970*)

the relation between gas removal and initial water discharge, the ejector can remove about 10 percent more gas than the vacuum pump. It should be noted, however, that the water level in the accumulator was higher for the ejector tests and this might have produced better performance of the ejector system.

Conclusions

1. The modified accumulator is considerably more efficient than the original accumulator used on prototype dredges. Up to 40 percent of the gas in the suction line can be removed with the modified accumulator before it enters the pump.

2. The water ejector removes a greater proportion of gas than the vacuum pump. It is more rugged, mechanically simple, and unaffected by water removal with air from the accumulator.
3. Liquid level in the accumulator must be controlled and kept at the highest possible level.
4. Pump speed has little effect on the gas removal system.
5. Prototype dredges operating in areas where a large percentage of gas may be present should be equipped with gas removal systems.
6. During normal dredging operations with little or no gas present in dredge material, the liquid level in the accumulator should be maintained at a certain level by a small excess vacuum over the pump suction vacuum.

REFERENCES

1. Herbich, J. B., and Vallentine, H. R., *Effect of Impeller Design Changes on Characteristics of a Model Dredge Pump,* Fritz Engineering Laboratory, Report 277-PR 33, Lehigh University, Sept. 1961.
2. Ippen, A., *The Influence of Viscosity on Centrifugal Pump Performance,* Paper no. A-45-57, ASME, New York, Nov. 27, 1945.
3. Herbich, J. B., *Characteristics of a Model Dredge Pump,* Fritz Engineering Laboratory, Report 277-PR 31, Lehigh University, Sept. 1959.
4. Peck, J. F., "Investigation Concerning Flow Conditions in a Centrifugal Pump, and the Effect of Blade Loading on Head Slip," *Proc.,* vol. 164, no. 1, Institution, Mechanical Engineers, England, 1951.
5. Stepanoff, A. J., *Centrifugal and Axial Flow Pumps,* John Wiley, New York, 1957.
6. Herbich, J. B., *Modifications in Design Improve Dredge Pump Efficiency,* Fritz Engineering Laboratory, Report no. 277.35, Lehigh University, Sept. 1962.
7. Shepherd, D. G., *Principles of Turbomachinery,* Macmillan, New York, 1956.
8. Tetlow, N., *A Survey of Modern Centrifugal Pump Practice,* Institution, Mechanical Engineers, England, 1942.
9. Ritter, W. R., Ginsberg, A., and Beede, W. L., *Performance Comparison of Two Deep Inducers as Separate Components and in Combination with an Impeller,* NACA, 1940.
10. Fischer, K., and Thoma, D., "Investigation of the Flow Conditions in a Centrifugal Pump," *Trans., ASME,* vol. 54, 1932.
11. Binder, R. C., and Knapp, R. T., "Experimental Determination of the Flow Characteristics in the Volutes of Centrifugal Pumps," *Trans., ASME,* vol. 58, 1936.
12. Herbich, J. B., and Christopher, R. J., "Use of High-Speed Photography to Analyze Particle Motion in a Model Dredge Pump," *Proc. IAHR Congress,* International Association for Hydraulic Research, London, England, 1963.
13. Krisam, F., *Influence of Volutes on Characteristics Curves of Centrifugal Pumps,* Fritz Engineering Laboratory, Translation T-5, Lehigh University, 1959.
14. Carslaw, C., "Communications on Flow Conditions in a Centrifugal Pump," *Proc., Inst. Mech. Engineers,* England, vol. 164, 1951.
15. Buckingham, E., "Model Experiments and the Form of Empirical Equations," *Trans., ASME,* vol. 37, New York, 1915.
16. *Hydraulic Models, ASCE Manual of Engineering Practice,* no. 25, 1942.
17. Lord Rayleigh, "The Principle of Similitude," *Nature,* vol. 95, 1915.

18. Langhaar, H. L., *Dimensional Analysis and Theory of Models,* John Wiley, New York, 1951.

19. Murphy, G., *Similitude in Engineering,* Ronald Press, New York, 1950.

20. Daily, J. W., and Harleman, D. R. F., *Fluid Dynamics,* Addison-Wesley, Reading, MA, 1966.

21. Hunsaker, J. C., and Rightmire, B. G., *Engineering Applications of Fluid Mechanics,* McGraw-Hill, New York, 1947.

22. De Groot, R., "Dredging Pipelines and Pumps," *Ports and Dredging,* no. 69, 1971.

23. Roorda, A., and Vertregt, J. J., *Floating Dredges,* De Technische Uitgeverij H. Stam N.V., Haarlem, the Netherlands, 1963.

24. Marnitz von, F., Blaum, E., and Marnitz, V., *Die Schwimmbagger,* Springer-Verlag, Berlin/Gottingen/Heidelberg, West Germany, 1963.

25. Herbich, J. B., "Discussion on *Dredging Fundamentals,* by J. Huston," *ASCE Journal of the Waterways and Harbors Division,* no. 6053, WW3, Aug. 1968.

26. Herbich, J. B., *Effects of Impeller Design Changes on Characteristics of a Model Dredge Pump,* ASME Paper no. 63-AHGT-33, 1963.

27. Herbich, J. B., *Coastal & Deep Ocean Dredging,* Gulf Publishing, Houston, TX, 1975.

28. Stepanoff, A. J., *Pumps and Blowers, Two-Phase Flow,* John Wiley, New York, 1964.

29. Hasegawa, Y., and Tokunaga, Y., "Performance Tests of Sand Pump," *Transportation Technical Research Institute,* vol. 7, no. 7, 1958 (in Japanese).

30. Anonymous, *Ports and Dredging,* 1986.

31. Herbich, J. B., Dredging Engineering Short Course, vol. 1, unpublished notes, 1985.

32. Hicks, T. G., *Pump Selection and Application,* McGraw-Hill, New York, 1957.

33. Burrill, L. C., "Sir Charles Parsons and Cavitation," *1950 Parsons Memorial Lecture, Trans., Institute of Marine Engineers,* vol. 63, 1951.

34. Knapp, R. T., Dailey, J. W., and Hammitt, F. G., *Cavitation,* McGraw-Hill, New York, 1970.

35. Herbich, J. B., and Cooper, R. L., II, "The Effect of Solid-Water Mixtures on Cavitation Characteristics of Dredge Pumps," *Proc., Fourth World Dredging Conference,* New Orleans, LA, 1971.

36. Mariani, V. R., and Herbich, J. B., "Effect of Viscosity of Solid-Liquid Mixture on Pump Cavitation," *Central Water and Power Research Station, Golden Jubilee Symposia,* vol. 2, no. 51, Poona, India, 1966.

37. Basco, D. R., "Particle Size and Density Effects on Cavitation Performance of Dredge Pumps," *Proc., Third Dredging Seminar,* Texas A&M University, Sea Grant Program Report no. TAMU-SG-71-109, 1971.

38. Basco, D. R., "Cavitation in Dredging Pumps," Dredging Short Course, Texas A&M University, College Station, TX, 1972.

39. Scheffauer, F. C., *The Hopper Dredge—Its History, Development and Operations,* U.S. Army Corps of Engineers, U.S. Government Printing Office, Washington, DC, 1954.

40. Herbich, J. B., and Miller, R. E., "Effect of Air Content on Performance of a Dredge Pump," *Proc., World Dredging Conference,* WODCON '70, Singapore, 1970.

41. Streeter, V. L., *Fluid Mechanics,* 3d ed., McGraw-Hill, New York, 1962.

42. Herbich, J. B., Adams, J. R., and Ko, S. C., *Gas Removal System. Part III, Model Study,* Lehigh University, Fritz Engineering Laboratory, Report no. 310.21, Feb. 1969.

43. Hoffman, R. T., U.S. Patents 2,795,873 and 3,119,344.

44. Herbich, J. B., and Isaacs, W. P., *Gas Removal System. Part I, Literature Survey and*

Formulation of Test Program, Lehigh University, Fritz Engineering Laboratory, Report no. 310.3, June 1964.

45. Adams, J. R., and Herbich, J. B., "Gas Removal Systems," *Proc., World Dredging Conference,* WODCON '70, Singapore, 1970.

46. Adams, J. R., and Gupta, R. P., *Gas Removal System. Part III: Model Study, Final Report,* Lehigh University, Fritz Engineering Laboratory Report, no. 310.22, Aug. 1970.

CHAPTER 4
DREDGING EQUIPMENT

All dredges can be classified as either mechanically operating or hydraulically operating.

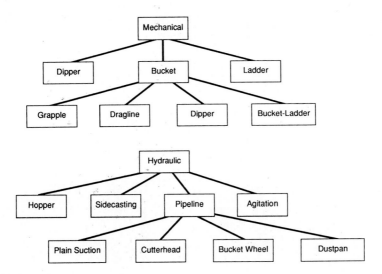

MECHANICAL DREDGES

Because of their simplicity and analogy with land-based excavating machines, mechanical dredges were the first to be developed. They can be further classified into the grapple dredge, the dragline, the dipper dredge, and the bucket-ladder dredge.

The grapple dredge consists of a derrick mounted on a barge and equipped with a "clamshell" bucket. It works best in very soft underwater deposits.

The dragline is an excavating tool consisting of a steel bucket which is suspended from a movable crane. After biting into the soil, it is dragged toward the crane by a cable.

The dipper dredge is the floating counterpart of the familiar land-based me-

FIGURE 4.1   Dipper dredge *Gaillard.* (*Courtesy, U.S. Army Corps of Engineers*)

chanically excavating shovel. Because of its great leverage and "crowding" action, it works best in hard compact material or rock. A dipper dredge is shown in Fig. 4.1.

The bucket-ladder dredge consists essentially of an endless chain of buckets. The top of the chain is thrust into the underwater deposit to be dredged so that each bucket digs its own load and carries it to the surface (Fig. 4.2). Since the

FIGURE 4.2   Continuous bucket dredge. (*Courtesy, IHC Holland*)

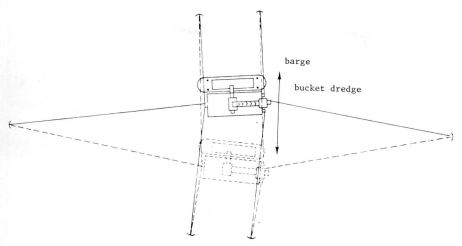

barge

bucket dredge

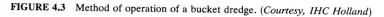

FIGURE 4.3 Method of operation of a bucket dredge. (*Courtesy, IHC Holland*)

work cycle is continuous, bucket-ladder dredges are more efficient than either the grapple or dipper dredge. Bucket-ladder dredges are particularly useful to sand and gravel suppliers since the end of the bucket-ladder can be terminated high above the supporting barge and the buckets can discharge their contents to vibrating screens. Thus, the different material sizes may be separated and stored on the barge, all by gravity.

The characteristic feature of a bucket dredge is the chain consisting of a large number of buckets and the links joining them. The chain is suspended from and driven by an upper "tumbler" and is guided and supported by a ladder. The drain is attached to a lower tumbler, and the lower end of the ladder is suspended from a hoisting gantry by means of a tackle. As each bucket reaches the lower tumbler, the soil is cut by the rim of the bucket and fills the bucket, which then travels up and discharges into a barge tied alongside the dredge. To achieve a fairly continuous operation, the dredge is swung from side to side with the aid of wires and anchors as shown in Fig. 4.3. Tugs are commonly used to move the empty barges to the dredge and to pull or push the loaded barges to a disposal or an unloading area.

Bucket dredges usually employ six anchors during operations (Fig. 4.4):

1. Two forward swing anchors
2. Two aft swing anchors
3. One bow anchor
4. One stern anchor

Wires connect anchors with winches; modern dredges are usually equipped with separate winches for the bow and forward swing wires and the warping lines.

Mechanical dredges are all characterized by their inability to transport the dredged material over long distances, lack of self-propulsion, and relatively low production. Their chief advantage lies in their ability to operate in restricted locations such as docks and jetties and to treat and dewater the dredged material in placer mining operations.

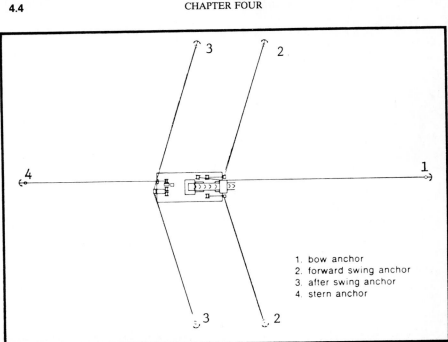

1. bow anchor
2. forward swing anchor
3. after swing anchor
4. stern anchor

FIGURE 4.4   Arrangement of anchors and winches. (*Courtesy, IHC Holland*)

The bucket dredge has been used extensively in placer mining of gold and more recently in tin mining in Malaysia. The digging depth of bucket dredges has gradually grown from about 50 ft (15.24 m) in 1905 to about 175 ft (53.3 m) in 1973 and the size of buckets from 7 ft^3 (0.20 m^3) to 54 ft^3 (1.5 m^3).[1]

HYDRAULIC DREDGES

Hydraulic dredges are self-contained units which handle both phases of the dredging systems. They not only dig the material but dispose of it either by pumping the material through a floating pipeline to a placement area or by storing it in hoppers which can be subsequently emptied over the disposal area. Hydraulic dredges are more efficient, versatile, and economical to operate because of this continuous, self-contained digging and disposal principle of operation.[2,3]

In a hydraulic dredge the material to be removed is first loosened and mixed with water by cutter heads or by agitation with water jets and then pumped as a fluid. The three basic units in a hydraulic dredge are the dredge pumps, the agitating machinery, and the hoisting and hauling equipment. The latter is used primarily to raise and lower the cutter and suction dragheads. Because of its relatively great importance and complexity, the dredge pump is discussed in Chap. 3.

Hydraulically operating dredges can be classified into four basic categories: hopper (trailing suction), pipeline (plain suction, cutterhead, dustpan), bucket wheel, and sidecasting.

Hopper Dredges

The self-propelled trailing-suction hopper dredge was originally developed in the United States (Fig. 4.5), and European manufacturers made refinements in recent decades (Fig. 4.6). The development has revolutionized the dredging industry by drastically reducing dredging costs. Today maintenance of European channels and ports is virtually dominated by trailing-suction hopper dredges, which are suitable for all but hard materials and are by far the best-suited dredges for offshore work. It consists of a ship-type hull with hoppers to hold material dredged from the bottom. The material is brought to the surface through a suction pipe and draghead. The configuration of the draghead varies with the type of material. Many improvements have been effected in the design of dragheads to make them suitable for even compact material. Hopper dredges have been built with hopper capacities ranging from several hundred cubic meters to 10,000 m^3 (13,080 yd^3). The notable improvements in the design achieved in recent years in the Netherlands are: (1) distribution system, (2) integral suction system, (3) submerged dredge pump, (4) active draghead with rotating cylinder, (5) modular draghead, (6) draghead winch control, (7) split-trail, (8) slick-trail hopper dredges, and (9) multipurpose dredges. The hoppers are usually unloaded through the bottom doors.

FIGURE 4.5 Small hopper dredge *Frontera I,* hopper capacity 500 yd^3. (*Courtesy, Ellicott Machine Corporation*)

FIGURE 4.6   Hopper dredge *DCI Dredge IX.* (*Courtesy, IHC Holland*)

Pump-out facilities are also provided in modern dredges; they are extremely mobile and do not require sheltered areas. The maximum dredging depth is 59.1 to 69 ft (18 to 21 m). By fitting a submerged pump at the draghead it has been possible to increase the dredging depth up to 131 ft (40 m). The effect of trailing-suction hopper dredges on the environment depends on the type of bed material. Dragheads can resuspend the sediments at the bottom in fine materials and the overflow water may also carry fine material, creating a turbidity plume, although some improvements have been made to reduce the turbidity.

Ship's bridge on a large hopper dredge (10,464 yd^3, or 8000 m^3 capacity) is shown in Fig. 4.7.

Distribution System in the Hoppers.   Turbulence in the hoppers maintains the dredged material in suspension and must be kept to a minimum to allow the material to settle quickly. The overflow weirs are kept at the end opposite to where the dredged material is discharged into the hopper to allow sufficient time for the particles to settle before overflowing. When the slurry flows from the discharge pipes into the hopper, significant turbulence is created since the slurry mixture falls into the hopper from some height and air is entrained, keeping the sediment particles in suspension. In a recent modification of the dredge hopper distribution system the discharge pipes were installed below the water level, or even below mid-depth, discharging sideways at the aft end of the hopper (Fig. 4.8).[4] Gratings were provided at the two sides to reduce the turbulence and air entrainment.

Draghead-Mounted Dredge Pump.   One of the recent and very important developments in trailing-suction dredges is the construction of a draghead-mounted

FIGURE 4.7 Ship's bridge, large hopper dredge *Cornelis Zanen* (8000 m³ hopper capacity). (*Courtesy, IHC Holland*)

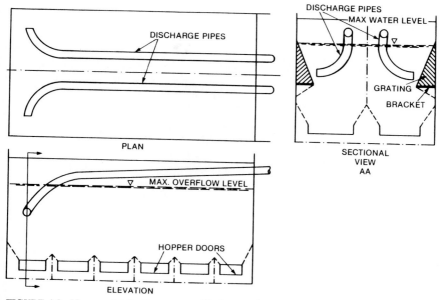

FIGURE 4.8 Hopper distribution system. (*Brahme and Herbich, 1977*)

(a)

(b)

FIGURE 4.9   A pump installed on a dragarm. (Ports and Dredging, *IHC Holland, 1986*)

dredge pump, enabling dredging in deeper water. A dragarm-mounted underwater pump is shown in Fig. 4.9.[5] Such installation increases the dredging depth, reduces chances of cavitation, and allows pumping at a higher specific gravity of the mixture; thus a smaller pump may be installed for a given production, reducing the cost of the dredge. When a pump is installed on the dragarm, adequate protection must be provided against choking. This is particularly important when dredging at shallow depths with smaller, higher-speed dredge pumps with higher minimum vacuum requirements.

The submerged pump was also installed on the draghead of a dredge *Maas* built in Belgium (Fig. 4.10)[6]; however, the majority of modern hopper dredges have dragarm-mounted pumps.

The reasons for installing a submerged pump are to increase the dredging depth and to prevent cavitation. Cavitation is defined as the formation and collapse of low-pressure vapor cavities in a flowing liquid, causing a drop of pressure head and resulting in damage to metal surface by pitting. Cavitation also causes noisy operation and mechanical vibration. A large zone of cavitation will obstruct the flow of solid-water mixtures and eventually may result in the failure of the pump to maintain the relation between head and rate of flow of the pump (Fig. 4.11).[7]

Normally, the dredge pump is installed at the lowest possible level of the ship's pump room since the attainable vacuum is limited to about 24.6 ft (7.5 m) of water by the cavitation characteristics of a centrifugal pump. This vacuum must provide adequate net positive suction head (NPSH, which is the total head available to the pump above the vapor pressure) to prevent the pump from cavitating. The NPSH is computed as follows:

FIGURE 4.10 Dredge Maas. (Ports and Dredging, *IHC Holland, 1978*)

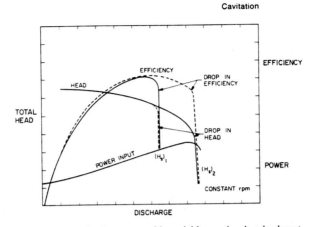

FIGURE 4.11 Performance with variable suction head: abrupt onset of cavitation. (*Herbich, 1975*)

$$\text{NPSH} = \frac{p_a}{\gamma} + \frac{p_s}{\gamma} + \frac{V_s^2}{2g} - \frac{p_v}{\gamma} \qquad (4.1)$$

where p_a = local barometric pressure in pounds per square inch (newtons per square meter)

γ = specific weight of liquid in pounds per foot (newtons per cubic meter)

p_s = pressure measured at the pump inlet in pounds per square inch (newtons per square meter)

V_s = velocity in the suction pipe at the pump inlet in feet per second (meters per second)

g = gravitational acceleration = 32.2 ft/s² (9.81 m/s²)

p_v = vapor pressure of liquid in pounds per square inch (newtons per square meter)

If available NPSH is less than that required by the pump, the pump will cavitate. It is thus very important to calculate the available NPSH for the proposed installation before it is built (see also Chap. 3).

The required NPSH can be obtained by model or field prototype testing and should be provided by the manufacturer. The only way to avoid cavitation is to install the dredge pump far enough under water. This has already been done in the case of cutter suction dredges where the pump could be easily installed on the cutter ladder, which is quite strong and rigid. In the case of the draghead, the design is more difficult. This was, however, realized in the case of some dredges. It was possible to achieve a specific gravity of the slurry mixture up to 1.4 even when the depth of dredging was increased. The dredge pump on dragheads has thus led to the possibility of achieving theoretically unlimited dredging depths without affecting the concentration and without causing cavitation in the pump.

California-type Draghead. A California-type draghead is shown in Figs. 4.12 and 4.13. Grating is used on many dragheads to prevent large objects from lodging themselves in the suction pipe or in the pump. Most gratings are rectangular and the ratio of grating area to suction pipe area is approximately 3 to 4.1. Dimensions for the California-type draghead are given in Fig. 4.14.

Another type of fixed draghead is the Ambrose draghead (Fig. 4.15). It has been found to be inefficient on firm sandy soils, being unable to adjust itself to provide an overall close contact with the bottom for different angles of the

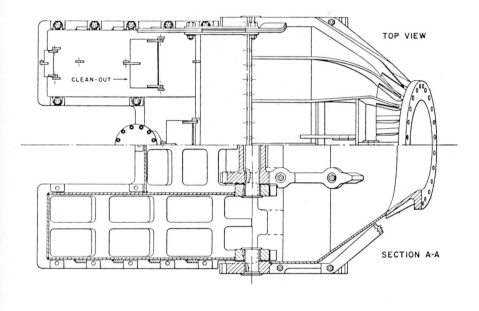

TOP VIEW

CLEAN-OUT

SECTION A-A

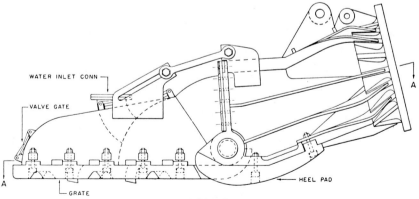

WATER INLET CONN

VALVE GATE

A

HEEL PAD

A

GRATE

SIDE VIEW

FIGURE 4.12 California-type draghead; sketch of dredge *Essayons*. (*Courtesy, U.S. Army Corps of Engineers*)

FIGURE 4.13 California-type draghead, dredge *Ham 309*. (*Courtesy, IHC Holland*)

dragarm. The draghead known as the "modified Ambrose" uses closure plates to reduce the entry of clear water.

A need to improve efficiency of dragheads led to adjustable types of dragheads.

Active Draghead. The ability to absorb the reactive forces generated during the dredging process is an essential feature of the freedom of movement of a trailing dredge. In the case of dredge anchor wires or spuds, the reactive forces produced during the dredging are absorbed. On a freely moving dredge such as a trailing-suction dredge, the thrust of the propeller(s) is the sole means by which these forces can be opposed. Since this thrust is of considerably smaller magnitude than the holding power of anchors or spuds, the digging power of a trailing-suction dredge is limited in comparison to that of a stationary dredge. As a result, the type of soil on which trailing-suction hopper dredges are used is largely restricted to silt and within certain limits of compacted sand and gravel (Fig. 4.16).[8]

The production achieved by trailing-suction hopper dredges in highly compacted soils with the standard dragheads is comparatively low. To improve the output in highly compacted sands, water jets have been used in the past with a fair degree of success. In the case of clayey soil, however, such jets are not quite satisfactory. Various attempts made to improve the output by fitting fixed knives or teeth to the dragheads were not successful since such teeth have imposed high and widely fluctuating stresses on the suction pipe. The dragheads modified in this manner have only a mediocre output.

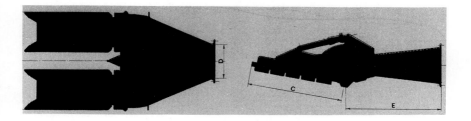

Pipe Diameter (mm)	A (mm)	C (mm)	E (mm)	Weight (kg)
450	1400	1185	1250	1600
500	1400	1185	1250	1650
550	1610	1545	1520	2550
600	1610	1880	1800	3750
650	1860	1880	1800	3750
700	2085	1840	1950	4650
750	2085	1840	1950	4700
800	2330	2300	2200	6650
850	2330	2300	2200	6700
900	2580	2760	2485	8250
1000	2880	2760	2785	9950
1100	3155	3220	3050	13000
1200	3425	3680	3300	15100

FIGURE 4.14 Dimensions of California-type draghead. (*Courtesy, IHC Holland*)

A new type of draghead which can produce economically acceptable output from a trailing-suction hopper dredge operating in clay has been developed. The draghead is designed as an "active rotary draghead" and it incorporates a rotating cylinder with a number of knives. The rotation of the cylinder and the forward movement of the dredge cause thin, wedge-shaped slices of soil to be removed (Fig. 4.17), thus eliminating the erratic dredging process experienced in the case of cutting knives. The draghead consists of an outer casing, inside of which a "cutter edge" is mounted on a horizontal shaft driven by hydraulic motors and revolving in the vertical plane.[9] The weight of the draghead, however, increases considerably. The action of the cutters causes the material to disintegrate, permitting it to be drawn up as slurry. A number of field tests were carried out on the active dragheads and the results indicated that they performed better than the conventional dragheads fitted with knives. The tests also served to confirm that the tensile stress exerted on the active draghead is smaller and displays far smaller fluctuations than the tensile stress on a conventional draghead fitted with knives.[8,9]

Venturi Draghead. In order to be able to remove the compacted noncohesive soils, a start was made on the development of the so-called Venturi draghead as early as

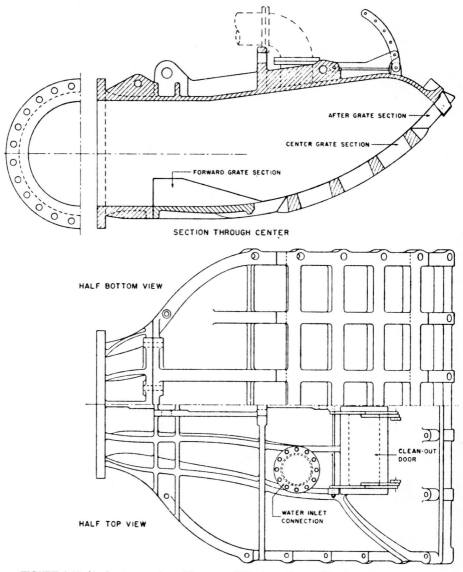

AFTER GRATE SECTION

CENTER GRATE SECTION

FORWARD GRATE SECTION

SECTION THROUGH CENTER

HALF BOTTOM VIEW

CLEAN-OUT DOOR

WATER INLET CONNECTION

HALF TOP VIEW

FIGURE 4.15 Ambrose-type drag. (*Courtesy, U.S. Army Corps of Engineers*)

1970.[10] Basically the Venturi draghead consists of three parts: (1) the pivoting part, the visor—the cross section of this part first reduces and then increases; (2) the fixed part which contains the water jets; and (3) an elbow piece fitted between the fixed part and the actual suction tube to prevent any bad transition to the visor. The operating principle is based on the creation of a negative pressure immediately above the seabed by converting part of the pressure energy into kinetic energy, creating a flow very close to the seabed. By avoiding sharp bends and transitions in the design of the draghead the conversion of pressure into a lower pressure plus velocity head

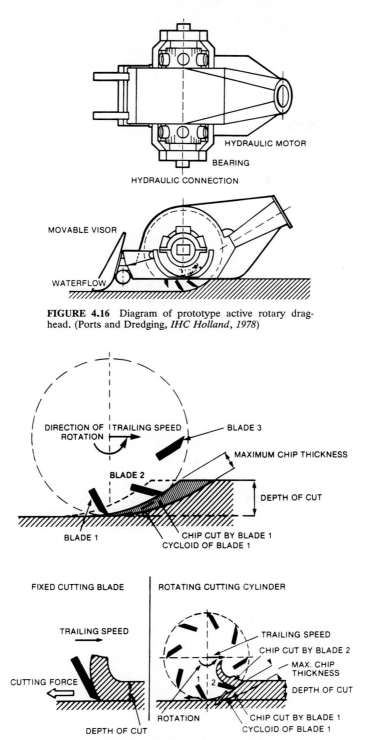

FIGURE 4.16 Diagram of prototype active rotary drag-head. (Ports and Dredging, *IHC Holland, 1978*)

FIGURE 4.17 Cutting action with active rotary draghead. (Ports and Dredging, *IHC Holland, 1976*)

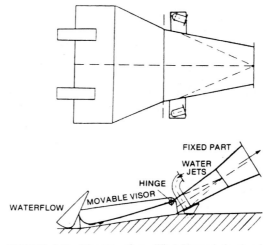

FIGURE 4.18   Diagram of modified Venturi draghead with water jets. (Ports and Dredging, *IHC Holland, 1986*)

takes place with the greatest possible efficiency, thus reducing the pressure drop across the entrance to the draghead to a minimum. By streamlining the entrance and by gradually narrowing it, a controlled acceleration is produced while the mixture flow is kept in contact with the seabed as long as possible. The change in the cross section is again given the most efficient shape for regaining the pressure drop that was necessary to accelerate the flow of water. This combination of Venturi tube with diffuser has given this draghead its name (Fig. 4.18). Prototype tests were carried out on the Venturi-type draghead and the results showed a 30 to 40 percent increase in production in the case of fine sand. In coarse sand, however, the output of the Venturi draghead, though somewhat better than the Berlin draghead, was slightly lagging behind the IHC draghead.[11] Dimensions of the IHC draghead are shown in Fig. 4.19.

Automatic Draghead Winch Control System.   An automatic draghead controller[12] regulates the movements of the suction pipe and draghead through-out the dredging cycle (Fig. 4.20). It is programmed to swing the pipe outboard, lower it, and in conjunction with the swell compensator maintain the correct pressure of the draghead on the bottom and the lateral position of the pipe hoist (Fig. 4.21). This is achieved through a number of watertight electronic sensors fitted to the winches, suction pipe gantries, swell compensator, and draghead gantries. Exhaustive model tests have been carried out which showed promising results. The automatic suction pipe controller affords continuous supervision and correction of sideways movement of the pipe during the dredging operation and protection against exceeding the maximum dredging depth. To this end it com-pletely controls the swell compensator and has enabled the vessel to continue op-erations in bad weather, while minimizing the risk of damage to the installation.

Sidecasting Dredges

Sidecasting dredges have the advantage of continuous operation since the ma-terial is dredged and disposed of some distance from the channel. There are a

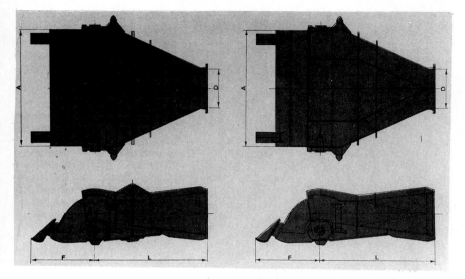

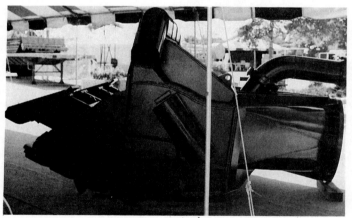

Pipe Diameter (mm)	A (mm)	F (mm)	L (mm)	Weight Undivided (kg)
450	1345	800	1300	1350
500	1345	800	1300	1400
550	1645	980	1600	2200
600	1645	980	1600	2250
650	1945	1280	1950	3500
700	2095	1310	2090	3900
750	2095	1310	2090	3950
800	2395	1590	2300	5500
850	2395	1590	2300	5550
900	2695	1610	2600	7000
1000	2995	1660	2840	8000
1100	3295	1780	3150	11700
1200	3595	1840	3450	13500

FIGURE 4.19 Dimensions of the IHC draghead. (*Courtesy, IHC Holland*)

few hopper dredges equipped with a sidecasting boom in the United States and Venezuela, and there are several small sidecasting dredges without hoppers (Fig. 4.22).

Cutterhead Dredge.

This is probably the most well-known dredging vessel as well as the most efficient and versatile. It is equipped with a rotating cutter apparatus surrounding the intake end of the suction pipe. These dredges, shown in Figs. 4.23 through 4.25, can efficiently dig and pump all types of alluvial materials and compacted deposits such as clay and hardpan. The larger, more powerful machines are used to dredge rock-like formations such as coral and the softer type of basalt and limestone without blasting. Some of the dredges have been known to excavate and transport boulders in sizes

FIGURE 4.20 Automatic draghead winch controller. (*Courtesy, IHC Holland*)

up to 30 in (762 mm) in diameter and to dredge limestone up to the unconfined compressive strength of 14,223 psi (1000 kg/cm^2).

The cutterhead dredge is generally equipped with two stern spuds, which are used to advance the dredge into the cut or excavating area. Newer cutterhead dredges have a spud carriage which increases the overall dredging efficiency. A well-designed 30-in (762-mm) dredge (size is given by the diameter of the discharge pipe) with 5000 to 8000 hp (3728 to 5964 kW) on the pump and 2000 hp (1491 kW) on the cutter will pump 2000 to 4500 yd^3/hr (1529 to 3440 m^3/hr) in soft material, and 200 to 2000 yd^3/hr (153 to 1529 m^3/hr) in soft to medium hard rock through pipeline lengths up to 15,000 ft (4572 m).

The cutterhead dredge is considered an American specialty. Bond[13] reports that nowhere else in the world has this type of machine been so highly developed and so widely used in submarine excavation.

The operator's control area on a large cutterhead dredge is shown in Fig. 4.26. The operator has all the information and controls within easy reach.

The cutter suction dredges are excellent machines for dredging hard compact material without pretreatment of the rock by drilling and blasting or the use of rock-breaking equipment. One of the important developments in the case of cutter suction dredges was the installation of a submerged dredge pump on the ladder of the dredge. This pump has certain advantages over the dredge pump in the hull: (1) it can increase the concentration of the dredged slurry in the pipe, and (2) it can dredge in deeper depths.

One serious problem with hydraulic dredges is the wear of the pumps. To deal with this problem two new types of pumps have been developed, both in the Netherlands and in the United States. The principal feature of the pump is a double-walled housing, the inner wall of which is replaceable and allowed to vir-

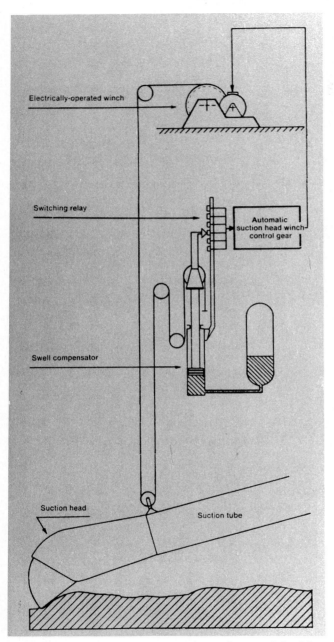

FIGURE 4.21   Swell compensator. (*Courtesy* Ports and Dredging, *IHC Holland*)

·(a)

(b)

FIGURE 4.22   Small sidecasting dredge *Schweizer* operating at Oregon Inlet. (*Courtesy, U.S. Army Corps of Engineers*)

tually wear out. The inside part is constructed of a highly wear-resistant material which is quite brittle and could not be used for the volute casing of the pump. The wear-resistant liner can also be manufactured from rubber or a synthetic material. The space between the inside and outside housing is filled with water. The double housing also provides an added factor of safety in case of an explosion in the pump.

FIGURE 4.23 Artist's rendering of self-propelled cutter suction dredge *Leonardo da Vinci*. (*Courtesy, IHC Holland*)

Cutterhead. The environmental concerns relating to cutterhead operations were raised in recent decades and caused rejection of dredging permits or work stoppage in several instances. It appears that the design of a cutterhead was developed on a trial-and-error basis rather than rational design.

FIGURE 4.24 Cutterhead dredge *Alameda*. Overall length 208 ft, width 50 ft, and depth 14 ft. Dredging depth 52 ft. Designed to move 40,000 yd^3 per day over distances over 4 miles. (*Courtesy, PACECO Company, California*)

FIGURE 4.25 Small cutter suction dredge. (*Courtesy, IHC Holland*)

FIGURE 4.26 Operator's control desk, self-propelled cutterhead dredge *Leonardo da Vinci*. (*Courtesy, IHC Holland*)

Both Slotta[14] and Breusers[15] conducted research on the mechanics of flow approaching the cutterhead. The cutterhead was developed initially to loosen up densely packed deposits and eventually to cut through soft rock. Common practice seems to be to employ the cutterhead whether it is needed or not. There are many instances where a cutter is not needed, for example in dredging silt, clay, or fine sand. In such a case the rotation of the cutterhead would produce a sediment cloud and increase the possible environmental impact. A variety of suction pipes have been tried by manufacturers and contractors; these range from simple cylindrical pipes cut at an angle from 45° to 90° to more elaborate suction nozzles. In some cases water jets are incorporated in the design of the suction pipe.

Apgar and Basco[16] presented results of a study of the flow field around a cylindrical suction pipe inlet. The analytical solution was compared with experimental model results. Since the analytical solution employing potential flow theory does not take into account viscous effects, the comparisons are only good for initial flow conditions when viscous effects are limited. For fully developed flow the viscous and boundary effects create separation zones and circulation patterns, considerably altering the theoretical streamline configuration.

One interesting application of the study is the concept of a hooded shield which reduces the environmental impact of cutterhead dredging. It was found that a major separation zone exists between eight and ten pipe diameters above the bottom and such a hood could lower the zone of separation, causing increased sediment entrainment, thus preventing turbid water from reaching the surface. A sketch of a hooded shield is shown in Fig. 4.27.

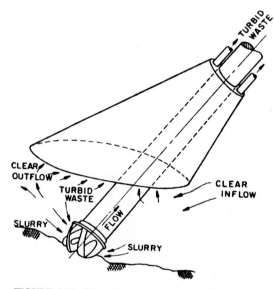

FIGURE 4.27   Hooded shield on cutterhead. (*Apgar and Basco, 1973*)

In designing the cutter there are a number of geometric variables which need to be considered. Terry[17] suggests several variables related to the digging-in function, which depends on the cone and face angle (Fig. 4.28), on angular displacement of the arm (Fig. 4.29), and on positioning the leading edge (Fig. 4.30). The face angle is usually equal to one-half of the cone angle and should be selected to produce a horizontal finished grade of the cut. The displacement angle of a cutter arm is measured from a selected point at the bottom part to a similar point at the top. A four-blade cutter has an angular displacement of 90°. The more arms, the lesser the force on the leading edge, shaft, and bearings, but this also creates a smaller opening for rocks to enter the suction pipe. The rake angle is defined in

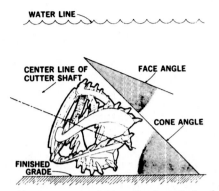

FIGURE 4.28 Definition sketch: cutterhead. (*Courtesy, WODCON IV*)

FIGURE 4.29 Angular displacement. (*Courtesy, WODCON I*)

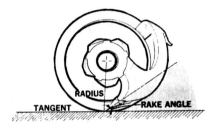

FIGURE 4.30 Rake angle. (*Courtesy, WODCON I*)

Fig. 4.30 as the angle between a tangent to the tip of a tooth and its outside surface. Terry suggests a rake angle of about 28° for best biting efficiency.

Cutters are used with and without teeth, depending on the hardness and compactness of the material to be dredged. The teeth may be part of the blade, detachable, or welded to the blade. The wear on the cutter is extremely high, particularly when working in hard materials. Terry[17] suggests that the leading edge of the cutter, wear shrouds, and weld-on edges be made of materials having the following characteristics:

1. Hardness: 500 Brinell or 51 Rockwell "C"
2. Yield strength: 200,000 psi (14,061 kg/cm^2)
3. V-notch (Charpy): impact resistance of 15 ft - lb at $-40°$F
4. Hardenability* of 5 in (12.7 cm)

Several types of cutter designs are used from the simplest straight-arm type to the more complicated basket type, typically with four to six blades. A cutter with replaceable pick points is shown in Fig. 4.31.

The approximate dimensions of the cutterhead are given in Fig. 4.32. Cutter diameter D_c may be expressed in terms of suction pipe diameter, or

---

*Hardenability is defined as the section thickness through which steel will fully harden upon quenching.

FIGURE 4.31 A cutter with replaceable pick points. (*Courtesy, Jan de Nul*)

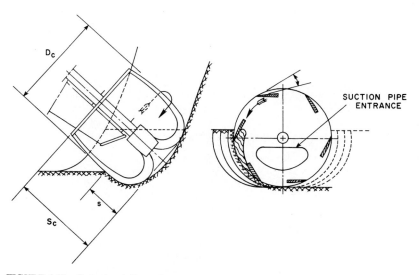

FIGURE 4.32 Cutterhead dimensions.

where C_c = coefficient equal to between 3.0 and 4.0.

$$D_c \approx C_c D_s \qquad (4.2)$$

The length of the cutter may be expressed in terms of cutter diameter, or

$$S_c = 0.75 \, D_c \qquad (4.3)$$

Speed of the cutterhead varies from about 15 to 25 rpm. The required horsepower may be approximately computed from

$$P = \frac{1}{75\eta} F_c U_s \qquad (4.4)$$

where P is in horsepower
$\quad F_c$ = cutting force at circumference per unit cut length in kilograms force per meter
$\quad U = \pi D_c n/60$, in meters per second
$\quad s$ = cut length, in meters
$\quad \eta$ = efficiency
$\quad 75$ = conversion factor (1 hp = 75 kgf/m/s)
$\quad n$ = speed in revolutions per minute

Cutters can be designed for a variety of dredged materials from peat through clay, silt, sand, gravel, and boulders to sedimentary rock such as limestone, dolomite, and carbonaceous rocks.

Cutterheads for dredging of sand and clay (Fig. 4.33) have the following features:

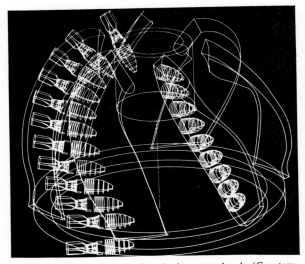

FIGURE 4.33 Sand and clay dredge cutterhead. (*Courtesy, ESCO Corporation*)

1. The hubs have up to six integrally cast sockets for use on dredges with spud carriages.
2. The ring may have as many as six teeth to provide maximum wear protection.
3. Special arm to hub joints create full openings at the hub, providing minimum overlap between point tips and the adjacent arm.
4. The arm openings are designed with maximum clearance.

The horsepower requirements to operate at the required revolutions per minute range from 400 to 5000 hp (298 to 3728 kW).

Cutterheads for relatively soft rock (Fig. 4.34) have the following features (in addition to those listed above):

1. Controlled arm openings to minimize the chance of rocks blocking the pumps.
2. Tooth roll angle is set especially for using pick points.
3. Tooth angle of attack is designed for optimum cutting in hard rock (Fig. 4.35).

A cutterhead installed on dredge *Leonardo da Vinci* is shown in Fig. 4.36.

There are several cutter edge designs available on the market. These are shown in Fig. 4.37 as the plain, plain serrated, offset serrated, pick point, and adapter edges. Many designs of teeth and adapters have been developed by various manufacturers. An example is shown in Fig. 4.38. Details of a cutterhead are shown in Fig. 4.39. Specifications for cutters are given in Figs. 4.40 through 4.42.

Transportable Cutterhead Pipeline Dredge. In recent years portability of cutterhead dredges has become popular, particularly for small units. Dredges up to 20 in (508 mm) in size have been built and in more recent years dredges of sizes larger than 20 in (508 mm) have been constructed in the portable range. However, the economic aspects of building bigger portable dredges versus the

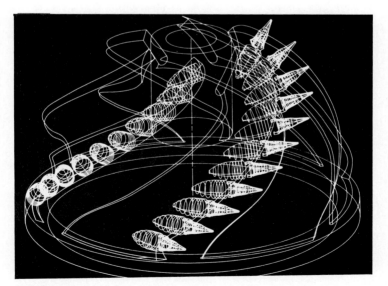

FIGURE 4.34 Light-duty rock dredge cutterhead. (*Courtesy, ESCO Corporation*)

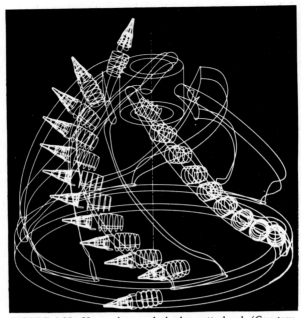

FIGURE 4.35 Heavy-duty rock dredge cutterhead. (*Courtesy, ESCO Corporation*)

FIGURE 4.36 A cutterhead installed on a massive ladder of *Leonardo da Vinci.*

Plain Edges are designed for sand, mud and loose aggregate material.

Plain Serrated Edges should be specified for hard-packed soil as well as soft and loose material.

Offset Serrated Edges are shaped for high production in clay, boulder clay, hard-packed sand, soft limestone, coral and more.

Pick Point Edges are used for hard limestone, coral or similar rocklike formations.

Adapter Edges broaden the work range of the replaceable edge cutter to include cutting hard, abrasive materials.

FIGURE 4.37 Replaceable cutter edge designs. (*Courtesy, Florida Machine and Foundry*)

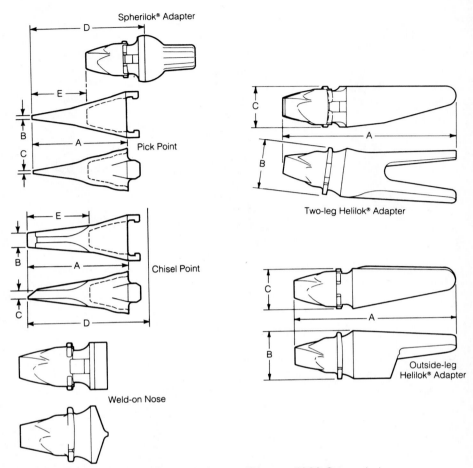

FIGURE 4.38 Replaceable teeth and adapters. (*Courtesy, ESCO Corporation*)

economic advantages which are normally associated with small portable dredges should be considered. Woodbury[18] discussed portable dredges in a paper presented at the 1968 World Dredging Conference.

According to Woodbury, there are two innovations which have improved modern portable dredge design. One is that recent designs use the forward-mounted underwater cutter drive, which reduces weight without sacrificing cutting ability. The second is mounting the swing drums and the ladder at the trunnion end. This arrangement eliminates the fair-leads sheaves at the upper end of the ladder and places the drum in view of the lever operator. It reduces the mechanical load on the trunnion and removes a piece of machinery from the hull. In regard to the portability of such a dredge one should consider the fact that most of the weight is concentrated in the central pontoon while the outer pontoons are relatively light but extremely bulky. Since transportation by ship is charged either by volume or weight, the bulkiness of the outer pontoons may

(a)

(b)

FIGURE 4.39 A cutterhead designed for cutting rock. (*Courtesy, Jan de Nul*)

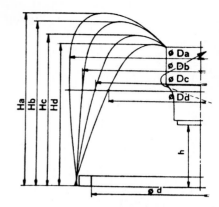

Suction tube diam.	Type A		Type B		Type C		Type D		Ring diam.	Height h
	Da	Ha	Db	Hb	Dc	Hc	Dd	Hd		
200	1000	670	900	640	800	595	700	520	830	245
250	1150	770	1035	735	920	680	805	635	955	280
300	1300	870	1170	830	1040	770	910	715	1080	320
350	1450	970	1305	930	1160	860	1015	800	1205	355
400	1600	1070	1440	1025	1280	950	1120	855	1330	390
450	1750	1170	1575	1120	1400	1040	1225	965	1455	430
500	1900	1275	1710	1215	1520	1125	1330	1050	1580	465
550	2050	1375	1845	1310	1640	1215	1435	1130	1700	505
600	2200	1475	1980	1410	1760	1305	1540	1215	1830	540
650	2350	1575	2115	1505	1880	1395	1645	1295	1950	575
700	2500	1675	2250	1600	2000	1485	1750	1380	2075	610
750	2650	1775	2385	1695	2120	1575	1855	1460	2200	650
800	2800	1875	2520	1790	2240	1660	1960	1545	2325	685
850	2950	1975	2655	1890	2360	1750	2065	1625	2450	725
900	3100	2075	2790	1985	2480	1840	2170	1710	2575	760
950	3250	2175	2925	2080	2600	1930	2275	1795	2705	800
1000	3400	2275	3060	2175	2720	2020	2380	1875	2830	835
Dimensions in mm										

FIGURE 4.40 Specifications for a cutter. (*Courtesy, IHC Holland*)

contribute to a high cost of transportation. To cut the weight of the main pontoon, dispense with the superstructure and simply provide covers for the pumps and the engines with easy access for servicing.

The lever room should have sufficient space for controls and instrumentation and it should provide the operator with all-around visibility for safe operation of the dredge (Fig. 4.43). The lever room may be removable for transportation by rail, road, and sea, with all the instruments and controls in one complete unit. One way of adjusting the weight of the whole dredge is by lengthening, shortening, or widening the outer pontoons to provide the best weight distribution and buoyancy for the whole dredge. The main pontoon may be connected to the outer pontoons with high-strength bolts. The bolts should be located in such a position that the pontoons may be assembled either in water or on land. The portability feature of the dredge is advantageous for an operator asked to dredge a channel

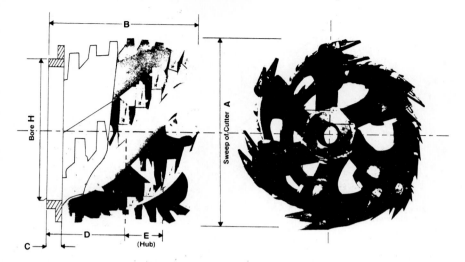

CUTTER SIZE	Use on Ellicott Dredge Series	BASIC DIMENSIONS								Approx. HP	Cutter Weight	Approx Ring Size Pass	Suits Suction Pipe In.&Mm	No. Blades	X-Sect Sq. Yd. Sq. M	FACTOR CODE
		A	B	C	D	E	F*	G	H							
43	770 970 1270	43.25	33	3.5	19.5	7.5	11.75	THREAD OR BORE TO SUIT	34.25	50/100	1500#	7.75	10-16	6	.877	.20
		1100	840	89	495	190	300		871		680 kg	197	254/406		.732	
54	970 1270 1600	54.0	40.88	4.0	24.0	9.0	14.72		42.88	100/200	2700#	9.75	14-20	6	1.35	.30
		1370	1018	102	610	228	374		1040		1225 kg	246	356/510		1.16	
85	3000	84.75	57.50	6.0	28.0	17.0	23.0		68.13	225/675	10.000#	15.25	20-27	6	3.02	.70
		2150	1458	152	712	430	585		1712		4536 kg	386.	510/680		2.53	
99	5000 7000 10,000	99.0	73.0	7.0	39.5	19.0	27.0		76.50	675/1200	16000#	18.0	24-30	6	4.31	1.00
		2510	1850	178	1000	481	685		1940		7260 kg	456	610/840		3.62	

INCHES – ENGLISH SYSTEM   MILLIMETERS – METRIC SYSTEM   *HEX

FIGURE 4.41 Specifications for a cutter. (*Courtesy, Ellicott Machine Corporation*)

of a certain width and depth on one job and then move on to a job requiring a much greater depth or much wider channel. In such cases the operator may purchase several pontoons and ladders of different lengths which can then be assembled as need arises. Figure 4.44 shows three sketches of hulls having 75-ft (22.9-m) lengths, one with a 70-ft (21.3-m) ladder at a maximum angle of 45°, which would provide for a cut 50 ft (15.4 m) deep and 150 ft (45.7 m) wide. A similar hull may be provided with only a 25-ft ladder with a maximum angle of 22.5°. This would cut a channel 10 ft (3.1 m) deep and 125 ft (38.1 m) wide. A smaller hull, 55 ft (16.8 m) in length with a 25-ft (7.6-m) ladder and swing reduced to 60°, will cut a channel 10 ft (3.0 m) deep and only 70 ft (23.1 m) wide.

Superstructure. The superstructure on the dredges is usually an A-frame or sometimes a square frame. Low-alloy, high-strength steel superstructures may be used for the assembly because of their light weight . Cable-supported A-frames as well as square frames may be used in view of the light weight of this type of mounting. In any design, care must be taken to provide a vibration-free unit which would not cause any problems of wear at the various connections.

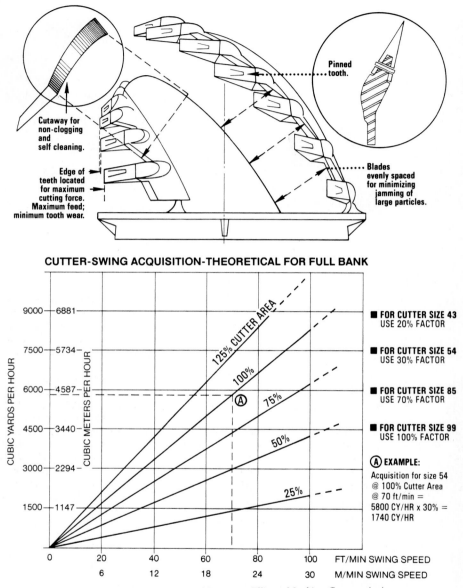

FIGURE 4.42 Cutter-swing acquisition. (*Courtesy, Ellicott Machine Corporation*)

Pump and Prime Mover. Pumps for portable dredges are usually purchased from the manufacturers. It would be expensive to design a special low-weight pump for a portable dredge. In selecting pumps for a portable dredge, it should be kept in mind that smaller pumps which would have to be operated at higher speeds have a higher wear rate. Although larger pumps running at lower speeds

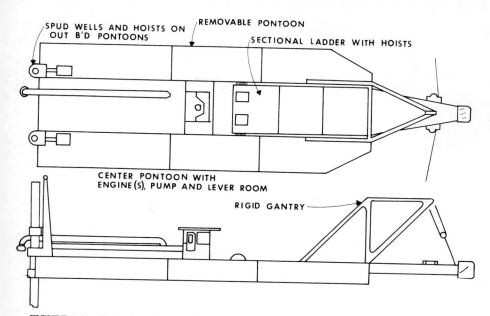

FIGURE 4.43 Typical portable dredge arrangement. (*Courtesy, World Dredging Conference*)

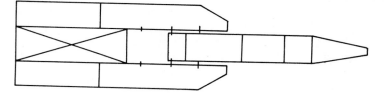

75' HULL 90° SWING 70' LADDER @45° CUT 50'x150'

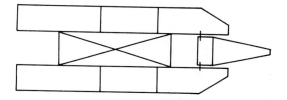

75' HULL 90° SWING 25' LADDER @22·5° CUT 10'x125'

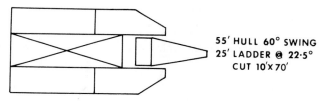

55' HULL 60° SWING
25' LADDER @ 22·5°
CUT 10'x70'

FIGURE 4.44 Hull and ladder combination. (*Courtesy, World Dredging Conference*)

may be satisfactory as far as wear, they will be heavy and may provide excessive weight for the main pontoon. In any design, one should consider the capacity and head developed by the pump per unit weight of the pump. Similar observations may be made for the prime mover, since one would desire maximum horsepower per unit weight.

In general, both the pump and motor should not be too heavy because removal and disassembly of the pump and the motor each time the dredge is moved to another location would be time consuming and cumbersome. Figure 4.44 shows the hull and ladder combinations to suit project depths and widths.

Power Requirements and Dredge Production.   Figure 4.45 indicates power requirements for dredges of sizes between 12 in (304 mm) and 24 in (609 mm) and distribution of power between the pump, the cutterhead, and the hoist. Although the actual horsepower requirements may vary from one category of dredge to another, this figure may be used as a rough guide for selection of power requirements of a cutterhead dredge. Figure 4.46 is a preliminary selection guide for dredges.

In summary, portable dredges which were introduced by several manufacturers in the past 15 years have their place in dredging. They have many uses, particularly when the operator or the owner has a number of locations where dredging may be required and it would be uneconomical to provide for a dredge in each location.

Typical dimensions of 14- to 33-in (355- to 838-mm) cutterhead dredges are shown in Figs. 4.47 through 4.50.

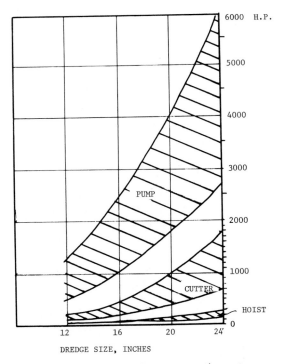

FIGURE 4.45 Typical dredge power requirements.

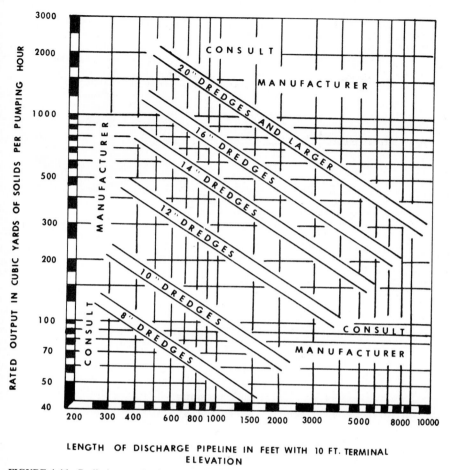

FIGURE 4.46   Preliminary selection guide for dredges.

Portable Dredges.   One of the major dredge pump manufacturers has developed portable dredges ranging from an 8-in (203-mm) model to a 20-in (508-mm) model. The summary of practical ranges of discharge by pipe sizes, ranges of shaft horsepower, normal digging depths, etc., is shown in Table 4.1.[19]

The large range of sizes of these standard dredges indicates their versatility to undertake a wide variety of earth-moving tasks, such as land reclamation, aggregate recovery, waterway improvement, new construction, industrial handling, and many more.

The Bucket Wheel Dredge

The bucket wheel dredge is a rotating wheel equipped with bottomless buckets (Figs. 4.51 and 4.52). In the American design of the bucket wheel (introduced in 1970) the soil is cut or loosened up and then directed into the interior of the wheel where it is conveyed into the suction line. The underwater bucket wheel unit

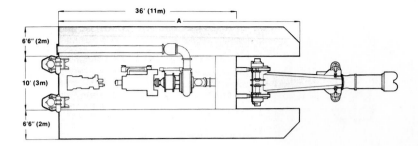

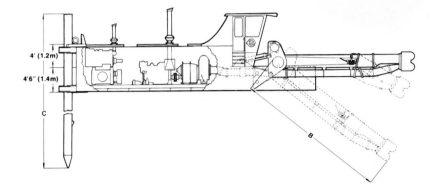

	A OVERALL HULL LENGTH FT. (m)	B TORQUE TUBE LENGTH FT. (m)	C SPUD LENGTH FT. (m)
20 FT. (6 m) **DIGGING DEPTH**	48' (14.6 m)	27' (8.2 m)	34' (10.4 m)
26 FT. (7.9 m) **DIGGING DEPTH**	50' (15.2 m)	34' (10.4 m)	40' (12.2 m)
33 FT. (10 m) **DIGGING DEPTH**	54' (16.5 m)	43' (13.1 m)	47' (14.3 m)

	MAXIMUM CHANNEL WIDTH 40° SWING EACH SIDE ₵	MINIMUM CHANNEL WIDTH
4 FT. (1.2 m) MINIMUM **DIGGING DEPTH**	97' (29.6 m)	60' (18.3 m) (Hull grounded)
26 FT. (7.9 m) **DIGGING DEPTH**	81' (24.7 m)	
33 FT. (10 m) **DIGGING DEPTH**	87' (26.5 m)	

FIGURE 4.47 Transportable dredge for 14- or 16-in discharge pipe. (*Courtesy, Ellicott Machine Corporation*)

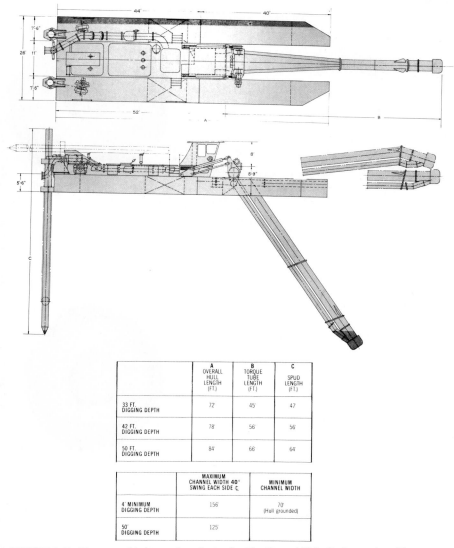

	A OVERALL HULL LENGTH (FT.)	B TORQUE TUBE LENGTH (FT.)	C SPUD LENGTH (FT.)
33 FT. DIGGING DEPTH	72'	45'	47'
42 FT. DIGGING DEPTH	78'	56'	56'
50 FT. DIGGING DEPTH	84'	66'	64'

	MAXIMUM CHANNEL WIDTH 40° SWING EACH SIDE ₵	MINIMUM CHANNEL WIDTH
4' MINIMUM DIGGING DEPTH	156'	70' (Hull grounded)
50' DIGGING DEPTH	125'	

FIGURE 4.48 Transportable heavy-duty dredge for 16-, 18-, and 20-in discharge pipe. (*Courtesy, Ellicott Machine Corporation*)

derwater bucket wheel unit uses a hydraulically driven wheel, roller-bearing support drive shaft, fabricated steel receiving hopper, suction pipe, and structural supports for attaching the module to the ladder. The new design is a dual wheel rather than a single bucket wheel (Fig. 4.53). The concentration of the cutting force on a much smaller cutting edge provides the bucket wheel with the capability of efficiently digging much harder or consolidated soil. Specifications for the dual wheel excavator are given in Fig. 4.54 and for the bucket wheel in Fig. 4.55.

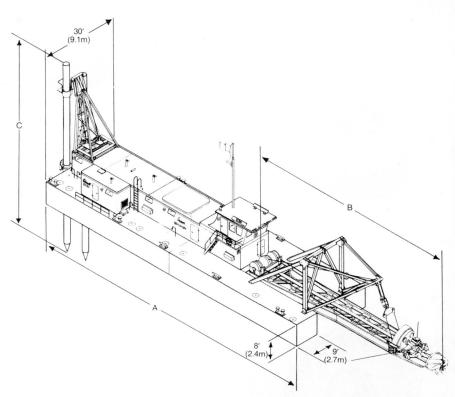

	A HULL LENGTH FT. (m)	**B** LADDER LENGTH FT. (m)	**C** SPUD LENGTH FT. (m)
58 FT. (17.7 m) DIGGING DEPTH (with wedge piece)	110' (34 m)	80.5' (24.5 m)	71.5' (22 m)

	MAXIMUM CHANNEL WIDTH 40° SWING EACH SIDE ¢	**MINIMUM CHANNEL WIDTH**
8 FT. (2.4 m) MINIMUM DIGGING DEPTH	210' (64 m)	87' (26.5 m) (Hull grounded)
58 FT. (17.7 m) MAXIMUM DIGGING DEPTH	175' (53 m)	142' (43 m) (Hull grounded)

FIGURE 4.49 Transportable heavy-duty dredge for 20-, 22-, and 24-in discharge pipe. (*Courtesy, Ellicott Machine Corporation*)

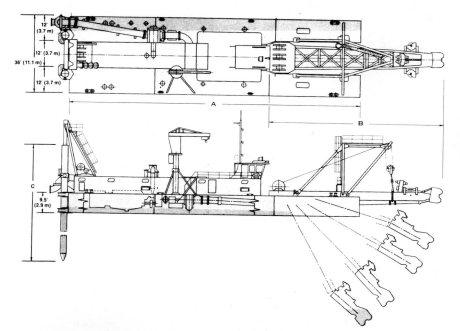

	A HULL LENGTH FT. (m)	B LADDER LENGTH FT. (m)	C SPUD LENGTH FT. (m)
50 FT. (15m) DIGGING DEPTH	136' (41.5 m)	82' (25 m)	67' (20.4 m)
58 FT. (17.7m) DIGGING DEPTH	136' (41.5 m)	84' (25.6 m)	75' (22.9 m)

	MAXIMUM CHANNEL WIDTH 40° SWING EACH SIDE ₵	MINIMUM CHANNEL WIDTH
8 FT. (2.4 m) MINIMUM DIGGING DEPTH	230' (70.1 m)	130' (39.6 m) (Hull grounded)
58 FT. (17.7 m) DIGGING DEPTH	200' (61 m)	

FIGURE 4.50 Transportable heavy-duty dredge for 26- or 33-in discharge pipe. (*Courtesy, Ellicott Machine Corporation*)

The Dustpan Dredge

The dustpan dredge is so named because its suction head resembles a large vacuum cleaner or dustpan. The dustpan dredge is a hydraulic, plain suction vessel. It consists essentially of a dredge pump which draws in a mixture of water and dredged materials through the suction head, which is lowered by winches to the face of the deposit to be removed. The suction head, which is about as wide as the hull of the dredge, is outfitted with high-velocity water jets for agitating and mixing the material. After sucking the mixture to the surface, the dredge pumps it to a disposal area, either at sea or shore, through a floating pipeline. Because it doesn't have a cutterhead, which loosens up hard

TABLE 4.1   Range of Portable Dredges[19]

Nominal sizes	8-in	10-in	12-in	14-in	16-in	24-in
Practical range of discharge pipeline sizes (inches I.D.)	8–10	10–12	12–16	14–18	16–22	20–24
Range of total connected maximum shaft horsepower	200–300	300–600	600–1000	1000–1500	1500–2500	2500 and up
Range of normal digging depths at 45° in feet	17–20	20–33	26–40	26–45	33–50	40–60
Number of standard models available	2	2	4	2	2	2

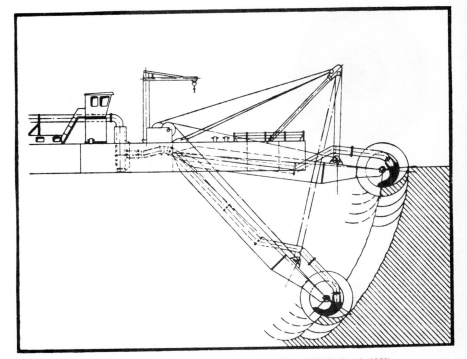

FIGURE 4.51   The bucket wheel dredge. (Ports and Dredging, *IHC Holland, 1979*)

FIGURE 4.52 Bucket wheel suction dredge *Scorpio*. (*Courtesy, IHC Holland*)

FIGURE 4.53 The dual wheel excavator. (*Courtesy, Ellicott Machine Corporation*)

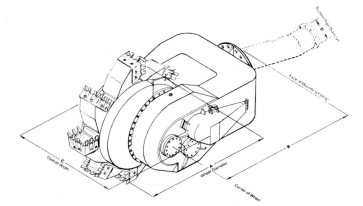

APPROXIMATE DIMENSIONS AND TECHNICAL DATA

Model	Dimensions					Weight (empty) kg Lbs Sheaves not included	Maximum R.P.M. of Wheel	Cutting Force (total) kgf Lbf	Cutting Force per unit effective length kgf/cm Lbf/inch	*Theoretical Production Capacity M3/Hr. Cu. yds./Hr.	Maximum Motor Shaft Power (total) kw H.P.
		A	B	C	D			WITH STANDARD BUCKETS			
WE50	mm	1270	787	1422	254	1000	30.0	1780	73	150	30
	inches	50	31	57.5	10	2250	30.0	3915	406	200	40
WE87	mm	2210	1702	1981	250-350	6800	15.9	4996	151	344	75
	inches	87	67	78	10-14	15,000	15.9	11,015	847	450	100
WE114	mm	2896	2845	3099	350-450	19,050	14.7	9557	198	650	186
	inches	114	112	122	14-18	42,000	14.7	21,070	1109	850	250
WE150	mm	3810	3683	3810	450-610	38,555	11.9	17,658	285	1150	373
	inches	150	145	150	18-24	85,000	11.9	39,830	1593	1500	500
WE184	mm	4674	5334	4445	559-686	68,040	10.0	26,343	335	1682	560
	inches	184	210	175	22-27	150,000	10.0	58,082	1874	2200	750
WE216	mm	5486	5969	4826	686-840	93,000	7.0	48,274	528	2300	746
	inches	216	235	190	27-33	205,000	7.0	106,425	2956	3000	1000
SPECIAL	Dimensions and other data for special models available on request										Up to 1500

*NOTE: Optional high capacity bucket assemblies allow approximately 50% higher capacities in less compact materials

FIGURE 4.54 Specifications for dual wheel excavator. (*Courtesy, Ellicott Machine Corporation*)

compact materials, the dustpan dredge is suited mostly for high-volume, soft-material dredging (Fig. 4.56).

The first dustpan dredge was developed to permit navigation on the Mississippi River during low water. A dredge was required which could operate in shallow water and be large enough to excavate the navigational channel in a reasonably short time. Dustpan dredge *Alpha* was completed in 1895 and operated until 1900 on the Mississippi River. *Alpha,* which was not self-propelled, had a wooden hull and a draft of only 59 in (1500 mm). It was equipped with a centrifugal pump with an open impeller 66 in (1676 mm) in diameter. Both the suction and discharge pipes were 30 in (762 mm) in diameter. The pump was powered by a vertical compound 400-hp (293-kW) steam engine. The width of the suction dustpan was 7½ ft (19.1 cm) and the normal working depth was 18 ft (5.5 m). The

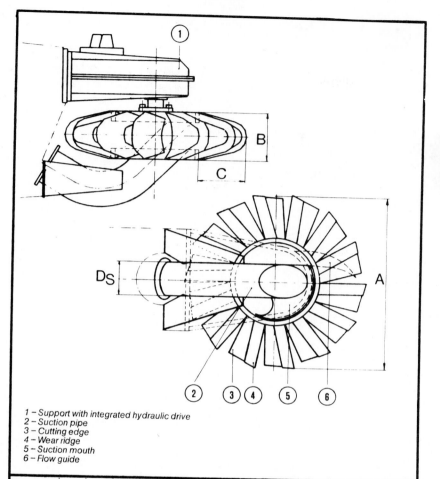

1 – Support with integrated hydraulic drive
2 – Suction pipe
3 – Cutting edge
4 – Wear ridge
5 – Suction mouth
6 – Flow guide

Model	D_S	A	B	C	Mass	Speed	Power	Theoretical production
		mm			kg	Hz	kW	m³/h
BW 1907	350	1900	500	450	8000	0.25	75	460
BW 2611	450	2550	700	600	11000	0.22	110	800
BW 3317	550	3290	800	820	17000	0.22	170	1300
BW 3527	600	3500	920	850	21000	0.22	265	1500
BW 3937	650	3910	1000	980	28000	0.22	370	2000
BW 4355	700	4260	1050	1080	36000	0.22	550	2400

FIGURE 4.55 Specifications for bucket wheel dredge module. (*Courtesy, IHC Holland*)

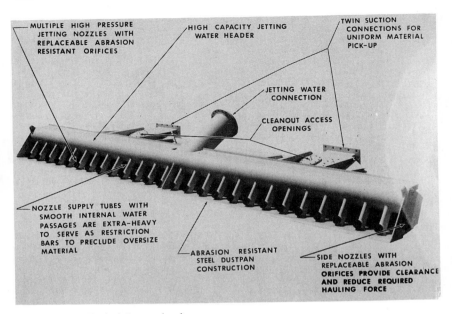

FIGURE 4.56 Typical dustpan head.

dustpan was equipped with water jets 2½ in (6.35 cm) in diameter at 9 psi pressure (62,052 N/m²). The normal dredging capacity was about 500 yd³/hr (382 m³/hr) with peak capacity of about 1070 yd³ (818 m³). Early history as well as a more complete description of dustpan dredges are given in *The Dustpan Dredge—An American Development and Its Future Possibilities.*[3]

The suction head of the dustpan dredge may consist of a rectangular box, open at the bottom and equipped with water jet nozzles.[20] Instead of a single suction pipe, it usually has two suction pipes connected to each half of the rectangular box. One of the successful dustpan-type dredges is the U.S. Army Engineers' dredge *Burgess* on the Mississippi River. The dredge is 224 ft (68.3 m) long, 52 ft (15.8 m) wide, and 9 ft (2.7 m) deep at the sides; its equipment is listed in Table 4.2.

The suction head on the *Burgess* is of special design and is quite different from either the cutterhead or draghead. It is 32 ft (9.8 m) wide with a rectangular opening 31 ft (9.4 m) wide and 16 in (0.41 m) high. The opening is fitted with equally

TABLE 4.2 Equipment for Dredge *Burgess*

Pump: centrifugal type
Volute: full cases cast from chrome, molybdenum alloy steel, liners from chrome cast iron
Impeller diameter: 84 in (2.13 m)
Total head: 80 ft (24 m)
Pump speed: 175 rpm
Specific speed: 2000
Power: 2500 hp (steam turbine)
Suction pipe diameter: 38 in (0.96 m)
Discharge pump diameter: 32 in (0.81 m)

spaced vertical members to prevent large objects from entering the suction line. The vertical members, which are connected to pipe cross members at the top, also convey water for jet nozzles located at the bottom of the members. The water is supplied from a 11,000-gpm (41,639-liters per minute) pump through a 16-in (0.41-m) water pipe at 45 psi (310,264 N/m^2). The multiple water jets dislodge the silt and sand, and they form a mixture of sand and water at the entrance to the suction pipe as seen in Fig. 4.57. The dustpan head is divided into two compartments which taper into two 26-in (660-mm) suction pipes. The two pipes join to form a wye with an outlet pipe diameter of 38 in (965 mm). The 38-in (965-mm) suction pipe is provided with a cleanout, and the pump inlet is of the same diameter as the suction pipe. The dustpan head is supported by a gantry with a multipart tackle. The joist line leads from the tackle to a single drum winch. Initially two spuds were located near the gantry but in recent years were replaced with anchors. There is one square spud used for maintaining the position of the dredge when the dustpan head and ladder are raised. The vertical spud is located directly in front of the bridge.

The arrangements on the discharge side of the pump are similar to a cutterhead pipeline dredge. The discharge pipe is connected to a swivel elbow at the stern and then to the floating pipeline. The floating pipelines in the case of the dustpan dredges used on the Mississippi are of special design. The pontoons are rectangular and usually line up with the direction of the prevailing current. Circular rails are provided on each pontoon and a pivot is mounted in the center of the rail on the deck. The pipe is supported on a saddle which has a fitting that mates with the pivot on the pontoon. The pontoon is free to pivot 360° around the pipe, but there is only a limited amount of freedom for the pontoon to respond to waves and the pipe sections are bolted with rigid flanges. Thus, this type of float-

FIGURE 4.58 Dustpan dredge *Jadwin* in operation on the Mississippi River. (*Courtesy, U.S. Army Corps of Engineers*)

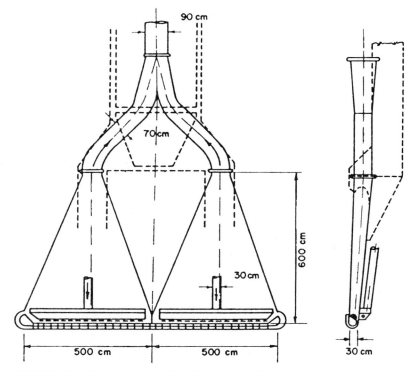

FIGURE 4.59 Czech dustpan dredge. (*Courtesy, Melzer*)

TABLE 4.3 Details of Czechoslovakian Dustpan Dredge

Length of dredge: 230 ft (70.1 m)
Width: 33 ft (10.1 m)
Depth: 9.2 ft (2.8 m)
Maximum draft of dredge: 5.2 ft (1.58 m)
Total hp: 2975 hp
Power for dredge pump: 1700 hp
Discharge pipe: 36-in diameter (0.91 m)
Delivery pipeline: 820–1960 ft (250–597 m)
Production rate: 3270–4578 yd³/hr (2500–3500 m³/hr)

Source: Melzer, 1968.

ing pipeline can only be used in rivers or sheltered waters. It cannot be employed in estuaries or bays where wave action may be expected.

Dustpan dredges have been developed and almost exclusively used in the United States. Figure 4.57 shows the dustpan employed on a U.S. Army Engineers' dustpan dredge *Jadwin*. Figure 4.58 shows the dustpan dredge in operation; the baffle plate in the foreground diverts the dredge material. A 28-in (711-mm) dustpan dredge manufactured in the United States is used on the Parana and other rivers in South America.

Melzer[21] reported on construction in Czechoslovakia of a dustpan dredge which has a dustpan head width of 30 ft (9.1 m) and is divided into two compartments (Fig. 4.59). The other pertinent information is given in Table 4.3.

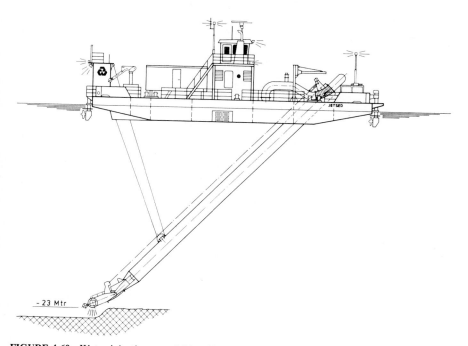

- 23 Mtr

FIGURE 4.60 Water-injection vessel (sketch) *Jetsed.* (*Estourgie, 1988*)

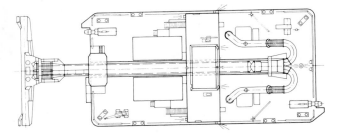

Principal particulars

Year of build	1987 Ravestein, Deest
Overall length hull	29.54 m
Overall length incl. discharge- and injectionpipe	37.00 m
Breadth, extreme outside	13.86 m
Depth, amidships at side	2.22 m
Draught at summer mark	1.40 m
Classification B.V.	I 3/3 Dredger/N.P.
	Sheltered Waters
	Deep Sea occasionally
Accommodation	4 persons
Propulsive power fwd. and aft	2 × 152 kW
Jet pump power SB and PS	2 × 375 kW
Total output of engines	1,107 kW
Diam. of discharge pipelines on board	800 mm
Max. dredging depth	23.00 m
Min. dredging depth	2.00 m

FIGURE 4.61   Specifications for water-injection vessel *Jetsed.* (*Estourgie, 1988*)

Agitation Dredges

Agitation dredging was practiced in ancient times and in the Middle Ages, as described in Chap. 1. A newly developed method in the Netherlands calls for the use of water jets to resuspend the sediment which is then carried away by prevailing water and turbidity currents.[22] Two water pumps discharge water through the nozzles as close as possible to the channel bed, as shown in Fig. 4.60. The water jet injection system with a capacity of up to 12,000 m³/hr can resuspend up to 5000 m³ of solids per hour according to the owner of the *Jetsed* (Figs. 4.61 and 4.62). The vessel is most efficient in silt and fine sand sediments, which typically accumulate in navigation channels. A density and turbidity current is set up which can transport the fine sediments out of the dredge area. During the first 6 months of operation the *Jetsed* completed several projects in the Netherlands and Germany totaling 1 million cubic meters of dredged sediment. The production is said to be up to 4000 m³/hr. Specifications for the dredge are shown in Fig. 4.61.

SURVEY BOATS

In Chap. 10, Weeks discusses the need for accurate surveys and indicates that the dredging contractors could increase the profitability of their operation by accurate and timely surveys of the channel being dredged. This section describes the requirement for high-speed hydrographic survey vessels which house the sophisticated modern instrumentation.

FIGURE 4.62 Water-injection vessel *Jetsed*. (*Estourgie, 1988*)

For condition surveys, lines are run along a channel centerline, along the channel quarterlines, and along the channel sidecut lines to provide a quick reference for the location of shoaling areas and for planning future dredging operations.[23] Pre- and postdredging surveys will include additional survey lines as required.

Heineman and Bechly[23] describe an air-cushion-assisted catamaran survey vessel to provide optimum performance characteristics. The survey vessel, called *Rodolf* (Fig. 4.63), is 48 ft (14.6 m) in length and has a beam of 24 ft (7.3 m) and

FIGURE 4.63 Survey vessel *Rodolf*. (*Courtesy, U.S. Army Corps of Engineers*)

a maximum draft (off cushion) of 5 ft 10 in (19.8 m). The maximum continuous speed with cushion is 35 mph (56 kph) and off-cushion it is 15 mph (24 kph) with the same load.

REFERENCES

1. Roorda, A., and Vertregt, J. J., *Floating Dredges,* De Technische Uitgeverij H. Stam, N.V., Haarlem, The Netherlands, 1963.
2. Scheffauer, F. C. (ed.), *The Hopper Dredge, Its History, Development and Operation,* U.S. Corps of Engineers, 1954.
3. Schmidt, F. J., "The Dustpan Dredge—An American Development and Its Future Possibilities," *Proc. World Dredging Conference, WODCON IV,* New Orleans, LA, 1971.
4. Brahme, S. B., and Herbich, J. B., *Dredging in India, Suggested Improvements in Techniques and Equipment,* Texas A&M University, TEES, Report CDS-204, June 1977.
5. Anonymous, "The Use of Underwater Pumps on Trailing-Suction Hopper Dredges," *Ports and Dredging,* E125, 1986.
6. Anonymous, "Trailing Dredger MAAS—Draghead Mounted Pump," *Ports and Dredging,* vol. 79, 1978.
7. Herbich, J. B., *Coastal and Deep Ocean Dredging,* Gulf Publishing Company, Houston, TX, 1975.
8. Anonymous, "Dredging Heavy Soil with an Active Draghead," *Ports and Dredging,* vol. 96, 1978.
9. Anonymous, "Tests with Active Draghead," *Ports and Dredging,* vol. 91, 1976.
10. Monster, G. A., and Kooijman, J., "New Generation of Dragheads for the Hopper Dredge," *Proc. Second International Symposium on Dredging Technology,* vol. 1, paper D-3, pp. 22–26, Nov. 1977.
11. Anonymous, "The IHC Modular Draghead. A New System for Dealing with All Types of Soil," *Ports and Dredging,* E124, 1986.
12. Anonymous, "Automatic Draghead Winch Controller," *Ports and Dredging,* vol. 96, 1978.
13. Bond, C. W., "Naval Architecture of Hydraulic Dredges," *SNAME Southwest Section Symposium,* paper no. 2, Tampa, FL, Oct. 2–22, 1966.
14. Slotta, L. S., "Flow Visualization Techniques Used in Dredge Cutterhead Evaluation," *Proc. World Dredging Conference, WODCON II,* Rotterdam, The Netherlands, Oct. 1968.
15. Breusers, H. N. C., Allersma, E., and van der Weide, J., "Hydraulic Model Investigations in Dredging Practice," *Proc. World Dredging Conference, WODCON II,* Rotterdam, The Netherlands, Oct. 1968.
16. Apgar, W. J., and Basco, D. R., *An Experimental and Theoretical Study of the Flow Field Surrounding a Suction Pipe Inlet,* Center for Dredging Studies Report no. 172, Texas A&M University, TAMU-SG-74-203, Oct. 1973.
17. Terry, L. E., "It's What's Up Front that Counts," *Proc. World Dredging Conference, WODCON I,* New York, pp. 91–113, 1967.
18. Woodbury, C. E., "The Transportable Hydraulic Cutterhead Pipeline Dredge," *Proc. World Dredging Conference, WODCON II,* Rotterdam, The Netherlands, pp. 734–770, 1968.

19. Anonymous, *Dragon Model Portable Dredges,* Ellicott Machine Corporation, MD, 1970.

20. Anonymous, "The Dustpan Dredge," report no. 90217, Ellicott Machine Corporation, MD, 1973.

21. Melzer, L., "Czech Floating Dredgers and Their Utilization," *Proc. World Dredging Conference, WODCON II,* Rotterdam, The Netherlands, pp. 771–799, 1968.

22. Estourgie, A. L. P., "A New Method of Maintenance Dredging," *IRO Journal,* May 1988.

23. Heineman, A., and Bechly, J., "High Speed Hydrographic Surveying," *Proc. 13th Dredging Seminar,* Center for Dredging Studies, Texas A&M University, pp. 124–144, 1981.

CHAPTER 5
SEDIMENT

Great strides have been made in recent years to improve dredging efficiency through the employment of better and more efficient equipment, deployment of modern, more reliable, instrumentation, and introduction of automated systems to the dredging process. Additional improvements can be achieved through better understanding of the engineering properties of the material to be dredged and through the appropriate selection of the cutter and draghead for different types of submarine soils.

Dredgeability depends on the flow and deformation of in situ material and disturbance by the mechanical action of the cutter.[1] Soil characteristics at the bottom of the channel are a function of many variables:

1. In situ density
2. Void ratio
3. Degree of consolidation
4. Layering
5. Cementation, etc.

Slurry flow capability is principally a function of concentration (by weight or volume), density and viscosity of the water-solids mixture, size and shape of the particles, whether cohesive or noncohesive, etc.

The volume of overflow solids from the hoppers depends to a large extent on grain size and concentration of the suspended sediments. The settlement and consolidation of dredged material in confined disposal areas depend on soil characteristics as well as active management techniques, which may include drainage and dewatering of the disposal site.

CLASSIFICATION OF SOILS

There are many basic classification methods that can be used to determine the dredgeability of soils in situ. The following is a system established by the International Association of Dredging Contractors[2] for identifying soil types and the parameters needed to determine their dredgeability:

1. Sand
   a. Weight
   b. Water content
   c. Specific gravity of grains
   d. Grain size

    e. Water permeability
    f. Frictional properties
    g. Lime content
    h. Organic content
 2. Silt
    a. Weight by volume
    b. Water content
    c. Grain size
    d. Water permeability
    e. Sliding resistance (shear strength)
    f. Plasticity
    g. Lime content
    h. Organic content
 3. Clay
    a. Weight
    b. Water content
    c. Sliding resistance
    d. Consistency ranges (plasticity)
    e. Organic content
 4. Peat: same parameters as clay
 5. Gravel: same parameters as sand

  Another system devised by the Permanent International Association of Navigation Congresses[3,4] is shown in Table 5.1. Classification by laboratory and in situ testing is shown in Table 5.2. The bold-outlined tests are of the first priority for assessing the soil characteristics for dredging purposes. The lightly-outlined tests are of second priority; the rest can be restricted to a few representative samples of each soil type found in situ. Table 5.3 outlines the testing procedures. Table 5.4 outlines the testing recommended for determination of engineering properties of rocks. The bold-outlined tests are of the first priority for assessing the soil characteristics for dredging purposes, the lightly-outlined tests are of second priority, and nonoutlined tests are of lesser importance. Table 5.5 provides the outline for sampling and testing procedures for dredging purposes.

  For underwater dredging of rock, the nature and strength of rock are determined from cores taken prior to dredging. Information on weathering should also be obtained since it can affect the dredging efficiency considerably. The strength of rock may be described by the following terms related to the unconfined compressive strength:

Term	Unconfined compressive strength	
	MN/m^2	kg/cm^2
Very weak	Less than 1.25	12.5
Weak	1.25–5.0	12.5–50
Moderately weak	5.0–12.5	50–125
Moderately strong	12.5–50.0	125–500
Strong	50–100	500–1000
Very strong	100–200	1000–2000
Extremely strong	Greater than 200	Greater than 2000

TABLE 5.1 General Basis for Identification and Classification of Soils for Dredging Purposes

Main soil type	Particle size identification range of size (mm)	Identification	Particle nature and plasticity	Strength and structural characteristics
Boulders Cobbles	Larger than 200mm Between 200-60mm	Visual examination and measurement (3)	Particle shape: Rounded Irregular Angular Flaky Elongated Flaky and elongated Texture: Rough Smooth Polished	N.A.
Gravels	Coarse 60-20 Medium 20-6 Fine 6-2mm	Easily identifiable by visual examination		Possible to find cemented beds of gravel which resemble weak conglomerate rock. Hard-packed gravels may exist intermixed with sand
Sands (4)	Coarse 2-0.6 Medium 0.6-0.2 Fine 0.2-0.06mm	All particles visible to the naked eye. Very little cohesion when dry		Deposits will vary in strength (packing) between loose, dense and cemented. Structure may be homogeneous or stratified. Intermixture with silt or clay may produce hard-packed sands
Silts (4)	Coarse 0.06-0.02 Medium 0.02-0.006 Fine 0.006-0.002mm	Generally particles are invisible and only grains of a coarse silt may just be seen with the naked eye. Best determination is to test for dilatancy (1). Material may have some plasticity, but silt can easily be dusted off fingers after drying and dry lumps powdered by finger pressure	Non-plastic or low plasticity	Essentially non-plastic but characteristics may be similar to sands if predominantly coarse or sandy in nature. If fine will approximate to clay with plastic character. Very often intermixed or interleaved with fine sands or clays. May be homogeneous or stratified. The consistency may vary from fluid silt through stiff silt into "siltstone"
Clays	Below 0.002mm Distinction between silt and clay should not be based on particle size alone since the more important physical properties of silt and clay are only related indirectly to particle size	Clay exhibits strong cohesion and plasticity, without dilatancy. Moist sample sticks to fingers, and has a smooth, greasy touch. Dry lumps do not powder, shrinking and cracking during drying process with high dry strength	Intermediate plasticity (Lean Clay) High plasticity (Fat Clay)	Strength / Shear Strength (2) V. soft May be squeezed easily between fingers. Soft Easily moulded by fingers. — Less 20 kN/m^2 Firm Requires strong pressure to mould by fingers. — 20-40kN/m^2 Stiff Cannot be moulded by fingers, indented by thumb. — 40-75 kN/m^2 Hard Tough, indented with difficulty by thumb nail — 75-150 kN/m^2 / Above 150 kN/m^2 Structure may be fissured, intact, homogeneous, stratified or weathered
Peats and Organic soils	Varies	Generally identified by black or brown colour, often with strong organic smell, presence of fibrous or woody material		May be firm or spongy in nature. Strength and structure may vary considerably in horizontal and vertical directions. Presence of gas should be noted

N.A.: Not applicable

[1]Dilatancy is the property exhibited by silt as a reaction to shaking. If a moistened sample is placed in an open hand and shaken, water will appear on the surface of the sample, giving a glossy appearance. A plastic clay gives no reaction.

[2]Defined as the undrained (or immediate) shear strength ascertained by the applicable *in situ* or laboratory test procedure.

[3]Although only visual examination and measurement are possible, an indication should be given with respect to the particles as well as to the percentages of different sizes.

[4]*Sands* and *silts* are terms denoting a particle size. Sands are not necessarily restricted to quartz sands but may include lime sands, iron ores, etc. Also silts denote a grain size, not a consistency. Therefore consistency terms such as *fresh harbor silts, muds,* etc., should not be used.

Source: Ref. 2.

TABLE 5.2 Classification of Soils for Dredging Purposes by in situ and Laboratory[1] Testing

Main soil type	Particle size distri- bution	Particle Shape	In situ density or bulk density	Specific gravity of the solid particles	Compact- ness (in situ)	Natural moisture content	Plastic and liquid wastes	Shear strength	Lime content	Organic content
(2) Boulders Cobbles	Visual in field	Visual inspec- tion	N.A.	Lab. test (on frag- ments)	N.A.	N.A.	N.A.	N.A.	N.A.	N.A.
Gravel	Lab. test	Lab. test	N.A.	Lab. test	In situ test	N.A.	N.A.	N.A.	(3) Lab. test	N.A.
Sands	Lab. test	Lab. test	(4) Lab. test on undis- turbed samples	Lab. test	In situ test	Lab. test	N.A.	N.A.	Lab. test	Lab. test
Silts	Lab. test	Lab. test	Lab. test on undis- turbed samples	Lab. test	In situ test or lab. test on undis- turbed samples	(5) Lab. test	Lab. test	Lab. test	Lab. test	Lab. test
Clays	(7) Lab. test	N.A.	Lab. test on undis- turbed samples	N.A.	In situ test or lab. test on undis- turbed samples	(6) Lab. test	Lab. test	(8) In situ and/or Lab. test	N.A.	Lab. test
Peats and organic soils	N.A.	N.A.	Lab. test on undis- turbed samples	N.A.	In situ test	Lab. test	Lab. test	In situ and/or Lab. test	N.A.	Lab. test

N.A.: Not applicable
[1]For testing procedures see Table 5.3.
[2]To be tested as rock.
[3]Applicable to dredged aggregates for construction purposes.
[4]Determination of max./min. dry density is also recommended.
[5]Silts often contain an appreciable amount of clay particles which have a strong influence on the soil characteristics. In such cases the tests for silts as well as for clays should be performed.
[6]Tests should be performed on samples in natural condition by preference using undisturbed samples.
[7]It may be useful to carry out particle size distribution on any sand/silt fraction within the clay sample but also expressing the percentages relative to the total sample.
[8]Tests should include sensitivity performed on representative samples.
Source: From Ref. 4.

It should be noted that the strength of a rock material determined by the uniaxial compression test is dependent on the moisture content of the core specimen, anisotropy, and the test procedure adopted.

The United Soil Classification System (USCS) has been generally used in the United States and can also meet the needs of the dredging industry.[5] In the USCS, the soil is classified according to:

1. Texture (grain size)
2. Plasticity
3. Engineering behavior

TABLE 5.3 In Situ and Laboratory Testing Procedures of Rocks for Dredging Purposes

Name of test	Purpose of test	Remarks	Lab (L) or in situ (S)	References
Visual inspection	Assessment of rock mass	Indicates in situ state of rock mass (1)	S or L	B.S. 5930 (1981)
Thin section	Identification	Aid to mineral composition	L	Geotechnical textbooks
Bulk Density	Volume/weight relationship	Wet and dry test	L	Int. Journal for Rock Mech. Min. Sci. (1979) 16, 141-156
Porosity	Measure of pores expressed as percentage ratio voids/total volume	To be calculated directly from wet and dry bulk density	L	Ditto
Carbonate content	Measurement of lime content	Useful for identification of limestone, chalks, etc.	L	A.S.T.M D 3155
Surface hardness	Determination of hardness	Graded according to Moh's scale from 0 (talc) to 10 (diamond)	L	Reference set commercially obtainable
Uniaxial compression	Ultimate strength under uniaxial stress	Test to be done on fully saturated samples. Dimensions of testpiece and direction of stratification relevant to stress direction are to be stated. Recommend 1:2 length/diameter ratio for cylindrical specimens	L	Int. Soc. for Rock Mech. Commission Committee on Lab. tests, publication 135 (Sept. 1978)
Brazilian split	Tensile strength (derived from uniaxial testing)	Ditto except length/diameter ratio recommendation	L	Ditto, Doc. No. 8 (March 1977)
Point load test	Strength indication	Easy and fast test but should be matched with uniaxial compressive strength test	L	Int. Journal for Rock Mech. Min. Sci. (1972) 9, 669-697
Protodiakonov	Indication of crushing resistance under dynamic load	Test has been devised for the harder type of rocks. Care should be taken with the execution and interpretation of test results on soft rocks, especially coarse-grained conglomerates	L	See note (2)
Standard penetration test	Strength indication	Applies to corals and highly weathered rocks	S	B.S. 1377 (1975) 103 et seq
Seismic velocity	Indication of stratigraphy and fracturing of rock mass	Useful in extrapolating laboratory and field tests to rock mass behaviour	S	A.S.T.M. Annual Book (1975) 340-347
Ultrasonic velocity	Longitudinal velocity	Tests on saturated core samples	L	A.S.T.M. Spec. Techn. Publication No. 402 (1966) 133-172
Static modulus of elasticity	Stress/strain rate.	Gives an indication of brittleness	L	Ditto
Drillability	Assessment of the rock mass	Measurement of drilling parameters including penetration rate, torque, feed force fluid pressure, etc., and statement of drill specification and technique	S	
Angularity	Determination of particle shape	May be by visual examination compared to standard specimens	L	B.S. 812 (Part 1) 1975

[1]Color photography for record purposes can be very useful.

[2]Concise references are not available for this test. A reference which gives a slight modification of the test procedure (in order to overcome some of the disadvantages of the original method such as rebonding of pulverized material) is: *The Strength, Fracture and Workability of Coal*, Evans, I., and Pomeroy, C. D., Pergamon Press, 1966.

[a]*Professor M. M. Protodiakonov's Strength Coefficient of Rocks*. Translation by the Foreign Technology Division of the Air Force Systems Command, Ohio (trans. 1981).

[b]*Methods for the Evaluation of the Fissurization and Strength of a Rock Mass* by M. M. Protodiakonov. Translation by the Council for Scientific and Industrial Research, Pretoria, 1965.

[c]*Methods of Evaluating the Cracked Stage and Strength of Rocks in Situ* by M. M. Protodiakonov, Department of Mines and Technical Surveys, Ottawa, Canada, 1965.

[d]"A critical appraisal of the Protodiakonov index," Misra, G. B., and Paithankar, A. G., Technical note, *International Journal of Rock Mechanics,* Min. Sciences and Geomech. Abstracts, vol. 13, pp. 249–251, 1976.

Source: From Ref. 4.

TABLE 5.4 Aid to Identification of Rocks for Engineering Purposes

Bedded rocks (mostly sedimentary)

Grain size (mm)	Grain size description		At least 50% of grains are of carbonate	At least 50% of grains are of fine-grained volcanic rock	
More than 20	RUDACEOUS — Coarse	CONGLOMERATE — Rounded boulders, cobbles and gravel cemented in a finer matrix		Fragments of volcanic ejecta in a finer matrix	
20		BRECCIA — Irregular rock fragments in a finer matrix	Calcirudite*	AGGLOMERATE — Rounded grains / VOLCANIC BRECCIA — Angular grains	SALINE ROCKS — Halite
6					
2	ARENACEOUS — Medium / Fine	SANDSTONE — Angular or rounded grains, commonly cemented by clay, calcitic or iron minerals	Calcarenite	TUFF — Cemented volcanic ash	Anhydrite
0.6		Quartzite — Quartz grains and siliceous cement			
0.2		Arkose — Many feldspar grains / Greywacke — Many rock chips			Gypsum
0.06	ARGILLACEOUS	MUDSTONE — SILTSTONE (Mostly silt) / Calcareous mudstone	Calcisiltite	Fine-grained TUFF	
0.002		SHALE (Fissile) — CLAYSTONE (Mostly clay)	Calcilutite	Very fine-grained TUFF	COAL
Less than 0.002			CHALK		LIGNITE
Amorphous or crypto-crystalline		Flint: occurs as bands of nodules in the chalk / Chert: occurs as nodules and beds in limestone and calcareous sandstone			

LIMESTONE and DOLOMITE (undifferentiated)

SILICIOUS	CALCAREOUS	SILICIOUS	CARBONACEOUS

Granular cemented — except amorphous rocks

SEDIMENTARY ROCKS

Granular cemented rocks vary greatly in strength, some sandstones are stronger than many igneous rocks. Bedding may not show in hand specimens and is best seen in outcrop. Only sedimentary rocks, and some metamorphic rocks, derived from them, contain fossils.

Calcareous rocks contain calcite (calcium carbonate) which effervesces with dilute hydrochloric acid

Obviously foliated rocks (mostly metamorphic)

Grain size description	
COARSE	GNEISS — Well developed but often widely spaced foliation sometimes with schistose bands / Migmatite — irregularly foliated: mixed schists and gneisses
MEDIUM	SCHIST — Well developed undulose foliation; generally much mica
FINE	PHYLLITE — Slightly undulose foliation; sometimes "spotted" / SLATE — Well developed plane cleavage (foliation) / Mylonite — Found in fault zones, mainly in igneous and metamorphic areas

CRYSTALLINE

SILICIOUS | mainly SILICIOUS

METAMORPHIC ROCKS

Most metamorphic rocks are distinguished by foliation which may impart fissility. Foliation in gneiss is best observed in outcrop, except by association. Any rock baked by contact metamorphism is described as a "hornfels" and is generally somewhat stronger than the parent rock

Most fresh metamorphic rocks are strong although perhaps fissile

Rocks with massive structure and crystalline texture (mostly igneous)

Grain size description					Grain size (mm)
COARSE	MARBLE / QUARTZITE / Granulite / HORNFELLS / Amphibolite / Serpentine	GRANITE[1] / Diorite[1,2] / GABBRO[3]	Pegmatite — These rocks are sometimes porphyritic and are then described, for example, as porphyritic granite	Pyroxenite / Peridotite	More than 20 / 20 / 6 / 2 / 0.6
MEDIUM		Micro-granite[1] / Micro-diorite[1,2] / Dolerite[4] — These rocks are sometimes porphyritic and are then described as porphyries			0.2 / 0.06
FINE		RHYOLITE[4,5] / ANDESITE[4,5] / BASALT[4,5] — These rocks are sometimes porphyritic and are then described as porphyries			0.002 / Less than 0.002
		Obsidian[5]	Volcanic glass		Amorphous or crypto-crystalline

ACID — Much quartz	INTER-MEDIATE — Some quartz	BASIC — Little or no quartz	ULTRA BASIC
Pale			Dark

Colour

IGNEOUS ROCKS

Composed of closely interlocking mineral grains. Strong when fresh; not porous

Mode of occurrence: 1. Batholiths; 2. Laccoliths; 3. Sills; 4. Dykes; 5. Lava flows; 6. Veins

TABLE 5.5 Sampling and Investigation Procedures for Dredging Purposes

Rock or soil type	Rotary (1) drilling	Shell & auger boring	Underwater (sea bed) devices	Undisturbed (2) sampling	Disturbed (2) representative samples	Dynamic penetration tests (3)	Static penetration test (e.g. Dutch, Swedish)	In situ vane testing	Geophysical methods
Rocks	Best method of obtaining core samples of intact rocks in in situ condition for examination and test	N.A.	Useful for obtaining core samples of limited penetration	Cores represent undisturbed samples of intrinsic rock	Cutting in drill fluid may be used for identification of non-recovered layers	Used only in soft or weathered rock and in corals	N.A.	N.A.	Useful to establish the likely geology over a large area will assist both to "set out" a borehole grid and to "fill in" details between borings and drillings. However, note should be taken that such methods still require careful interpretation. Very useful where relatively simple soil/rock conditions exist (i.e. soft alluvium over rock). Where only slight changes in strata density occur, great care needed in interpretation
Boulders Cobbles	May be used to penetrate and obtain core samples	Chiselling required to penetrate strata	N.A.	Cobbles retained as undisturbed samples	N.A.	N.A.	N.A.	N.A.	
Gravels	N.A.	Method employed for site investigation in order to obtain representative & undisturbed samples and to carry out field (in situ) tests	N.A.	Not practicable to retain gravel as an undisturbed sample unless in cemented condition	Obtained from borings in tins or bags. Must be "representative" (i.e. only from a single horizon or stratum). Essential for identification of various strata	Used with cone gives reasonable in situ compactness estimate	Very difficult to penetrate coarse gravel	N.A.	
Sands	N.A.		Various devices are available to obtain representative samples, but generally of limited penetration	Patent samplers available, difficult to sample in undisturbed condition		Useful for in situ compactness estimate at the same time as sample is obtained	Useful method for determining in situ properties and "hard" strata levels. In areas with very wide soil variation may be useful to supplement borehole information	N.A.	
Silts	N.A.			If cohesive in nature can use clay undisturbed core samplers, otherwise see Sands		Can very well be used, but interpret with care		Used for estimate of shear strength but great care needed in interpretation	
Clays	N.A.			Variety of undisturbed core samplers available				Very useful for shear strength evaluation in alluvial clays	
Peats, etc.	N.A.			Variety of undisturbed core samplers available				Used for estimate of shear strength but great care needed in interpretation	

Source: From Ref. 4.

Texture

The following grain sizes are recognized in the USCS:

Component	Size range
Cobbles	Above 3 in (76.2 mm)
Gravel	3 to 4 in (76.2 to 101.6 mm)
Coarse	3 to 3/4 in (76.2 to 19.1 mm)
Fine	3/4 in to no. 4 (19.1 to 4.76 mm)
Sand	No. 4 (4.76 mm) to no. 200 (0.074 mm)
Coarse	No. 4 (4.76 mm) to no. 10 (2.0 mm)
Medium	No. 10 (2.0 mm) to no. 40 (0.42 mm)
Fine	No. 40 (0.42 mm) to no. 200 (0.074 mm)
Fines (silt or clay)	Below no. 200 (0.074 mm)

In addition, the shape of the grain size curve has an important effect on the properties of sands and gravels. This can be described with two coefficients, the coefficient of curvature, C_c, and the coefficient of uniformity, C_u, defined as follows:

$$C_c = \frac{(D_{30})^2}{D_{60} \, D_{10}} \tag{5.1}$$

$$C_u = \frac{D_{60}}{D_{10}} \tag{5.2}$$

where D_{60} = the grain size at which 60 percent of the soil is finer
D_{30} = the grain size at which 30 percent of the soil is finer
D_{10} = the grain size at which 10 percent of the soil is finer

If C_c is between 1 and 3, the grain size distribution curve will be smooth, and if C_u exceeds 4 for gravels, or 6 for sands, there will be a wide range of sizes. When both of these criteria are met, the soil is said to be well graded (designated W); otherwise, it is poorly graded (designated P). The identification procedure is outlined in Fig. 5.1.

No practical significance can be attached to the shape of the grain size curve for silts and clays.

Plasticity

Knowledge of the mineralogy and particle size distribution of clays is not sufficient to describe their behavior. It is also necessary to conduct the liquid limit (LL) and plastic limit (PL) tests of Atterberg.

The liquid limit is the moisture content of a remolded soil above which it acts as a fluid and below which it acts as a plastic substance. The liquid limit test is carried out by thoroughly mixing the clay with water, placing it in a brass cup, and finding the number of bumps required to close a groove cut in the pat of clay in the cup.

Atterberg defined the plastic limit of a clay as the water content below which it ceased to behave as a plastic material and became friable or crumbly. The test to determine this water content consists of finding the water content at which it ceases to be possible to roll out the clay in an unbroken thread about 1/8 in (3.18 mm) in diameter.

The difference between the liquid and plastic limits is the range of water content over which the remolded clay behaves as a plastic material and is referred to as the plasticity index (PI) (Figs. 5.2 and 5.3). Thus:

$$\text{PI} = \text{LL} - \text{PL} \tag{5.3}$$

THE SOIL GROUPS

Major Divisions

Soils are primarily divided into coarse-grained, fine-grained, and highly organic soils. On a textural basis, coarse-grained soils are those that have 50 percent or more by dry weight of the constituent material retained on the no. 200 sieve, and fine-grained soils are those that have more than 50 percent by dry weight passing the no. 200 sieve. Highly organic soils are, in general, readily identified by visual

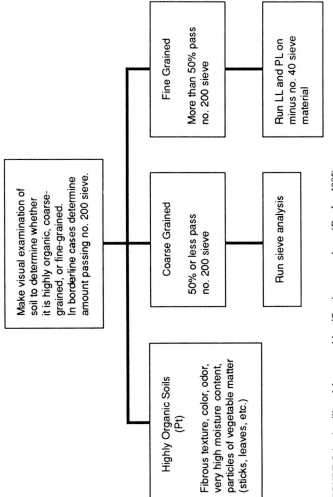

FIGURE 5.1 Auxiliary laboratory identification procedure. (*Dunlap 1985*)

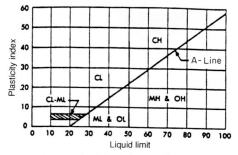

FIGURE 5.2   Plasticity index as a function of liquid limit. (*Dunlap 1985*)

examination. Peat and other highly organic soils are designated by the symbol Pt and are not subdivided.

Coarse-Grained Soils.   In general, there is no clear-cut boundary between gravelly and sandy soils, and as far as behavior is concerned, the exact point of division is relatively unimportant. For purposes of identification, however, coarse-grained soils are classed as gravels and given the symbol G if the greater percentage of the coarse-grained material is larger than the no. 4 sieve; they are classed as sands and given the symbol S if the greater portion of the coarse fraction is finer than the no. 4 sieve. The gravel (G) and sand (S) are further subdivided into four secondary groups, as shown in Fig. 5.4.

Well-graded material with little or no fines (less than 5 percent passing the no. 200 sieve) is given the symbol W and must satisfy both of the following requirements:

1. $C_u > 4$ for gravels; $C_u > 6$ for sands
2. C_c between 1 and 3 for both sand and gravels

The groups following these criteria would be GW and SW.

Poorly graded material with little or no fines (less than 5 percent passing the no. 200 sieve) is given the symbol P. A poorly graded material is one in which the gradation does not meet all the requirements for a well-graded material. The groups following this criterion would then be GP and SP.

Coarse material with nonplastic fines or fines with low plasticity (greater than 12 percent passing the no. 200 sieve) is given the symbol M. The groups following this criterion would then be GM and SM.

Coarse material with plastic fines (greater than 12 percent passing the no. 200 sieve) is given the symbol C. The groups following this criterion would then be GC and SC.

In the last two subgroups, use is made of the plasticity chart (Fig. 5.2). If the intersection of the liquid limit and plasticity index falls below the A line and the hatched zone, it is designated with the symbol M; if the intersection is above the A line and the hatched zone, it is designated with the symbol C.

Fine-Grained Soils.   The fine-grained soils (i.e., more than 50 percent passes the no. 200 sieve) are subdivided into groups based on whether they have a relatively low liquid limit (designated with the symbol L) or a relatively high liquid limit

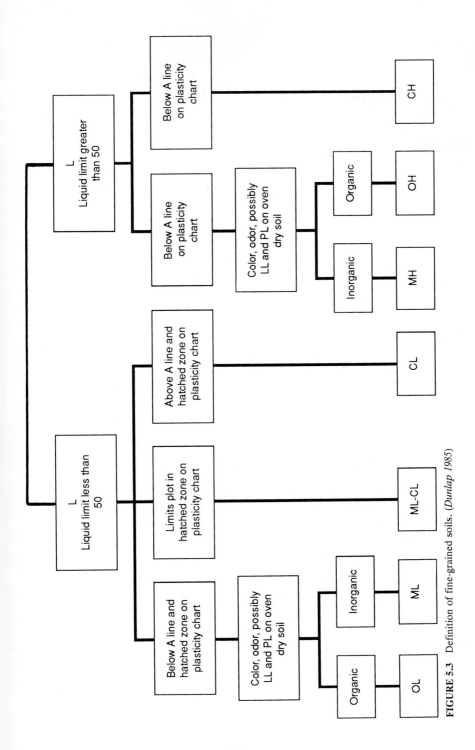

FIGURE 5.3 Definition of fine-grained soils. (*Dunlap 1985*)

```
Gravel (G)
Greater percentage of coarse fraction
retained on no.4 sieve
```

| Less than 5% pass no. 200 sieve | Between 5% and 12% no. 200 sieve | More than 12% no. 200 sieve |

```
Examine grain-size curve
```

```
Borderline, to have double symbol appropriate to grading and plasticity characteristics, e.g., GW-GM
```

```
Run LL and PL on minus no. 40 sieve fraction
```

| Well-graded | Poorly graded | | Below A line and hatched zone on plasticity chart | Limits plot in hatched zone on plasticity chart | Above line and hatched zone on plasticity chart |

| GW | GP | | GM | GM-GC | GC |

5.12

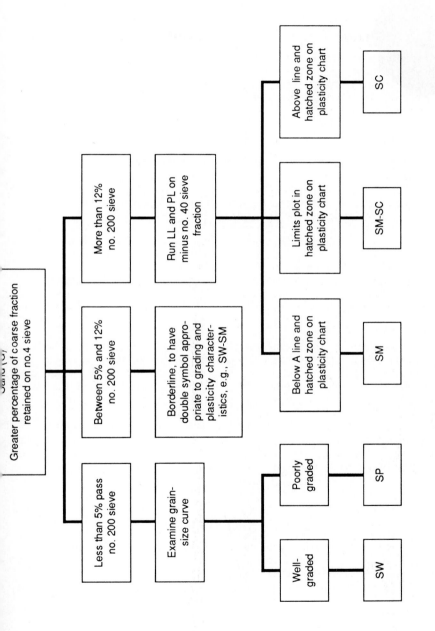

(b)

FIGURE 5.4 Definition of (a) gravel and (b) sands.

5.13

(designated with the symbol H). The dividing line is a liquid limit of 50. The two groups are further subdivided according to their composition.

Inorganic silts and very fine sandy soils are given the symbol M. The groups following these criteria would be designated as ML and MH. Organic silts and clay are given the symbol O. The groups following these criteria would be designated OL and OH.

As before, use is made of the plasticity chart. If the intersection at the liquid limit and plasticity index falls below the A line, the symbol designated will be either M (silt) or O (organic material present); if the intersection plots above the A line, the symbol designated will be C for clay.

Coarse-Grained Soils

GW and SW Groups. These groups comprise well-graded gravelly and sandy soils having little or no plastic fines (less than 5 percent passing the no. 200 sieve). The presence of the fines must not noticeably change the strength characteristics of the coarse-grained fraction and must not interfere with its free-draining characteristics (Fig. 5.4).

GP and SP Groups. Poorly graded gravels and sands containing little or no plastic fines (less than 5 percent passing the no. 200 sieve) are classed in GP and SP groups. The materials may be called uniform gravels, uniform sands, or nonuniform mixtures of very coarse material and very fine sands, with intermediate sizes lacking (sometimes called skip, gap, or step graded). This last group often results from borrow pit excavation in which gravel and sand layers are mixed.

GM and SM Groups. In general, the GM and SM groups comprise gravels or sands with fines (more than 12 percent passing the no. 200 sieve) having low or no plasticity. The plasticity index and liquid limit of soils in the group should plot below the A line on the plasticity chart. The gradation of the material is not considered significant and both well- and poorly graded materials are included.

GC and SC Groups. In general, the GC and SC groups comprise gravelly or sandy soils with fines (more than 12 percent passing the no. 200 sieve) which have fairly high plasticity. The liquid limit and plasticity index should plot above the A line on the plasticity chart.

Fine-Grained Soils

ML and MH Groups. In these groups the symbol M has been used to designate predominantly silty materials. The symbols L and H represent low and high liquid limits, respectively, and an arbitrary dividing line between the two is set at a liquid limit of 50. The soils in the ML and MH groups are sandy, clayey, or inorganic silts with relatively low plasticity. Also included are loess-type soils and rock flours.

CL and CH Groups. In these groups the symbol C stands for clay, with L and H denoted low or high liquid limits, with the dividing line again set at a liquid limit of 50. The soils are primarily inorganic clays. Low plasticity clays are classified as CL and are usually lean, sandy, or silty clays. The medium and high plasticity clays are classified as CH. These include fat, gumbo, and some volcanic clays.

OL and OH Groups.   The soils in the OL and OH groups are characterized by the presence of organic odor or color, hence the symbol O. Organic silts and clays are classified in these groups. The materials have a plasticity range that corresponds with the ML and MH groups.

Highly Organic Soils

The highly organic soils are usually very soft and compressible and have undesirable construction characteristics. Particles of leaves, grass, branches, or other fibrous vegetable matter are common components of these soils. They are not subdivided and are classed into one group with the symbol Pt. Peat, humus, and swamp soils with a highly organic texture are typical soils of the group.

SOIL INVESTIGATION IN SITU

The contractor must have sufficient soil information to be able to prepare a bid for accomplishing a dredging task. The owner must complete an investigation which should be submitted as part of the Request for Bids (RFB). There are two main types of investigations that can be conducted to achieve a fair degree of confidence in estimating the soil conditions *in situ:*

1. Desk study
2. Field sampling

A desk study can be conducted if there is sufficient historical and local information available to establish geotechnical and geological conditions at the site. This information may be based on earlier surveys, site investigations or records of previous dredging contracts, or construction work at the same location. The validity of information must be scrutinized for accuracy and applicability. If bids are for a maintenance dredging project, where the material is generally fine grained, a desk study based on factual information may be all that is required. There will never be any concerns that rock may be present above the required dredging depth since the channel was dredged before.

For channel deepening, site investigation using established methods must be conducted. Field work, which will generally include drilling and boring, should be conducted by a contractor familiar with river, estuarine, or marine environments according to one of the various standards. The most common standards are the American Society of Testing Materials (ASTM), the British Standards (BS), and the German Standards (DIN).

Although it is desirable to obtain as much information as possible on the levels of configuration of deposits, there are economic limits to any drilling and sampling problems. Obviously the more, the better since one of the most common reasons for litigation is "the changed condition" argument. As Ottmann and Lahuec[6] state: "All borings are very expensive, but those which cost the most are those which were not done!" It is very important from the contractor's and owner's points of view to have sufficient knowledge of the geotechnical characteristics of the proposed dredging project to avoid financial consequences.

Every dredging project and every dredging site are unique, and every site investigation plan must be tailored to the needs of a specific project. Open water investigations are more difficult to accomplish because of the possible presence of waves, currents, and tides. Mobilization costs of equipment are much higher than on land.

Spigolon and Fowler[7a] discuss the important soil properties that are significant for the various types of equipment employed in the several stages of a dredging operation. Table 5.6 outlines the significant soil properties for various types of equipment used in dredging. For example in hydraulic suction dredging the compactness of granular soils, water content, mass density, grain size distribution, plasticity of fines, and organic content are significant. In hydraulic disposal, grain size distribution, plasticity of fines, organic content, sedimentation rate in water, and bulking factor are important.

TABLE 5.6 Soil Properties Significant in Dredging

Dredging equipment	Soil properties*											
	1	2	3	4	5	6	7	8	9	10	11	12
Hydraulic excavation:												
Plain suction	x			x	x	x	x	x				
Trailing draghead suction	x			x	x	x	x	x				
Dustpan suction		x		x	x	x		x				
Mechanical excavation:												
Bucket, shovel, backhoe, dragline	x	x	x	x	x	x	x	x				
Cutter blades—rotary/fixed	x	x	x	x	x	x	x	x				
Hydraulic removal:												
Suction pipeline				x	x	x	x	x	x	x		
Mechanical removal:												
Bucket, shovel, backhoe, dragline					x	x	x	x				
Bucket-ladder, bucketwheel					x	x	x	x				
Hydraulic transport:												
Pumped slurry in pipeline				x	x	x	x	x	x	x		
Mechanical transport:												
Hopper (own hold)						x	x	x			x	x
Barge, self-propelled or towed						x						x
Land-based trucks, belts, etc.						x						x
Hydraulic disposal:												
Pipeline slurry—land disposal						x	x	x			x	x
Pipeline slurry—water disposal						x	x	x			x	x
Mechanical disposal:												
Bottom discharge—hopper or barge						x	x					x
Grabs/scrapers for emptying barges						x	x					x

*1. Compactness of granular soils
2. Consistency of cohesive soils
3. Sensitivity of cohesive soils
4. Water content
5. Mass density
6. Grain size distribution
7. Plasticity of fines
8. Organic content
9. Particle shape and hardness
10. Rheologic prop. of slurry
11. Sedimentation rate in water
12. Bulking factor

Source: From Ref. 7.

Drilling methods include removal of soils down to the maximum dredging depth. This can be done by excavating pits or trenches (which require heavy equipment or dredges) or by drilling holes and recovering either disturbed or relatively undisturbed samples. Drilling methods include wash borings, augering, or rotary drilling. A pipe casing is employed to confine a drilling hole and water is used in the wash borings. In the augering method, a hollow stem auger employs an auger flight around a pipe casing advancing the auger and casing at the same time. Rotary drilling uses drilling mud to maintain the below-bottom hole. All drilling methods rely on either boats, jack-up rigs, or bottom supported rigs, etc., to provide firm attachment to the bottom soils.

Sampling methods include surficial samplers, projectile or impact tube samplers, thin-wall tube samplers, or vibrating core samplers.

Grab Sampling. A grab sampler consists of a scoop or bucket container that bites into the soft sediment deposit and encloses the sample. Grab samplers are easy and inexpensive to obtain and may be sufficient to characterize sediment for routine maintenance dredging. Grab sampling may indicate relatively homogeneous sediment composition, segregated pockets, or coarse- and fine-grained sediment and/or mixtures. Figure 5.5 shows several surficial samplers that are currently in use.

Projectile tube samplers may be forced into the bottom soils as a projectile using free fall, a propellant, or an explosive charge. The depth of penetration is a function of the weight of the sampling tube, propelling force, explosive charge, diameter of the tube, wall thickness, and the soil characteristics. Successful soil samples are obtained if the soil enters the tube instead of being pushed aside during penetration. Tube sampling includes a tube sampler and the split barrel sample spoon.

A tube sampler is an open-ended tube that is thrust vertically into the sediment deposit to the depth desired. The sampler is withdrawn from the deposit with the sample retained within the tube. Differences among tube samplers relate to tube size, tube wall thickness, type of penetrating nose, head design including valve, and type of driving force. Tube samplers (also called harpoon samplers) are available with adjustable weights in the range of from 17 to 77 lb (7.7 to 34.9 kg) and with fixed weights in excess of 90 lb (40.8 kg). The amount of weight required depends upon deposit texture and required depth of penetration.[8]

The split barrel sample spoon (also known as a split-spoon sampler) is capable of penetrating hard sediments, provided sufficient force is applied to the driving rods. The sampler is thrust into the deposit by the hammering force exerted on rods connected to the head. During retrieval, the sample is retained within the barrel by a flap. The nose and head are separated from the barrel in order to transfer the sample to a container. Refer to EM 1110-2-1907 for more information on soil sampling.

Thin-wall tube samplers may be pushed slowly in clay and clayey silts to obtain relatively undisturbed samples.

The Phleger tube (gravity corer) is shown in Fig. 5.6.

The vibratory core samples consist of a fairly lightweight sampling tube which can be suspended over the side of a vessel or platform in relatively shallow water depths (Fig. 5.7).[7b] Penetration is achieved by a light vibratory force applied to the top of the device. Penetration can be achieved in a water depth of about 20 ft (6 m) in loose to dense granular soils and soft to stiff cohesive clays.

Sampler		Weight	Remarks
Peterson		39–93 lb	Samples 144-in^2 area to a depth of up to 12 in, depending on sediment texture.
Shipek		150 lb	Samples 64-in^2 area to a depth of approximately 4 in.
Ekman		9 lb	Suitable only for very soft sediments.
Ponar		45–60 lb	Samples 81-in^2 area to a depth of less than 12 in. Ineffective in hard clay.
Drag Bucket		Varies	Skims an irregular slice of sediment surface. Available in assorted sizes and shapes.

FIGURE 5.5   Surficial sampling equipment. (*Office of Chief of Engineers 1983*)

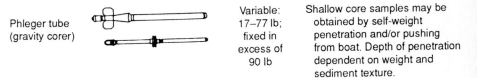

Phleger tube (gravity corer)		Variable: 17–77 lb; fixed in excess of 90 lb	Shallow core samples may be obtained by self-weight penetration and/or pushing from boat. Depth of penetration dependent on weight and sediment texture.

FIGURE 5.6   Phleger tube gravity corer. (*Office of Chief of Engineers 1983*)

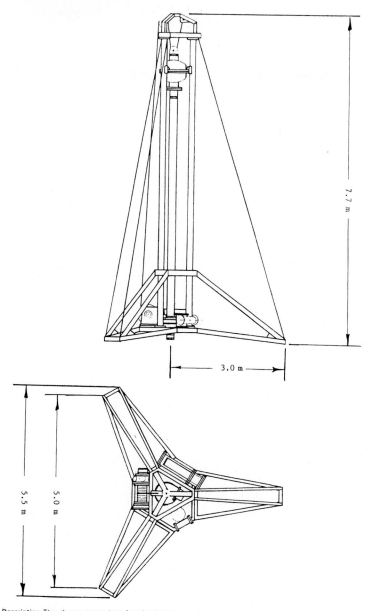

7.7 m

3.0 m

5.5 m

5.0 m

Description: The vibrocorer consists of a twin vibrator motor housed in a pressure vessel driving a core barrel of 102 mm outside diameter with a vibration force of 6 tonnes at 50 Hz. The standard system weighing in the order of 3½ tonnes uses a 6 m barrel but smaller units with correspondingly lighter frames are available. A base mounted winch on the vibrocorer providing up to 12 tonnes withdrawal force enables full barrel retraction prior to recovery on the main lift wire. A penetrometer

with a chart recorder and analog display gives a precise measure of penetration rate and depth. The power requirements is 30 kva 415 v 3 ph 50 Hz.

Sample: The samples are retrieved in a clear plastic liner tube of 83 mm internal diameter.

Operational depth: The system has been tested to depths in excess of 1,800 metres and is designed for use to 2,000 metres.

FIGURE 5.7   A vibratory core. (*Courtesy, British Geological Survey*)

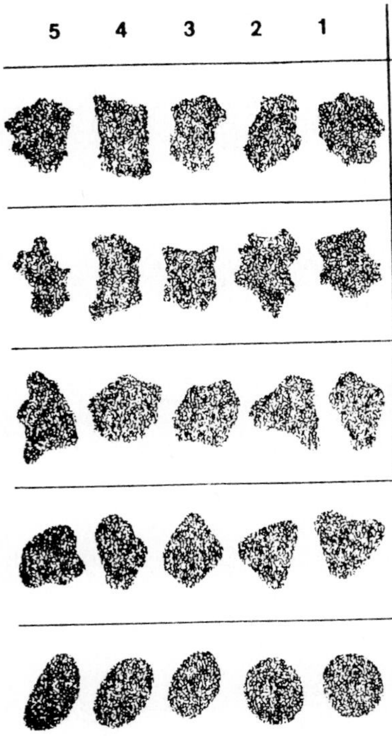

5	4	3	2	1	Names of classes	Degrees of roundness	Description
					E Angular	0 – 0.15	Sharp corners sharply defined. Large embayments with numerous equally sharply defined embayments.
					D Sub-angular	0.15 – 0.25	Incipient rounding of corners. Large embayments preserved, small embayments smoother and less numerous.
					C Sub-rounded	0.25 – 0.40	Corners well rounded, large embayments weakly defined, small embayments few in number and gently rounded.
					B Rounded	0.40 – 0.60	Original corners are gently rounded, large embayments are only suggested, small embayments absent.
					A Well-rounded	0.60 – 1.00	Original corners and large embayments are no longer recognizable. Uniformly convex cutline (subordinate planar sections possible).

FIGURE 5.8 Classification and designation of grain shapes according to Russel and Taylor. (*Courtesy,* Terra et Aqua)

Laboratory soil testing methods include tests

1. To determine *in situ* water content.

2. To perform plasticity analyses on the fine-grain fraction of the soil (passing no. 40 sieve).

3. To determine specific gravity.

4. To determine the grain size of the sediments, including the shape of individual grains. The shape of sediment pumped in hydraulic dredging affects the wear of the pump and pipeline. The wear is a function of sediment size, shape, specific gravity, and hardness of individual particles as well as slurry velocity. Figure 5.8 provides a classification and designation definition of grain shapes by Russell and Taylor.[9]

5. To obtain *in situ* density by conducting a Standard Penetration Test (SPT). Relative density of sands according to standard penetration tests is shown in Table 5.7. The SPT uses a 140-lb (63.5-kg) hammer falling freely 30 in (76 cm) on a standard 2-in (5.1-cm) O.D. thick wall split-tube sampler.

The number of hammer blows needed to drive the sampler a distance of 12 in (30.5 cm) is called the *blow count.* The Cone Penetration Test (CPT) is also used; it has a sleeved cone with a 1.4-in (35.7-mm) diameter and a 60° apex angle. The cone rods are forced into the soil vertically at the rate of 0.8 in/s (20 mm/s) and

TABLE 5.7 Relative Density of Sands According to Results of Standard Penetration Tests

No. of blows/ft	Relative density
0–4	Very loose
4–10	Loose
10–30	Medium
30–50	Dense
Over 50	Very dense

the penetration resistance of the cone and the forcing effort on the sleeve are measured separately.

A cone penetration test provides not only the information regarding compactness of the soil but also gives some indication of the type of soil, as shown in Fig. 5.9. The correlation between the standard penetration values and cone penetration tests is given by Verbeek[9] (Table 5.8).

6. For organic content.

7. For grain shape and hardness.

8. For rheological properties, many fine grain-water mixtures form nonnewtonian fluids with a variable viscosity. This will assist in determining the "navigable depth" of the channel.

9. For determination of bulking factors required for sizing confined disposal areas.

Major projects may justify additional soil analyses and tests such as unconfined compression tests and direct shear tests.

Verbeek[9] recommends that samples should be taken from all over the site to obtain an indication of the soil conditions in situ. For dredging projects the distance between borings should be between 165 ft (50 m) and 655 ft (200 m). An equation for calculating the required number of boreholes is

$$N = 3 + \left(\frac{A^{0.5} \, d^{0.33}}{50} \right) \tag{5.4}$$

where A = dredging area
d = average depth to be removed by dredging

MATERIALS IN DISPOSAL AREAS

Hydraulic dredging typically mixes 20 percent of solids with 80 percent water; this mixing process produces bulking. The sediment being pumped into a disposal area has a larger volume than the sediment in situ.

This increase in volume is accompanied by an increase in the void ratio and water content of the soil.[10,11] In case of dense sands this expansion may be minimal, while for loose sand the dredging process may even serve to densify the material. However, in the case of silts and clay soils, the bulking of the soil may be quite substantial, particularly for consolidated clays. This phenomenon is partially caused by the

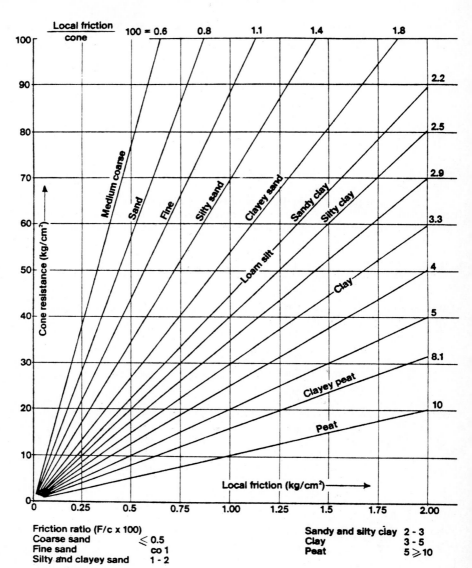

FIGURE 5.9 Core penetration resistance as a function of local friction. (*Courtesy*, Terra et Aqua)

TABLE 5.8 Correlation between Standard Penetration N Value and Cone Penetration Tests

	Sand		
Soil condition	N value (SPT according to Terzaghi and Peck)	Cone resistance (Dutch cone penetration test) in bars	Relative density (Dr)
Very loose	< 4	< 25	< 0.15
Loose	4–10	25–50	0.15–0.35
Medium dense	10–30	50–100	0.35–0.65
Dense	30–50	100–200	0.65–0.85
Very dense	> 50	> 200	> 0.85

	Clay		
Soil condition	N value (SPT according to Terzaghi and Peck)	Unconfined compression strength in bars	Torvane cohesion in bars
Very soft	< 2	< 0.25	< 0.13
Soft	2–4	0.25–0.5	0.13–0.25
Plastic	4–8	0.5–1	0.25–0.5
Stiff	8–15	1–2	0.5–1
Very stiff	15–30	2–4	1–2
Hard	> 30	> 4	> 2

Relationship between N value and cone resistance	
Soil	Cone resistance/n_{30}
Gravel	5.5–8
Coarse sand	4–5.5
Fine sand	2.5–4
Clayey sand	6
Sandy loam	5–6
Sandy clay	3–4
Clay	2

Source: Courtesy, *Terra et Aqua.*

aforementioned factors and partially by the absorption of water by the clay. The *bulking factor* of a particular soil is the dimensionless factor expressed by the ratio of the volume of the soil in a containment area after dredging to that volume of the soil in situ. The following equations are presented from Ref. 12:

$$B = \frac{V_c}{V_i} \tag{5.5}$$

$$B = \frac{\gamma_{d,i}}{\gamma_{d,c}} \tag{5.6}$$

$$B = \frac{w_c G_s + 100}{w_i G_s + 100} \tag{5.7}$$

$\qquad$ = bulking factor
V_c = volume in containment area
V_i = volume in situ
$\gamma_{d,i}$ = dry density in situ
$\gamma_{d,c}$ = dry density in containment area
w_c = water content in the containment area
w_i = water content in situ
G_s = specific gravity of the solids

Equation 5.5 is valid only for a soil under saturated conditions. In order to calculate the unit volume of a soil in situ, the water content and specific gravity must be determined. In order to predict the volume occupied by the dredged material in the containment area, the bulking factor for the soil type must be accurately known. If any long-term prediction is expected, the settlement and consolidation characteristics of the soil type must also be considered.[13,14] In the majority of cases cited in the literature, government agencies and private contractors involved in dredging operations rely heavily on practical experience to predict bulking or sizing factors. There has been much dissatisfaction or uncertainty expressed about those factors commonly employed for clay or silty-clay soils. These factors have historically resulted in undersizing volumes by as much as 50 percent or oversizing them by as much as 100 percent.[15] This problem is partially related to the wide range of clay soil characteristics and the close relationship between the behavior of clay and the dredging system used to move it. Once dredged material is in suspension, its settlement characteristics are a function of water salinity, turbulence, and solids concentration as well as the properties of the soil. Increasing the salinity tends to intensify flocculation up to a limiting concentration, above which increased concentration has little effect. Those clay particles suspended in fresh water tend to remain in suspension until all water motion ceases and then settle very slowly to the bottom where they accumulate as sediments. As water salinity approaches 14‰, the clay particles flocculate and settle out of suspension much faster than in fresh water. This is because the abundant number of positively charged ions in salt water tend to change the surface charge of some of the clay particles from negative to positive. These clay particles with positive surface charges tend to aggregate with clay particles having negative surface charges, forming flocculants which rapidly settle out of suspension.[16]

As turbulence increases, so does flocculation because of the increased opportunity for collisions between particles; turbulence also has a limiting value above which further increases tend to break up the flocculated soil particles. For solids concentrations less than 2.7 percent by weight, an increase in concentration tends to cause an increase in flocculation. On the other hand, for typical dredge slurries which range from 10 to 30 percent solids by weight, increases in concentration tend to reduce particle movement, increase excessive pore-water pressure, and reduce flocculation.[11] Prediction of bulking factors for clay is further complicated by the fact that some clay particles tend to remain in clods, depending on the dredging method. These clods are transported as a "bed load" in the pipeline, exiting the line as clay balls. These clay balls do not contribute significantly to the bulking characteristics of clay, thus the percentage of clay so transported directly influences the bulking factor.[12,17,18]

Another problem associated with the use of containment areas is that a horizontal sorting of the dredged material by particle size results. The larger, heavier particles tend to settle near the discharge line in a fan-shaped distribution, while the fine silts and clays tend to remain in suspension longer and settle nearer the discharge line. This horizontal sorting appears to be limited to an area with a radius of 295 to 656 ft (90 to 200 m) from the discharge line, depending on particle size distribution, discharge velocity, and containment area topography.

Commonly used bulking factors are described in detail in Chap. 8.

Laboratory Determination of Initial Bulking Factor

The following is a step-by-step laboratory procedure:[19,20]

1. The wax is trimmed from the undisturbed samples as each is prepared for testing. The sample itself is then trimmed into a uniform cylindrical shape with a knife and a small carpenter's square.
2. The sample is carefully measured with a metric scale and its volume in cubic centimeters is calculated from the measurements.
3. The weight of the sample to the nearest decigram is obtained and its specific volume obtained by dividing the volume from step 2 by this weight.
4. The sample is then chopped and mixed into a homogeneous mass using a knife and a glass cutting plate.
5. For those samples tested in fresh water, 13.1‰ salt water, and 28.6‰ salt water, three 200-gram portions are taken from the sample and placed aside in separate, covered 1000-ml beakers.
6. The in situ volume of each gram sample is calculated by multiplying its weight by the specific volume obtained in step 3.
7. The portions of each sample are then allowed to slake in their separate beakers with 500 ml of the water in which they were to be tested. This slaking is continued for a period of approximately 24 hr.
8. After slaking, a variable speed mixer and a special four-bladed plastic impeller-shaped blade are used to mix the soil particles and water into a homogeneous slurry. Water is added during the mixing process, increasing the slurry volume to about 975 ml. A slurry density of 1200 grams per liter by weight is desired to simulate that density typically found in dredge slurries. The impeller blade provides a strong vortex within the beaker and serves to raise the soil particles into suspension. A mixing time of 5 to 6 min is required to ensure all soil particles, with the exception of the clay balls, are in suspension.
9. The slurry is then poured into a 1000-ml graduated cylinder. About 25 ml of the proper salinity water is used to rinse the adhering soil particles from the impeller blade and beaker into the graduated cylinder, raising the volume within the cylinder to exactly 1000 ml.
10. The slurry is then allowed to stand undisturbed while the soil particles settle out of suspension. The level of the interface between the suspended material and the supernatant liquid is observed and recorded ½ hr after beginning the test and hourly thereafter until the rate of settlement has decreased to 10 ml/hr or less. The level of suspended material is recorded thereafter at 24-hr intervals from the start of the test until the amount of settlement is undetectable for a 48-hr period.[9,11]
11. When the sedimentation test for a particular sample is completed, the super-

natant liquid is drawn off and its salinity determined by a conductivity meter. The sample and cylinder are weighed together, then placed in an oven to dry at 140°F (60°C).

12. The cylinders are removed from the oven and weighed at 24-hr intervals until two consecutive weights are the same. By compensating for the weight of the small amount of salt water left in the cylinder prior to placement in the oven, the dried salt on top of the soil after removal from the oven, and the cylinder weight, the dry weight of the sample and moisture content of the sedimented material are determined. This moisture content is used to calculate the void ratio of the sediment material.

Effect of Water Salinity and Time

The fine soil particles tend to remain in suspension in fresh water while flocculation and relatively rapid settlement occur in saline water (13.1‰ and 28.6‰ salt water). Although the bulking factors in fresh water are larger than those in salt water, the indications are that the fresh water curves may approach the salt water curves over a long period of time. Typical test results are shown in Figs. 5.10 and 5.11.[13]

The size of the test cylinder has a bearing on the bulking factor determination. Too small a testing cylinder (60 ml) will give higher bulking factors; a minimum size of 1000 ml is recommended (Fig. 5.12).

Effect of Physical Properties of Soil on Bulking Factor

The relationships between the physical properties of the soil and the bulking factor were examined by DiGeorge and Herbich.[19] The effect of the fines (silt and

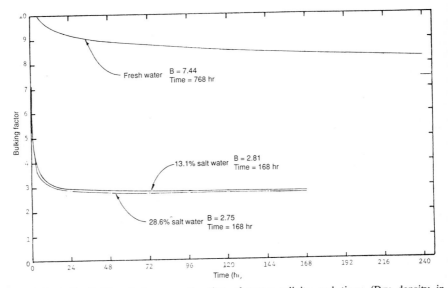

FIGURE 5.10 Bulking factor as a function of water salinity and time. (Dry density in situ = 1.826 gm/ml, water content in situ = 17.0%.) (*DiGeorge and Herbich 1978*)

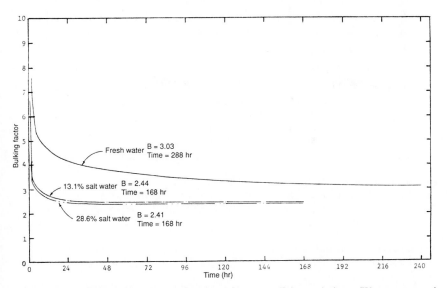

FIGURE 5.11 Bulking factor as a function of water salinity and time. (Water content in situ = 18.0%.) (*DiGeorge and Herbich 1978*)

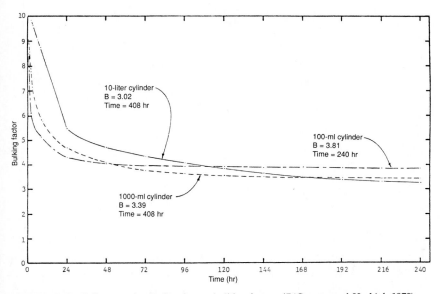

FIGURE 5.12 Influence of cylinder size on bulking factor. (*DiGeorge and Herbich 1978*)

clay content) is shown in Fig. 5.13. While there is considerable scatter, a relationship does seem to exist. A linear regression equation was developed which resulted in:

$$\text{Bulking factor} = 1.897 + 0.013 \quad (\% \text{ fines}) \tag{5.8}$$

The correlation coefficient was quite low ($r^2 = 0.25$), in part because of the flatness of the line, but the average error in using this equation is only about 11 percent, well within any state-of-the-art predictions of sizing factor.

Figure 5.14 shows the relationship between the liquid limit (LL) and bulking factor. The following equation was obtained:

$$\text{Bulking factor} = 0.005 \, (\text{LL}) + 2.66 \tag{5.9}$$

(correlation coefficient = 0.05).

The effect of the plasticity index (PI) on the bulking factor is shown in Fig. 5.15. The following equation was obtained from linear regression:

$$\text{Bulking factor} = 0.0043 \, (\text{PI}) + 2.80 \tag{5.10}$$

(correlation coefficient = 0.03).

The effect of the liquidity index (LI) on the bulking factor is shown in Fig. 5.16, and the equation is

$$\text{Bulking factor} = 0.31 \, (\text{LI}) + 2.87 \tag{5.11}$$

(correlation coefficient = 0.03).

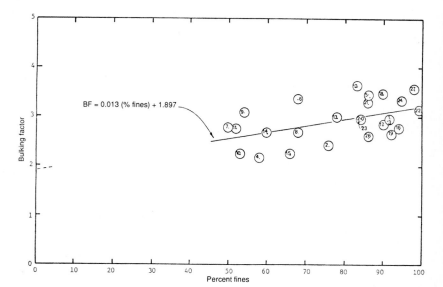

FIGURE 5.13 Bulking factor as a function of percent soil particles passing no. 200 sieve. (*DiGeorge and Herbich 1978*)

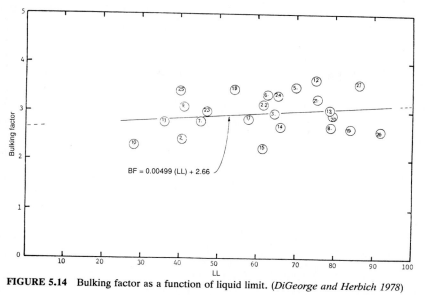

FIGURE 5.14 Bulking factor as a function of liquid limit. (*DiGeorge and Herbich 1978*)

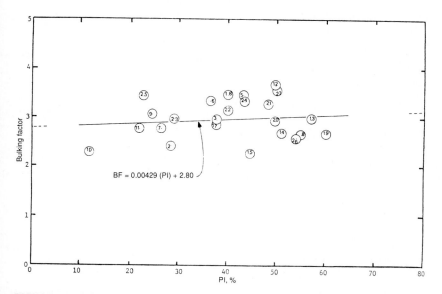

FIGURE 5.15 Bulking factor as a function of plasticity index (PI). (*DiGeorge and Herbich 1978*)

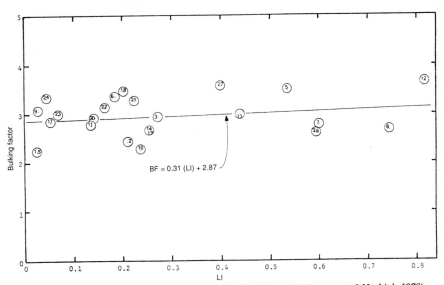

FIGURE 5.16 Bulking factor as a function of liquidity index. (*DiGeorge and Herbich 1978*)

The following conclusions may be drawn from this study of bulking factors:

1. Salinity has a definite effect on the bulking factor. A threshold salinity, below which flocculation will not occur, was not determined from these tests; however, it seems apparent that in performing bulking factor tests, water of the same salinity as expected in situ should be used.
2. The bulking factors obtained by this method appear to be a function of cylinder geometry. Possibly the amount of drainage along the walls of the cylinder is an important geometrical effect. A minimum size of the cylinder recommended is 1000 cm^3.
3. The bulking factors determined were larger than usually experienced. Placement of a large single lift of material rather than smaller multiple lifts may have contributed to the large bulking factors.
4. The bulking factors increased as the fines content (silt and clay) increased.
5. The bulking factors seem to generally increase as the Atterberg limits increased. The relationship found between Atterberg limits and bulking factor was, however, very weak.

PROPERTIES OF ROCK

Dredging of rock, now possible with specially designed heavy-duty cutterhead dredges, deserves a separate comment. The strength of rock is generally described by the Unconfined Compressive Strength (UCS) test as follows:[3]

Term	Unconfined compressive strength (kg/cm²)
Very weak	Less than 12.5
Weak	12.5 to 50
Moderately weak	50 to 125
Moderately strong	125 to 500
Strong	500 to 1000
Very strong	1000 to 2000
Extremely strong	Greater than 2000

Rock characteristics at a dredging site have a major influence on the performance of a dredge. The rock characteristics must be determined prior to commencement of work to determine whether a dredge can handle the rock and, if so, how the output would be affected by rock characteristics.

The rock properties are a combination of:

1. *Rock material properties*, which are established by laboratory tests on undisturbed rock specimens, examples of which are: compressive strength and tensile test values (without taking into account the fissured condition of the rock in situ)
2. *Rock mass properties,* which are established by in situ measurements of the rock as they occur in their geological condition, intersected by planes of weakness (i.e., discontinuities)

It is therefore important to perform an adequate investigation of the soil conditions.

The two primary objectives of a geotechnical investigation are to:

1. Identify all the rock types present within the dredging area and to determine their extent and volumes
2. Define the physical and mechanical properties of those materials in terms of parameters which affect the dredging process

To achieve these objectives, rotary drilling is generally employed to obtain cores. Rotary drilling also provides continuous samples of rock that are necessary for examination and testing.

The total core recovery and its condition are heavily dependent upon the capability and experience of the drilling equipment operator.

The diameter of cores recovered by this method usually lies in the range of 2–3.2 in (50–80 mm). There is, however, a relationship between the core diameter and the quality of the recovered samples. Core diameters at the upper end of this range, or greater, should be used, particularly in the case of weak or friable rock, because it generally produces better and more representative samples.

Often trial pits are used onshore, in addition to boreholes, to investigate materials to a maximum depth of about 16 ft (5 m). These pits have an advantage in that the materials can be examined in far greater detail than drilled cores which provide only comparatively small samples.

The investigation should cover the entire area where the dredging is to take place. This is best achieved by placing the outermost boreholes just outside the dredging area, and the boreholes are then preferably made within this area in a regular pattern to permit drawing of profiles.

The density of borings (number of borings and distances between each one) is always difficult to establish and is decided upon by the owner's geologist's interpretation of the probable geological uniformity or nonuniformity of the area.

The samples from the boreholes, adequately packed, sealed, and labeled, should be sent as soon and as carefully as possible to the laboratory for classification and testing. Special care must be taken during transportation so that the samples do not dry out. A report on the site investigation is required to interpret the findings of the subsurface exploration and the laboratory testing.

When an engineering interpretation of the various geological data present at the site is required, the report is best prepared in two distinct parts. The first part should contain (among other things) a description of the site, tables and diagrams giving the test results, field and laboratory report forms, and the data sheets, all of which provide a detailed record of the data obtained. The second part of the report should contain the analysis, conclusions, and recommendations of the laboratory staff and the geologist.

For rotary-cored boreholes, it is good practice to have an engineering geologist on site so that the rock cores may be logged and described in their fresh condition. The borehole logs should contain:

1. A description of each stratum together with its thickness
2. The depth and level of each change in stratum
3. The depth of the start and finish of each core run
4. The core recovery for each run, usually expressed as percentage total core recovery
5. The fraction state, expressed in terms of one or more of the following:
   a. Solid core recovery
   b. Rock quality designation expressed as percentage of cores longer than 3.9 in (10 cm)
   c. The degree of weathering, cementation, cleavage, and fissures
   d. The direction of the fissures and the bedding plane
6. A record of tests carried out

The results of the laboratory tests should also be incorporated in the borehole logs. This additional information will lead to the final engineering and geological rock descriptions in the borehole logs.

Laboratory tests on rock samples fall into one of three groups:

1. *Index testing,* to provide basic information on the physical characteristics of the rock, such as density and porosity
2. *Strength testing,* to determine the strength and elastic properties of rocks in compression and tension
3. *Materials testing,* to establish the durability and chemical properties of the rock

The analysis of the geotechnical investigation results consists of the interpretation of the geology of the site, the derivation of the engineering properties of the rocks, and a detailed engineering assessment, which would lead to the final recommendations for the design of structures and dredged slopes and the proposed dredging plan.

The entire success of the soils investigation is dependent upon the quality of the geological interpretation and the ability of those undertaking the investigation to understand the geological environment in which they are working. Furthermore, it is essential that all those who have to communicate information to others employ the same technical language in describing materials to be dredged. This calls for a uniform system of classification, particularly at the international level.

A step toward a uniform nomenclature has been made in recent years with the publication and increasing use of two reports by the Permanent International Association of Navigational Congresses (PIANC) entitled *Classification of Soils and Rocks to Be Dredged.*[3,4] These reports do not attempt to classify materials in terms of their dredgeability; instead they concentrate on providing a widely understood basis for judging dredgeability by setting out a systematic method of classifying materials in terms of their basic characteristics and indicating appropriate methods for their identification. Table 5.9 provides the general basis for identification and classification of rocks for dredging purposes.[3] Igneous, sedimentary, and metamorphic rocks are described, and guidelines are given for identifying the various types of rocks. Table 5.10 describes the in situ and laboratory testing procedures of rocks for dredging purposes.

DISTRIBUTION OF DREDGED MATERIAL IN THE UNITED STATES

Figures 5.17 through 5.20 show the general distribution of dredged material types according to USCS, AASHO, and FAA classification systems published by Bartos.[21] Figure 5.21 gives the dredged material classification graph for the nation and Figure 5.22 gives grain size distribution curves for typical samples from regions A to E. Table 5.11 shows the ranges of classification test data determined for the dredged material. The information presented in this report is indicative of the types of dredged material found in each of the study regions.

The samples of dredged material taken from within the Gulf States study region fell into seven of the USCS classification groups. Figure 5.17 shows that less than 33 percent of the samples were classified as sandy material and the remaining two-thirds (67 percent) of the samples were classified in one of the fine-grained designations, mostly CH, which indicates inorganic fines of high plasticity. In the South Atlantic study region the dredged material samples ranged from poorly graded gravels (GP) to plastic and organic clays (CH and OH). Three-fourths of all the samples were classified as sands and gravels. Only 24.5 percent of the samples were fine-grained. In the North Atlantic study region the samples consisted of 27 percent organic clay, 26 percent poorly graded sand, and the remaining samples were evenly distributed among 10 different classifications of the USCS system. Forty-seven percent of the samples were coarse-grained and 53 percent were fine material, mostly CL and OH. In the Great Lakes study region the predominant types of dredged samples were poorly graded sand (SP) and clay of high plasticity (CH). Slightly more than half of the samples, 52.9 percent, were coarse-grained material, and the remaining 47.1 percent were fine-grained. The majority of the fine-grained material was highly plastic (CH). There were no samples of organic dredged material. In the Pacific Coast study region the material ranged from well-graded sand (SW) to organic fines (OH). The predominant type of dredged material was poorly graded sand (SP). Approximately 75.5 percent of the samples were coarse-grained. Figure 5.20 shows the division of samples into four categories and is intended to show the fractions of the samples that were coarse, plastic, nonplastic, or organic.

TABLE 5.9 General Basis for Identification and Classification of Rocks[1] for Dredging Purposes

Group	Examples of rock type	Origin	Identification	Remarks
Igneous	Granites, dolerites, basalts, etc.	Formed by the solidification (crystallization) of original molten material (magma) extruded from within the earth's crust.	All exhibit a crystalline form although the individual crystals may be invisible to the naked eye. Complex system of rocks. All igneous rocks are hard although may be altered by various natural causes such as weathering. Because of stress rocks may possess systems of joints and fissures.	Full identification of rocks may be complex. Hand examination will give approximate classification based on rock-type name. Laboratory examination may be required using rock slices to confirm the more difficult cases.
Sedimentary	Sandstone, limestones, marls, chalk, corals, conglomerates, etc.	Derived from preexisting formations, by weathering and desintegration, often being reconsolidated in hard strata. Occurring as sequence of deposits in beds.	Often recognizable by bedded structure. In general terms the older the formation, the harder the rock although a considerable variation in hardness, color, and other characteristics is likely. In many sedimentary rocks the individual particles forming the body of the material may be seen (e.g., sandstone) and a rough grading given in description.	Engineering properties of rock for dredging purposes generally must be carried out in laboratory using Test Procedures suggested in Table 5.3. While for practical purposes it may not be necessary to identify a rock by name, it is of inestimable value in analyzing the project as a whole.

| Metamorphic | Gneisses, marbles, etc. | Includes an igneous or sedimentary rock which has been altered by heat or pressure. | Wide range in degree of metamorphism with some rocks still close to original condition, other rocks completely structure obscured. Rock is normally very hard with glassy surface. | Degree of weathering in rock is of extreme importance and will alter the engineering properties of even the hardest igneous rocks. |

[1] Rock may be defined in the engineering sense as the hard and rigid deposits forming part of the earth's crust as opposed to deposits classed as soil. Geological rock embraces both soft and hard naturally occurring deposits, excluding topsoil.

Source: From Ref. 3.

TABLE 5.10 General Characteristics of Soils and Rocks for Dredging Purposes (Rocks Unweathered* and Unblasted)

| Rock/Soil Type | Excavation Characteristics | | | | | | | Suitable as reclamation material | Suitability to pipeline transportation | Often observed bulk density before excavation |
	Dipper dredger	Bucket dredger	Suction dredger	Cutter dredger	Trailer dredger	Grab dredger				
ROCK*										
I. Igneous	N.A.	N.A.	N.A.	N.A.	N.A.	N.A.		N.A.	N.A.	2.0 -2.8
II. Sedimentary	Possible in soft rock but difficult	Possible in soft rock but difficult	N.A.	Difficult to fair in softer rocks	N.A.	Possible in softer rocks but very difficult		Very good	Fair, large fragments may block pipes	1.9 -2.5
III. Metamorphic	N.A.	N.A.	N.A.	N.A.	N.A.	N.A.		N.A.	N.A.	2.0 -2.8
• Weathering of rocks will alter form and strength considerably and may allow direct dredging without blasting, etc.										
Boulders	Fair	Very slow, may require slinging	N.A.	N.A.	N.A.	Difficult but large units cope		Not acceptable	N.A.	N.A.
Cobbles or Cobbles with gravel	Fair	Fair	Difficult	Difficult	Difficult	Fair		Bad to Good	Poor	N.A.
Gravel	Easy	Fair	Difficult to Fair	Fair	Difficult te Fair	Fair		Good	Fair	1.75-2.2
Sandy gravel	Easy	Fair to Easy	Fair	Fair to Easy	Fair to Easy	Fair to Easy		Very good	Fair to good	2.0 -2.3

Soil type	1	2	3	4	5	6	7	8	Density
Medium sand	Easy but low production	Easy	Easy	Easy	Fair to Easy but high overflow losses likely	Easy	Very good	Good	1.7 -2.3
Fine sand	Easy but low production	Easy	Easy	Easy	Fair to Easy but high overflow losses likely	Easy	Good	Very good	1.7 -2.3
Extra Fine Sand	Easy but low production	Easy	Easy	Easy	Fair to Easy but high overflow losses likely	Easy	Good	Very good	1.7 -2.3
Silty Fine Sand	Easy but low production	Easy	Easy	Easy	Fair to Easy but high overflow losses likely	Easy	Good	Very good	1.7 -2.3
Cemented fine sand	Fair	Fair	N.A.	Fair to Easy	Difficult	Difficult-	Good	Bad to good	1.7 -2.3
Silt	N.A.	Easy	Difficult to Fair	Easy	Fair to Easy but high overflow losses	Fair	Bad	Very good	1.6 -2.0
Firm or stiff gravelly or sandy clays (i.e. boulder clays)	Fair	Difficult to Fair	N.A.	Difficult to Fair	N.A.	Difficult to Fair	Good	Only possible after disintegration	1.8 -2.4
Soft silty clays (i.e. alluvial clays)	N.A.	Fair to Easy	N.A.	Easy	Fair	Easy	Bad	Fair	1.2 -1.8. (fresh harbour sediment 1.15-1.6)
Firm or stiff Silty clays	Fair to Easy	Easy	N.A.	Fair to Easy	Difficult to Fair	Fair	Bad to fair	Only possible after disintegration	1.5-2.1
Peats	N.A.	Easy	N.A.	Easy if no gas encountered	Fair	Easy	Unacceptable	Very good	0.9 -1.7

N.A.: Not applicable

Note: This table only gives a rough indication and should be used with caution.

The feasibility to use a certain type of dredging equipment depends not only on the soil type but also on site conditions, the size, strength of construction and power supply of that piece of equipment, etc.

The qualifications used above (bad, poor, fair, easy, very good, etc.) are meant to show the degree of suitability but should not be related to the output or even less as indicative of the cost per excavated unit.

Source: Ref. 3.

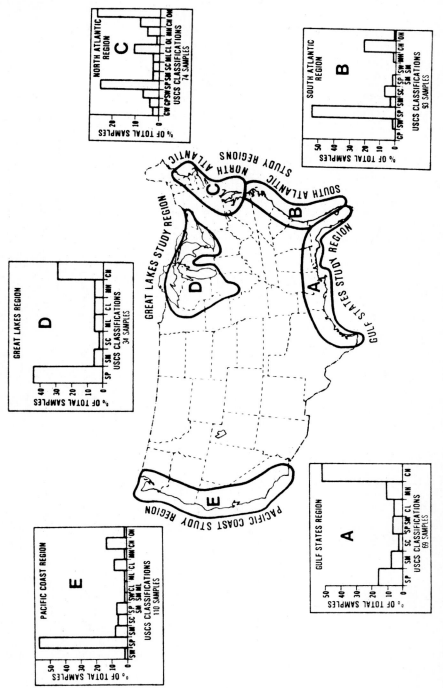

FIGURE 5.17 Regional distribution of dredged material types according to USCS classifications. (*Bartos 1977*)

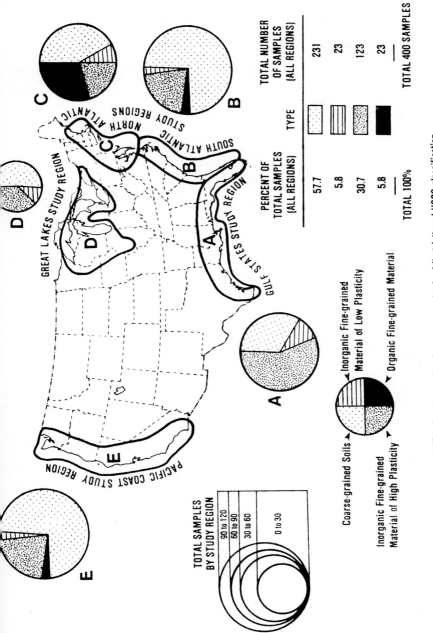

NOTE: Samples assigned to groups on basis of first letter of USCS classification

FIGURE 5.18 Types of dredged material samples by region. (*Bartos 1977*)

TYPE	PERCENT OF TOTAL SAMPLES (ALL REGIONS)	TOTAL NUMBER OF SAMPLES (ALL REGIONS)
(dotted)	57.7	231
(vertical lines)	5.8	23
(speckled)	30.7	123
(black)	5.8	23
	TOTAL 100%	TOTAL 400 SAMPLES

Coarse-grained Soils

Inorganic Fine-grained Material of Low Plasticity

Inorganic Fine-grained Material of High Plasticity

Organic Fine-grained Material

TOTAL SAMPLES BY STUDY REGION

90 to 120
60 to 90
30 to 60
0 to 30

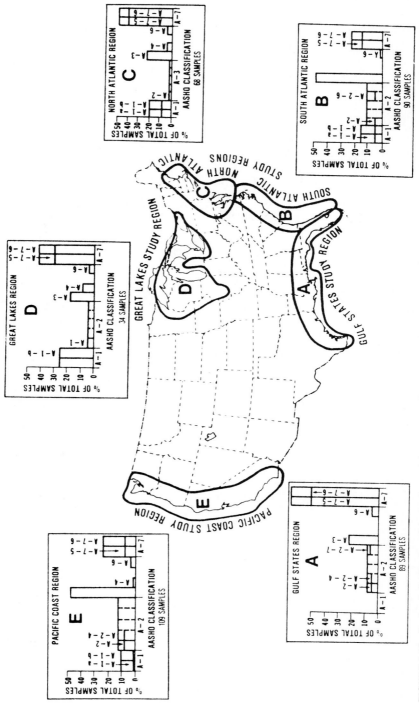

FIGURE 5.19 Regional distribution of dredged material types according to AASHO classifications. (*Bartos 1977*)

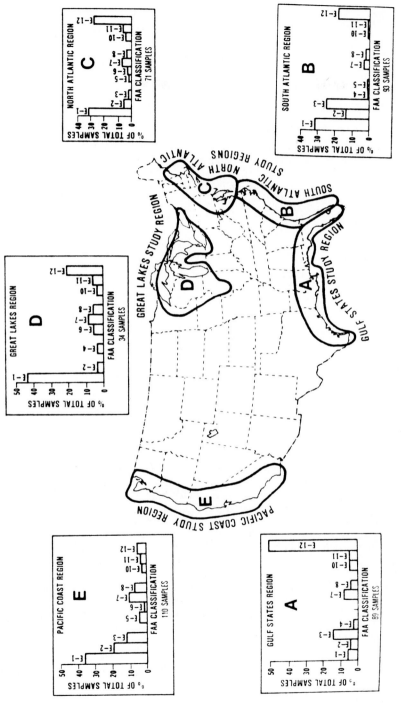

FIGURE 5.20 Distribution of FAA classifications by region. *(Bartos 1977)*

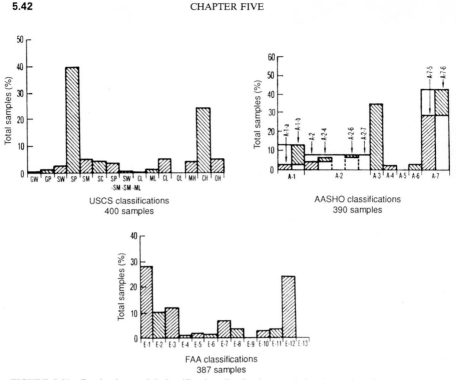

FIGURE 5.21   Dredged material classification distribution graph for the nation. (*Bartos 1977*)

DREDGEABILITY OF SOILS

Recent research deals both with granular and cohesive materials. In evaluating the dredgeability of compacted sand underwater, the pore water plays an important role.[22] Shear strain caused by the cutting action in densely compacted sand causes an increase in the pore volume of the sand layer. This results in the decreased pressure of the pore water and causes higher effective stresses, an increase in the shear resistance, and thus an increase in the cutting force required. A method for calculating the mean volumetric strain rates during the deformation cycle (caused by cutting action) in compacted sand has been developed.[22] Further research is needed before dredgeability of soils can be accurately predicted. Reliable field tests are desirable but are very difficult to conduct and evaluate.[23]

The cutting of soil under hydrostatic pressure was investigated by Lobanov and Joanknecht[24] for different types of soil (including cohesive materials) and rock (such as limestone). It was found that the increase in cutting force in clays at greater depths is caused by increases in cohesion under hydrostatic pressure. The total cutting force in clay results from two distinct processes:[25]

1. The actual cutting
2. The transportation of the clay cuttings over the blade

When the cutting angle and blade angle coincide, the total draft force is proportional to the cutting velocity, cutting width, and the square root of the cutting

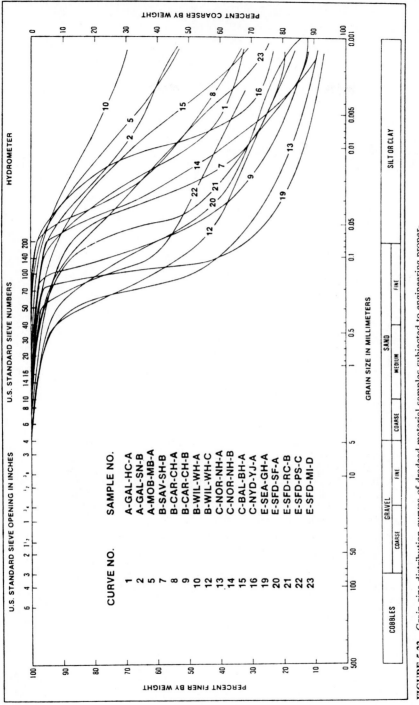

FIGURE 5.22 Grain-size distribution curves of dredged material samples subjected to engineering properties testing. (*Bartos 1977*)

5.43

TABLE 5.11 Ranges of Classification Test Data Determined for Dredged Material*

Region	Total no. samples	Type of material**	Grain size†			Percent passing no. 200 sieve	Atterberg limits			Organic content (%)
			D_{10} (mm)	D_{60} (mm)	D_{90} (mm)		LL	PL	PI	
A	89		81 < 0.001–0.24	89 < 0.001–0.42	89 0.0065–0.80	89 1–99 63	66 32–202 104	65 17–71 35	65 9–144 69	60 0.17–10.64 3.95
B	93		90 < 0.001–0.47	89 < 0.001–7.50	90 0.0057–12.00	93 1–100 26	34 21–273 100	33 15–90 35	33 1–183 68	9 0.13–9.61 5.76
C	74		46 < 0.001–5.00	74 0.0019–78.00	20 0.008– > 78.00	74 0.5–99 50	38 29–152 89	38 17–82 41	38 9–98 48	10 0.32–9.74 4.53
D	34		34 < 0.001–0.46	34 0.007–1.10	34 0.031–7.00	34 0.5–99 46	18 21–161 72	18 19–69 34	18 2–92 38	34 0.09–13.45 3.67
E	110		109 < 0.001–0.45	110 0.0053–2.70	110 0.027–10.30	110 0–99 27	33 28–99 55	33 17–43 25	33 5–57 30	10 0.28–6.53 2.77
Nation	400		360 < 0.001–78.00	396 < 0.001–78.00	397 0.0057– > 78.00	400 0–100 40	189 21–273 88	187 15–90 35	187 1–183 55	123 0.09–13.45 3.95

*Conclusions drawn on basis of data shown apply only to samples tested for this study. Data entries for each region are shown in the following format:

    xx Number of samples
    xx-xx Range of values
    xx Average value, if meaningful

**Legend for material types is as follows:

Sand and gravel (> 50% retained on no. 200 sieve).

Silt (low plasticity fines).

Clay (high plasticity fines).

Organic material (soil with organic matter present).

Note: For the purpose of this table, silts plot below the A line and clays plot above the A line on a plasticity chart.

†D_{10} = Grain size at 10% passing.
 D_{60} = Grain size at 60% passing.
 D_{90} = Grain size at 90% passing.

Source: From Ref. 21.

depth. When the blade angle is smaller than the cutting angle, the total cutting force is decreased.

REFERENCES

1. Mauriello, L. T., and Denning, R. A., "Assessing and Controlling Hydraulic Dredge Performance," *Proc. World Dredging Conference, WODCON 1968,* Rotterdam, The Netherlands, pp. 465–485, Oct. 1968.

2. International Association of Dredging Contractors, The Hague, The Netherlands.

3. Permanent International Association of Navigation Congresses, *Classification of Soils to Be Dredged,* bulletin no. 11, vol. I, 1972.

4. Sargent, J. H., et al., *Classification of Soils and Rocks to Be Dredged,* supplement to bulletin no. 47, Permanent International Association of Navigation Congresses, 1984.

5. Dunlap, W. A., "Soil Classification," Short Course Notes, Center for Dredging Studies, Texas A&M University, 1985.

6. Ottmann, F., and Lahuec, G., "Dredging and Geology," *Terra et Aqua,* no. 11, 1972.

7a. Spigolon, S. J., and Fowler, J., "Site Investigation for Dredging Operations," *Proc. 23d Annual Dredging Seminar,* Center for Dredging Studies, Texas A&M University, Virginia Beach, VA, pp. 83–107, Oct. 1990.

7b. Pheasant, J., "A Microprocessor Controlled Seabed Rockdrill/Vibrocover," *Underwater Technology*, British Geological Survey, U. K., pp. 10–14, Spring 1984.

8. Office of Chief of Engineers, Corps of Engineers, *Dredging and Dredged Material Disposal,* EM1110-2-5025, 1983.

9. Verbeek, P. R. H., "Soil Analysis and Dredging," *Terra et Aqua,* no. 28, 1984.

10. Huston, J., *Hydraulic Dredging,* Cornell Maritime Press Inc., Cambridge, MD, 1970.

11. Lacasse, S. E., Lambe, W. T., Marr, A. W., and Neff, T. L., "Void Ratio of Dredged Material," *Proc. Conference on Geotechnical Practices for Disposal of Solid Waste Material,* University of Michigan, Ann Arbor, June 13–15, ASCE, New York, pp. 153–168, 1977.

12. Johnson, L. D., "Mathematical Model for Predicting the Consolidation of Dredged Material in Confined Disposal Areas," U.S. Army Engineers, Dredged Material Research Program, technical report D-76-1, Jan. 1976.

13. Bowles, J. E., *Foundation Analysis and Design,* 2d ed., McGraw-Hill, New York, 1977.

14. Richards, A. F., Hirst, T. J., and Parks, J. M., "Bulk Density-Water Content Relationships in Marine Silts and Clay," *Journal of Sedim. Petrol.,* vol. 44, no. 4, pp. 1004–1009, 1974.

15. Lacasse, S. F., Lambe, W. T., and Marr, A. W., *Sizing of Containment Areas for Dredged Material,* U.S. Army Engineers, Dredged Material Research Program, technical report D-77-21, Oct. 1977.

16. Gustafson, J. F., "Beneficial Effects of Turbidity," *World Dredging and Marine Construction,* vol. 8, no. 13, pp. 44–52, Dec. 1972.

17. Mehta, A. J., and Partheniades, E., *Depositional Behavior of Cohesive Sediments,* technical report no. 16, Department of Coastal and Oceanographic Engineering, University of Florida, Gainesville, FL, Mar. 1973.

18. VanBaardewijk, A. P. H., "The Influence of the Conditions of Soil on the Dredging Output," *Proc. World Dredging Conference, WODCON II,* Rotterdam, The Netherlands, pp. 465–485, 1968.

19. DiGeorge, F. P., and Herbich, J. B., *Laboratory Determination of Bulking Factors for*

Texas Coastal Fine-Grained Materials, Center for Dredging Studies report no. 218, Texas A&M University, Aug. 1978.

20. DiGeorge, F. P. III, Herbich, J. B., and Dunlap, W. A., "Bulking Factors for Texas Coastal Fine-Grained Sediments," *9th World Dredging Conference, WODCON IX,* Vancouver, B.C., Oct. 1980.

21. Bartos, M. J., *Classification and Engineering Properties of Dredged Material,* U.S. Army Engineers Waterways Experiment Station, Vicksburg, MS, technical report D-77-18, Sept. 1977.

22. Anonymous, "Cutting Sand Underwater. A Theoretical Model," *Ports and Dredging,* no. 121, 1985.

23. Van Leussen, W., and Nieuwenhuis, J. D., "Soil Mechanics Aspects of Dredging," *Geotechnique,* vol. 34, no. 3, pp. 359–381, 1984.

24. Lobanov, V. A., and Joanknecht, L. W. F., "The Cutting of Soil Under Hydrostatic Pressure," *Proc. World Dredging Conference, WODCON IX,* pp. 327–339, 1980.

25. Joanknecht, L. W. F., and Lobanov, V. A., "Linear Cutting Tests in Clay," *3d International Symposium on Dredging Technology,* Bordeaux, France, paper E-2, pp. 315–332, Mar. 1980.

CHAPTER 6
PIPELINE TRANSPORT OF SOLIDS

Transportation of solids by pipeline is not a new system. Records indicate that jet pumps were used to transport gold-bearing gravel in California in the late 1850s. Many proposals demonstrating the economic feasibility of pipelines compared with other forms of surface transportation have been prepared, but considerations other than the cost per ton probably have made the systems unattractive since only a few pipelines for transporting solids have been built in this country. However, many such installations exist in England, Poland, France, Russia, and Australia.

Table 6.1 presents a summary of industrial installations of pipeline transportation abroad and in this country.[1]

Existing pipelines in the United States include a 6-in pipeline transporting limestone slurry across the Snake River in Idaho, completed in 1953; a 12-in line transporting 235 tons per day (tpd) of anthracite silt from the Susquehanna River to a preparation plant 275 ft above the river, completed in 1953; and a line at Bonanza, Utah, handling 50 tpd of gilsonite pumping 72 miles from a mine across a mountain range to a refinery in Colorado.[2] Abroad, Anaconda Company has a 14-mile-long, 6-in copper concentrate gravity pipeline in Chile transporting about 800 tpd since January 1959. There are plans to extend the line 60 miles to the coast. It is expected that the projected 6-in line will be able to transport about 2000 tpd at concentrations of 40 to 60 percent solids entirely by gravity.

Other projects include an iron ore slurry line in northwest Tasmania.[3] A 9-in pipeline extends 53 miles and provides a vital link between the mine concentrating facility and the pelletizing plant. It takes about 14 hr for the slurry, having a solids content of about 60 percent, to travel this distance. Total conveyance is estimated at 2,000,000 tons of iron ore concentrates per year.[3] The ore deposits are located at an elevation of 1200 ft, and the plant is at sea level. The pumping station is located at the mine, and operating pressures are 2500 psi at entrance and 1000 psi at outlet. Pipe wall thickness varies from ½ in at inlet to ¼ in at outlet.

PLACER MINING BY DREDGING METHODS

Placer mining using dredging equipment is the most commonly used method to recover sand and gravel near a construction site. A large source of water is re-

TABLE 6.1 Industrial Installations of Pipeline Transportation

Location	Material transported	Particle diameter	Pipe diameter (in)	Length of pipe (miles)	Volume transported
Poland	Lump coal	Up to 2 in	10	1.26	Not specified
Germany	Coal	Not specified	16	Not specified	3,000,000 tons per yr
France	Coal	0.1 in	15	5.7	6000 tons per day, 15% solid
Russia	Coal	Not specified	12	Not specified	4800 tons per day, 38% solid
Bonanza, Utah, to Grand Junction, Colo.	Gilsonite	Not specified	6	72	850 tons per day, 49% solid
Cadiz, Ohio, to East-lake, Ohio	Coal	0.047 in	10	108	3200 tons per day, 42% solid
United States	Iron ore tailing	−100 mesh	10	1.65	Not specified
United States	Fly ash	−325 mesh	10	1.5	Not specified
United States	Gold slime	−200 mesh	9	Not specified	Not specified

Source: From Zandi and Govatos, 1967.

quired. The pumping operation performs much of the washing requirement. This system of using a dredge for open-cast mining is also used when the terrain precludes the use of overland machinery, for example, on rough, rugged, or marshy land. Solids pumping can be efficiently used whenever grinding of the ore to a fine mesh is part of the refining process.

In Sierra Leone, West Africa, rutile (a raw material from which titanium dioxide is extracted) is mined by a 20-in suction dredge. A reservoir supplies the water for the dredge and during the dry season it is supplemented by supplies from the Jong River, 6 miles away. Mineral sands in Australia are mined in a similar fashion along the East coast between Broken Bay and Wide Bay.

One example, which required a dredge in an earth-moving project, occurred on the Cumberland River above Carthage, Tennessee, during construction of the Cordell Hull lock and dam. The first stage of excavation was accomplished by conventional earth-moving equipment; however, because of the proximity of a canal, the earth was firm to a depth of only about 4 or 5 ft. The disposal area was over a large hill from the excavation site, requiring a long, steep haul. The excavation problem was solved by importing a portable 14-in cutterhead dredge which pumped at an average rate of 350 yd^3/hr.

Characteristics of the Transported Material

The characteristics of transport materials are a function of the geometric, kinematic, physical, and chemical properties of the solids.

Transport = f (distance x, geometric characteristics e, kinematic characteristics k, physical characteristics l, chemical characteristics m, and time t)

$$e = \text{geometric characteristics} = f(d_g, s, V_t, d) \qquad (6.1)$$

where d_g = grain diameter
s = shape of particles
V_t = terminal settling velocity
d = pipe diameter

$$k = \text{kinematic characteristics} = f(V, \Delta p) \qquad (6.2)$$

where V = flow velocity
Δp = pressure change between two points along the pipe a distance (x) apart

$$l = \text{physical characteristics} = f(\rho_s, \rho_w, \text{physical strength } s', v) \qquad (6.3)$$

where ρ_s = particle density
ρ_w = water density
s' = physical strength
v = kinematic viscosity (may not be a constant)
$= f(\tau, du/dy)$

where τ is a shear stress.

$$n = \text{chemical characteristics} = f(\text{chemical variables}) \qquad (6.4)$$

The hydraulic transportation of mixtures is largely dependent upon the combination of inertia and resistance forces. It is not possible to define the solid particle by a single characteristic dimension such as a diameter since the solids have different shapes. Wiedenroth[4] suggested that for irregularly shaped particles their volume should be determined, and spherical particle diameters of equivalent volume should be used in the calculations.

Properties of Sediment. Sediments may be divided into these categories:

1. Cohesive such as silt and clays—median diameter $(d_m) < 0.0625$ mm
2. Noncohesive such as sand, gravel, cobbles, etc.—median diameter $(d_m) > 0.0625$ mm

Noncohesive sediments generally consist of discrete particles, and consequently their movement in pipe flow depends almost exclusively on the physical properties of particles, such as size, shape, and density. The movement of cohesive sediments depends on many other factors associated with Brownian motion, electrostatic force effects, flocculation, etc. In addition, eroded cohesive sediments may or may not move cohesively in pipe flow.

In the United States and overseas there are many classifications of sediment with regard to its size. Recently the Permanent International Association of Navigation Congresses recommended a unified, international classification of soils.

TABLE 6.2 PIANC Classification of Soils

Main soil type[1]	Particle size identification		Identification	Strength and structural characteristics
	Range of size	B.S. Sieve[2]		
		Granular (noncohesive)		
Boulders Cobbles	Larger than 200 mm Between 200–60 mm	6	Visual examination and measurement.	Not applicable
Gravels	Coarse 60–20 mm Medium 20–6 mm Fine 6–2 mm	3–¾ in ¾–¼ in ¼–no. 7	Easily identifiable by visual examination.	Possible to find cemented beds of gravel which resemble weak conglomerate rock. Hard-packed gravels may exist intermixed with sand.
Sands[3]	Coarse 2–0.6 mm Medium 0.6–0.2 mm Fine 0.2–0.06 mm	7–25 25–72 72–200	All particles visible to the naked eye. Very little cohesion when dry.	Deposits will vary in strength (packing between loose, compact, and cemented). Structure may be homogeneous or stratified. Intermixture with silt or clay may produce hardpacked sands.
		Cohesive		
Silts[3]	Coarse 0.06–0.02 mm Medium 0.02–0.006 mm Fine 0.006–0.002 mm	Passing no. 200	Generally particles are invisible and only grains of a coarse silt may just be seen with naked eye. Best determination is to test for dilatency.[4] Material may have some plasticity, but silt can easily be dusted off fingers after drying and dry lumps powdered by finger pressure.	Essentially nonplastic but characteristics may be similar to sands if predominantly coarse or sandy in nature. If finer will approximate to clay with plastic character. Very often intermixed or interleaved with fine sands or clays. May be homogeneous or stratified. The consistency may vary from fluid silt through stiff silt onto "siltstone."
Clays	Below 0.002 mm. Distinction between silt and clay should not be based on particle size alone since the more important physical properties of silt and clay are only related indirectly to particle size.	Not applicable	Clay exhibits strong cohesion and plasticity, without dilatency. Moist sample sticks to fingers and has a smooth, greasy touch. Dry lumps do not powder; shrinking and cracking during drying process with high dry strength.	Strength — Shear strength[5] V. Soft: May be squeezed easily between fingers. — Less 0.17 kg/cm² Soft: Easily molded by fingers. — 0.17–0.45 kg/cm² Firm: Requires strong pressure to mold by fingers. — 0.45–0.90 kg/cm² Stiff: Cannot be molded by fingers, indented by thumb. — 0.90–1.34 kg/cm² Hard: Tough, indented with difficulty by thumb nail. — Above 1.34 kg/cm² Structure may be fissured, intact, homogeneous, stratified, or weathered.
		Organic		
Peats and organic soils	Not applicable	Not applicable	Generally identified by black or brown color, often with strong organic smell; presence of fibrous or woody material.	May be firm or spongy in nature. Strength may vary considerably in horizontal and vertical directions.

TABLE 6.2 PIANC Classification of Soils (*Continued*)

[1]Soil may be defined in the engineering sense as any naturally occurring loose or soft deposit forming part of the earth's crust. The term should not be confused with "pedological soil" which includes only the topsoil capable of supporting plant growth, as considered in agriculture.

[2]Or national equivalent sieve size or numbers.

[3]There may be some justification for including a range of "extra fine" sand and "extra coarse" silt over the particle size ranges (0.1–0.06 mm) and (0.06–0.04 mm), respectively. It is recommended that whenever possible in borehole description or verbal discussion such further identification of these soils be used. However, to avoid the chance of confusion, if the classification "fine" sand or "coarse" silt is used without further qualification, it will be taken that the particle size ranges fall within those given in Table 6.1.

[4]Dilatency is the property exhibited by silt as a reaction to shaking due to the higher permeability of silt. If a moistened sample is placed in the open hand and shaken, water will appear on the surface of the sample giving a glossy appearance. A plastic clay gives no reaction.

[5]Defined as the undrained (or immediate) shear strength ascertained by the applicable in situ or laboratory test procedure.

[6]Although only visual examination and measurement are possible, an indication should be given with respect to the size of the "grain" as well as to the percentages of the different sizes.

Literature

BSI: "Site Investigations," *BS Code of Practice CP2001,* British Standards Institution, London, 1957.

BSI: *Methods of Using Testing Soils for Civil Engineering Purposes,* BS 1377, British Standards Institution, London, 1961.

FIDC-FIBTP: *Conditions of Contract (International) for Works of Civil Engineering Construction,* 1st ed., Aug. 1957.

Sargent, J. H.: "Investigations for Dredging Projects," Terra et Aqua 3/4 (1973), pp. 10–15.

Source: From Oosterbaan, 1973.

Since dredging in many cases is an international operation, it would be to everybody's advantage to use this classification, which is reproduced from a paper by Oosterbaan[5] (Table 6.2). The success of many dredging operations depends to a large extent on the proper prior estimates of dredging and pumping efficiency, which in turn is affected by the nature of the soil. Soil samples should be secured from the area to be dredged to the depth required and analyzed, and the dredging estimates should be made only after this information is on hand. Table 6.2 also describes a visual or manual classification of both cohesive and noncohesive sands in the field.

It appears that the terminal velocity parameter is ideally suited for determination of particle resistance to flow. Quick measurements of fall velocity may be obtained in a visual accumulation tube shown in Fig. 6.1*a*. Figure 6.1*b* shows the recording drum of the accumulation tube device. The results of many terminal velocity measurements are summarized in Fig. 6.2. The terminal velocity is obtained by equating buoyant and drag forces with weight forces or

$$\frac{\pi}{6} d_g^3 \, \rho_w g + C_D \rho_1 \frac{V_t^2}{2} \frac{\pi}{4} d_g^2 = \frac{\pi}{6} d_g^3 \rho_s g \tag{6.5}$$

or

$$C_D = \frac{4}{3} \left(\frac{\rho_s - \rho_w}{\rho_w} \right) \frac{g d_g}{V_t^2} \tag{6.6}$$

Small Particles. For small particles ($d_{50} < 0.15$ mm) drag forces are due mostly to viscosity, and the conditions are laminar. Under laminar conditions

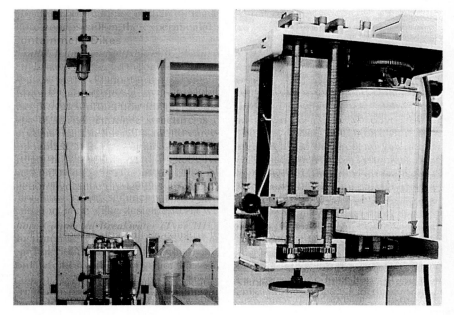

FIGURE 6.1   (*a*) Visual accumulation tube; (*b*) recording drum of visual accumulation tube device.

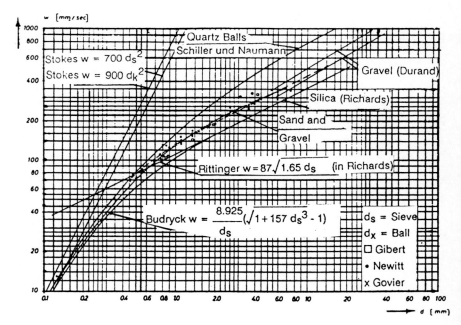

FIGURE 6.2   Terminal velocity. (*Courtesy, World Dredging Conference*)

$$C_D = \frac{24}{N_R} = \frac{24\mu}{V_t d_g \, \rho_w} \qquad (6.7)$$

or

$$V_t = \frac{g d_g^2}{18v}\left(\frac{\rho_s - \rho_w}{\rho_w}\right) \qquad (6.8)$$

Equation (6.8) is known as Stokes' Law and it applies only as long as the viscosity is the principal variable affecting the forces acting on a particle settling in a fluid. The equation is valid for $N_R < 0.1$ and may be used for $N_R < 1.0$ as an approximation.

The radius of particles for which the Stokes' Law does not apply[6] may be computed from

$$r = \sqrt[3]{\frac{9}{2}\frac{v^2}{(\rho_s - \rho_w)/\rho_w}} \qquad (6.9)$$

Rubey[7] points out that for a quartz sphere in water (at a temperature of 16°C) Stokes' Law does not apply when $\tau > 0.007$ cm. Studies by Richards[8] have shown that for quartz particles greater than 0.2 mm in diameter the settling velocity cannot be estimated even approximately by Eq. (6.9). For particles greater than 1.55 mm in diameter the settling velocity varies as the square root rather than the square of the diameters.

Intermediate Size Particles. Richards[8] conducted tests with quartz particles (S.G. = 2.65) under laminar, transitional, and turbulent conditions and derived an expression for terminal velocity

$$V_t = \frac{8.925}{d_g[\sqrt{1 + 95\,(2.65 - 1)d_g^3} - 1]} \qquad (6.10)$$

where V_t is given in mm/s and d_g is in mm.

The range of application of this equation is for particles between 0.15 and 1.5 mm and Reynolds numbers of between 10 and 1000.

Large Particles. For particles larger than 1.5 mm the terminal velocity may be given as

$$V_t = c\sqrt{d[(\rho_s - \rho_w)/\rho_w]} \qquad (6.11)$$

where c = experimental constant.

Worster and Denny[9] published the data shown in Tables 6.3 and 6.4 on terminal velocities for coal and gravel particles.

Drag coefficients as a function of Reynolds number are given in standard texts on fluid mechanics and are reproduced here as Fig. 6.3. Drag coefficients as a function of Reynolds number for selected ocean nodules recovered from the Pacific Ocean floor were determined by Herbich[6] and are shown in Fig. 6.4.

TABLE 6.3 Terminal Velocities for Gravel of Specific Gravity 2.67

Mesh size (in)	Single particle		30% concentration	
	ft/s	cm/s	ft/s	cm/s
1/16	0.3	9.1	0.2	6.1
1/4	1.0	30.5	0.7	21.3
1/2	2.0	61.0	1.4	42.7
1	3.5	106.7	2.4	73.1

Source: Worster and Denny, 1955.

TABLE 6.4 Terminal Velocities for Coal of Specific Gravity 1.5

Mesh size (in)	Single particle		30% concentration	
	ft/s	cm/s	ft/s	cm/s
1/16	0.15	4.6	0.10	3.0
1/4	0.50	15.2	0.35	10.7
1/2	1.00	30.5	0.70	21.3
1	1.70	51.8	1.20	36.6

Source: Worster and Denny, 1955.

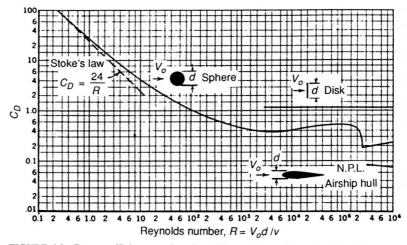

FIGURE 6.3 Drag coefficient as a function of Reynolds number—after Prandtl.

Composition of Solid-Water Mixtures

The composition of mixture is usually given as the ratio of solids to the total amount of the mixture either by volume or by weight. It is also convenient to use the specific gravity of mixtures particularly when discussing dredge pump characteristics.

Concentration of solids by volume (C_v) is equal to the ratio of volume of solids and volume of mixture, or

$$C_v = \frac{\text{S.G.}_m - \text{S.G.}_f}{\text{S.G.}_s - \text{S.G.}_f} \tag{6.12}$$

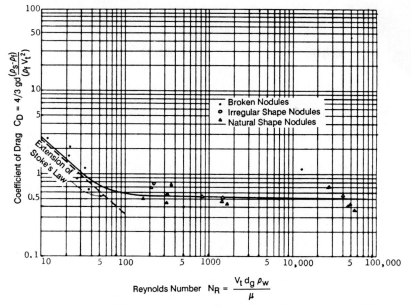

FIGURE 6.4   Drag coefficient as a function of Reynolds number for ocean nodules.

where S.G.$_m$ = specific gravity of mixture
S.G.$_f$ = specific gravity of water
S.G.$_s$ = specific gravity of solids

Concentration of solids by weight (C_w) is equal to the ratio of weight of solids and weight of mixture, or

$$C_w = \frac{\text{S.G.}_s(\text{S.G.}_m - \text{S.G.}_f)}{\text{S.G.}_m(\text{S.G.}_s - \text{S.G.}_f)} \qquad (6.13)$$

Specific gravity for water (S.G.$_w$) is equal to 1 and for sea water the specific gravity (S.G.$_{sw}$) may be taken as 1.025; thus, the concentration of solids by volume in sea water is

$$C_v = \frac{\text{S.G.}_m - 1.025}{\text{S.G.}_s - 1.025} \qquad (6.14)$$

and by weight

$$C_w = \frac{\text{S.G.}_s(\text{S.G.}_m - 1.025)}{\text{S.G.}_m(\text{S.G.}_s - 1.025)} \qquad (6.15)$$

or

$$C_w = \frac{\text{S.G.}_s}{\text{S.G.}_m} C_v$$

or

$$S.G._m = S.G._s \frac{C_v}{C_w}$$

or

$$S.G._m = \frac{1 - C_v}{1 - C_w} S.G._f \qquad (6.16)$$

Figure 6.5 presents specific gravity of mixture as a function of concentration of solids by volume and by weight for clay, sand, coal, and flyash.

Regimes of Sediment Flow

There are four regimes of solid-water mixtures that flow in a pipeline for a given mixture composition and pipe diameter.[10] These are qualitatively shown in Fig. 6.6. The flow regimes are (1) as a homogeneous suspension, (2) as a heterogeneous flow with all solids in suspension, (3) as a moving bed, saltation (with or without suspension), and (4) as a flow with a stationary bed.

These regimes overlap and no distinct boundaries between them exist. The lower regime in Fig. 6.6 represents the type of flow in which the particles are so small (and consequently their vertical fall velocities are very small) that vertical distribution is almost uniform. The fall velocities of particles are insignificant when compared with vertical motion of fluid.

In the heterogeneous flow regime the particles are also in suspension, but the particle vertical distribution is not uniform since the concentration of particles is greater at the bottom than near the top of the pipe. The most economical trans-

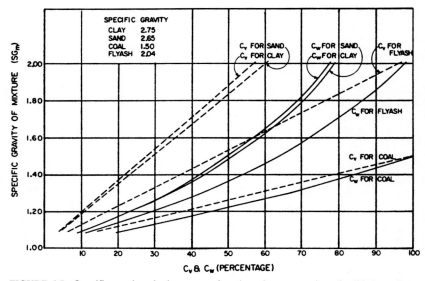

FIGURE 6.5 Specific gravity of mixture as a function of concentration of solids by volume and by weight.

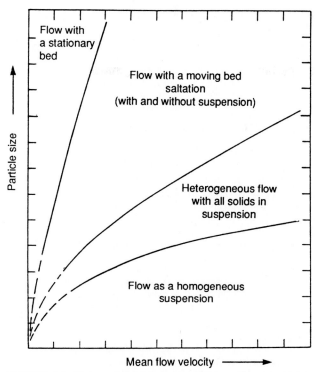

FIGURE 6.6 Flow regimes for a given fluid, sediment and pipe size—qualitative only. (*From Shen, 1970*)

portation of sediment in pipes is in this regime. The weight of material transported per unit of power required is at a maximum.

Flow with a moving bed regime (ripples and dunes) will form at the bottom boundary and the mixture above the moving bed will travel at a substantially higher velocity. This causes additional head losses and may produce an uneconomical operation.

In the flow with a stationary bed, regime transport of material will only occur above the stationary bed which will form a new bottom boundary.

Durand and Condolios[11] suggested the following classification of particles with reference to the flow regime:

1. Homogeneous suspension—particles smaller than 40 μ in size

2. Suspension maintained by turbulence—particle size between 40 μ and 0.15 mm

3. Suspension and saltation—particle size between 0.15 and 1.5 mm

4. Saltation—particle greater than 1.5 mm in size

The above classifications refer to particles having specific gravity equal to 2.65 and which are subject to forces of sufficient magnitude to produce movement in water.

Homogeneous Two-Phase Flow.   Basically there are a number of forces acting on a particle transported by fluid:

1. Weight or a downward vertical force
2. Buoyancy or an upward vertical force
3. Force caused by the current in the horizontal direction of motion
4. Resisting drag force in the horizontal direction

   In order to maintain the particle in suspension in turbulent flow and prevent the downward movement of sediment due to gravity, there must be an exchange of fluid containing greater sediment concentration at a lower level with the fluid at an upper level, which has a lesser sediment concentration. This exchange is usually referred to as turbulent mixing. Generally, in dredging, there will be a vertical sediment concentration gradient in the pipe since a range of particles from very small to fairly large will be encountered in most operations. The homogeneous flow will be observed with small particles for which the fall velocities are insignificant. Figure 6.7a shows the uniform distribution of sediment in a homogeneous flow. Examples of homogeneous flow include pumping of clays, drilling muds, thorium oxide, and other very fine particles.

Studies on Flow of Solid-Liquid Mixtures.   Literature survey indicates that extensive studies were conducted by many investigators. The more important are those of O'Brien and Folsom,[12] Durand,[13] Condolios and Chapus,[14] Newitt et al.,[15] Zandi and Govatos,[16] and Bonnington.[17] A good review of theory and/or applications is contained in books by Stepanoff,[18] Bain and Bonnington,[19] and Graf,[20] and in articles by Shen and the ASCE Committee on Sedimentation,[10] Babcock,[21] and Colorado School of Mines.[22]

   O'Brien and Folsom[12] studied flow of sand in small pipes 2 to 3 in in diameter. The median sizes of sand used were between 0.17 mm and 1.7 mm, or 0.17 mm $< d_{50} <$ 1.7 mm. Their experiments covered sand-water mixtures up to solid concentrations of 26 percent by volume. They defined the "critical velocity" as the velocity at which the head loss in feet of mixture differs appreciably from head loss (in feet of water) for flow of water at the corresponding flow conditions. At velocities greater than the critical velocity homogeneous flows may be expected. Durand[13] also concluded that for homogeneous flow the head losses for sand-water mixtures are similar to head losses for water, provided the head loss is expressed in feet of mixture.

   Newitt, et al.[15] made the same conclusions as O'Brien and Folsom[12] and Durand[13] and suggested that

$$\frac{i_m - i}{C_v i} = \frac{\rho_s}{\rho_w} - 1 \tag{6.17}$$

where i_m = gradient of piezometric head line for mixture
      i = gradient of piezometric head line for water
      C_v = sediment concentration by volume
      ρ_s = density of the sediment
      ρ_w = density of water

   However, as reported by Shen,[10] the numerical value of Eq. (6.17) is approximately 3, which is greater than $(\rho_s/\rho_w) - 1$ for sand with a mean particle size of

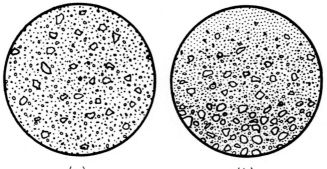

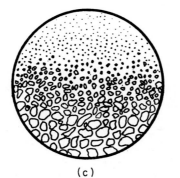

(a) (b)

(c)

FIGURE 6.7 Sediment distribution in a pipeline. (*a*) Homogeneous flow. Uniform distribution of sediment. Excessive power requirements. (*b*) Heterogeneous flow. Nonuniform distribution of sediment. Optimum power. (*c*) Flow with a moving bed. Segregated distribution of sediment. Bed moving at lower velocity than material above. Excessive power requirements.

0.02 mm. Similarly the left-hand side of Eq. (6.17) is less than $(\rho_s/\rho_w) - 1$ for sand with mean particle size of 0.096 mm for velocities greater than 5.5 ft/s.

Several investigators, including Homayounfar,[23] Zandi,[24] and Herbich and Vallentine,[25] have found that head losses for coal-water and silt-clay-water mixtures were lower than the head losses for water at the corresponding conditions. Shen suggests that the homogeneous flow with sediments slightly larger than the fine sediments required for truly homogeneous flow may be assumed to have approximately the same head losses as the homogeneous flows.

Heterogeneous Regime and Saltation Flow. The heterogeneous flow regime is most important from an economic point of view. In this regime the sediment is nonuniformly distributed, as shown in Fig. 6.7*b*. Head losses will be higher in flow with a moving bed, as depicted schematically in Fig. 6.7*c*.

Durand and Condolios[11] conducted extensive experimental studies on the transportation of solids in the heterogeneous regime and the equations presented are valid only for flow with no net deposition of solids at the bottom of the pipe. They defined the limiting deposit velocity as

$$V_c = F_L \left[2gd \left(\frac{\rho_s}{\rho_w} - 1 \right) \right]^{1/2} \tag{6.18}$$

where F_L = constant = $f(C,d_g)$.

F_L is given in Fig. 6.8a for the range of concentration of solids by volume (C_v) as a function of grain diameter. It will be noted that for particles having a grain size greater than 2 mm the concentration has no effect, and F_L is approximately equal to 1.34.

Limiting deposit velocities for sand (S.G. = 2.62) and coal (S.G. = 1.5) are shown in Table 6.5.

Durand and Condolios[11] presented an equation for flow based on test results as follows:

$$\frac{i_m - i}{C_v i} = 81 \left[\frac{gd \ (\rho_s/\rho_w - 1)}{V^2} \left(\frac{1}{\sqrt{C_D}} \right) \right]^{1.5} \tag{6.19}$$

Newitt, et al.[14] conducted an experiment in a smaller size pipe (1 in) with several sediments ranging from plastics to gravel and manganese dioxide. They suggested Eq. (6.20) for heterogeneous flow:

$$\frac{i_m - i}{C_v i} = 1100 \left(\frac{\rho_s}{\rho_w} - 1 \right) \frac{gd}{V^2} \frac{V_t}{V} \tag{6.20}$$

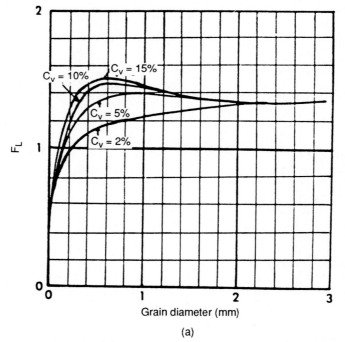

FIGURE 6.8(a) Variation of parameter F_L as a function of grain diameter. (*From Durand, 1953*)

TABLE 6.5 Limiting Deposit Velocity for Particles Greater than 1 mm in Diameter

| | Limiting Deposit Velocity | |
| | Sand | Coal |
Pipe diameter d (mm)	(m/s)	(m/s)
150	2.95	1.65
250	3.75	2.15
440	5.00	2.85
900	7.15	4.00

Source: Durand, 1953.

where V_t = particle settling velocity.

For saltation flow (for particle $d_{50} > 0.001$ in)

$$\frac{i_m - i}{C_v i} = 66 \left(\frac{\rho_s}{\rho_w} - 1 \right) \frac{gd}{V^2} \qquad (6.21)$$

The transition zone between homogeneous and heterogeneous regimes is given as

$$V_{th} = (1800 \, gdV_t)^{1/3} \qquad (6.22)$$

where V_{th} = transition velocity from homogeneous to heterogeneous regime.

The transition zone between heterogeneous and saltation region is given as

$$V_{ts} = 17 \, V_t \qquad (6.23)$$

where V_{ts} = transition velocity from heterogeneous to saltation regime.

Worster[9] presented the flow equations based on work by Turtle:

$$\frac{C_c - C_c'}{g} = 4 \left(\frac{V_i}{V} \right)^{0.173} \frac{C_c}{\sqrt{g}} \left[\frac{C_v d \, g(\rho_s/\rho_w - 1)}{V^2} \right]^{0.413} - 4 \qquad (6.24)$$

where C_c = Chezy friction factor for water
C_c' = Chezy friction factor for mixture

which applies when

$$\frac{C_c - C_c'}{1/\sqrt{g}} < 13$$

Wilson[26] proposed an equation for friction head loss in feet of slurry based on data obtained on pumping molybdenum tailings in Colorado.

$$h_f' = L \left(\frac{fV^2}{2gd} + C_1 \frac{C_w V_t}{V} \right) \qquad (6.25)$$

where f = Darcy-Weisbach friction factor
C_1 = constant

The velocity (in ft/s) which produces a minimum friction loss for given conditions is

$$(V_{hf})_{min} = C_2 \sqrt[3]{(C_w V_t g d)/f} \qquad (6.26)$$

where C_2 = constant.

Wilson also presented Eq. (6.27), which gives a guide as to whether particles are suspended in flowing water or whether they will settle to form a bed at the bottom of the pipe.

$$C_3 = \frac{V_t}{\sqrt{(h_f g d)/4L}} \qquad (6.27)$$

when $C_3 > 1$, most of the particles with terminal velocity (V_t) will stay in suspension and when $C_3 \leq 1$, most of the particles with terminal velocity (V_t) will settle out.

Zandi and Govatos[16] in an effort to separate experimental data for the heterogeneous regime from the saltation flow developed an index number (N_I)

$$N_I = \frac{V^2 \sqrt{C_d}}{C_v d g \left[(\rho_s - \rho_w)/\rho_w\right]} \qquad (6.28)$$

The critical value of N_I indicates the separation of the two flow regimes, that is,

$$(N_I)_{critical} = 40 \qquad (6.29)$$

The saltation flow occurs for $N_I < 40$ and the heterogeneous regime for $N_I > 40$.

Babcock[21] using Blatch's data[27] concluded that N_I should be equal to 10 for separation of the heterogeneous and moving bed regimes. Additional experimental verifications are required.

Modified Durand Equation (by Zandi).   Zandi analyzed Durand's data which were selected on the basis of the index number (N_I); only those data points which were in the heterogeneous regime were selected for analysis. Zandi points out that

$$N_I = \frac{1}{C_v} \left[\frac{V^2 \sqrt{C_d}}{gd\,(S.G._s - S.G._w)/S.G._w} \right] = \frac{\psi}{C_v} \qquad (6.30)$$

and

$$\phi = \frac{J - J_w}{C_v J_w} = K\,(\psi)^m \qquad (6.31)$$

where both K and m are coefficients shown in Table 6.6.

Durand's reanalyzed data are plotted in Fig. 6.8b as ϕ versus ψ. Note that better correlation of all values of ψ is achieved if ψ is divided into two separate ranges, for $\psi > 10$ and for $\psi < 10$ as indicated in Table 6.6.

Blockage of Pipe.   In some cases it may be desirable to design for solids transport in the moving bed regime or such flows may be the result of insufficient power.

TABLE 6.6 Values of Coefficient K and m in Eq. (6.31)

Range of ψ	K	m
$10 < \psi$	6.3	-0.354
$\psi < 10$	280.0	-1.93

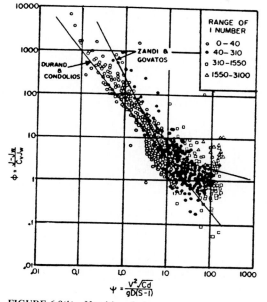

FIGURE 6.8(b) Head loss in heterogeneous flow. (*After Zandi and Govatos, 1966*)

Many designs call for no settlement of sediment (heterogeneous regime), but under certain conditions it may be economically desirable to permit some settlement resulting in partial blockage of the cross-sectional pipe area. The partial blockage occurs when the rate of sediment supply to the pipe exceeds the transporting capacity of the water. The sediment will deposit at the bottom of the pipe until an equilibrium condition is reached when the cross-sectional area is sufficiently reduced to provide sufficient transporting capacity.

Craven[28] conducted studies in 2- and 5.5-in diameter pipes, with three approximately uniform sands having median diameters of 0.25, 0.58, and 1.62 mm, respectively, to determine blockage characteristics. He determined that for relatively high values of relative transport rate (Q_s/Q) Darcy's hydraulic gradient (i) was proportional to the two-thirds power of Q_s/Q, as shown in Eq. (6.32):

$$i = \frac{dh}{dx} = C_3 \left(\frac{Q_s}{Q}\right)^{2/3} \tag{6.32}$$

where Q_s = absolute rate of sediment transport
$\quad C_3$ = constant
$\qquad$ = $1/1.65[(\gamma_s - \gamma_w)/\gamma_w]$ for 0.58- and 1.62-mm sands
$\qquad$ = $0.6/1.65[(\gamma_s - \gamma_w)/\gamma_w]$ for 0.25-mm sand

Vallentine[29] conducted studies with nonuniform sands of median diameters 0.53 mm (no. 1) and 1.05 mm (no. 2) in 2- and 6-in pipes to determine the effect of nonuniformity of sands on pipe blockage.

Darcy's equation may be written as in Eq. (6.33):

$$i = f\frac{l}{4R}\left(\frac{V^2}{2g}\right) = f\frac{l}{8Rg}\frac{Q^2}{A^2} = \frac{fQ^2}{8gd^5}\psi(B) \qquad (6.33)$$

where B = blockage
$\quad \psi(B)$ = function of blockage

Vallentine suggests that a good approximation to the blockage function is

$$B = \phi\left[\frac{Q}{d^{2.5}}\sqrt{\frac{\rho}{\gamma_s - \gamma_w}}\left(\frac{Q_s}{Q}\right)^{-1/3}\right] \qquad (6.34)$$

Figure 6.9 is a plot of blockage as a function of the right-hand side of Eq. (6.34) for the two nonuniform sands used by Vallentine and a uniform sand used by Craven. Vallentine also presents a blockage chart (Fig. 6.10) which may be used for design purposes.

Example 6.1. Estimate blockage, if any, for the following conditions:

$$Q = 4 \text{ ft}^3/\text{s}$$

$$d = 1.25 \text{ ft}$$

$$\frac{Q_s}{Q} = \text{bed load} = 1 \text{ percent}$$

Uniform sand: Enter Fig. 6.10 at $Q = 4$ ft^3/s, draw a horizontal line until a curve for $Q_s/Q = 1$ percent is reached; enter the diagram at $d = 1.25$ ft and draw a horizontal line until it intersects a vertical line from the point where the 4 ft^3/s and 1 percent lines meet. Read the value of the blockage, $B = 48$ percent.

Nonuniform sand: For nonuniform sand use Fig. 6.9. Forty-eight percent blockage for uniform sands corresponds to 36 percent blockage for nonuniform sands.

Head Losses. Head losses in a pipeline are one of the most important considerations from an economic point of view since power required is proportional to the head loss. Since the characteristics of flow in different regimes are different, it is important to determine the likely flow regime first so that appropriate head loss equations are used for estimating head losses. The relationships between head loss and mean velocity of mixture for different sediment concentrations are shown in Fig. 6.11. These typical curves are for a given fluid, sediment characteristics, and pipe size. The deposition regime is to the left of the dashed line at low flow velocities, and the nondeposition regime is to the right at higher velocities of flow. For water in fully turbulent flow the

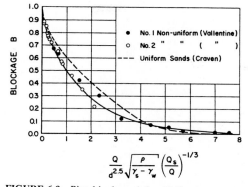

$$\frac{Q}{d^{2.5}}\sqrt{\frac{\rho}{\gamma_s - \gamma_w}}\left(\frac{Q_s}{Q}\right)^{-1/3}$$

FIGURE 6.9 Pipe blockage (*after Vallentine, 1955*).

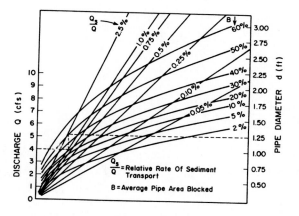

FIGURE 6.10 Blockage chart (*after Vallentine, 1955*).

head loss is proportional to the second power of velocity. Note that there is a minimum head loss for each concentration of mixture occurring at certain velocity. This velocity has been given various names by different researchers, such as limiting deposit velocity (Durand),[13] economical velocity (Blatch),[27] critical velocity, etc.

Clear Water. As discussed in Chap. 2, an energy equation may be written between two locations along the pipeline. Consider two locations, 1 and 2, spaced distance L apart:

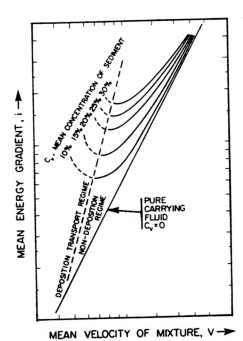

FIGURE 6.11 Typical head loss and sediment concentration curves for a given fluid, sediment, and pipe size (*after Shen, et al., 1970*).

$$\frac{p_1}{\gamma} + \alpha\left(\frac{V_1^2}{2g}\right) + Z_1 = \frac{p_2}{\gamma} + \alpha\left(\frac{V_2^2}{2g}\right) + Z_2 + h_l \qquad (6.35)$$

where α = kinetic energy coefficient
h_l = head loss expressed in foot pounds per pound or feet

The kinetic energy coefficient depends on the variation in velocity over the cross section of the pipe. The coefficient for laminar flow in a circular pipe is equal to 2, and in turbulent flow it varies from 1.01 to 1.15. Since most slurry flows are in turbulent range, α is usually neglected; however it should be pointed out that in some applications the flow may be in laminar range, and a proper value of the coefficient should be used:

Therefore, neglecting α

$$h_l = \left(\frac{p_1 - p_2}{\gamma}\right) + \left(\frac{V_1^2 - V_2^2}{2g}\right) + Z_1 - Z_2 \qquad (6.36)$$

In horizontal pipe, for a constant discharge Eq. (6.36) reduces to

$$h_l = \frac{p_1 - p_2}{\gamma} \qquad (6.37)$$

The equation relating the head loss with velocity, usually called the Darcy-Weisbach equation, indicates that the head loss varies almost directly with velocity head and length of pipe and inversely with pipe diameter, or

$$h_l = f\left(\frac{l}{d}\right)\frac{V^2}{2g} \tag{6.38}$$

where f = Darcy-Weisbach dimensionless friction factor
$= \psi\,(N_R,\ \epsilon/d)$

$$N_R = \text{Reynolds number} = (Vd\rho)/\mu \tag{6.39}$$

For laminar flow in a circular pipe

$$f = \frac{64}{N_R} \tag{6.40}$$

The laminar flow exists for N_R less than 2000 and turbulent flow exists for N_R between 2000 and 10,000. The flow for $2000 < N_R < 10,000$ is sometimes taken as transitional flow.

The relationship between friction factor N_R and the relative roughness ϵ/d has been presented by Moody[30] and is shown in Fig. 6.12. Several distinct regions are readily apparent in this diagram. For $N_R < 2000$ data for pipe roughness fall on the line defined by Eq. (6.40) (note that f is independent of ϵ/d in this region). In the transitional range the friction factor is dependent on both N_R and ϵ/d and changes rapidly with N_R. For large N_R's, in excess of 10^5 to 10^7 (depending on ϵ/d value), the friction factor is independent of N_R but highly dependent on relative roughness ϵ/d. The diagram may be used to find friction factors for water at different temperatures and for different pipe diameters.

Other losses include those in elbows, through valves, at entrances, etc. Such losses are usually expressed as a function of velocity head, that is,

$$h_l = K\frac{V^2}{2g} \tag{6.41}$$

Values of coefficient K are given in Table 6.7 and in Fig. 6.13.

Flow of Solids-Water Mixtures

Estimating head losses for solids-water mixtures is complicated by the fact that solids tend to settle out because of gravity, and the settling rate not only depends on geometric and physical properties of the mixture but also on the flow condition in a pipe. For very fine particles the brownian movement may become important, affecting the settling rates. The particles may also flocculate and change the settling characteristics. Slurries on dredging projects are often non-homogeneous and the density and viscosity of the mixture are difficult to evaluate and may change rapidly with time. Some silt-clay mixtures exhibit Bingham Body properties and their flow behavior is quite different from the newtonian flow. It is therefore not surprising that no simple theory and no clear-cut methods are available for estimating slurry head losses for engineering purposes.

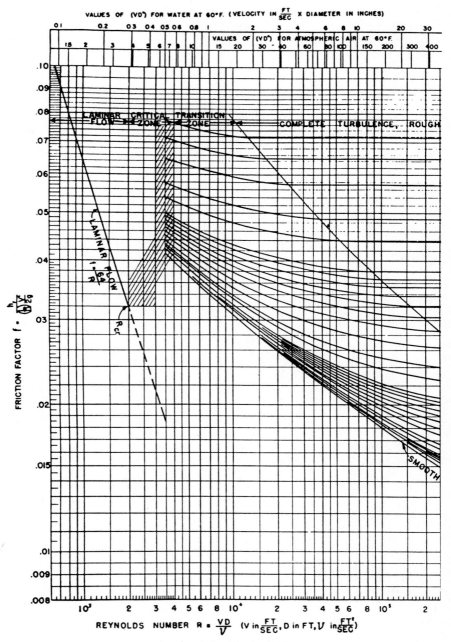

FIGURE 6.12   Friction factors for pipes (*after Moody, 1944*).

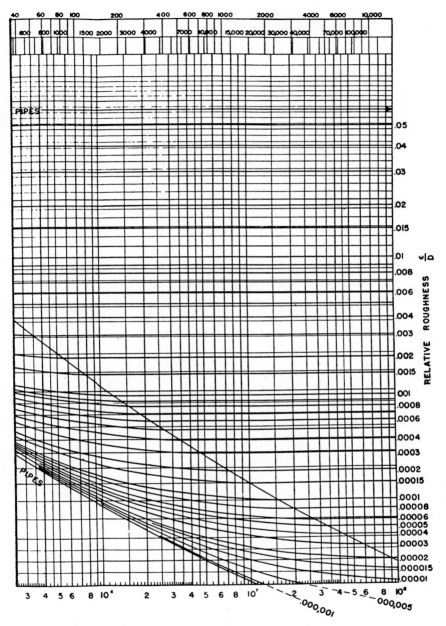

FIGURE 6.12 (*Continued*)

TABLE 6.7 Minor Losses*

Fitting	K	L/D
Globe valve, wide open	10.00	350
Angle valve, wide open	5.00	175
Close return bend	2.20	75
T, through side outlet	1.80	67
Short-radius elbow	0.90	32
Medium-radius elbow	0.75	27
Long-radius elbow	0.60	20
45° elbow	0.42	15
Gate valve, wide open	0.19	7

Flow of Fluids through Valves, Fittings, and Pipe, Crane Co., Tech. Paper 409, p. 20, May 1942. Values based on tests by Crane Co. and at the University of Wisconsin, the University of Texas, and A&M College of Texas.

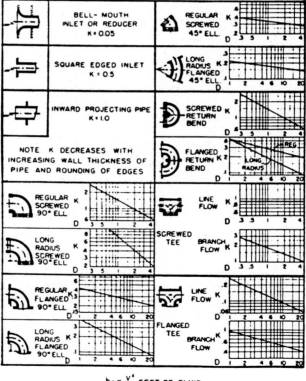

$$h = K \frac{v^2}{2g} \text{ FEET OF FLUID}$$

FIGURE 6.13 (*a*) Resistance coefficients for valves and fittings (*from Hydraulic Institute, 1961*).

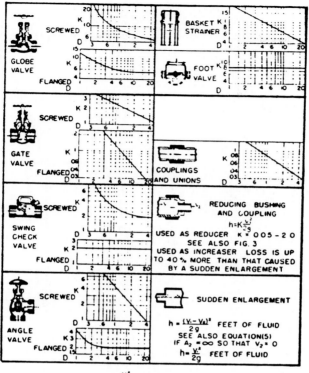

$$h = K \frac{V^2}{2g} \quad \text{FEET OF FLUID}$$

FIGURE 6.13 (*Continued*) (*b*) Resistance coefficients for values and fittings (*from Hydraulic Institute, 1961*).

Example 6.2. Determine frictional head losses in a 24-in diameter steel pipe over a distance of 2 miles.

Given   Water pumped at temperature 80°F
        Welded steel pipe $\epsilon = 0.03$
        Average pipe velocity = 15 ft/s

To find the solution, compute the following

$$\text{Reynolds number} = N_R = \frac{Vd}{v}$$

$$= \frac{15 \times 2}{0.93 \times 10^{-5}}$$

$$= 3.225 \times 10^6$$

$$\frac{\epsilon}{d} = \frac{0.03}{2} = 0.015$$

Friction factor (from Fig. 6.13) $f = 0.045$

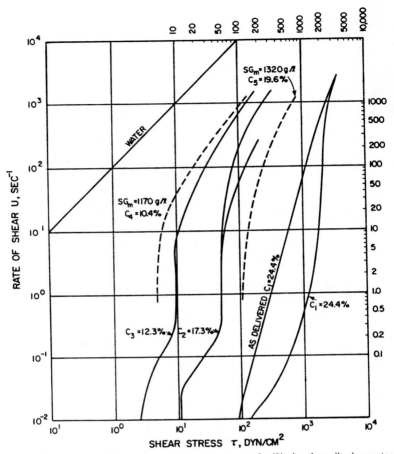

FIGURE 6.14 Shear stress as a function of shear rate for Weehawken silt-clay-water mixture (*from Fortino, 1959*).

Frictional head loss $h_f = f \dfrac{L}{d} \dfrac{V^2}{2g}$

$$= 0.045 \frac{2 \times 5280}{2(15^2/64.4)}$$

$$= 830.1 \text{ ft of water}$$

Example 6.3. What additional losses would occur if there were two 90° elbows and a fully open gate valve in the pipeline of Ex. 6.2?

k for 90° elbow = 0.9

$$h_l = k \frac{V^2}{2g} = 0.9 \frac{15^2}{64.4} = 3.14 \text{ ft of water}$$

k for a gate valve fully open $= 0.19$

$$h_l = k\frac{V^2}{2g} = 0.19\frac{15^2}{64.4} = 0.66 \text{ ft of water}$$

Total head loss   elbow $h_l = 3.14$ ft of water

elbow $h_l = 3.14$ ft of water

gate valve $h_l = 0.66$ ft of water

Total head loss $= 6.94$ ft of water

Example 6.4. Bingham Body Solids-Water Mixture.

Given   Suction pipe diameter 30 in

Discharge pipe diameter 28 in

Specific gravity of mixture $= 1.17$

Rate of shear-stress relationship for the silt-clay-water mixtures as shown in Fig. 6.14 (from Ref. 31)

Determine.   Reynolds numbers for the flow in suction and discharge pipes.

Solution: Suction pipe

Assume velocity in the suction line equal to 15 ft/s

Discharge $Q = (\pi/4)(30/12)^2\, 15 = 73.63 \text{ ft}^3/\text{s}$

The shear rate $= (40)/(\pi r^3) = [4(73.63)]/[\pi(15/12)^3] = 48.00 \text{ s}^{-1}$

The shear stress (from Fig. 6.14) corresponding to the shear rate of 48.0 s^{-1} is 10.1 dyn/cm^2. The absolute viscosity $\mu = 10.1/48.0 = 0.21$ poises. The kinematic viscosity $v = (0.21/1.170) = 0.179$ stokes. The Reynolds number in the suction line $N_R = (Vd/v) = [15(12)2.54(30)2.54]/(0.179) = 1.94 \times 10^5$

Discharge pipe   The discharge velocity $= (30/28)^2 15 = 17.22 \text{ ft/s}$

The shear rate in the discharge pipe $= (4Q)/(\pi r^3) = [4(73.63)]/[\pi(14/12)^3] = 59.03 \text{ s}^{-1}$

The shear stress (from Fig. 6.14) corresponding to the shear rate of 59.07 s^{-1} is 10.3 dyn/cm^2

The absolute viscosity $\mu = (10.3)/(59.03) = 0.174$ poises

The kinematic viscosity $v = (0.174)/(1.170) = 0.149$ stokes

Reynolds number in the discharge line, $N_R = [17.22(12)2.54(28)2.54]/0.149 = 2.50 \times 10^5$

Example 6.5. Compute the head loss per unit length of pipeline using any of the recognized equations for flow of solids-water mixtures in pipes. It is designed to deliver 60 lb of sand per second.

Given   S.G. sand $= 2.65$

Median diameter of sand $= 0.08$ in

Desired concentration $(C_w) = 40\%$ by weight

Terminal settling velocity (V_s) (estimated) $= 0.95$ ft/s

Drag coefficient (C_D) (estimated) = 0.4
Temperature = 70°F

Solution
Note: Select a 4-in diameter pipe

1. Determination of limit deposit velocity (Durand).

$$V_c = F_L \left[2gd \left(\frac{S.G._s - S.G._w}{S.G._w} \right) \right]^{1/2}$$

$$= 1.3 \, [64.4 \, (0.33) \, (1.65)]^{1/2}$$

$$= 7.698 \text{ ft/s}$$

2. Determination of transition velocity (Newitt).
   a. Transition between homogeneous and heterogeneous regime.

$$V_{th} = (1800 \, g \, d \, V_s)^{1/3}$$

$$= [1800 \, (32.17) \, (0.33) \, (0.95)]^{1/3}$$

$$= 26.28 \text{ ft/s}$$

   b. Transition between heterogeneous and saltation region.

$$V_{ts} = 17V_t = 16.15 \text{ ft/s}$$

3. Determination of transition velocity (Zandi). From heterogeneous to saltation regime.

N_I = 40 at transition velocity

$$N_I = \frac{V^2 \sqrt{C_d}}{C_v dg \, [(\rho_s - \rho_w)/\rho_w]}$$

$$V = \sqrt{\frac{N_I C_v dg \, [(\rho_s - \rho_w)/\rho_w]}{\sqrt{C_d}}}$$

$$= \sqrt{\frac{[40(0.201)(0.33)(32.17)(1.65)]}{\sqrt{0.4}}}$$

$$= 14.92 \text{ ft/s}$$

4. Determine unit pipe head losses.[11]

Solids transport rate = $60 \text{ lb/s} \left(\frac{1}{2.65} \right) \left(\frac{1}{62.4} \right) = 0.363 \text{ ft}^3/\text{s}$

$$C_w = \frac{2.65 \, (S.G._m - 1)}{S.G._m (2.65 - 1)} = 0.4$$

$$\therefore S.G._m = 1.33$$

$$C_v = \frac{S.G._m}{S.G._s} \, C_w = \frac{1.33}{2.65} \, 0.4 = 0.201 = 20.1\%$$

Volumetric flow rate (Q)

$$Q = \frac{0.363}{0.201} = 1.806 \ \text{ft}^3/\text{s}$$

Velocity (V)

$$V = \frac{Q}{A} = \frac{1.806}{(\pi/4)(0.33)^2} = 21.2 \ \text{ft/s}$$

Flow equation

$$\frac{i_m - i}{C_v i} = 81 \left[\frac{gd \ (\rho_s/\rho_w - 1)}{V^2} \left(\frac{1}{\sqrt{C_D}} \right) \right]^{1.5}$$

$$= 81 \left[\frac{32.17 \ (0.33) \ 1.65}{(21.1)^2} \left(\frac{1}{\sqrt{0.4}} \right) \right]^{1.5}$$

$$= 1.239$$

$$i_m = 0.201 \ (i) \ 1.239 + i$$

$$i_m = 1.249 \ i$$

Reynolds number (N_R)

$$N_R = \frac{Vd}{\nu} = \frac{21.1 \ (0.33)}{1.05 \times 10^{-5}} = 6.63 \times 10^5$$

Relative roughness (ϵ/d)
Assume steel pipe, $\epsilon/d = 0.00012$
$f = 0.0142$ (from Fig. 6.12)
Unit head loss (h_f)

$$h_f = f \frac{l}{d} \frac{V^2}{2g} = 0.0142 \ \frac{l}{0.33} \ \frac{(21.1)^2}{64.4} = 0.297 \ \text{ft/ft}$$

$$i_m = 1.249 \times 0.297 = 0.372 \ \text{ft/ft}$$

Piping

Steel pipe has generally been used on dredging projects in the past; in recent years high-density polyethylene pipe has been available on the market.

Steel Pipe. Steel pipe comes in various sizes; standard sizes for dredging applications are 6, 8, 10, 12, 14, 16, 18, 20, 24, 26, 30, 32, 34, 36, 40, 42, 44, 46, and 48 in. Tables 6.8 and 6.9 provide details of steel pipe including outside diameter (O.D.), wall thickness, weight per unit length, class, and schedule number.

TABLE 6.8 Steel Pipe Specifications, Diameter 2 to 18 In

Nominal Size	OD	Wall Thickness, Inches	Weight Lbs./Ft. Plain End	Nominal T&C Weight, Lbs./Ft. Standard Pipe Coupling	Class	Sched. No.
2	2.375	0.154	3.65	3.68	Std.	40
		0.218	5.02		XS	80
		0.344	7.46			160
		0.436	9.03		XXS	
2½	2.875	0.203	5.79	5.82	Std.	40
		0.276	7.66		XS	80
		0.375	10.01			160
		0.552	13.69		XXS	
3	3.500	0.216	7.58	7.62	Std.	40
		0.250	8.68			
		0.281	9.66		XS	
		0.300	10.25			80
		0.438	14.32			160
		0.600	18.58		XXS	
3½	4.000	0.226	9.11	9.20	Std.	40
		0.250	10.01			
		0.281	11.16		XS	80
		0.318	12.50			
		0.636	22.85			
4	4.500	0.156	7.24			
		0.172	7.95			
		0.188	8.66			
		0.203	9.32			
		0.219	10.01			
		0.237	10.79	10.89	Std.	40
		0.250	11.35			
		0.281	12.66			
		0.312	13.96			
		0.337	14.98		XS	80
		0.438	19.00			120
		0.531	22.51			160
		0.674	27.54		XXS	
5	5.563	0.156	9.01			
		0.188	10.79			
		0.219	12.50			
		0.258	14.62	14.81	Std.	40
		0.281	15.85			
		0.312	17.50			
		0.344	19.17			
		0.375	20.78		XS	80
		0.500	27.04			120
		0.625	32.96			160
		0.750	38.55			
		0.875	43.81			
		0.938	46.33		XXS	
6	6.625	0.156	10.78			
		0.172	11.85			
		0.188	12.92			
		0.203	13.92			
		0.219	14.98			
		0.250	17.02			
		0.280	18.97	19.18	Std.	40
		0.312	21.04			
		0.344	23.08			
		0.375	25.03		XS	80
		0.432	28.57			
		0.500	32.71			
		0.562	36.39			120
		0.625	40.05			
		0.719	45.35			160
		0.864	53.16		XXS	
		1.000	60.07			
		1.125	66.08			
		1.188	68.98			

Nominal Size	OD	Wall Thickness, Inches	Weight Lbs./Ft. Plain End	Class	Sched. No.
8	8.625	0.188	16.94		
		0.203	18.26		
		0.219	19.66		
		0.250	22.36		20
		0.277	24.70		30
		0.312	27.70		
		0.322	28.55	Std.	40
		0.344	30.42		
		0.375	33.04		
		0.406	35.64		60
		0.438	38.30		
		0.500	43.39	XS	80
		0.562	48.40		
		0.594	50.95		100
		0.625	53.40		
		0.719	60.71		120
		0.812	67.76		140
		0.875	72.42	XXS	
		0.906	74.69		160
		1.000	81.43		
		1.125	90.11		
		1.250	98.46		
10	10.750	0.188	21.21		
		0.203	22.87		
		0.219	24.63		
		0.250	28.04		20
		0.279	31.20		30
		0.307	34.24		
		0.344	38.23		
		0.365	40.48	Std.	40
		0.438	48.24		
		0.500	54.74	XS	60
		0.562	61.15		
		0.594	64.43		
10	10.750	0.625	67.58		
		0.719	77.03		100
		0.812	86.18		
		0.844	89.29		120
		1.000	104.13		140
		1.125	115.64		160
		1.438	143.01		
12	12.750	0.188	25.22		
		0.203	27.20		
		0.219	29.31		
		0.250	33.38		20
		0.281	37.42		
		0.312	41.45		
		0.330	43.77		30
		0.344	45.58		
		0.375	49.56	Std.	
		0.406	53.52		40
		0.438	57.59		
		0.500	65.42	XS	
		0.562	73.15		60
		0.625	80.93		
		0.688	88.63		80
		0.750	96.12		
		0.844	107.32		100
		1.000	125.49		120
		1.125	139.67		140
		1.312	160.27		
		1.500	180.23		160
		1.594	189.92		
		1.625	193.07		
		1.750	205.59		
		2.000	229.62		
		2.375	263.16		
		2.500	273.68		

Nominal Size	OD	Wall Thickness, Inches	Weight Lbs./Ft. Plain End	Class	Schedule No.
14	14.000	0.188	27.73		
		0.203	29.91		
		0.210	30.93		
		0.219	32.23		
		0.250	36.71		20
		0.281	41.17		
		0.312	45.61		
		0.344	50.17		
		0.375	54.57	Std.	30
		0.406	58.94		
		0.438	63.44		40
		0.459	67.78		
		0.500	72.09	XS	
		0.562	80.66		
		0.594	85.05		60
		0.625	89.28		
		0.688	97.81		
		0.750	106.13		80
		0.812	114.37		
		0.938	130.85		100
		1.094	150.79		120
		1.250	170.21		140
		1.406	189.11		
		1.500	200.25		160
		2.000	256.32		
16	16.00	0.188	31.75		
		0.203	34.25		
		0.219	36.91		
		0.250	42.05		
		0.281	47.17		
		0.312	52.27		20
		0.344	57.52		
		0.375	62.58	Std.	30
		0.406	67.62		
		0.438	72.80		
		0.469	77.79		
		0.500	82.77	XS	40
		0.562	92.66		
		0.625	102.63		
		0.656	107.50		60
		0.688	112.51		
		0.750	122.15		
		0.812	131.71		
		0.844	136.61		80
		1.031	164.82		100
		1.219	192.43		120
		1.438	223.64		140
		1.594	245.25		160
		1.618	248.52		
		2.000	299.04		
18	18.000	0.219	41.59		
		0.250	47.39		
		0.281	53.18		
		0.312	58.94		
		0.344	64.87		20
		0.375	70.59	Std.	
		0.406	76.29		30
		0.438	82.15		
		0.469	87.81		
		0.500	93.45	XS	
		0.562	104.67		40
		0.625	115.98		
		0.688	127.21		
		0.750	138.17		60
		0.812	149.06		
		0.938	170.92		80
		1.156	207.96		100
		1.375	244.14		120
		1.500	264.33		
		1.562	274.22		140
		1.652	288.43		

Source: Courtesy, Bartow Steel, Inc.

Polyethylene Pipe. Polyethylene pipe is made from a high-density, extra high molecular weight compound. Polyethylene pipe comes in sizes from 1½ to 36 in in diameter. Specifications for pressure rating from 255 to 50 psi are given in Table 6.10.

Pankow[32] reported on a laboratory study of polyethylene pipe in a dredging test loop at the U.S. Army Waterways Experiment Station. Tests were conducted with homogeneous and heterogeneous slurries and a field experiment performed with 20-ft-long sections of 30-in (I.D.) polyethylene pipe. The pipe handled 2.6 million cubic yards of dredged material over about a 6-month period.

TABLE 6.9 Steel Pipe Specifications, Diameter 20 to 48 In

Nominal Size	OD	Wall Thickness, Inches	Weight Lbs./Ft. Plain End	Class	Schedule No.
20	20.000	0.219	46.27		
		0.250	52.73		
		0.281	59.18		
		0.312	65.60		
		0.344	72.21		
		0.375	78.60	Std.	20
		0.406	84.96		
		0.438	91.51		
		0.469	97.83		
		0.500	104.13	XS	30
		0.562	116.67		
		0.594	123.11		40
		0.625	129.33		
		0.688	141.90		
		0.750	154.19		
		0.812	166.40		60
		1.031	208.87		80
		1.281	256.10		100
		1.375	273.51		
22	22.000	0.375	86.61	Std.	20
		0.406	93.63		
		0.438	100.86		
		0.469	107.85		
		0.500	114.81	XS	30
		0.562	128.67		
		0.625	142.68		
		0.688	156.60		
		0.750	170.21		
		0.812	183.75		
		0.875	197.41		60
		1.125	250.81		80
		1.219	270.55		
24	24.000	0.250	63.41		
		0.281	71.18		
		0.312	78.93		
		0.344	86.91		
		0.375	94.62	Std.	20
		0.406	102.31		
		0.438	110.22		
		0.469	117.86		
		0.500	125.49	XS	
		0.562	140.68		30
		0.625	156.03		
		0.688	171.29		40
		0.750	186.23		
		0.812	201.09		
		0.875	216.10		
		0.938	231.03		
		0.969	238.35		60
		1.219	296.58		90
		1.312	317.91		
26	26.000	0.250	68.75		
		0.281	77.18		
		0.312	85.60		
		0.344	94.26		
		0.375	102.63	Std.	
		0.406	110.98		
		0.438	119.57		
		0.469	127.88		
		0.500	136.17	XS	20
		0.562	152.68		
		0.625	169.38		
		0.656	177.56		
		0.688	185.99		
		0.750	202.25		
		0.875	234.79		
		1.188	314.81		
30	30.000	0.250	79.43		
		0.281	89.19		
		0.312	98.93		
		0.344	108.95		
		0.375	118.65		
		0.406	128.32		
		0.438	138.29		
		0.469	147.92		
		0.500	157.53		
		0.562	176.69		
		0.625	196.08		
		0.656	205.59		
		0.688	215.38		
		0.750	234.29		

Nominal Size	OD	Wall Thickness, Inches	Weight Lbs./Ft. Plain End
32	32.000	0.250	84.77
		0.281	95.19
		0.312	105.59
		0.344	116.30
		0.375	126.66
		0.406	136.99
		0.438	147.64
		0.469	157.94
		0.500	168.21
		0.562	188.70
		0.625	209.43
		0.656	219.60
		0.688	230.08
		0.750	250.31
34	34.000	0.250	90.11
		0.281	101.19
		0.312	112.25
		0.344	123.65
		0.375	134.67
		0.406	145.67
		0.438	157.00
		0.469	167.95
		0.500	178.89
		0.562	200.70
		0.625	222.78
		0.656	233.61
		0.688	244.77
		0.750	266.33
36	36.000	0.250	95.45
		0.281	107.20
		0.312	118.92
		0.344	131.00
		0.375	142.68
		0.406	154.34
		0.438	166.35
		0.469	177.97
		0.500	189.57
		0.562	212.70
		0.625	236.13
		0.656	247.62
		0.688	259.47
		0.750	282.35
40	40.000	0.312	132.25
		0.344	145.69
		0.375	158.70
		0.406	171.68
		0.438	185.06
		0.469	198.01
		0.500	210.93
		0.562	236.71
		0.625	262.83
		0.688	288.86
		0.750	314.39
42	42.000	0.312	138.91
		0.344	153.04
		0.375	166.71
		0.406	180.35
		0.438	194.42
		0.469	208.02
		0.500	221.61
		0.562	248.72
		0.625	276.18
		0.688	303.55
		0.750	330.41
44	44.000	0.344	160.39
		0.375	174.72
		0.406	189.03
		0.438	203.78
		0.469	218.04
		0.500	232.29
		0.562	260.72
		0.625	289.53
		0.688	318.25
		0.750	346.43

Nominal Size	OD	Wall Thickness, Inches	Weight Lbs. Ft. Plain End
46	46.000	0.344	167.74
		0.375	182.73
		0.406	197.70
		0.438	213.13
		0.469	228.06
		0.500	242.97
		0.562	272.73
		0.625	302.88
		0.688	332.95
		0.750	362.45
		0.812	391.88
		0.875	421.69
		0.938	451.42
		1.000	480.60
48	48.000	0.375	190.74
		0.406	206.37
		0.438	222.49
		0.469	238.08
		0.500	253.65
		0.562	284.73
		0.625	316.23
		0.688	347.64
		0.750	378.47
		0.812	409.22
		0.875	440.38
		0.938	471.46
		1.000	501.96

WELL CASING THREADED & COUPLED		
OD	WALL	WEIGHT
5 9/16	.258	14.81
6 5/8	.280	19.18
8 5/8	.277	25.55
8 5/8	.322	29.35
10 3/4	.279	32.75
10 3/4	.307	35.75
10 3/4	.365	41.85
12 3/4	.330	45.45
12 3/4	.375	51.15

Source: Courtesy, Bartow Steel, Inc.

TABLE 6.10 Polyethylene Pipe

IPS* PIPE SIZE	O.D. SIZE (in.)	255 psi DR 7.3		200 psi DR 9		160 psi DR 11		130 psi DR 13.5		110 psi DR 15.5		100 psi DR 17		80 psi DR 21		65 psi DR 26		50 psi DR 32.5		IPS* PIPE SIZE
		MIN WL/AVG ID (in.)	WEIGHT LB/FT	MIN WL/AVG ID (in.)	WEIGHT LB/FT	MIN WL/AVG ID (in.)	WEIGHT LB/FT	MIN WL/AVG ID (in.)	WEIGHT LB/FT	MIN WL/AVG ID (in.)	WEIGHT LB/FT	MIN WL/AVG ID (in.)	WEIGHT LB/FT	MIN WL/AVG ID (in.)	WEIGHT LB/FT	MIN WL/AVG ID (in.)	WEIGHT LB/FT	MIN WL/AVG ID (in.)	WEIGHT LB/FT	
½"	1.900	—	—	—	—	.173/1.533	.41	—	—	—	—	—	—	—	—	—	—	—	—	½"
2"	2.375	.325/1.686	.91	.264/1.815	.76	.216/1.917	.64	—	—	—	—	—	—	—	—	—	—	—	—	2"
3"	3.500	.479/2.485	1.98	.389/2.675	1.65	.318/2.826	1.39	.259/2.951	1.15	.226/3.021	1.02	.206/3.063	.93	—	—	—	—	—	—	3"
4"	4.500	.616/3.194	3.27	.500/3.440	2.74	.409/3.633	2.30	.333/3.794	1.90	.290/3.885	1.67	.265/3.938	1.54	.214/4.046	1.26	—	—	—	—	4"
†5"	5.563	.762/3.948	5.00	.618/4.253	4.18	.506/4.490	3.50	.412/4.690	2.91	.359/4.802	2.57	.327/4.870	2.35	.265/5.001	1.93	.214/5.109	1.58	—	—	†5"
6"	6.625	.908/4.700	7.09	.736/5.065	5.93	.602/5.349	4.97	.491/5.584	4.13	.427/5.720	3.64	.390/5.798	3.34	.316/5.957	2.74	.255/6.084	2.23	.204/6.193	1.80	6"
7"	7.125	—	—	—	—	—	—	—	—	—	—	.420/6.237	3.87	.340/6.406	3.17	.274/6.544	2.58	.220/6.661	2.09	7"
8"	8.625	1.182/6.119	12.01	.958/6.594	10.05	.785/6.963	8.43	.639/7.270	7.00	.556/7.446	6.16	.508/7.550	5.66	.411/7.754	4.64	.332/7.921	3.79	.265/8.063	3.05	8"
10"	10.750	1.473/7.627	18.66	1.194/8.219	15.62	.978/8.679	13.10	.797/9.062	10.89	.694/9.279	9.59	.633/9.410	8.80	.512/9.665	7.21	.413/9.874	5.87	.331/10.048	4.75	10"
12"	12.750	1.747/9.046	26.25	1.417/9.746	21.97	1.160/10.293	18.43	.945/10.749	15.31	.823/11.005	13.47	.750/11.160	12.36	.608/11.463	10.14	.490/11.711	8.26	.392/11.919	6.67	12"
14"	14.000	1.918/9.934	31.64	1.556/10.701	26.49	1.273/11.301	22.20	1.037/11.802	18.44	.903/12.086	16.24	.824/12.253	14.91	.667/12.586	12.22	.538/12.859	9.96	.431/13.086	8.05	14"
16"	16.000	2.192/11.353	41.34	1.778/12.231	34.61	1.455/12.915	29.00	1.185/13.488	24.09	1.032/13.812	21.21	.941/14.005	19.46	.762/14.385	15.97	.615/14.696	13.02	.492/14.957	10.51	16"
18"	18.000	2.466/12.772	52.31	2.000/13.760	43.79	1.636/14.532	36.69	1.333/15.174	30.48	1.161/15.539	26.85	1.059/15.755	24.65	.857/16.183	20.19	.692/16.533	16.48	.554/16.826	13.29	18"
20"	20.000	2.740/14.191	64.57	2.222/15.289	54.05	1.818/16.146	45.30	1.481/16.860	37.64	1.290/17.265	33.13	1.176/17.507	30.42	.952/17.982	24.92	.769/18.370	20.34	.615/18.696	16.41	20"
22"	22.000	—	—	2.444/16.819	65.41	2.000/17.760	54.82	1.630/18.544	45.56	1.419/18.992	40.09	1.294/19.257	36.81	1.048/19.778	30.19	.846/20.206	24.62	.677/20.565	19.87	22"
24"	24.000	—	—	2.667/18.346	77.85	2.182/19.374	65.24	1.778/20.231	54.22	1.548/20.718	47.72	1.412/21.007	43.82	1.143/21.577	35.92	.923/22.043	29.29	.738/22.435	23.62	24"
†26"	26.000	—	—	2.889/19.875	91.35	2.364/20.988	76.58	1.926/21.917	63.63	1.677/22.445	56.02	1.529/22.759	51.40	1.238/23.375	42.13	1.000/23.880	34.39	.800/24.304	27.74	†26"
†28"	28.000	—	—	—	—	2.545/22.605	88.79	2.074/23.603	73.76	1.806/24.171	64.94	1.647/24.508	59.62	1.333/25.174	48.86	1.077/25.717	39.89	.862/26.173	32.20	†28"
†30"	30.000	—	—	—	—	2.727/24.219	101.94	2.222/25.289	84.68	1.935/25.898	74.56	1.765/26.258	68.45	1.429/26.971	56.13	1.154/27.554	45.78	.923/28.043	36.92	†30"
†32"	32.000	—	—	—	—	2.909/25.833	115.99	2.370/26.976	96.35	2.065/27.622	84.88	1.882/28.010	77.86	1.524/28.769	63.83	1.231/29.390	52.10	.985/29.912	42.04	†32"
†34"	34.000	—	—	—	—	3.090/27.450	130.92	2.519/28.660	108.80	2.194/29.349	95.83	2.000/29.760	87.91	1.619/30.568	72.06	1.308/31.227	58.79	1.046/31.782	47.44	†34"
†36"	36.000	—	—	—	—	—	—	—	—	2.323/31.075	107.40	2.118/31.510	98.56	1.714/32.366	80.79	1.385/33.064	65.93	1.108/33.651	53.18	†36"

Wear in the pipe was between 1.1 and 29.9 percent and the wear patterns were more closely associated with the material type (i.e., coarse sands and gravel) than with the amount dredged. Pankow[32] indicates that the use of high-density polyethylene pipe can provide cost savings in installation, maintenance, freedom of design, and extended life of pipeline systems. The pipe will not rot, rust, or corrode; conduct electricity; or support growth of, nor be affected by algae, bacteria, or fungi and is resistant to marine biological growth. Specific gravity of the pipe is 0.955 to 0.957, and it should be joined by heat fusion. Pipe flexibility allows it to conform with uneven topography.

The pipe is hydraulically smooth, and the C coefficient in Hazen-Williams formula is estimated to be about 155. The Hazen-Williams is defined as

$$H = (0.2083)\left[\frac{100}{c}\right]^{1.85}\left[\frac{Q^{1.85}}{d^{4.8655}}\right]$$

where H = friction head, in feet of fresh water per 100 ft of pipe
  c = a constant accounting for pipe roughness
  Q = rate of flow in gallons per minute
  d = inside diameter of pipe

Floating Pipelines.   The conventional pontoon-supported discharge pipeline with sections of the pipe joined by ball joints cannot be operated successfully in waves over 3 ft (1 m) as discussed in Chap. 7. The discharge pipe is placed inside the

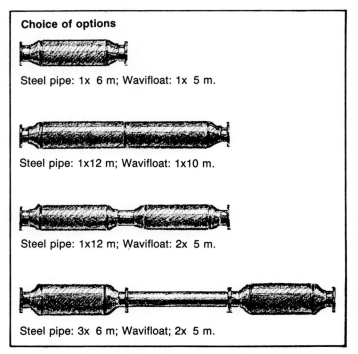

Choice of options

Steel pipe: 1x 6 m; Wavifloat: 1x 5 m.

Steel pipe: 1x12 m; Wavifloat: 1x10 m.

Steel pipe: 1x12 m; Wavifloat: 2x 5 m.

Steel pipe: 3x 6 m; Wavifloat; 2x 5 m.

FIGURE 6.15   Flotation collars. (*Courtesy, Wavin Co.*)

buoyant collar to provide the desired flotation for the pipeline. Flotation collars are available in various lengths; most commonly used lengths are 18 ft (6 m) and 36 ft (12 m). Figure 6.15 shows the various combinations of steel and flotation collar. Figure 6.16 shows the discharge line deployed on a dredging project.

Floating discharge lines have also been used when dredging in areas subject to wave action. Floating armored pipe is also available for pumping abrasive slurries in pipelines subject to swells and/or tidal variations (Fig. 6.17). Figure 6.18 provides information on bending angles as a function of pipe length for different pipe diameters. Figure 6.19 gives values of net weight of complete floating armored pipe and Table 6.11 summarizes pertinent information on floating armored pipelines. The pipeline will be buoyant up to slurry specific gravity of 1.8.

FIGURE 6.16   A floating discharge line deployed on a dredging project.

1 Discharge hose with flanges

2 Suction hose with flanges

3 Armoured hose with flanges

FIGURE 6.17   Floating armored pipe. (*Courtesy, Eddelbuttel & Schneider K.G.*)

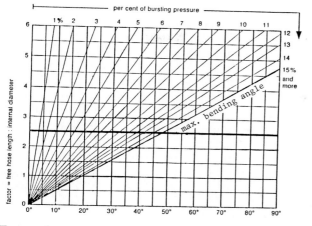

A safety factor of 3-4 is built in so that the operating pressure is 25 to 35 per cent of bursting pressure.

The heavy line on the diagram shows that where there is a hose length : i.d. factor of 2.5 or more, the hose may be bent to 90 deg. or more when not in use. (i.e. with no inner pressure).

Formulae

Maximum permissible distortions under normal working conditions

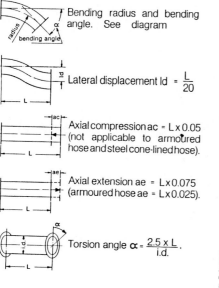

Bending radius and bending angle. See diagram

Lateral displacement $ld = \dfrac{L}{20}$

Axial compression $ac = L \times 0.05$ (not applicable to armoured hose and steel cone-lined hose).

Axial extension $ae = L \times 0.075$ (armoured hose $ae = L \times 0.025$).

Torsion angle $\alpha = \dfrac{2.5 \times L}{i.d.}$.

L   = original length   (in centimetres or in inch)
i.d. = inner diameter   (in centimetres or in inch)
p   = pressure   (atm. or psi)

When the cord angle β is exactly 54°44' a free moving hose will expand not only in diameter but also in length.

The accompanying graph shows this feature.

When the cord angle β is greater than 54°44' a free moving hose under working pressure conditions needs to be correctly designed if it is to maintain its inner diameter and only extend in length.

If the angle β is smaller than 54°44' the diameter of the hose will always increase and the hose length will shorten.

Note: if a delivery hose is used with additional joints or struts it can **never** extend. The elasticity of the cord threads allows only for an increase in diameter. The load on the struts or joints may be calculated very simply by means of the following formula

$$\dfrac{i.d.^2 \times \pi \times p}{4} = \text{axial tension force.}$$

FIGURE 6.18 Bending angles as a function of pipe length for different pipe diameters. (*Courtesy, Eddelbuttel & Schneider K.G.*)

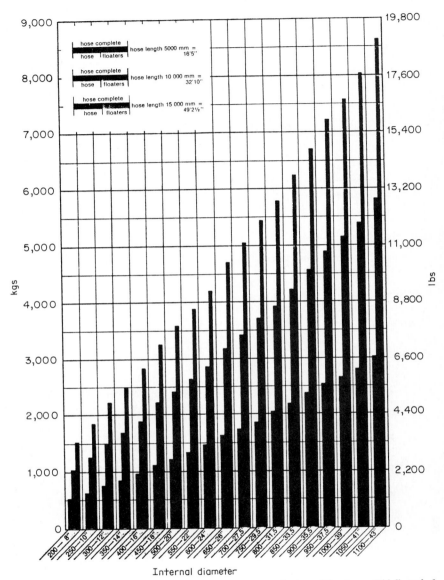

FIGURE 6.19 Net weight of complete floating armored pipe. (*Courtesy, Eddelbuttel & Schneider K.G.*)

TABLE 6.11 Floating Armored Pipe Dimensions

i.d. of hoses		o.d. of floaters		flange dimensions		working/bursting pressure		tensile breaking strength	
mm	inches	mm	inches	mm	inches	bar	psi	kgs	lbs
200	8	500	20	335/ 435 x 385 x 8 x 23/54	13⅜ /17⅛ x 15⅛ x 12 x ⅞ / 2	27/110	370/1540	35,000	77,000
250	10	580	22¾	385/ 485 x 435 x 12 x 23/54	15⅛ /19⅛ x 17⅛ x 12 x ⅞ / 2	20/94	280/1310	46,140	101,500
300	12	650	26	435/ 535 x 485 x 12 x 23/54	17⅛ /21⅛ x 19⅛ x 12 x ⅞ / 2	20/80	280/1120	56,540	124,400
350	14	725	28½	485/ 585 x 535 x 12 x 23/54	19⅛ /23⅛ x 21⅛ x 12 x ⅞ / 2	17/70	240/ 980	67,340	148,160
400	16	800	31½	535/ 655 x 595 x 16 x 23/54	21⅛ /25¾ x 23½ x 12 x ⅞ / 2	15/63	210/ 880	79,160	174,160
450	18	865	34	590/ 710 x 650 x 20 x 23/64	23¼ /28 x 25½ x 20 x ⅞ / 2½	20/80	280/1120	127,230	279,900
500	20	940	37	640/ 760 x 700 x 20 x 23/64	25¼ /30 x 27½ x 20 x ⅞ / 2½	18/75	250/1050	147,260	323,970
550	22	1010	39¾	690/ 810 x 750 x 20 x 27/64	27¼ /32 x 29½ x 20 x 1 / 2½	17/70	240/ 980	166,300	365,870
600	24	1105	43½	740/ 860 x 800 x 20 x 27/64	29⅛ /34 x 31½ x 20 x 1 / 2½	15/65	210/ 910	183,780	400,300
650	26	1180	46½	800/ 920 x 860 x 24 x 27/71	31½ /36¼ x 34 x 24 x 1 / 2¾	19/77	270/1070	255,500	562,100
700	27½	1255	49½	850/ 990 x 920 x 24 x 27/75	33½ /39 x 36¼ x 24 x 1 / 3	18/73	250/1020	280,900	618,060
750	29½	1330	52½	900/1040 x 970 x 24 x 27/75	35½ /41 x 38¼ x 24 x 1 / 3	16/66	225/ 920	291,570	641,470
800	31½	1405	55½	950/1090 x 1020 x 24 x 30/75	37⅜ /43 x 40 x 24 x 1⅛ / 3	15/63	210/ 880	316,670	696,670
850	33½	1490	58½	1000/1140 x 1070 x 24 x 30/75	39⅜ /45 x 42 x 24 x 1⅛ / 3	15/60	210/ 840	340,470	749,030
900	35½	1565	61¾	1060/1200 x 1130 x 28 x 33/80	41¾ /47½ x 44½ x 28 x 1⅜ / 3⅛	17/70	240/ 980	445,320	979,700
950	37½	1640	64½	1110/1250 x 1180 x 28 x 33/80	43¾ /49½ x 46½ x 28 x 1⅜ / 3⅛	16/66	225/ 920	467,820	1029,200
1000	39	1715	67½	1160/1320 x 1240 x 28 x 33/80	45¾ /52 x 49 x 28 x 1⅜ / 3⅛	15/63	210/ 880	494,800	1088,560
1050	41	1795	70¾	1210/1370 x 1290 x 28 x 33/80	47⅝ /54 x 51 x 28 x 1⅜ / 3⅛	15/61	210/ 850	528,190	1162,030
1100	43	1870	73½	1265/1425 x 1345 x 32 x 33/87	49⅞ /56 x 53 x 32 x 1⅜ / 3⅜	17/70	240/ 980	665,230	1463,510

Source: Courtesy, Eddelbuttel & Schneider K.G.

REFERENCES

1. Zandi, I., and Govatos, G., "Heterogeneous Flow of Solids in Pipelines," *Journal, Hydraulic Division,* ASCE HY. 3, pp. 145–159, May 1967.

2. Nardi, J. J., "Pumping Solids Through a Pipeline," *Mining Engineering,* Sept. 1959.

3. Scholes, W. A., "First Cross-Country Ore Slurry Line Completed in Tasmania," *Pipe Line Industry,* pp. 56–59, Nov. 1968.

4. Wiedenroth, W., "An Examination of the Problems Associated with the Transportation of Sand-Water-Mixtures in Pipelines and Centrifugal Pumps," *Proc., World Dredging Conference, WODCON II,* Rotterdam, the Netherlands, 1968.

5. Oosterbaan, N., "On the PIANC Classification of Soils to be Dredged," *Terra & Aqua,* no. 5, International Association of Dredging Companies, the Netherlands, 1973.

6. Herbich, J. B., "Deep Ocean Mineral Recovery," *Proc., World Dredging Conference, WODCON II,* Rotterdam, the Netherlands, 1968. ·

7. Rubey, W. W., "Settling Velocities of Gravel, Sand and Silt Particles," *American Journal of Science,* vol. 25, no. 148, pp. 325–338, Apr. 1933.

8. Richards, R. H., "Velocity of Galena and Quartz Falling in Water," *Trans., AIME,* vol. 38, pp. 230–234, 1908.

9. Worster, R. C., and Denny, D. F., "Hydraulic Transport of Solid Material in Pipes," *Proc., AIME,* England, vol. 169, pp. 563–586, 1955.

10. Shen, H. W., and The Committee on Sedimentation, Hydraulics Division, "Sediment Transportation Mechanics: J. Transportation of Sediment in Pipes," *Journal, Hydraulics Division,* ASCE, vol. 96, HY7, July 1970.

11. Durand, R., and Condolios, E., "Experimental Investigation of the Transport of Solids in Pipes," Deuxième Journées de l'Hydraulique, Compte Rendu, Société Hydrotechnique de France, Grenoble, France, June 1952.

12. O'Brien, M. P., and Folsom, R. G., "The Transportation of Sand in Pipelines," University of California, *Publications in Engineering,* vol. 3, 1937.

13. Durand, R., "Basic Relationship of the Transportation of Solids in Pipes—Experimental Research," *Proc., International Association for Hydraulic Research,* University of Minnesota, Sept. 1953.

14. Condolios, E., and Chapus, E. E., *Chemical Engineering,* June 2 and July 8, 22, 1963.

15. Newitt, D. M., Richardson, J. F., Abbott, M., and Turtle, R. B., "Hydraulic Conveying of Solids in Horizontal Pipes," *Trans., Institution of Chemical Engineers,* England, vol. 33, 1955.

16. Zandi, I., and Govatos, G., "Solid Transportation in Pipelines Heterogeneous Regimes," *Hydraulics Division Conference ASCE,* 1966.

17. Bonnington, S. T., BHRA *Technical Note no. 7098,* England, 1961.

18. Stepanoff, A. J., *Pumps and Blowers; Two-Phase Flow,* John Wiley, New York, 1965.

19. Bain, A. G., and Bonnington, S. T., *The Hydraulic Transport of Solids by Pipeline,* Pergamon Press, Elmsford, NY, 1970.

20. Graf, W. H., *Hydraulics of Sediment Transport,* McGraw-Hill, New York, 1971.

21. Babcock, H. A., "Head Losses in Pipeline Transportation of Solids," *Proc., World Dredging Conference, WODCON I,* pp. 261–304, 1967.

22. Anonymous, Colorado School of Mines Research Foundation Inc., "Transportation of Solids in Steel Pipelines," Colorado, 1963.

23. Homayounfar, F., "Flow of Multi-component Slurries," M.S. Thesis, University of Delaware, 1965.

24. Zandi, I., "Decreased Head Losses in Raw-Water Conduits," *Journal, American Water Works Association,* vol. 59, no. 2, Feb. 1967.

25. Herbich, J. B., and Vallentine, H. R., Unpublished head loss data for silt-clay-water mixtures, 1962.

26. Wilson, W. E., "Mechanics of Flow with Non-Colloidal Inert Solids," *Trans., ASCE,* vol. 107, 1942.

27. Blatch, N. S., "Water Filtration at Washington, D.C.," Discussion, *Trans., ASCE,* vol. 57, p. 400, 1906.

28. Craven, J. P., "A Study of Transportation of Sands in Pipes," Ph.D. Dissertation, State University of Iowa, 1951.

29. Vallentine, H. R., "Transportation of Solids in Pipelines," *Commonwealth Engineer,* Australia, pp. 349–354, Apr. 1, 1955.

30. Moody, L. F., "Friction Factors for Pipe Flow," *Trans., ASME,* vol. 66, 1944.

31. Fortino, E. P., "Viscosity and Pumpability of Bottom Material in a Dredge Channel," (Draft), U.S. Army Engineer District, Philadelphia, May 1, 1959.

32. Pankow, V. R., "Dredging Applications of High Density Polyethylene Pipe," *Proc. 19th Dredging Seminar,* Center of Dredging Studies, TAMU-SG-880-103, Texas A&M University, Sept. 1987.

SEDIMENT TRANSPORT IN PIPES

R. E. Schiller, Jr.

Professor Emeritus of Ocean and Civil Engineering
Texas A&M University

INTRODUCTION

Determination of sediment transport in pipes involves the use of at least three flow equations involving three flow regimes. The objective is to compute slurry head loss for a range of slurry concentrations and present these results in the form of a velocity versus head loss graph and also in the form of a velocity versus horsepower graph.

Computation and presentation of these graphs are moderately complicated but may be readily carried out by the use of an overlay to Lotus 1-2-3© on a personal computer. All equations that are used in this overlay are ones that do not involve an iterative process and have been derived from existing equations that require an iterative solution, by regression analysis, when necessary. So, after all required data are available, the data may be entered in the appropriate write unprotected cells in the Lotus 1-2-3 spread sheet and all computations are then automatic if recalculation is set to automatic. If recalculation is set to manual, the F9 key must be pressed for recalculation to take place.

This overlay program entitled "Pipe sediment transport using Lotus 1-2-3" is available on 5-1/4- or 3.5-in disks for the IBM PC and compatibles and requires Lotus 1-2-3, version 2.0, 2.01, 2.2, 2.3, or 3.1 or spread sheets capable of importing Lotus .WK1 files. Wysiwyg (from Lotus 1-2-3 release 2.3) was used to enhance Tables 6.13 through 6.16, and print Figs. 6.26 and 6.28. Harvard

Graphics© Release 2.31 was used to convert the Lotus graphs to logarithmic graphs and to enhance the printing and lettering of Figs. 6.25 and 6.27. All of the graphs in this section, with the exception of those printed using Lotus Wysiwyg, were constructed using Harvard Graphics.

Harvard Graphics, unlike some other spread sheet graphics programs or some stand-alone graphics programs, allows the truncation of logarithmic cycles at the upper end of logarithmic cycles. Figure 6.25 (produced by use of Harvard Graphics) is an example of truncation of the logarithmic cycles where the velocity scale stops at 35 ft/s and the head loss scale stops at 200 ft. Certain other graphics programs do not allow truncation and the velocity scale would extend to 100 ft/s and the head loss scale would extend to 1000 ft in this case. Harvard Graphics also allows truncation of logarithmic cycles at the lower end of the cycles but requires that the graduations be scaled and entered on the screen by the operator. Figure 6.22 is an example of a logarithmic graph that has the lower and upper logarithmic cycles truncated for the X coordinate.

SEDIMENT PROPERTIES

Sediments may be broadly divided into cohesive materials (silts and clays) with a median particle diameter of less than 0.063 mm and noncohesive materials (sand, gravel, cobbles, etc.) with a median particle size greater than 0.063 mm. The division point of 0.063 mm is based on the ϕ scale classification proposed by W. C. Krumbein (1934), where ϕ is the $-\log$ to the base 2 of the grain size in mm.

Grain size (mm)*	Krumbein ϕ scale
0.063	4.0
0.088	3.5
0.125	3.0
0.177	2.5
0.250	2.0
0.354	1.5
0.500	1.0
0.707	0.5
1.000	0.0
1.414	-0.5
2.000	-1.0

*Grain size values used for visual accumulation tube analysis.

Note: Changing $-\log_2$ by -1 doubles the grain size in mm. Intermediate values of grain size such as 0.177, 0.354, etc., are the geometric mean of the adjacent values, or $0.354 = \sqrt{((0.25)(0.5))}$.

The division point of 0.063 mm is of practical field use since individual grain particles larger than 0.063 mm are visible to the naked eye (20/20 vision assumed). In addition, particles smaller than 0.063 mm in size follow Stokes law, whereas larger particles do not. Noncohesive sediments generally consist of discrete particles and transport of these sediments depends on the physical properties of the individual particles, such as grain size, density, and particle shape.

There are many soil classification systems (with regard to size) in use around the world. One is the Permanent International Association of Navigation Congresses (PIANC) classification which is nearly identical, insofar as grain size distribution values are concerned, to the system used by the United States Geological Survey. The PIANC classification system divides soils into the following:

Boulders	Larger than 200 mm
Cobbles	Between 200–60 mm
Gravels	Between 60–2 mm
Sands	Between 2–0.06 mm
Silts	Between 0.06–0.002 mm
Clays	Smaller than 0.002 mm

Gravels and the coarser sand sizes (greater than 1 to 2 mm in size) are generally analyzed by sieving. The finer sand sizes (less than 1 mm in size) are generally analyzed by the use of the visual accumulation tube. A special longer length visual accumulation tube is available to analyze sands up to 2 mm in size. For materials in the silt and clay size range (using an application of Stokes Law), the grain size distribution may be found by the use of pipette analysis, bottom withdrawal tube analysis, or hydrometer analysis.

For most natural sands, the particle size distribution will be found to be very close to a log-normal probability distribution. The results of a grain size analysis are usually plotted on log-probability paper with the percentage finer being plotted on the probability scale and the grain size being plotted on the logarithmic scale. A typical plot of grain size test results on log-normal paper is shown in Fig. 6.20. The median particle size (d_{50}) may be estimated by reading the particle size at the 50 percent probability value on the logarithmic probability paper.

Particle shape affects the terminal settling velocity of the sediment particle,

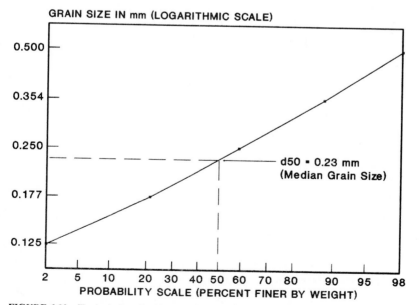

FIGURE 6.20 Typical plot of grain size test results plotted on log-normal probability distribution paper.

which in turn affects the sediment transport rate of sand size material. Particle shape may be expressed by some type of shape factor. A Waddell's sphericity value of 0.670 was used to determine values of the terminal settling velocity from data for quartz particles (SG_s = 2.65) in 20°C water after Graf et al. (1966). An expression for terminal velocity (V_{ss}) versus grain size (d) was then found using regression analysis. The equation gives terminal velocity in ft/s, or:

$$V_{ss} = (134.14 * (d_{50} - 0.039) \, \hat{} \, 0.972)/(25.4 * 12) \qquad (6.42)$$

where V_{ss} = the terminal velocity in ft/s
  d_{50} = the median grain size in mm

Note that Eq. (6.42) is valid for sand size materials only (materials with d_{50} between 0.063 mm and 2.0 mm).

Equation (6.42) was used in the Lotus 1-2-3 overlay to determine terminal velocity for the median soil particle.

Density of the sediment particles is an important property and should be known. Some sea shells have a specific gravity of approximately 2.9, whereas a beach sand (quartz and feldspar particles) has a specific gravity of about 2.65. In addition, beach sand particles are generally angular in shape, whereas shells are plate shaped. For visual accumulation tube analysis, any sea shells must be removed from the sample (generally with a 10% solution of hydrochloric acid) since the visual accumulation tube apparatus is calibrated for angular or rounded particles having a specific gravity of 2.65.

SEDIMENT TRANSPORT IN PIPES

Flow Regimes

Flow of sediment-laden water in pipes may be broken down into four different flow regimes. Figure 6.21 shows a schematic representation of concentration and velocity for various regimes of flow. Figure 6.21 thus shows that the flow regimes are based on the sediment concentration distribution in the pipe. The boundary velocities between the various regimes of flow may be determined by various equations. Figure 6.22 shows flow regimes as delineated by the boundary velocities for a 27-in diameter steel pipe.

1. *Pseudohomogeneous flow:* Flow where V is greater than V_H (the transition velocity between pseudohomogeneous and heterogeneous flow) and the sediment concentration varies only a small amount from top to bottom of the pipe.

2. *Heterogeneous flow (no deposit):* Flow where V is greater than V_c (the transition velocity between flow with a stationary bed and heterogeneous flow) and less than V_H and the concentration varies from top to bottom of the pipe but no particles remain on the bed of the pipe.

3. *Heterogeneous flow (flow with a moving bed):* Heterogeneous flow with particles settling to the bottom but continuing to move along the pipe. V greater than V_c and less than V_H.

4. *Flow with a stationary bed:* Slurry flow continues to take place; however, some particles settle out and remain as a stationary deposit on the bottom. V less than V_c.

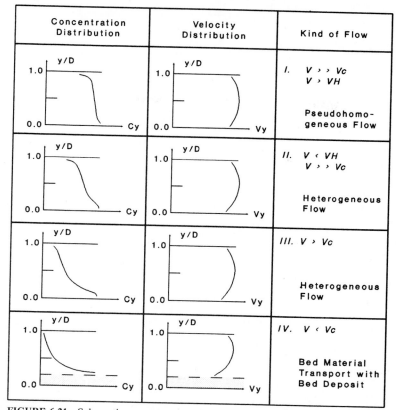

FIGURE 6.21 Schematic representation of concentration and velocity distribution in a pipe for various regimes of flow.

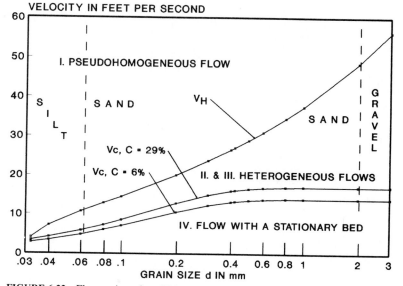

FIGURE 6.22 Flow regimes for a 27-in diameter steel pipe with a range of grain sizes and sediment concentrations.

Transition Velocities

V_c, or the transition velocity between flow with a stationary bed and heterogeneous flow, may be found from the following equations:

$$F_L = (1.3 * C^{0.125}) * (1 - \exp - 6.9 * d_{50} \text{ mm}) \qquad (6.43)$$

$$V_c = F_L * (2 * g * D(SG_s - SG_f))^{1/2}$$

where F_L = coefficient based on grain size and sediment concentration (after Durand, et al., 1956)
V_c = critical velocity in ft/s
C = sediment concentration by volume
g = gravitational acceleration (32.2 ft/s²)
D = pipe diameter in feet
SG_s = specific gravity of sediment particles
SG_f = specific gravity of the fluid

In the overlay procedure, values of sediment concentration C were assumed, and values of specific gravity of the slurry were then computed using the relationship

$$SG_m = C * (SG_s - SG_f) + SG_f \qquad (6.44)$$

where SG_m = specific gravity of the slurry.
V_H, or the transition velocity between heterogeneous and pseudohomogeneous flow, may be found by the expression

$$V_H = (1800 * g * V_{ss} * D)^{1/3} \qquad (6.45)$$

Head Loss Determination

Current U.S. engineering practice employs the use of the Darcy-Weisbach equation for the determination of head loss of clear water in pipes

$$h_f = f * (L/D) * (V^2/2g) \qquad (6.46)$$

where h_f = head loss of clear water for pipe length L
f = friction factor = $\phi(N_r, k/D)$
L = pipe length
D = pipe diameter
N_r = Reynolds number = $V * D/\nu$ $\qquad (6.47)$
κ = pipe roughness
ν = kinematic viscosity

For turbulent flow in smooth pipes the friction factor is given by

$$1/\sqrt{f} = 2.0 * \log(N_r * \sqrt{f}) - 0.8 \qquad (6.48)$$

Normally, pipes carrying sediments will be polished smooth by the movement of the sediment through the pipe and the friction factor may be defined by the smooth pipe equation. The above expression may be solved only by iteration.

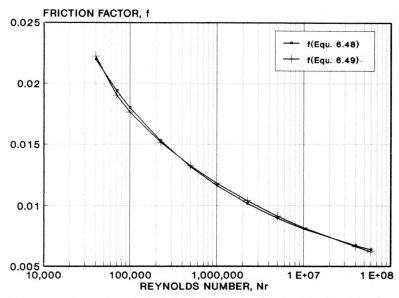

FRICTION FACTOR, f

FIGURE 6.23 Comparison of the friction factor f for smooth pipe (after Moody) and as calculated by the regression analysis Eq. (6.49).

For use in the Lotus 1-2-3 overlay program, a direct solution expression (using values from the above expression) was found by regression analysis. This expression is

$$f = 0.10408 * (N_r - 22,000)^{-0.157} \qquad (6.49)$$

and is valid for Reynolds numbers from 40,000 to 60,000,000. Figure 6.23 shows a comparison between values from Eqs. (6.48) and (6.49). If the pipe in question is not a smooth pipe, the values of f may be read from a Moody diagram and entered into the Lotus 1-2-3 overlay program after disabling the worksheet protection by the sequence /WGPD. This program should be saved using a different file name than those used in the supplied overlay program since the equations for f for a smooth pipe will be lost if the new computations are saved into the old file.

Minor losses have been taken care of in the computational procedure by adding an equivalent length of pipe to replace the minor losses, such as entrance loss, bend losses, etc. A tabulation of minor loss coefficients appears in Table 6.12. The Darcy-Weisbach equation, which gives the head loss for pipe friction, is

$$h_f = f * (L/D) * (V^2/2g) \qquad (6.46)$$

Minor losses may be found by the expression

$$h_m = K * (V^2/2g) \qquad (6.50)$$

where K is the minor loss coefficient and may be found in Table 6.12. If the two expressions are equated, an expression for the equivalent length may be found and is

TABLE 6.12 Minor System Head Loss Coefficients K for Dredging

System component	K = 2 * g * h/V ^ 2	Reference
Suction Entrance*		
Cutter with flare opening		Unknown
Plain-end suction	1.0	Crane
Rounded suction	0.05	Crane
Dragheads	Varies	WES report
Nozzle	5.5	Salzman
Oval	1.0	Salzman
Funnel	0.1	Salzman
Pear	0.02	Salzman
Elbows		
Long-radius suction	0.6	Crane (also depends
45° elbows	0.4	on pipe diameter)
90° elbows	0.9	
Stern swivel	1.0	Estimated
Ball joints		
Straight	0.1	
Medium cocked	0.4–0.6	
Fully cocked (17°)	0.9	
Wedge joints		Unknown
End section	1.0	

*Depends on the distance from suction opening to bottom, suction angle, etc. The above values are from a report by D. R. Basco, 1974.

$$L_e = (K * D)/f \qquad (6.51)$$

The length that appears under the column heading of Le in Tables 6.13 to 6.16 (pp. 6.49 to 6.52) is the sum of the equivalent length as determined by Eq. (6.51) and the total length of pipe in place. The effect of difference in diameter between the suction pipe and the discharge pipe has not been introduced into the Lotus 1-2-3 overlay program, since the effect is generally minor.

Graf (1971) and Herbich (1975) have summarized the flow equations for the different flow regimes for sediment transport in pipes. Please consult these references for a complete listing of the flow equations. The following equations were used to determine head loss in this treatment:

Flow with Deposit

$$V/(4 * g * R_h)\char`^1/2 = V_c/(D * g)\char`^1/2 \qquad (6.52)$$

where R_h is the hydraulic radius for flow with a bed deposit, and

$$R_h = ((V/V_c)\char`^2) * D/4 \qquad (6.53)$$

$$\frac{C * V * R_h}{((SG_s - SG_f) * g * d\char`^3)\char`^1/2} = 10.29 * \left(\frac{(SG_s - SG_f) * d}{S * R_h}\right)\char`^{-2.52} \qquad (6.54)$$

where $S = (h_f/L)_m$ = head loss of the slurry divided by the length L of the pipe.

Equations (6.52), (6.53), and (6.54) may be combined into the following expression for use with Lotus 1-2-3:

$$(h_f)_m = (((SG_s - SG_f) * d_{50})/((V/V_c)^2 * D/4))/((((C * V * ((V/V_c)^2 *$$
$$(D/4))/(((SG_s - SG_f) * g * (d_{50})^3)^0.5)/10.39)^- (1/2.52)) * L$$

$$(6.55)$$

Heterogeneous Flow

$$((h_f/L)_m - (h_f/L)_f)/(C * (h_f/L)_f) = \Phi \qquad (6.56)$$

$$\Phi_N = 1100 * (SG_s - SG_f) * (V_{ss}/V) * (g * D/V^2) \qquad (6.57)$$

where Φ = dimensionless sediment-transport parameter.
    Equations (6.56) and (6.57) may be combined into the following expression for use with Lotus 1-2-3:

$$(h_f)_m = ((((C * 1100 * (SG_s - SG_f) * V_{ss} * g * D)/(V^3)) * (h_f)_f) + (h_f)_f) \qquad (6.58)$$

Pseudohomogeneous Flow

$$((h_f/L)_m - (h_f/L)_f)/(C(h_f/L)_f) = \Phi \qquad (6.59)$$

$$\Phi = (SG_s - SG_f) \qquad (6.60)$$

    For use with Lotus 1-2-3, Eqs. (6.59) and (6.60) may be combined into the following expression:

$$(h_f)_m = (C * (SG_s - SG_f)(h_f)_f) + (h_f)_f \qquad (6.61)$$

Horsepower Determination

Horsepower may be calculated by using the following expression:

$$HP = (Q * \gamma * h_f)/550 \qquad (6.62)$$

where Q = discharge rate of slurry in ft^3/s
    γ = unit weight of slurry in lb/ft^3
    h_f = head loss in feet for pipe considered

    The unit weight γ may be found by multiplying the unit weight of water by the specific gravity of the mixture or:

$$\gamma = SG_m * 62.4 \qquad (6.63)$$

Applications

A quick determination of head losses and horsepower requirements for a given pipe size, water temperature, pipe length, sediment grain size, and total minor loss K-factors may be carried out using the overlay program entitled "Pipe Sediment Transport Using Lotus 1-2-3." The results may be presented in tabular or graphical form on the computer monitor or may be printed using the Lotus 1-2-3 print procedure. Enhanced logarithmic graphs may be produced using Harvard Graphics in connection with Lotus 1-2-3.

Reproductions of the tabular layouts are shown on Tables 6.13 to 6.16 (pp. 6.49 to 6.52) using Wysiwyg to enhance the printouts. A reproduction of the velocity versus head loss graph is shown in Fig. 6.24. A printout of the graphs using the Lotus 1-2-3-Harvard Graphics procedure is shown on Figs. 6.25 and 6.27. Figures 6.26 and 6.28 show the graphs as they print out using files PSHLPGS and PSHPPGS and Wysiwyg from Lotus 1-2-3 Release 2.3.

To use the overlay program, first call up Lotus 1-2-3 and then file PSHLPGS or PSHPPGS from the overlay program. File PSHLPGS will give a velocity versus head loss table and graph [log(velocity) versus log(head loss)] for a range of sediment concentrations whereas file PSHPPGS will give a velocity versus horsepower table and graph of log(velocity) versus log(horsepower) for a range of sediment concentrations. The reason for a graph of log(velocity) versus log(head loss) is that the graphical relationship is generally logarithmic. The log values may be converted to actual numerical values by using the expression

$$V \text{ or } h_f \text{ or } HP = 10 \char94 (\log \text{ of number}) \tag{6.64}$$

Files PSHLHGS and PSHPHGS will give graphs that are intended for use with Harvard Graphics.

After calling up the desired file, the diameter (D), the water temperature (Temp), the length (L), the median grain size of the sediment (d_{50}), and the total minor head loss coefficient K (see Table 6.12 for K values) may be changed to suit the new circumstances. If recalculation is not set on automatic, the recalculation key (F9) should be pressed to recalculate the tabulated values. The

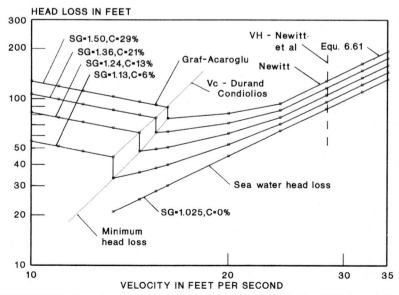

FIGURE 6.24 Slurry head loss in feet for D = 27 in, L = 1000 ft, T = 60°F, d_{50} = 0.42 mm, and K = 3.0. Logarithmic plot. Graf-Acaroglu, Newitt, and Eq. (6.61) relationship.

TABLE 6.13 Sea Water Slurry Head Loss for Smooth Pipes for Various Velocities and Sediment Concentrations Using Lotus 1-2-3 Release 2.3. File: A:\PSHLHGS. File Intended for Use with Harvard Graphics Release 2.31 or Earlier

g =	32.20	ft/s^2	D =	2.25	ft.				
VH =	28.22	ft/s	Temp =	60	Deg F	Visc =	1.217	*10^−5	
AREA =	3.976	ft^2	L =	1000	ft.			ft^2/sec	
			d50 =	0.42	mm				
Vss =	0.172	ft/s	K(tot)=	3.00					
CNST1=	1100.00		FL =	0.869	0.954	1.010	1.053		
d50 =	0.00138	ft.	Vc =	13.34	14.63	15.50	16.16		
			SG =	1.025	1.127	1.239	1.365	1.499	
			C =	0.000	0.063	0.132	0.209	0.292	

V,ft/s	Nr	f	Le	hf	hf	hf	hf	hf
10.00	1.85E+06	0.0107	1629		55.3	82.9	106.6	128.0
11.00	2.03E+06	0.0106	1639		51.5	77.2	99.3	119.2
12.00	2.22E+06	0.0104	1648		48.2	72.3	93.0	111.7
13.34	2.47E+06	0.0102	1659		44.6	66.8	86.0	103.2
13.34	2.47E+06	0.0102	1659	20.9	33.2	66.8	86.0	103.2
14.63	2.71E+06	0.0101	1669	24.9	36.1	62.4	80.2	96.4
14.63	2.71E+06	0.0101	1669	24.9	36.1	48.3	80.2	96.4
15.50	2.87E+06	0.0100	1675	27.8	38.3	49.7	76.9	92.3
15.50	2.87E+06	0.0100	1675	27.8	38.3	49.7	62.6	92.3
16.16	2.99E+06	0.0099	1679	30.1	40.1	51.1	63.3	89.5
16.16	2.99E+06	0.0099	1679	30.1	40.1	51.1	63.3	76.5
20.00	3.70E+06	0.0096	1703	45.2	53.1	61.8	71.5	81.9
24.00	4.44E+06	0.0093	1723	63.9	70.4	77.5	85.5	94.1
28.22	5.22E+06	0.0091	1742	87.1	96.0	105.8	116.6	128.4
32.00	5.92E+06	0.0089	1757	110.7	122.1	134.5	148.3	163.3
35.00	6.47E+06	0.0088	1768	131.4	144.8	159.6	176.0	193.7

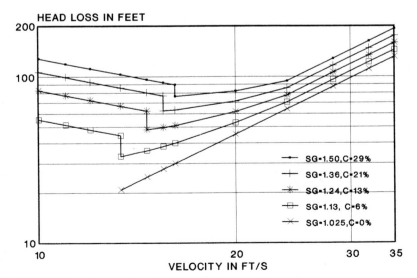

FIGURE 6.25 Graph of data from Table 6.13 using Harvard Graphics.

TABLE 6.14 Sea Water Slurry Head Loss for Smooth Pipes for Various Velocities and Sediment Concentrations Using Lotus 1-2-3 Release 2.3. File: A:\PSHLPGS. File Intended for Use with Lotus PrintGraph, Lotus Wysiwyg, or Lotus Allways.

g =	32.20	ft/s^2	D =	2.25	ft.			
VH =	28.22	ft/s	Temp=	60	Deg F Visc =		1.217	*10^-5
AREA =	3.976	ft^2	L =	1000	ft.			ft^2/sec
			d50 =	0.42	mm			
Vss =	0.172	ft/s	K(tot)=	3.00				
CNST1=	1100.00				FL =	0.869	0.954	1.010	1.053
d50 =	0.00138	ft.			Vc =	13.34	14.63	15.50	16.16
			SG =	1.025	1.127	1.239	1.365	1.499
			C =	0.000	0.063	0.132	0.209	0.292

V,ft/s	Nr	f	Le	hf	hf	hf	hf	hf
10.00	1.85E+06	0.0107	1629		55.3	82.9	106.6	128.0
11.00	2.03E+06	0.0106	1639		51.5	77.2	99.3	119.2
12.00	2.22E+06	0.0104	1648		48.2	72.3	93.0	111.7
13.34	2.47E+06	0.0102	1659		44.6	66.8	86.0	103.2
13.34	2.47E+06	0.0102	1659	20.9	33.2	66.8	86.0	103.2
14.63	2.71E+06	0.0101	1669	24.9	36.1	62.4	80.2	96.4
14.63	2.71E+06	0.0101	1669	24.9	36.1	48.3	80.2	96.4
15.50	2.87E+06	0.0100	1675	27.8	38.3	49.7	76.9	92.3
15.50	2.87E+06	0.0100	1675	27.8	38.3	49.7	62.6	92.3
16.16	2.99E+06	0.0099	1679	30.1	40.1	51.1	63.3	89.5
16.16	2.99E+06	0.0099	1679	30.1	40.1	51.1	63.3	76.5
20.00	3.70E+06	0.0096	1703	45.2	53.1	61.8	71.5	81.9
24.00	4.44E+06	0.0093	1723	63.9	70.4	77.5	85.5	94.1
28.22	5.22E+06	0.0091	1742	87.1	96.0	105.8	116.6	128.4
32.00	5.92E+06	0.0089	1757	110.7	122.1	134.5	148.3	163.3
35.00	6.47E+06	0.0088	1768	131.4	144.8	159.6	176.0	193.7

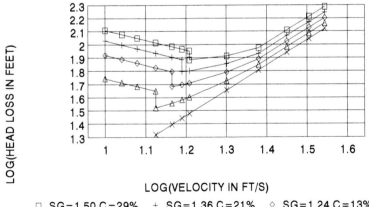

□ SG=1.50,C=29%   + SG=1.36,C=21%   ◇ SG=1.24,C=13%
△ SG=1.13,C=6%   × SG=1.025,C=0%

FIGURE 6.26 Graph of data from Table 6.14 using Lotus Wysiwyg.

TABLE 6.15 Sea Water Slurry Horsepower Requirement for Various Velocities and Sediment Concentrations Using Lotus 1-2-3 Release 2.3. File: A:\PSHPHGS. File Intended for Use with Harvard Graphics Release 2.31 or Earlier

g =	32.20	ft/s^2	D =	2.25	ft.			
VH =	28.22	ft/s	Temp=	60	Deg F	Visc =	1.217	*10^-5
AREA =	3.976	ft^2	L =	1000	ft.			ft^2/sec
			d50 =	0.42	mm			
Vss =	0.172	ft/s	K(tot)=	3.00				
CNST1=	1100.00		FL =	0.869	0.954	1.010	1.053	
d50 =	0.00138	ft.	Vc =	13.34	14.63	15.50	16.16	
			SG =	1.025	1.127	1.239	1.365	1.499
			C =	0.000	0.063	0.132	0.209	0.292

V,ft/s	Nr	f	Le	HP	HP	HP	HP	HP
10.00	1.85E+06	0.0107	1629		281	464	656	866
11.00	2.03E+06	0.0106	1639		288	475	672	887
12.00	2.22E+06	0.0104	1648		294	485	687	907
13.34	2.47E+06	0.0102	1659		302	499	706	932
13.34	2.47E+06	0.0102	1659	129	225	499	706	932
14.63	2.71E+06	0.0101	1669	168	268	510	723	954
14.63	2.71E+06	0.0101	1669	168	268	395	723	954
15.50	2.87E+06	0.0100	1675	199	302	431	733	968
15.50	2.87E+06	0.0100	1675	199	302	431	597	968
16.16	2.99E+06	0.0099	1679	225	329	461	630	978
16.16	2.99E+06	0.0099	1679	225	329	461	630	836
20.00	3.70E+06	0.0096	1703	418	540	691	880	1108
24.00	4.44E+06	0.0093	1723	709	860	1041	1263	1527
28.22	5.22E+06	0.0091	1742	1136	1377	1668	2026	2451
32.00	5.92E+06	0.0089	1757	1638	1986	2406	2922	3534
35.00	6.47E+06	0.0088	1768	2126	2578	3123	3792	4587

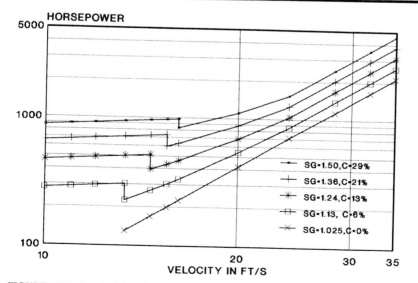

FIGURE 6.27 Graph of data from Table 6.15 using Harvard Graphics.

TABLE 6.16 Sea Water Slurry Horsepower Requirements for Various Velocities and Sediment Concentrations Using Lotus 1-2-3 Release 2.3. File: A:\PSHPHGS. File Intended for Use with Lotus PrintGraph, Lotus Wysiwyg, or Lotus Allways.

g =	32.20	ft/s^2	D =	2.25	ft.				
VH =	28.22	ft/s	Temp=	60	Deg F	Visc =	1.217	*10^−5	
AREA =	3.976	ft^2	L =	1000	ft.			ft^2/sec	
			d50 =	0.42	mm				
Vss =	0.172	ft/s	K(tot)=	3.00					
CNST1=	1100.00				FL =	0.869	0.954	1.010	1.053
d50 =	0.00138	ft.			Vc =	13.34	14.63	15.50	16.16
			SG =	1.025	1.127	1.239	1.365	1.499	
			C =	0.000	0.063	0.132	0.209	0.292	

V,ft/s	Nr	f	Le	HP	HP	HP	HP	HP
10.00	1.85E+06	0.0107	1629		281	464	656	866
11.00	2.03E+06	0.0106	1639		288	475	672	887
12.00	2.22E+06	0.0104	1648		294	485	687	907
13.34	2.47E+06	0.0102	1659		302	499	706	932
13.34	2.47E+06	0.0102	1659	129	225	499	706	932
14.63	2.71E+06	0.0101	1669	168	268	510	723	954
14.63	2.71E+06	0.0101	1669	168	268	395	723	954
15.50	2.87E+06	0.0100	1675	199	302	431	733	968
15.50	2.87E+06	0.0100	1675	199	302	431	597	968
16.16	2.99E+06	0.0099	1679	225	329	461	630	978
16.16	2.99E+06	0.0099	1679	225	329	461	630	836
20.00	3.70E+06	0.0096	1703	418	540	691	880	1108
24.00	4.44E+06	0.0093	1723	709	860	1041	1263	1527
28.22	5.22E+06	0.0091	1742	1136	1377	1668	2026	2451
32.00	5.92E+06	0.0089	1757	1638	1986	2406	2922	3534
35.00	6.47E+06	0.0088	1768	2126	2578	3123	3792	4587

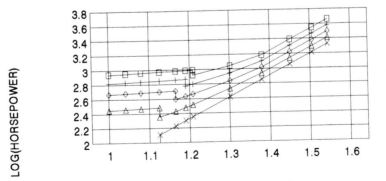

□ SG=1.50,C=29%   + SG=1.36,C=21%   ◇ SG=1.24,C=13%
△ SG=1.13,C=6%   × SG=1.025,C=0%

FIGURE 6.28 Graph of data from Table 6.16 using Lotus Wysiwyg.

resulting velocities in column A may not be in the ascending sequence. If the resulting velocities are not in ascending sequence (after recalculation), the velocity values should be changed so that all velocity values in column A are in ascending sequence. The F9 recalculation key should again be pressed if necessary to recalculate the tabular values for the rearranged velocity values.

The recalculated graph may be viewed by pressing /GV. If the new graph does not show a set of smooth curves, the velocity values are probably not in sequence and the velocity values should be altered so they are in sequence and key F9 punched again. The graph should be saved by hitting /GS and selecting a graph file name and then pressing Enter. The above procedure should be used to get both a velocity versus head loss set of results and a velocity versus horsepower set of results.

Lotus Graphs may be printed using Wysiwyg in connection with Lotus 1-2-3. Allowing about 16 lines of space will produce a graph about 3 in high. If a 10-point font is used for the spread sheet, about a 14-point font should be used for a graph of this size. Enhanced logarithmic graphs may be produced by using Harvard Graphics version 2.3. Lotus 1-2-3 files PSHLHGS and PSHPHGS should be used in connection with Harvard Graphics.

In this example the dredge pump should be operating with a flow velocity of about 16 ft/s with the system filled with sea water before starting to pump dredged material. The head loss is 30 ft (see Fig. 6.25) and the horsepower required is about 130 (see Fig. 6.27). Once enough dredged material is picked up to cause the slurry specific gravity to approach 1.50 (generally considered to be an optimum operating condition), it will take about 1 min at a velocity of 16 ft/s for the pipe to fill with this slurry. At the same time the head loss will rise to about 77 ft and the horsepower requirement will rise to about 830. The actual required horsepower will have to be computed using the efficiency of the operating system. When the head loss is 30 ft of sea water ($SG_f = 1.025$), the pressure drop over 1000 ft is

$$\Delta p = 30 * 62.4 * 1.025/144 = \underline{13.3} \text{ lb/in}^2 \qquad (6.65)$$

When the head loss is 77 ft of slurry ($SG_m = 1.50$), the pressure drop over 1000 ft is

$$\Delta p = 77 * 62.4 * 1.50/144 = \underline{50.0} \text{ lb/in}^2 \qquad (6.66)$$

Special notes: The overlay programs are computed using sea water as the transport fluid. The overlay may also be used if the transport fluid is fresh water since there is little difference in the results between the two fluids. No macro compatibility problems should be encountered when using the overlays in other spread sheet programs that will import .WK1 files, since the overlay files contain no macro commands.

BIBLIOGRAPHY

Acaroglu, E. R., "Sediment Transport in Conveyance Systems," Ph.D. Thesis, Cornell University, 1968.

Acaroglu, E. R., and Graf, W. H., "Sediment Transport in Conveyance Systems," Part 2, *Bull. Intern. Assoc. Sci. Hydr.,* vol. XIII, no. 3, 1968.

Alger, George R., and Simons, Daryl B., "Fall Velocities of Irregular Shaped Particles," *Journal of the Hydraulics Division,* ASCE, vol. 94, no. HY3, pp. 721–739, May 1968.

Basco, D. R., "Systems Engineering and Dredging—The Feedback Problem," *TAMU-SG-74-205*, report no. CDS 173, June 1974.

Condolios, E., "Transport of Materials in Bulk or in Container by Pipelines," United Nations Publ. No. 66-VIII-1, 1976.

Craven, J. P., "The TRansportation of Sand in Pipes, "Proceedings of the 5th Hydraulic Conference," *Engineering Bulletin no. 34*, State University of Iowa, Iowa City, Iowa, 1953.

Durand, R., "Basic Relationship of the transportation of Solids in Experimental Research," *Proc. Parer, International Association for Hydraulic Research*, University of Minnesota, September 1953.

Durand, R., and Condolios, E., "Experimental Investigation of the Transportation of Solids in Pipes," *Le Journals d'Hydraulique*, Societe Hydrotechnique de France, Grenoble, June 1952.

Graf, W. H., *Hydraulics of Sediment transport*, McGraw-Hill, New York, 1971.

Herbich, John B., *Coastal & Deep Ocean Dredging*, Gulf Publishing Company, Houston, Texas, 1975.

King, Cuchlaine A. M., *Beaches and Coasts*, 2d ed., Edwqard Arnold, London, 1972.

Newitt, D. M., Richardson, J. F., Abbott, M., and Turtle, R. B., "Hydraulic Conveying of Solids in Horizontal Pipes," *Transactions of Chemical Engineers*, vol. 33, 1955.

Lotus 1-2-3 Release 2.2 Reference, Lotus Development Corporation, 55 Cambridge Parkway, Cambridge, MA 02142, 1989.

"Report K, Operators Manual on the Visual Accumulation Tube Method for Sedimentation Analysis of Sands," Saint Anthony Falls Hydraulic Laboratory, Minneapolis, Minn., 1958.

Sagman, S. W., and Sandler, J. G., *Using Harvard Graphics*, Que Corporation, 1989.

Shen, H. W., Karaki, S., Chamberlain, A. R., and Albertson, M. L., "Sediment Transportation Mechanics: Transportation of Sediment in Pipes," *Journal of the Hydraulics Division*, ASCE, vol. 96, HY7, pp. 1503–1538, July 1970.

Simons, D. G., and Senturk, F., *Sediment Transport Technology*, Water Resources Publications, Fort Collins, Col., 1977.

Turner, Thomas M., *Fundamentals of Hydraulic Dredging*, Cornell Maritime Press, Centreville, Md., 1984.

Vennard, J. K., and Street, R. L., *Elementary Fluid Mechanics*, 6th ed., John Wiley, New York, 1982.

LIST OF SYMBOLS

Symbol	Definition or description	Dimension
A	Cross-section area of flow	L^2
C	Volumetric concentration of solids and equal to the ratio of volume of solids and volume of mixture	
D	Pipe diameter	L
d	Soil particle size	L
d_{50}	Median soil particle size	L
F_L	A coefficient to determine V_c	
f	Darcy friction factor	
f_m	Darcy friction factor of sediment-water mixture	
g	Gravitational acceleration	L/T^2
HP	Horsepower $= Q * \gamma * h/550$ (U.S. units)	
h_f	Pipe friction loss	L
h_m	Minor losses	L
K	Minor pipe loss coefficient	
L	Pipe length	L
L_e	Equivalent pipe length	L
ln	Log to the base e	
$\ln_2$	Log to the base 2	
log	Log to the base 10	
p	Wetted perimeter	L
p	pressure	F/L^2
Q	Discharge rate of water	L^3/T
N_r	Reynolds number $= V * D/\nu$	
R	Hydraulic radius $= A/p$	L
R_h	Hydraulic radius for flow with a bed deposit	L
S	Slope of energy grade line	
SG_s	Specific gravity of solids	
SG_f	Specific gravity of fluid	
SG_m	Specific gravity of sediment-water mixture	
V	Mean velocity of flow in pipe	L/T
V_c	Transition velocity, heterogeneous flow to flow with a stationary bed	L/T
V_H	Transition velocity, pseudohomogeneous flow to heterogeneous flow	L/T
V_{ss}	Terminal or fall velocity of soil particle in water	L/T
$\wedge$	Exponential symbol	
y	Flow depth	L
$*$	Multiplication symbol	
γ	Unit weight	F/L^3
Φ	Dimensionless sediment transport parameter	
ϕ	Krumbein phi scale ($-$ log to the base 2 of the grain size in mm)	
ϕ	Function of symbol	
κ	Pipe roughness	L
ν	Kinematic viscosity	L^2/T

CHAPTER 7
DREDGING METHODS

The Corps of Engineers' Engineering Manual EM 1110-2-5025 provides an inventory of dredging equipment and disposal techniques used in the United States and provides guidelines for developing a dredging project. The development of a dredging project involves the study and evaluation of many factors to assure that dredging and disposal are carried out in an effective, efficient, economical, and environmentally compatible manner. The following are some of the factors that should be considered in the planning and design phase:

1. Analysis of dredging locations and quantities
2. Dredging environment (depths, waves, currents, distance to potential disposal area, etc.)
3. Evaluation of physical, geological, chemical, and biological characteristics of sediments to be dredged
4. Identification of social, environmental, and institutional factors
5. Evaluation of dredge plant requirements
6. Evaluation of potential disposal alternatives
7. Hydrographic surveys of proposed project
8. Field investigations of sediments to be dredged
9. Performance of required laboratory tests (chemical characterization, sedimentation, engineering properties, bioassay, bioaccumulation, etc.)
10. Evaluation of in situ density of sediments to be dredged
11. Evaluation of long-term dredging and disposal requirements for project
12. Coordination of project plans with engineering, construction-operation, and planning elements of the district
13. Evaluation of potential productive uses
14. Coordination of project plans with other agencies and public and private groups

RIVER DREDGING

One of the most successful dredges for maintaining navigational channels in large rivers is a dustpan dredge.

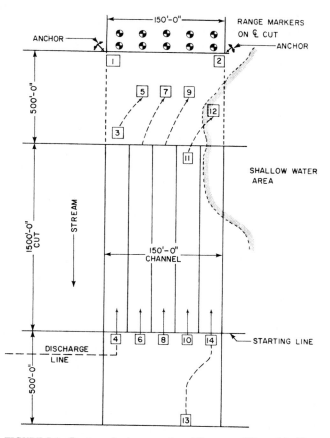

FIGURE 7.1   Dustpan dredge operation. (*Courtesy, Ellicott Machine Corp.*)

Operation of the Dustpan Dredge

Since dustpan dredging is principally a low-water season operation, it is necessary to survey river channels prior to the end of the high-water season to determine the location and depths at the crossings and sand bar formations and plan dredging operations accordingly. Prior to actual dredging at any given location, range markers have to be set out at the upper end of the centerline of the cuts to be dredged to maintain alignment. Dredging operation is controlled primarily by hauling winches, but the dredge can move from one location to another under its own propulsion.[1]

An operation of the dustpan dredge is illustrated in Fig. 7.1. The procedure is:

1. Dredge is moved in position 1, the spud is dropped, and the starboard hauling anchor with attached hauling wire is set with a floating marker.

2. Dredge is moved to position 2, the positioning spud is dropped, the port anchor is set with a floating marker, and the cables are crossed to facilitate maneuverability.

3. Dredge is moved to position 3 using either the cables or its own propulsion; then it is moved back downstream on the hauling cables to position 4.

4. The suction head is lowered to the required depth, dredge pump and jetting pumps are turned on, and the dredging commences. The dredge is moved forward by hauling cables. The rate of movement depends on the materials being dredged, the depth of the dredging cut, the currents, and the wind. In shallow cuts the advance may be quite rapid, about 800 ft/hr (244 m/hr).

5. When the upstream end of the cut is reached, the suction head is raised and the dredge is moved to position 5 and then dropped back to position 6.

6. The dredge next moves upstream using the cables and dredging commences at a rate commensurate with the depth of cut, type of material, etc.

7. After reaching the end of the cut, the procedure is repeated to dredge other cuts until the 150-ft- (45-m-) wide channel is dredged.

The suction head may have to be lowered or raised if obstacles such as boulders, logs, or tree stumps are encountered. Experience with dustpan dredges indicates that the best results are obtained when cuts do not exceed 6 ft in depth.

Typical discharge pipe arrangements are shown in Figs. 7.2 and 7.3. The pontoon closest to the dredge should be held in the same line as the pipe to prevent the pontoon from hitting the aft end of the dredge when swinging the dredge. The pontoon line should not swing more than 60° upstream on the starboard side of the cut and 80°

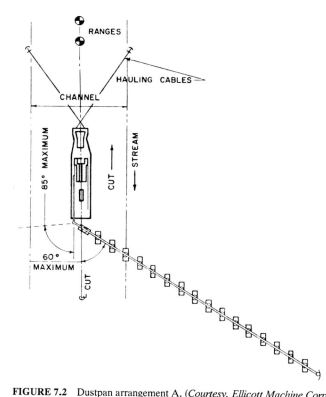

FIGURE 7.2 Dustpan arrangement A. (*Courtesy, Ellicott Machine Corp.*)

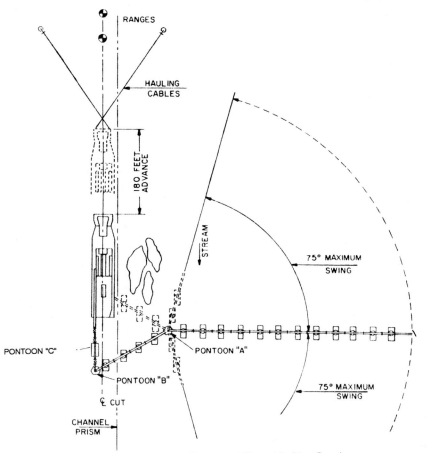

FIGURE 7.3 Dustpan arrangement B. (*Courtesy, Ellicott Machine Corp.*)

upstream on the port side of the cut. The dustpan dredge would normally discharge the dredged material in the river away from the channel. If this is not desirable, the discharge pipeline could be arranged as shown in Fig. 7.3.

In this case a swivel elbow is located on anchored pontoon A, placed some 100 to 120 ft (30.5 to 36.6 m) below the obstruction in the channel (such as an island) and some 120 to 140 ft (36.6 to 42.7 m) from the centerline of the dredge channel. The floating discharge line is assembled as shown in the drawing and is free to swing 150° by moving a baffle plate at the end of the discharge line. The second free-floating pontoon is located behind the dredge (pontoon B) and it is connected with pontoon A through a pipeline floating on at least four or five pontoons and a ball joint. Another pontoon (C) is located between pontoon B and the dredge to provide additional flexibility.

The dredge can advance approximately 180 to 200 ft (55 to 61 m); before further advance, the forward spud on the dredge must be dropped and additional pontoons added between pontoon C and the dredge. Some possible channel dredging conditions are shown in Fig. 7.4. Typical dustpan dredge arrangements are shown in Fig. 7.5.

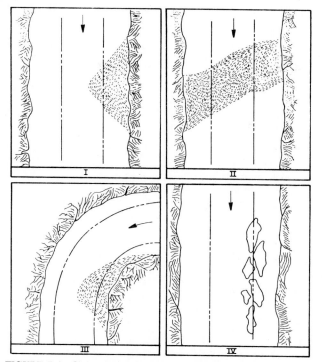

FIGURE 7.4   Channel dredging conditions showing shoaling areas. (*Courtesy, Ellicott Machine Corp.*)

ESTUARY AND COASTAL DREDGING

Operation of a Cutterhead Dredge

In actual operation a dredge swings from side to side using the port spud as a pivot, as indicated in Fig. 7.6. The action of the dredge is controlled by swing cables attached to swing anchors. To advance, after the swing of the dredge has stopped, the starboard side of the centerline of the cut is the same distance it was from this point when the port spud was raised. The port spud is lowered and the starboard spud lifted to advance the dredge.

The length and diameter of the steel spuds depend on the size of the dredge and the depth of the channel. There are two types of arrangements for the spuds:

1. Spuds located outside of the dredge's stern, as shown in Fig. 7.7*a*; one spud is called the working, the other walking.
2. One spud located in a spud carriage, and the other outside of the dredge's stern, as shown in Fig. 7.7*b*.

Turner[2] shows that the spud carriage is more efficient than the walking and working spud arrangement, as indicated in Fig. 7.7*b*. The employment of the spud carriage increases the overall dredge efficiency from 50 to 75 percent. The

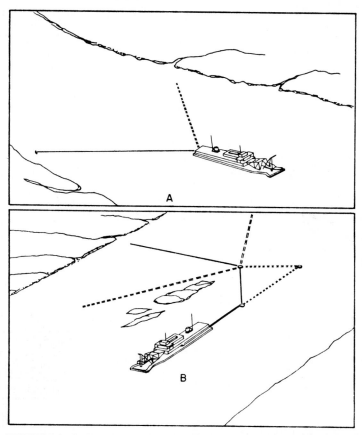

FIGURE 7.5   Dustpan dredge arrangements. (*Courtesy, Ellicott Machine Corp.*)

spud carriage system is also more effective since it allows the operator of the dredge to adjust the advance of the cutter into the material at any time and at any position of the cutter. The spud carriage is activated by a hydraulic cylinder. When excessive forces on the cutterhead occur, the excessive pressure in the cylinder can be relieved by a safety valve. When this occurs, the cutter is moved back from the dredged face and the dredge moves backward. Figure 7.8 shows very massive steel spuds used on dredge *Leonardo da Vinci*.

Optimum Operation of the Cutterhead Dredges

The production of a cutterhead dredge depends on the volume of the material cut away in the case of rock or the volume of material dislodged as in the case of sand and gravel per unit of time.[3]

With sandy soil the production

$$P_r = f(\text{PI},\mu,\rho,\text{MC},C_c,C_u,\text{shape of grains}) \qquad (7.1)$$

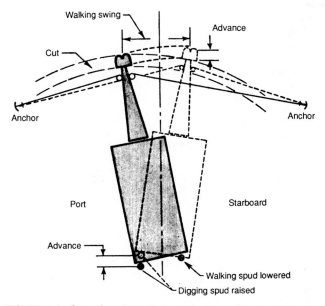

FIGURE 7.6 Operation of a cutterhead dredge, walking spud.

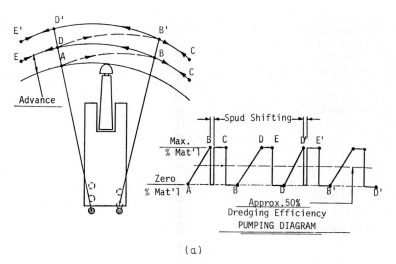

(a)

FIGURE 7.7 Operation of a cutterhead dredge, spud carriage. The advancing operation of a dredge is a major factor affecting dredge efficiency. These diagrams assume single-level swinging. (a) Walking-working spud arrangement.

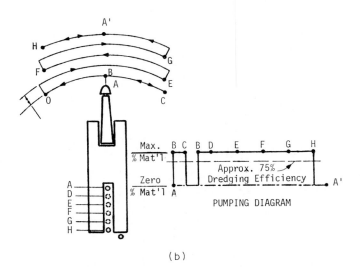

(b)

FIGURE 7.7 (*Continued*) Operation of a cutterhead dredge, spud carriage. The advancing operation of a dredge is a major factor affecting dredge efficiency. These diagrams assume single-level swinging. (*b*) spud carriage.

(*a*)

FIGURE 7.8 Spuds installed on cutterhead dredge *Leonardo da Vinci*. (*a*) Spuds and connection with a floating discharge pipeline.

(b)

(c)

FIGURE 7.8 (*Continued*) Spuds installed on cutterhead dredge *Leonardo da Vinci*. (a) Spuds and connection with a floating discharge pipeline; (b) spuds and a spare spud (horizontal); (c) spare spud.

where

$$PI = \text{plasticity index} = LL - PL \qquad (7.2)$$
$$LL = \text{liquid limit}$$
$$PL = \text{plastic limit (Atteberg)}$$
$$\mu = \text{viscosity of mixture}$$
$$\rho = \text{density}$$
$$MC = \text{moisture content}$$
$$C_c = \text{curvature coefficient}$$

$$= \frac{(D_{30})^2}{(D_{60})(D_{10})}$$

$$D_{30} = 30 \text{ percent of soil is finer}$$
$$D_{60} = 60 \text{ percent of soil is finer}$$
$$D_{10} = 10 \text{ percent of soil is finer}$$

$$C_u = \frac{D_{60}}{D_{10}} = \text{uniformity coefficient}$$

$1 < C_c < 3$ denotes a smooth grain size distribution; for gravel ($C_u > 4$) and for sand ($C_u > 6$), there will be a wide range of sizes.

In addition, production depends on the cutterhead:

$$P_r = f_1(\omega, n_o \text{shape of blade,size,type of cutting edge}) \qquad (7.3)$$

where ω = speed of rotation
n_o = number of blades

and on swing winches:

$$P_r = f_2(s_r, \text{direction of moving})$$

where s_r = rate of swing, and on the dredge itself

$$P_r = f_3(V, d_c)$$

where V = ship's speed
d_c = depth of cut

or

$$P_r = f_4(PI, \mu, \rho, MC, C_c, C_u, \text{shape of grain}, \omega, n_o, \text{shape of blade,size,}$$
$$\text{type of cutting edge}, s_r, \text{direction of swing}, V, d_c) \qquad (7.4)$$

It is obvious then that the cutterhead design is probably one of the most difficult since it depends on a complicated flow pattern and a large number of geometric variables. There is no simple relation between the production and all the variables involved. Field experience indicates that the production for a given type of soil and a given cutter is a function of the depth of cut, ship speed, and cutter speed. These three factors are varied within the limits of swing and cutter motors to develop optimum production. Figure 7.9 shows the production as a function of the flow rate for different specific gravities of solid-liquid mixtures. Note that the production for any given soil is independent of the flow rate.

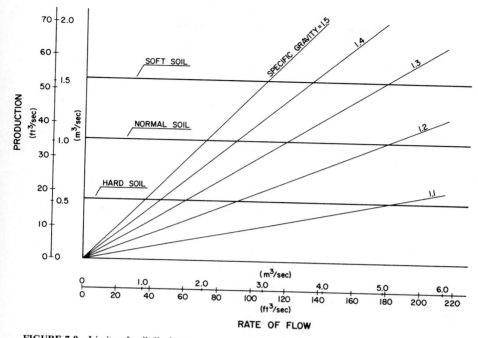

FIGURE 7.9 Limits of soil displacement. (*Courtesy, IHC Holland*)

Vacuum Pressure Limitation. Maximum production is a function of a number of variables:

$$(P_r)_{max} = f(y_d, Z, d_s, h_v, Q, S.G., C_f, C_s) \tag{7.5}$$

where y_d = dredging depth
Z = depth of pump inlet below water line
d_s = diameter of suction pipe
h_v = vacuum head
Q = discharge of mixture
S.G. = specific gravity of wet sediment
C_f = pipe friction coefficient
C_s = soil constant

$$P_r = Q(\gamma - 1) \tag{7.6}$$

$$Q = \left(\frac{\pi}{4}\right) d_s^2 V \tag{7.7}$$

Maximum allowed vacuum head (h_v) is equal to

$$h_v = y_d(S.G. - 1) - Z\gamma_s + C_f\left(\frac{S.G.}{2g}\right)V^2 + C_s\frac{(S.G. - 1)}{V} \tag{7.8}$$

Example 7.1. Assume that the diameter of the suction pipe is 2.8 ft (0.85 m), C_f = 2.5, and C_s = 430 ft²/s (40 m²/s), h_v = 26.2 ft (8 m) of water, Z = 1.0 ft (0.3 m), S.G. = 2.0, and g = 32.2 ft/s² (9.8 m/s²).

The maximum production may be computed from Eq. 7.2 with the aid of Eqs. 7.3 and 7.4. The results of computations for various dredging depths and number of mixture specific gravities are shown in Fig. 7.10. It should be noted that optimum production rate occurs at about 100 ft³/s (2.8 m³/s) and that pumping of higher specific gravities of mixture is not possible because of vacuum limitations. Figure 7.11 indicates that mixture specific gravities of over 1.4 are only possible at shallow dredging depths of less than 32.8 ft (9.8 m). This also points out that maximum production reading on production meters does not necessarily mean that optimum production is being achieved. The use of jet pumps or pumps on the ladder are recommended for dredging at greater depths.

Power Limitation. Power available is also one of the factors limiting the optimum production. Assume in this example that the efficiency of the pump does not vary with the discharge and mixture density and consider three pumping distances (L) of 1968, 3280, and 5741 ft (600, 1000, and 1750 m). Let the suction pipe diameter be 2.6 ft (0.8 m), hydrostatic head (h_t) 13.1 ft (4 m), the power available 3000 hp, pump efficiency (η) 75 percent, and vacuum head (h_v) 26.2 ft (8 m) of water.

$$\eta(hp) = C_1 L(\gamma - 1) + C_2 L(Q^3) + C_3(Q)^3 + C_4(h_v + h_t\gamma)Q \qquad (7.9)$$

where L = pipe length (pumping distance) and C_1, C_2, and C_3 = constants depending on pipe diameter and type of mixture pumped.

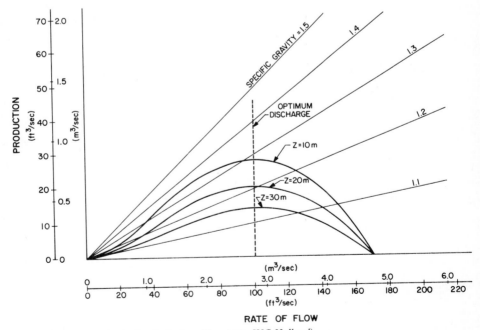

FIGURE 7.10 Cavitation limitation. (*Courtesy, IHC Holland*)

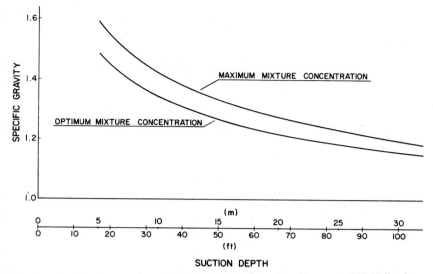

FIGURE 7.11 Optimum and maximum mixture concentration. (*Courtesy, IHC Holland*)

Let

$$C_1 = 0.2 \ t/s \quad C_2 = 0.0025 \ t/(s^2 \cdot m^9) \quad C_3 = 1.25 \ t/(s^2 \cdot m^6) \quad C_4 = 1 \ t/m^3$$

Using the above values Hadjidakis[3] prepared Fig. 7.12, which indicates that maximum production occurs for a given discharge and decreases for a greater or a lower discharge and that the minimum resistance occurs at a critical flowrate.

Critical Velocity Limitation. It may be assumed that discharge of mixture depends on the type of soil pumped and the pipe diameter, and the critical discharge (Q_c) is proportional to the square root of gravitational acceleration and pipe diameter to the 2.5 power, or

$$Q_c \propto (g)^{0.5} D^{2.5} \tag{7.10}$$

where Q_c is that discharge at which the sediment will not settle in the pipe, or critical velocity, or

$$Q_c = C\sqrt{gD^5}$$

and

$$V_c = C_1\sqrt{gD}$$

Thus the critical velocity is a function of the square root of the pipe diameter. It is a common practice to select a pipe diameter which would give a discharge greater than the critical value. The pipeline resistance is shown in Fig. 7.13 as a function of discharge and specific gravity of mixture. The minimum resistance is at a critical velocity.

Figure 7.14 shows the limit of critical discharge for medium sand in a pipe having a 2.6-ft (80-cm) diameter.

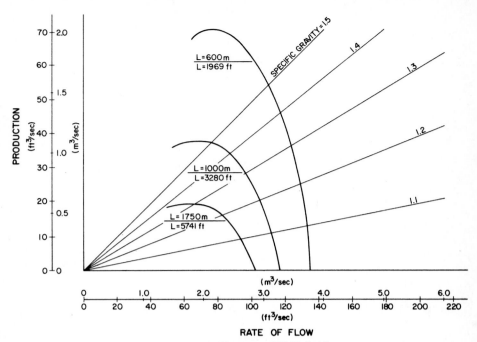

FIGURE 7.12 Limits of pumping power. (*Courtesy, IHC Holland*)

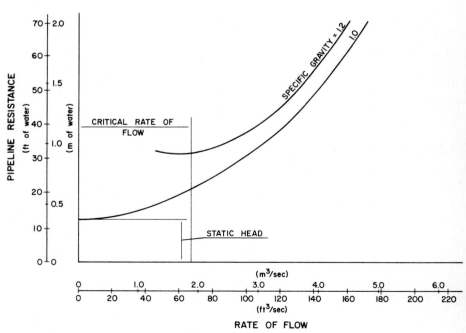

FIGURE 7.13 Critical discharge. (*Courtesy, IHC Holland*)

7.14

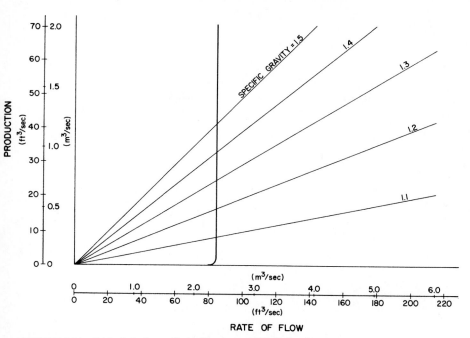

FIGURE 7.14 Critical discharge limit. (*Courtesy, IHC Holland*)

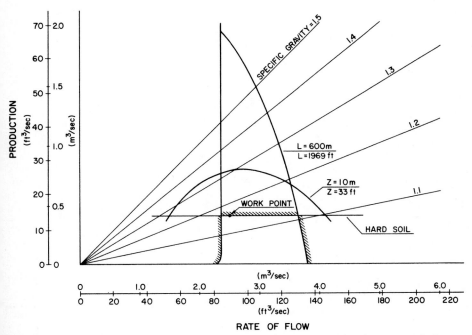

FIGURE 7.15 Production determined by the power of the cutter and swing motor winches. (*Courtesy, IHC Holland*)

7.15

Three practical examples are given by Hadjidakis:

1. A cutterhead dredge operates in hard soil. The dredging depth is low and discharge pipe is short. The production is limited as shown in Fig. 7.15 but can be improved by:
   a. Selecting an optimum combination of cutter speed, depth of cut, and rate of movement
   b. Selecting a different cutter or a smaller cutter of the same design
   c. Installing a more powerful cutter and bigger swing winch motors
2. A cutterhead dredge is working at a great depth in normal soil (Fig. 7.16). The discharge pipe is quite short. The production in this case is limited by the cavitation characteristics of the pump but may be improved by:
   a. Installation of a flow and density meter, or a production meter, to permit the operator to achieve the optimum production for given conditions
   b. Increasing suction pipe diameter and thus reducing head losses in the suction system
   c. Lowering the pump to reduce the suction head, inside the hull, or placing the pump halfway down the ladder
   d. Placing a jet pump in the suction line
3. A cutterhead dredge is operating at an average depth in normal soil with a long distance line (Fig. 7.17). The production is determined by the horsepower of the pump, and it can only be improved by:
   a. Installing booster pumps in the discharge line
   b. Installing a bigger motor

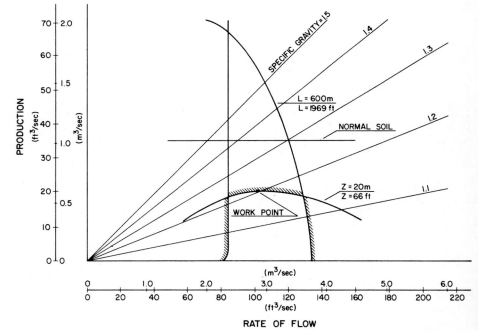

FIGURE 7.16 Production determined by cavitation characteristics. (*Courtesy, IHC Holland*)

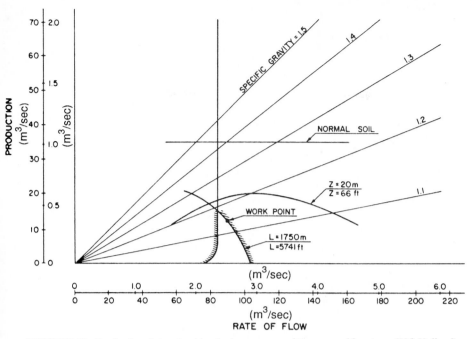

FIGURE 7.17 Production determined by the horsepower of the pump. (*Courtesy, IHC Holland*)

FIGURE 7.18 Cutterhead dredge *Wm. A. Thompson* and booster barge *Mullen* operating at St. Paul Barge Terminal. (*Courtesy, St. Paul District, U.S. Army Corps of Engineers*)

7.17

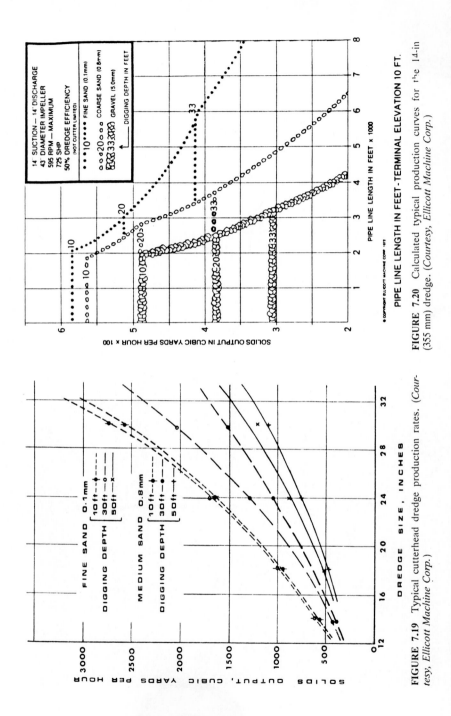

FIGURE 7.20 Calculated typical production curves for the 14-in (355 mm) dredge. (*Courtesy, Ellicott Machine Corp.*)

FIGURE 7.19 Typical cutterhead dredge production rates. (*Courtesy, Ellicott Machine Corp.*)

7.18

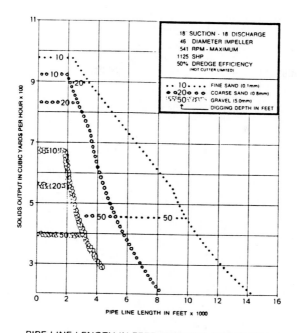

PIPE LINE LENGTH IN FEET-TERMINAL ELEVATION 10 FT.

FIGURE 7.21 Calculated typical production rates for 18-in (457-mm) dredge. (*Courtesy, Ellicott Machine Corp.*)

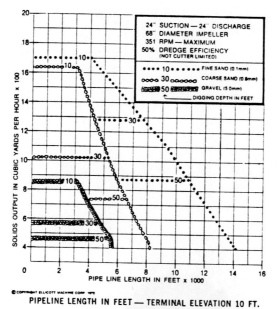

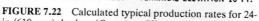

PIPELINE LENGTH IN FEET — TERMINAL ELEVATION 10 FT.

FIGURE 7.22 Calculated typical production rates for 24-in (610-mm) dredge. (*Courtesy, Ellicott Machine Corp.*)

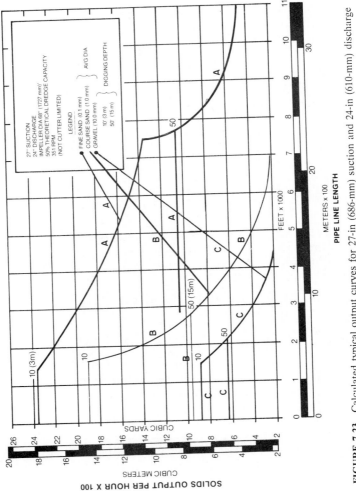

FIGURE 7.23 Calculated typical output curves for 27-in (686-mm) suction and 24-in (610-mm) discharge cutterhead dredge. (*Courtesy, Ellicott Machine Corp.*)

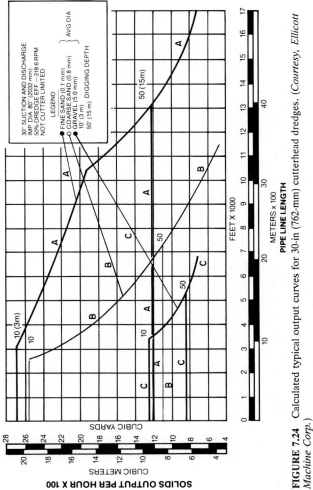

FIGURE 7.24 Calculated typical output curves for 30-in (762-mm) cutterhead dredges. (*Courtesy, Ellicott Machine Corp.*)

A cutterhead dredge operating with a booster pump in the discharge line is shown in Fig. 7.18. Typical cutterhead production rates for fine and medium sand and dredging depths from 10 to 50 ft (3.05 to 15.2 m) are shown in Fig. 7.19. Solids output curves for a series of cutterhead suction dredges built by American manufacturers are shown in Figs. 7.20 through 7.24 (as shown on pages 7.18 through 7.21).

OPERATING CHARACTERISTICS OF CUTTERHEAD DREDGES

A survey was made to evaluate the operating characteristics of cutterhead dredges in this country and overseas.[4] The survey was made to

1. Determine the physical characteristics of the dredges
2. Find out how many cutterhead dredges are equipped with modern instrumentation
3. Determine the average crew size on the dredges
4. Determine the estimated percentage of maintenance time required
5. Find out whether cutterhead dredges presently owned can operate in waves and swells
6. Determine the type of pipelines used
7. Compare the dredging practices of the United States and Canada with those of Europe and Asia

The survey covered cutterhead dredges from 6 to 42 in (152 to 1067 mm) in size. Some of the responses from U.S. contractors follow:

QUESTION   Size of crew on an 8-hr shift. Give answers for short pipeline operation and for long pipeline operation.

ANSWER   Average size crew for short pipeline: 9.0
Range in crew size for short pipeline: 2–37
Average size of crew for long pipeline: 14.4
Range in crew size for long pipeline: 2 to 55

QUESTION   Estimate percentage maintenance time required on your dredges.

ANSWER   Smaller size, 12 to 18 in (304 to 457 mm): 14.4%
Medium size, 20 to 27 in (508 to 686 mm): 12.9%
Larger size, over 28 in (711 mm): 11.8%

QUESTION   Estimate the net operating time in calm water and waves up to 3 ft (0.9 m).

ANSWER

Calm waters	%	hr/day
Ave.	81.5	17.9
Range	20–100	8–24
Waves		
Ave.	57.9	15.9
Range	4–80	4–24

QUESTION   Can you operate the dredge in the following size swells and wind waves?

ANSWER   2 to 3 ft (0.61 to 0.91 m):

        waves (swell)   Yes—58%   No—42%

        wind waves   Yes—79%   No—21%

    3 to 4 ft (0.91 to 1.2 m):

        waves (swell)   Yes—35%   No—65%

        wind waves   Yes—40%   No—60%

    over 5 ft (1.52 m):

        waves (swell)   Yes—10%   No—90%

        wind waves   Yes—11%   No—89%

QUESTION   Can you operate a floating pipeline in the following size swells and wind waves?

ANSWER   2 to 3 ft (0.61 to 0.91 m):

        waves (swell)   Yes—74%   No—26%

        wind waves   Yes—84%   No—16%

    3 to 4 ft (0.91 to 1.2 m):

        waves (swell)   Yes—34%   No—66%

        wind waves   Yes—44%   No—56%

    over 5 ft (1.52 m):

        waves (swell)   Yes—11%   No—89%

        wind waves   Yes—15%   No—85%

QUESTION   Do you operate with spuds?

ANSWER   Yes—96%   No—4%

QUESTION   Do you operate with wires and anchors in exposed areas?

ANSWER   Yes—64%   No—36%

QUESTION   Have you used a floating buoyant pipeline (which floats without supporting pontoons)?

ANSWER   Yes—10%   No—90%

QUESTION   Have you generally used ball joints in floating pipelines?

ANSWER   Yes—73%   No—27%

QUESTION   Have you used rubber sleeves instead of ball joints in floating pipelines?

ANSWER   Yes—67%   No—33%

    A summary of comparisons between the United States and Canada and overseas countries is given in Tables 7.1 and 7.2.

Greater Dredging Depths

There are two possible ways of increasing dredging depth on a cutterhead dredge: (1) locate a dredge pump on the ladder, and (2) introduce a jet pump in a suction line. Both methods have been tried and have been found to work satisfactorily.

    As pointed out in Chap. 3, cavitation limits the operation of the pump; however, the available NPSH may be increased by placing the pump on the ladder below the water surface. There are a number of manufacturers who can supply a ladder pump.

TABLE 7.1 Summary—Equipment, Instrumentation, Crew Size, and Maintenance

	U.S. & Canada		Overseas		Notes
	Number	%	Number	%	
Self-propelled cutterhead dredges		2.1		18.2	There are more self-propelled dredges overseas.
Dredges with pump on the ladder		10.4		30.0	To increase digging depth, dredges equipped with ladder pumps are becoming more popular, particularly overseas.
Swell-compensating device on the ladder		0		8.1	Swell-compensating devices are not in use at all in the United States and Canada, but 8% of foreign dredges are equipped with the device.
Instrumentation					
Magnetic flow meter		2.2		77.2	Overseas dredges are much more instrumented than U.S. and Canadian dredges.
Density meter		2.3		69.6	
Total production meter		6.6		68.2	
Size of crew on an 8-hr shift					No significant differences.
Short pipeline	9.0		9.6		
Range	2–37		1–35		
Long pipeline	14.4		10.8		
Range	2–55		2–40		
Percentage maintenance time required					
12 to 18 in (304 to 457 mm) in size		14.4		15.0	More maintenance time allowed for larger dredges overseas.
20 to 27 in (508 to 686 mm) in size		12.9		15.7	
Over 28 in (711 mm)		11.8		17.3	

TABLE 7.2 Summary—Operational Characteristics of Cutterhead Dredges in Open Water

	U.S. & Canada		Overseas		Notes
	Number	%	Number	%	
Dredge operation in waves (swell)					U.S. and Canadian dredges can operate under slightly higher wave (swell) conditions.
2–3 ft (0.6–0.9 m)		58		73	
3–4 ft (0.9–1.2m)		35		19	
Over 5 ft (1.5 m)		10		0	
Dredge operation in wind waves					U.S. and Canadian dredges can operate under slightly higher wind wave conditions.
2–3 ft (0.6–0.9 m)		74		83	
3–4 ft (0.9–1.2 m)		40		32	
Over 5 ft (1.5 m)		11		4	
Operation of a floating pipeline in waves (swell)					U.S. and Canadian contractors operate floating pipelines under slightly higher wave (swell) conditions.
2–3 ft (0.6–0.9 m)		74		77	
3–4 ft (0.9–1.2 m)		34		27	
Over 5 ft (1.5 m)		11		8	
Operation of a floating pipeline in wind waves					U.S. and Canadian contractors operate floating pipelines in slightly higher wind waves.
2–3 ft (0.6–0.9 m)		84		79	
3–4 ft (0.9–1.2 m)		44		29	
Over 5 ft (1.5 m)		11		8	

	Yes (%)	No (%)	Yes (%)	No (%)	Notes
Use of floating buoyant pipeline	10	90	26	74	Greater use of floating buoyant pipelines overseas.
Use of ball joints in floating pipelines	73	27	74	26	No significant differences.
Use of rubber sleeves in floating pipelines	67	33	69	31	No significant differences.

EFFECT OF A LADDER PUMP ON THE CAVITATION CHARACTERISTICS OF A CUTTERHEAD DREDGE[5]

Giulio Venezian

Associate Professor, Physics Department,
Southeast Missouri State University

A cutterhead dredge system of fixed geometry is analyzed to determine the head required to achieve a given discharge of mixture for various concentrations. These relationships determine the characteristics of the pumping system needed for the operation of the dredge in the given configuration.

Other parameters calculated are the power required, the weight of solids delivered per unit time, and the net positive suction head available at the pump.

The optimal design would have the best efficiency of the pump coinciding with the highest weight of delivered solids per unit of energy expended. There are various limitations on this optimization: the maximum power available, the minimum permissible suction head for the pump, the maximum concentration of solids that can be carried, and the characteristics of the cutterhead and seafloor, which combine to give a relationship between discharge rate and concentration of sand.

Of these four limitations, only the suction head limitation can be avoided easily. The power limitation will be taken as an absolute one, limited by capital cost, size, and weight. The maximum permissible concentration of solids that can be pumped is unknown, but in the context of this chapter it will be assumed to be 35 percent, which is the approximate limit of validity of the Durand-Gibert[6] relations used in the analysis. While higher concentrations can undoubtedly be pumped, the head requirements increase rapidly so that this limitation becomes similar to the power limitation. Finally the characteristics which determine the concentration for a given flow rate are not known at this time. This is an area where further studies are needed if a rational design of a dredging system is ever going to be carried out.

DESCRIPTION OF THE SYSTEM

An idealized dredging system is shown in Fig. 7.25. It consists of a suction pipe, one or more pumps, and a discharge line. The system is basically the one analyzed by Basco[7] except that the system geometry is kept fixed and some details of the analysis have been changed.

In the configuration shown, l_s is the length of the suction line, l_d is the length of the discharge line, d is the water depth, and e is the elevation of the pump and discharge above the water surface. An alternative configuration is also considered, in which an auxiliary pump is placed on the ladder at a distance l_a from the inlet.

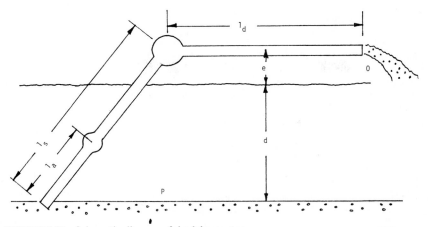

FIGURE 7.25 Schematic diagram of dredging system.

In analyzing the system, it is assumed that the fluid in the lines is a slurry containing a concentration C of solids (volume of solids per volume of mixture) while the fluid surrounding the pipe is clear fluid of specific weight γ_1.

The pressure at a point P has the same elevation as the inlet is thus γd, and if it is assumed that slurry is drawn into the inlet starting from rest at point P, the Bernoulli equation between P and the discharge point 0, where the pressure is atmospheric is

$$\gamma_1 d + \Delta p_p = \frac{1}{2}\rho_m V_D^2 + \Delta p_s + \Delta p_d + \gamma_m(d + e) \qquad (7.11)$$

where Δp_p is the pressure increase across the pump, Δp_s and Δp_d are the pressure drops due to frictional losses in the suction and discharge lines, and V_D is the velocity of the fluid at the discharge.

If the flow rate is given, the velocities in the suction and discharge lines can be found, and hence the pressure drops can be calculated for any given concentration. The Bernoulli equation then gives the pressure rise required across the pump to achieve the desired flow rate, and the power can then be calculated.

The stagnation pressure at the suction inlet of the pump is given by

$$p_{ss} = \gamma_e d - \gamma_m(d + e) - \Delta p_s$$

The available positive suction head relative to the vapor pressure of the liquid (expressed in feet of mixture) is

$$h_s = \frac{p_{ss} + p_a - p_v}{\gamma_m} \qquad (7.12)$$

where p_a is the local atmospheric pressure and p_v the vapor pressure of the liquid.

LOSSES

The literature on losses in a pipe when slurry is flowing traditionally expresses the losses as a head of clear liquid, so that $\Delta p = \gamma_1 h_L$.

The formulation used here will follow the results of Durand[6] and Gibert.[8] The head loss in a horizontal pipe carrying mixture $(h_L)_m$ is expressed in terms of the corresponding head loss for the same conditions carrying clear fluid $(h_L)_f$ as follows:

$$(h_L)_m = (1 + C\phi)(h_L)_f \tag{7.13}$$

where C is the concentration and ϕ is a function which does not involve the concentration and is given by

$$\phi = 121 \left[\frac{DV_s}{V^2} \sqrt{\frac{g(s-1)}{d}} \right]^{3/2} \tag{7.14}$$

Here D is the pipe diameter, d the median particle diameter, s the specific gravity of the sediment (relative to the conveying fluid), and V_s is the settling velocity of the particles.

This equation applies only when there is no settling of the sediment to the bottom of the pipe, a condition which holds only when the velocity is larger than a critical velocity V_c which Durand expressed as

$$V_c = F_L[2gD(s-1)]^{1/2} \tag{7.15}$$

where F_L depends on the concentration and median sediment diameter.

Gibert found that the relation could be applied to a partially blocked pipe by assuming that the blockage would decrease the effective diameter of the pipe and thus the critical velocity until the point would be reached when the velocity became equal to the critical velocity for that blockage and no further blockage would occur. Expressing the effective diameter as 4 times the hydraulic radius (R_h), this equilibrium blockage would be reached when

$$\frac{V^2}{4R_h} = \frac{V_c^2}{D} \tag{7.16}$$

For inclined pipes, Gibert found that ϕ should be replaced by $\phi(\cos\theta)^{3/2}$ in Eq. 7.13. Worster and Denny[9] proposed that ϕ should be replaced by $\phi\cos\theta$, and this relation was used in the calculations presented here.

It should be noted that in the case of an inclined pipe the question of partial blockage does not arise, unless the slope is very small.

CALCULATIONS AND RESULTS

Calculations were performed for a system described as follows:

Length of suction line: 100 ft (30.5 m)
Length of discharge line: 2000 ft (610 m)

Suction pipe diameter: 33 in (838 mm)

Discharge line diameter: 30 in (762 mm)

Median particle diameter: 0.0197 in (0.5 mm)

Settling velocity: 0.21 ft/s (0.064 m/s)

Concentrations were varied from 0.02 to 0.25, and digging depths of 40 and 70 ft (12.2 and 21.3 m) were considered.

Figure 7.26*a* and 7.26*b* shows the head versus discharge curves for different concentrations for digging depths of 40 and 70 ft (12.2 and 21.3 m), respectively. The horsepower requirement is also indicated on the curves. There is no significant variation of the curves with digging depths, indicating that the frictional loss is dominant for this geometry. The head loss increases with concentration and the power requirements increase both with flow rate and concentration. There is thus a limitation on the system imposed by the maximum power available.

Figure 7.27*a* shows curves of available NPSH at the pump superposed on the head versus discharge curves for a digging depth of 70 ft (21.3 m). These curves are nearly parallel to the concentration curves, indicating that there is a maximum concentration that can be conveyed by the system.

Of course the NPSH alone is not a suitable criterion, since the NPSH required by a pump to avoid cavitation varies with flow rate and rotational speed. Nevertheless, the curves do indicate that the available NPSH decreases with concentration.

The efficiency of the system in terms of energy cost per unit weight of solids delivered increases with concentration; therefore, if the NPSH limitation can be removed, the efficiency of the system can be improved. Figure 7.27*b* shows the effect of a ladder pump on available NPSH. The calculations were done with an auxiliary pump halfway up the ladder, assuming that the pressure increase across

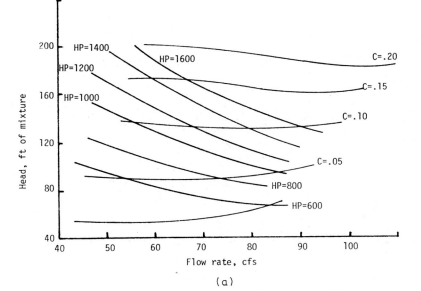

(a)

FIGURE 7.26 Head as a function of discharge curves for different concentrations. (*a*) Dredging depth = 40 ft (12.2 m).

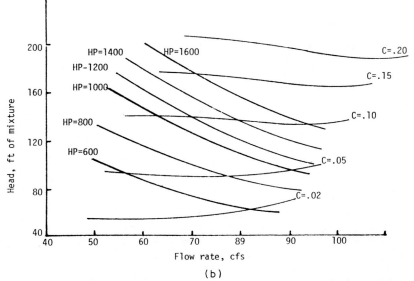

(b)

FIGURE 7.26 (*Continued*) Head as a function of discharge curves for different concentrations. (*b*) Dredging depth = 70 ft (21.3 m).

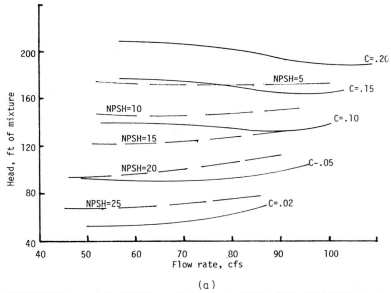

(a)

FIGURE 7.27 Available NPSH at a digging depth of 70 ft. (21.3 m) (*a*) Single pump.

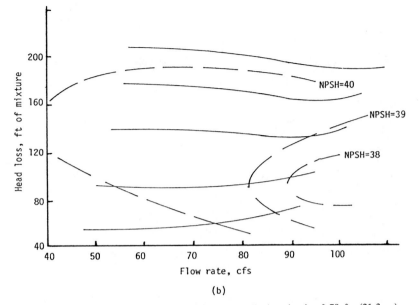

FIGURE 7.27 (*Continued*) Available NPSH at a digging depth of 70 ft. (21.3 m) (*b*) with auxiliary pump on ladder.

the auxiliary pump is 20 percent of the total head. As it can be seen from the resulting curves, the available NPSH at the main pump is virtually constant over the range of values computed. The auxiliary pump thus eliminates the NPSH limitation.

It must be pointed out, however, that the Durand-Gibert formulation does not hold for large concentrations, so the increase in efficiency with concentration has a limit also. Moreover, the concentration itself is not a controllable variable since it depends on the inlet conditions and the characteristics of the sediment. Optimization of the process is still an operator function, although calculations of the type described here can be of value in the design of the dredging system and indicating desirable operating conditions.

OCEAN DREDGING*

Hopper (Trailing Suction) Dredges

A seagoing hopper dredge has the molded hull and shape of an ocean vessel and functions in a manner similar to that of the plain suction type of dredge. The bottom material is raised by dredge pumps through dragarms, which are connected to the ship by trunnions and are designed to trail over the side of the vessel. The lower ends of the dragarms are equipped with dragheads, which are designed to pick up the maximum amount of bottom sediments. The dragarms are raised and lowered by hoisting tackle and winches while sailing at low speed. The pumps lift

---

*The following material has been written by John B. Herbich.

the mixture through the dragheads to the surface where it is discharged into hoppers. As pumping continues, the solid particles settle in the hoppers and the excess water passes overboard through overflow troughs. After the hoppers have been filled, the dragarms are raised and the dredge proceeds at full speed to the disposal site and empties the loaded hoppers through bottom doors. The doors then close and the dredge returns to the dredging area to continue the cycle.

American dredges operate with dragarms trailing at a ground speed of 2 to 3 knots (3.7 to 5.6 km/hr). Hopper dredges range in size from approximately 180 to 550 ft (55 to 168 m) in length and have hopper capacities between 500 and 8000 yd³ (382 and 6116 m³). They are equipped with twin propellers and twin rudders to provide the required maneuverability. Dredging depths vary from 10 to over 100 ft (3 to over 30.5 m). Some of the hopper dredges built overseas have hopper capacities up to 14,780 yd³ (11,300 m³) and can dredge from up to 135 ft (40 m).

Dredges of this type are necessary for maintenance work and improvement in exposed harbors and navigational channels where traffic and operating conditions rule out the use of cutterhead dredges or are employed to haul sands for reclamation projects. These dredges can also be equipped for "agitation dredging" where soft or free-flowing alluvial materials are picked up and then discharged directly overboard through a suspended discharge pipe, thus eliminating storage in the hoppers. The material is then carried out of the dredging area by currents and stream action. This method of dredging only applies to clean materials and is generally ruled out if contaminated materials are present.

Variables Affecting Hopper Efficiency

There are a number of variables affecting hopper efficiency, which may be separated into three groups:

1. Geometric variables
2. Kinematic and dynamic variables
3. Fluid properties

The geometric variables include the beam (B), length (L), and draft (d) of the hopper; shape of the cone (whether circular, triangular, etc.) (s); the angles of the cone (α_1, α_2); whether single slopes or composite sloped; number of baffle plates (n); and size of baffle openings (d_b). A simplified hopper showing the geometric variables is presented in Fig. 7.28.

The kinematic variables include the settling velocity of particles (V_t); discharge velocity of mixture into hopper (V_d); bed load transport velocity (V_{tr}); mixing velocity (V_m); velocity of mixture induced by ship pitching (V_p); heaving (V_h); rolling (V_r); and acceleration due to gravity (g).

The fluid properties include density (ρ) and viscosity (μ). Note: Viscosity may not be a constant, but a variable, if the fluid exhibits properties of a nonnewtonian fluid.

Thus the efficiency (η) of the hopper may be a function of a great number of variables:

$$\eta = f(B,L,d,s,\alpha_1,\alpha_2,n,d_b,V_t,V_d,V_p,V_h,V_r,V_m,V_{tr},g,\rho,\mu) \qquad (7.17)$$

It is obvious that the list of variables is formidable and the problem of designing an efficient hopper for dredging operations is quite complicated. It is not sur-

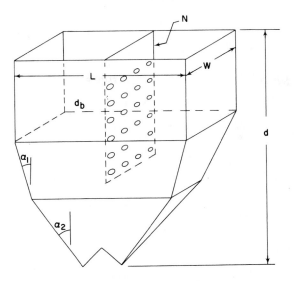

FIGURE 7.28 Hopper definition sketch.

prising to note that no standard designs have been developed since no basic methods employing hydrodynamic principles are available.

Physical Considerations

The solid particles settle in the hopper by gravity. If the solid particles are large (stone, gravel, or coarse sand), the settling velocity is high, and if they are small (sand, silt, or clay), the settling velocity is low. In the former case, the particles will settle quickly and very few solids will be found in the overflow, while in the latter case the settling will be a slow process and a large percentage of solids will be lost in the overflow. Since the large percentage of solids in the overflow not only causes an inefficient operation but also may silt up the channel being dredged or may contribute to sedimentary pollution, the reduction in overflow losses becomes very important.

Theoretically the solution for an efficient operation is to use large hoppers equipped with small inverse funnels for filling the hopper and a low rate of discharge into the hoppers. Of course, the idea is to keep all solid-water mixture and eddy velocities as low as possible. However, the dredging time would certainly be unreasonably increased and any physical restrictions in the hopper, such as funnels, would probably become clogged with tree roots, debris, etc., in many dredging operations. The vertical baffles are generally used, however.

It was earlier assumed that the filling of a hopper is a gravitational phenomenon. However, in the last stages of filling, the velocities near the load bed in the hopper may become large enough to produce sediment movement along the bed and scouring action. The Froude phenomenon is then accompanied by the Reynolds phenomenon, which complicates the problem, particularly if one seeks solutions through model testing.

In addition, the dredging operation (pumping rate and concentration of solids) is seldom of the steady-state type, and the fluid mechanics involved are further complicated by ship motion and possibly by the ship's vibration.

Model Tests

Model hopper tests were conducted several years ago in the Netherlands.[10] Although the study was limited to the mechanics of filling the hopper and overflowing, it does provide some guidelines for the designer. A model test facility is shown in Fig. 7.29.

Figure 7.30 presents typical results of tests. In this figure the loss of sand (as percentage of sand pumping rate) is plotted as a function of sand retained (in percentage of hopper capacity) for several systems tested. Figure 7.31 presents the total volume of sand retained in the hopper as a function of pumping time in minutes for several geometric configurations tested.

Bucket-Wheel Dredges

American and Dutch designers have developed a dredging wheel which replaces the cutter in cutterhead dredges (Figs. 4.51 through 4.53).[11] In the dredging wheel the buckets are bottomless. By placing the buckets close together—overlapping—a tunnel is created, the inner limitation of which is the suction mouth itself. Immediately after excavation, therefore, the soil is within the sphere of hydraulic suction. The process sequence is mechanical excavation followed by suction. A bucket-wheel produces a substantially higher output of solids than does a conventional cutter suction dredge. The bucket cutting edges are fitted with hard-

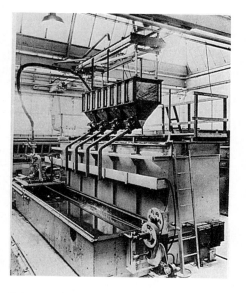

FIGURE 7.29 Model hopper test facility. (*Courtesy, U.S. Army Corps of Engineers*)

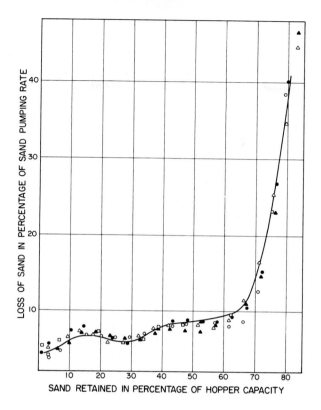

FIGURE 7.30 Instantaneous losses in overflow system III. Pumping rate 1060 ft³ (30 m³) of mixture per hour. (*Courtesy, U.S. Army Corps of Engineers*)

wearing blades and offer a choice of quick-change blades on adapters. The new Dutch (IHC Holland) range consists of six wheel sizes with outputs varying from 602 to 3140 yd³/hr (460 to 2400 m³/hr). The dredging wheels are supplied as complete units inclusive of drive and adapter. These can be installed on new dredges or replace cutter units on existing dredges.[12]

In the American design of the bucket wheel the material which is dug is directed into the interior of the wheel where it is immediately conveyed up the suction line. Modulation of the solids rate can be achieved by the dredge. The submerged dredge pump has certain advantages over the dredge pump in the hull: (1) it can increase the concentration of the dredge slurry in the pipe, and (2) it allows dredging in greater depths. One serious problem with hydraulic dredges is the wear of the pumps. To deal with this problem two new types of pumps have been developed; one in the Netherlands and the other in the United States. The principal feature of the Dutch pump is a double-walled housing, the inner wall of which is replaceable and allowed to virtually wear out. A highly important aspect of the design is the provision of means to compensate for differences in pressure inside and outside the housing during operation. This is achieved by filling the

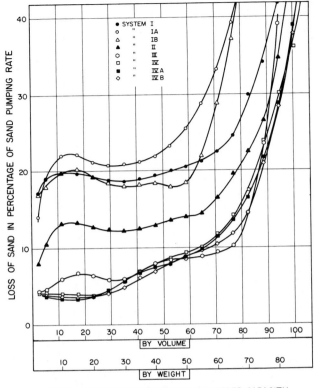

FIGURE 7.31 Instantaneous losses in overflow for systems tested. Pumping rate 1060 ft³ (30 m³) of mixture per hour. (*Courtesy, U.S. Army Corps of Engineers*)

space between the walls with water so as to provide a cushion that will transfer stresses from the inner wall to the outer wall, which is of welded steel construction.[13] This pump is ideally suited to conditions in which extremely high resistance to wear is demanded. The cost of the double-walled pump is almost double that of a single-walled pump but the maintenance required is very low (Fig. 7.32). In the American design there are two types of pumps to deal with the problem of wear. In one case the volute is lined; in the other case the volute is unlined.

Effect of Waves on Operation of Cutterhead Dredges

Most of the cutter suction dredges are designed to operate in moderate swells of 3.3 ft (1 m) or so. The feasibility of cutter suction dredging in swell conditions is increasingly occupying the attention of those concerned with the technological

(a)

(b)

FIGURE 7.32 Double-wall dredge pump. The smaller part is designed to allow inspection and changing of the impeller. The larger casing is only removed for the purpose of fitting a new inner housing (*Courtesy, IHC Holland*)

aspects of dredging. The main problem is that in order to achieve optimum output, the cutter must be kept fairly accurately in position and its uncontrolled movements must be kept within certain tolerances. When the soil to be dredged is hard, these tolerances should be small to achieve good output. Excessive freedom of movement results in heavy impact between the cutter and the hard material at the bottom, resulting in damage to the cutter and the ladder. Various devices have been thought of as an alternative to the present method of cutter dredging. These are (1) stabilizing the floating platform by increasing the size of the pontoon, introducing catamaran construction, and semi-submersible design, (2) introducing an active or passive spring system, (3) a fully submerged dredge, and (4) a self-elevating platform. A walking cutter suction dredging platform is the latest in the new development of cutter dredges. This type of equipment can continue to work in waves up to 14.8 ft (4.5 m) in height with a 7.2-s period.

The design consists of two pontoons linked by two lattice girder bracing members. The cutter ladder is suspended between the two pontoons. The legs are located at the extremities of the pontoons, which can be moved longitudinally. There are four main legs and four auxiliary legs. Under survival conditions all eight legs are lowered. The cutter installation is similar to that provided on heavy-duty cutter dredges. The disposal of the dredged material is accomplished through a suspended pipe 29.5 in (750 mm) in diameter. The disposal material is carried by a dynamically positioned dredged material carrier.

The Dutch have also designed a multipurpose semi-submersible self-elevating cutter dredge. The main advantage of such platforms is that the cutter ladder is independent of the wave motion, the position of the cutter can be accurately determined at any moment, and production can be kept at a constant rate. Moreover, there is no anchoring problem when dredging is at a stationary position. The platform is designed to work in 8.2-ft (2.5-m) waves. The dredge was developed and constructed at great expense but is no longer in operation.

Herbich and Lou[14] discussed the requirements for a seagoing cutterhead dredge:

1. An offshore dredge must be capable of operation in open seas. This requires a minimum motion response to waves and high maneuverability and control. According to Turner,[12] an offshore dredge which can operate safely and efficiently in seas with a wave height of 6.6 ft (2 m) would establish a 95 percent on-the-job record along the continental United States coasts.

2. Since an offshore dredge must move frequently from site to site, it must have high mobility and low drag during transit.

3. It should be able to dig to a great depth, say, up to 100 ft (30.5 m). Since the maximum depth that a dredge can dig is limited by the barometric pressure, this would require a pump being installed well below the ocean surface.

4. It should have a high operating efficiency, low costs, and a large capacity of 2000 yd^3/hr (1529 m^3/hr) or higher.

A standard design cutterhead dredge built to operate in calm waters cannot function offshore or in estuaries where waves are over 2 to 3 ft (0.61 to 0.91 m) in height. The main difficulty is caused by the excessive movement of the ladder, which may be forced into the sand bed or bounced off the bottom, creating excessive shock loads on both ladder and trunnion. Another problem relates to a connection between the discharge line and the floating pipeline. The dredge and the floating pipeline have different dynamic responses to wave action and this creates large vertical and horizontal forces on the connection. The ball joints, which have been used on some projects, tend to jam in the extreme positions and eliminate the flexibility for which they were designed.

Seagoing cutterhead dredges should have:

1. A specially designed hull to minimize rolling, pitching, yawing, swaying, and heaving motions of the platform, yet still be able to achieve high mobility and maneuverability

2. An efficient swell-compensating device on the ladder

3. A well-designed connection between the discharge pipe on the dredge and the submerged pipeline

Four modifications appear to have merit:

1. A semi-submersible or catamaran-type twin-hull to provide a stable dredging platform
2. An instantaneously responsive self-compensating device on the ladder or an articulated ladder described by Turner[12]
3. A neutrally buoyant submerged pipeline
4. Rudderless side-thruster self-propulsion.

Although semi-submersible dredges provide a very stable platform, a seagoing cutterhead dredge appears to be most promising because of its versatility and easy adaptation to a variety of dredging jobs offshore. The dredge must be seaworthy with a capability to survive in 15-ft- (4.5-m-) high waves and work efficiently in waves up to 6 ft (1.8 m) in height. A semi-submersible catamaran hull appears to be most suitable for development of a seagoing cutterhead dredge.

Accuracy of Dredging Processes[15]

Accuracy depends on the types of equipment used, the sediments encountered, control of the dredge's position, whether the work to be performed is new or maintenance dredging, and in the latter case, the previous dredging work. It can be strongly influenced by such local conditions as tides and currents and by the accuracy of pre- and postdredging surveys.

Type of Equipment. A dredging plant is diverse; types used in the United States for dredging major navigational channels can be categorized as fixed relative to the channel bottom or independent.

Fixed Equipment. Fixed equipment includes cutter suction dredges with a spud pole (pilelike leg) firmly implanted in the channel bottom and the dredge itself rotating about this axis. The distance from the cutterhead to the spud pole is fixed mechanically, as is the depth of the cutter beneath the surface. In theory, the only variable is the angular motion of the dredge itself as it pivots about the fixed spud pole. This motion is usually controlled by cables attached to an anchor.

It is imperative that the dredge operator knows the exact location of the spud pole and the relative position of the cutterhead to the spud pole as well as the transverse position of the cutterhead to the channel centerline at all times. The position of the spud pole can be assumed to be known with an accuracy of less than 5 ft (1.5 m). Spud placement has little effect on the dredging operation, except in terms of the horizontal cut. Therefore, the position of the spud pole in normal maintenance dredging is not of prime concern. Electronic positioning equipment or gyrocompasses are used to measure the deflection of the dredge from the channel heading. With the use of these two control methods, the accuracy of the channel width can be controlled within 5 to 10 ft (1.5 to 3.0 m).

Other fixed-spud plants are grab dredges and dipper dredges. They differ from the cutter suction dredge in that their hulls do not move relative to the spud pole. The bucket used for the excavation is moved relative to the hull. The movement of the excavating bucket can be controlled more precisely than that of the cutterhead.

Independent Equipment. Hopper dredges are the most common among independent equipment. Since they are not fixed relative to the channel bottom, they have no fixed point of reference for determining the accuracy of operations. The head attached to the suction pipe acts very much like a household vacuum cleaner, with the exception that the hopper dredge drags its suction head ("draghead") while the cutterhead is pushed forward in dredging operations. Unlike the cutter suction dredge, the hopper dredge removes a very thin layer of the bottom material as it dredges. It must therefore traverse the area to be dredged many times before the channel is substantially deepened. Electronic positioning and track plotters are used to indicate those areas that have been traversed. Continued surveys are required to assist the operator in controlling the dredging operation. Occasional passes outside the dredging area or in areas that have already been dredged entail very little damage or lost effort because only 2 or 3 vertical in (5 to 7.6 cm) of material are removed.

The accuracy of hopper dredges is much more dependent on the physical shape and dimension of the area to be dredged than that of the fixed dredges, which can control their location more precisely. The accuracy that can be achieved also depends on the type of soil encountered.

Another example of independent dredges is the dustpan dredge which is propelled by cables attached to anchors using propulsion devices to aid steering. This dredge, like a cutter suction dredge, excavates a substantial depth of material at one time. It therefore moves much more closely and, because of the anchoring system, has much better control of its horizontal position. It relies very heavily on electronic positioning. The vertical control of the dustpan is such that accuracy within 1 ft (0.3 m) can be achieved and horizontal control within 3 ft (0.9 m).

Type of Bottom. The accuracy of the dredging process is affected by a change in the material encountered. Cohesive or hard soils lead to the development of trenches. Different materials may dictate the dredging method. Certain hard materials such as a sandstone or limestone cannot be dredged with hopper or dustpan dredges, since these types of dredges generally require the material to be loose and free-flowing. Such hard materials can be dredged with a cutterhead dredge.

Achieving Design Prism. Again, the accuracy of the equipment remains unchanged, but the quantity of no-pay material dredged to ensure achieving the design prism can vary significantly. Dredging narrow shoals adjacent to the slope with a hopper may result in dredging 200 to 300 percent of the pay quantity. Rock often must be cut 3 to 4 ft (0.9 to 1.2 m) below grade to preclude strikes during a bar survey.

Environmental Conditions. Landlocked channels present fewer problems than open channels for controlling the dredging process and for pre- and postdredging surveys. Exposed waters allow less control of both operations. Tides, especially those of some magnitude, affect dredging and surveying, particularly if the gauge is far from the site. Where the location of the tide-gauge station is known to be unrepresentative of the dredging site, significant no-pay overdepths may be dredged to ensure that the required depths will be measured in the postdredging survey. Currents have a marked effect on hopper dredge accuracy, influencing not only the movement of the ship but the attitude of the ship in relation to the channel.

New Work versus Maintenance. The accuracy of dredging is little affected by the change from new work to maintenance. However, there may be substantial variations in the no-pay volume, depending on the type of material, side slopes, previous dredging, consistency of the material in the cut, underlying material, and other variables.

Examples of Accuracy of Dredging Operations

Cutterhead Dredge. A typical section of maintenance dredging by a cutterhead dredge is shown in Fig. 7.33. A channel such as the one shown is dredged in two

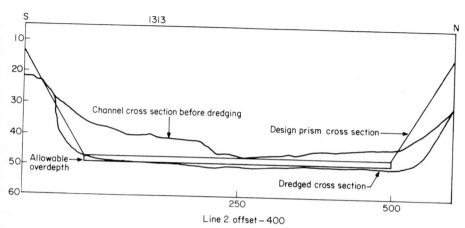

FIGURE 7.33 Sample maintenance dredging work by cutterhead dredge.

passes. How the maintenance dredging is accomplished depends principally on how the original channel was dredged, the type of soil encountered within design depth (silt or sand), and the virgin material encountered just below grade. If the channel was previously well overdredged, maintenance dredging will be easy. This is shown clearly in the cross section, particularly on the right side (250 to 500 ft, 76 to 152 m) of the channel. Here the maximum swing speed of the dredge is the limiting factor for production, because of the low bank, so if no hard (virgin) material is encountered, the total pay quantity can be removed with dispatch. The channel is dredged too widely on the right side because of the previous overdredging. Almost all maintenance material must be removed to achieve grade in the corners (toes) of the cut, as can be seen on the left side, where higher banks were encountered and where the virgin slope is close to the required prism.

New Work in Protected Waters. Figure 7.34 shows a typical section for new work in protected waters performed by a cutterhead dredge. Positioning was very accurate; stakes could be set out because there was no traffic, and the water was calm and shallow. The soil consisted mainly of soft to medium clays, which is ideal for cutting slopes, as can be seen in the cross section. However, the channel is overdug on one side because of the inaccuracy of width indication and spud position. No electronic positioning system was used on this dredge, but the cross section indicates very accurate dredging to the required depth, which can be attributed to good tidal information and (in this case) small tidal differences.

New Work in Unprotected Water. The cross section pictured in Fig. 7.35 is for a hydraulic dredging project 10 mi (16 km) offshore. Use of an electronic positioning system was essential to this project, but accuracy can vary between 7 and 50 ft (2.1 and 15.2 m). This explains why the channel is overexcavated on both sides. Depth control is very difficult in open waters because of the movement of the dredge hull and consequent problems controlling the position of the draghead.

Maintenance Work in Sand and Silty Sand. The irregular bottom made by a hopper dredge in progress is illustrated in Fig. 7.36. For width accuracy, the dredger relies on electronic positioning and thus experiences the problems noted for cutterhead dredges. Moreover, this type of dredge is free-sailing, which necessi-

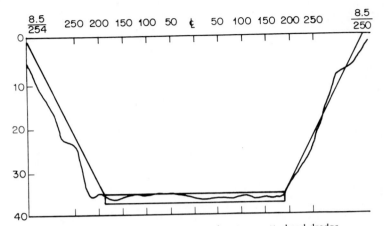

FIGURE 7.34 Sample new work in protected waters, cutterhead dredge.

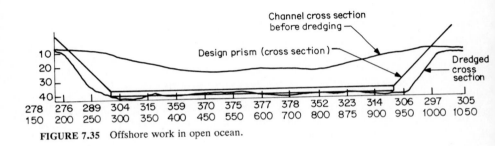

FIGURE 7.35 Offshore work in open ocean.

tates maneuvering the ship while dredging. Assuming that experienced personnel are operating the dredge, the important factors in control of the ship are currents (especially cross-currents), winds, and swells since these factors influence position. Another factor important to accurate work with a hopper dredge is frequent surveying since a hopper dredge gradually brings a large area to the required depth.

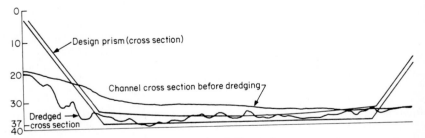

FIGURE 7.36 Maintenance dredging in sand and silty sand.

Maintenance Work in Silt and Soft Clay. If the material is soft, overdredging is likely, as is shown in the cross section of Fig. 7.37. Soft clay was encountered in the corners, which is much more difficult to remove with a hopper dredge than with other types of dredges.

Implications

The overdredged depths actually left by dredging are likely to be greater than those allowed by the pay-overdepth specified. Estimates compiled by Lacasse[16] of overdredging to achieve design prism indicate it may represent 10 to 15 percent of the total volumes dredged. As implied by the examples, the pay-overdepth specified is an incentive to dredging accuracy. Although each case will be different, accurate pre- and postdredging surveys may yield multiple benefits—better depth information for navigation, for example, and increasingly precise pay-overdepth specifications.

Sand Transfer at Harbor Entrances

The importance of the coastal zone of the United States continues to increase. The growth of population in the coastal states is an important factor for the increase in demand for residential, commercial, and recreational utilization of the coastal zone. An increased number of small boat owners has created demand for new and improved small-craft harbors. The progressive increase in size and draft of vessels, serving shipping interests, necessitates new and improved coastal harbor entrance channels and inland waterway systems.

For many years, the principal source of sand for beach rehabilitation and nourishment purposes was from lagoonal and inland deposits. However, during the past decade it has become increasingly difficult to obtain suitable sand from such sources in sufficient quantities and at economical costs primarily because of public and private developments in the estuarine areas. Also in many instances,

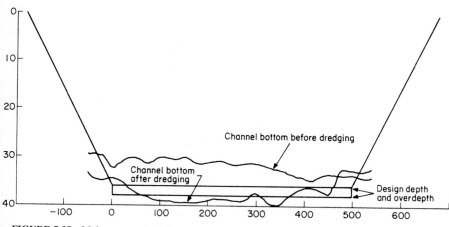

FIGURE 7.37 Maintenance dredging in silt and soft clay.

bottom materials dredged from estuaries, lagoons, and bays, because of their finer nature, are unsuitable for long-term beach stabilization purposes. Even if suitable material exists in such tideland areas, with the advent of environmental constraints its use is prohibited in most cases because of potential ecological imbalance or damage that may result. Consequently, increasing attention is being given to the development of techniques to use sand dredged from navigation channels and from the offshore zone for beach nourishment purposes.[17]

Types of Protected Entrances

The type of entrance structure, along with orientation and layout, is important to the consideration of viable alternatives and selection of the best-suited method of mechanically transferring sand from the inlet complex to adjacent shores. Figure 7.38 illustrates four structure configurations[17] normally used at harbor entrances and the littoral material impoundment zone for each. As implied in each config-

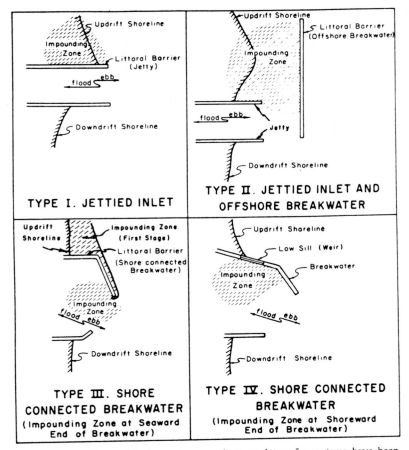

FIGURE 7.38 Types of harbor entrances where sand transfer systems have been used. (*Courtesy, U.S. Army Corps of Engineers*)

uration, the accretion zone is predicated on conditions that alongshore littoral transport is dominant in one direction.

Parallel Jetties (Type I).   For a parallel jettied inlet, type I in Fig. 7.38, transfer operations of the impounded littoral drift to downdrift shores can be effectively accomplished by use of dragline or comparable land-based dredging plant and land vehicles if the haul distance from the impoundment zone to downdrift shores is reasonable or compatible with economics. Also, in this type of operation there are other factors to be considered such as the use and related effects of the land-haul vehicles on public roads between the borrow and disposal sites. Transfer of the impounded littoral drift on parallel jetties can also be accomplished by a fixed or semifixed hydraulic dredging plant with the discharge pipeline extending across the entrance channel (generally submerged to not interfere with navigation) to downdrift shores. Another technique of sand bypassing for the type I parallel jettied inlet involves the use of a conventional hydraulic dredge to excavate sand impounded on the updrift side of the updrift jetty and pumped through a discharge line to the downdrift side of the harbor. The impoundment zone is exposed to the open ocean and subjected to high wave energy, thus techniques and procedures for the dredging operations must be adjusted to this physical environment.

There are many type I, parallel jettied, inlets in the United States where the longshore littoral material movement is allowed to follow the updrift impounded fillet shoreline, thence around the outer end of the updrift jetty and to become deposited in the entrance channel.

The material is then removed by mechanical or hydraulic means to maintain the desired channel dimensions to serve navigation interests. Transfer of the littoral drift from the entrance channel to the shores adjacent to the harbor has been accomplished in a number of ways. Physical and environmental factors applicable to the site-specific project generally control or limit the choice of alternative operational procedures that are viable and economically competitive to accomplish objectives. For a project located in a low wave energy environment, a number of operational alternatives are viable and competitive. However, in a high wave energy location, the entrance channel dredging is limited to a plant that can operate in those conditions and is generally limited to a seagoing hopper-type dredge. There are several alternative operational techniques for transferring the sand from the bins of the hopper dredge to the shores adjacent to the harbor. These include, but are not limited to:

1. Dumping of the hopper bin into one or more prepared and suitably protected rehandling basin(s). The material is then excavated by a conventional hydraulic pipeline dredge and pumped through a floating and onshore line to the desired discharge zone on the adjacent shore.

2. Equipping the seagoing hopper dredge with a hydraulic self-unloading system which can be coupled to the shore discharge line positioned on a moored or anchored barge which, depending on site-specific requirements, may contain equipment that functions as an integral part of the discharging operations.

3. An operational technique, which must be compatible with local physical conditions, is to dump the hopper dredge load in the nearshore zone, but as far shoreward as possible, and rely on wave-induced currents to transport the dumped material to the shore zone.

For these three alternatives, it can be stated that the rehandling and direct pumpout techniques are effective and well proven. However, the nearshore dumping procedure and reliance on wave-induced currents to transport the material to the shore zone is still in the experimental stage.

Parallel Jetties with Offshore Breakwater (Type II). Several harbors in the United States have the type II entrance structures, as shown in Fig. 7.38. The length and location of the offshore breakwater serve to minimize wave action at the channel entrance and create a trap for the downdrift movement of littoral transport. It also provides protection for the dredge plant during bypassing operations. This type of harbor entrance structure is considered to be a very effective plan to serve navigation interests and to carry out sand bypassing operations. Since the impounded littoral materials are in a sheltered area and the dredging operations are not normally subjected to excessive wave action, the type or system used for the sand bypassing is a matter of economics. However, the conventional cutter-suction hydraulic dredge has proved to be very cost effective for sand bypassing operations for this type II harbor entrance structure.

Shore-Connected Breakwater (Type III). Another type of harbor entrance is that illustrated as type III in Fig. 7.38 and involves a shore-connected breakwater system. In Southern California there are seven harbors of this type. The shore-connected breakwater type of harbor is most effective in an area where sand transport is negligible. However, in those locations where longshore littoral material transport does exist, shoaling problems can be expected in the entrance channel. One of the best examples of a harbor with a shore-connected breakwater, built in unidirectional sand transport regime, is located at Santa Barbara, California. This harbor was built at a time when adequate consideration was not given to longshore sand transport problems. Initially an offshore detached breakwater was constructed wherein the breakwater provided a sheltered anchorage to small craft from storm waves. After construction of the offshore breakwater was completed in the 1920s, sand began accreting from shore to the detached breakwater, thus reducing the available anchorage area. It was decided by the harbor owners to obviate the problem by connecting the updrift end of the detached breakwater to shore—in effect creating a fish-hook shape configuration. Upon completion of this connection, a fillet rapidly grew on the updrift side of the harbor. Within a few years the fillet reached capacity and the sand began to move along the face of the breakwater and around the downdrift tip of the breakwater back toward the shore. If the growth of this spit had been left unchecked, the harbor would have been completly closed. To preclude this, a hydraulic dredge is used to keep the entrance channel clear and to keep the spit from migrating shoreward. Prior to 1972 the harbor maintenance was performed by the City of Santa Barbara and an average of 150,000 yd^3 (114,700 m^3) of sand per year was dredged from the entrance channel area and placed on downdrift shores. Subsequent to 1972, the federal government assumed maintenance responsibility for the harbor and over a period of 8 years the annual average bypassing had been approximately 185,000 yd^3 (141,450 m^3). Alternative types of sand bypassing operations are being studied with the view of reducing costs of transferring the impounded littoral materials to downdrift shores.

Shore-Connected Breakwater with Weir System (Type IV). The type IV schematic layout harbor entrance system shown in Fig. 7.38 has been called the *weir-jetting* system in the United States. A properly designed and constructed weir-jetting system is a viable solution and will allow effective mechanical or hydraulic bypassing of sand to downdrift shores. A number of these systems have been constructed on the Atlantic and Gulf coasts of the United States.

A weir-jetty system is a modified version of the typical two-jetty system. In regions where the longshore sand transport is predominantly in one direction, the jetting on the updrift side of the inlet is provided with a low silt section over which sand is carried by wave action. The sand moves into a deposition area on the channel side of the jetting. The deposition area, usually located adjacent to the updrift jetty, is protected from all but the most severe wave action by the weir and jetty. A conventional hydraulic pipeline dredge can operate safely in the deposition area to bypass sand under a wide range of prevailing wave conditions at the inlet site.

The updrift jetty usually comprises a relatively sand-tight landward section that controls the platform of the updrift beach, a weir section with a crest elevation near the midtide level, and a seaward section with a typical jetty cross section. In the United States, jetties on exposed coasts are usually of rubble-mound construction. The downdrift jetty usually has a typical cross section without a weir.

The weir and the deposition basin essentially add a sand bypassing function to a basic jettied entrance system. The need to provide a deposition area and the need for a relatively low crested weir results in some design compromises. Weir-jetty systems often have an arrowheadlike configuration (or modification thereof) to provide room for a deposition basin apart from the normal navigation channel between the jetties. The weir section modifies the hydraulic behavior of the inlet. During flood flow, a portion of the tidal prism entering the inlet flows across the weir. The amount of flow exiting the inlet across the weir at ebb tide is significantly less than the amount entering on flood tide. This ebb flow dominance between the entrance jetties assists in keeping the navigation channel free of sediments.

BEACH REPLENISHMENT

Many of the beaches in this country and overseas have been eroding either because of artificial structures or by natural processes. Since population has been increasing in the coastal areas and since recreation along the seashore has become more popular, there is more awareness of beach erosion in some areas and the need to replenish the sand on the beaches.

Beach replenishment is not a new concept; beaches along the New Jersey coast (Ocean City) and along Florida's Atlantic Coast (Pompano Beach, Fort Lauderdale, Miami Beach) have all been replenished on a regular basis for many decades. In recent years it has become increasingly more difficult to find the right type of sand (preferably it should be coarser than the sand which was eroding) and at a reasonable cost for beach-fill purposes. This is due to the depletion of previously used, nearby sources and the increased cost of transporting the sand from further-away areas.

A typical operation would consist of a dredge, either cutterhead or hopper, to dredge suitable material from offshore deposits (½ to 2 mi away) and transport it by a direct pipeline to the shore in the case of a cutterhead dredge or pump the dredged material from the hopper through a pipeline in the case of a hopper dredge. Since the cutterhead dredge can only operate in relatively calm waters, the location, wave climate, and season will dictate whether a cutterhead dredge would be employed. Two examples of cutterhead dredge operation are beach replenishment by a large cutterhead dredge operating about 1 to 1½ mi (1.6 to 2.4

FIGURE 7.39   Beach replenishment from offshore sources at Pompano Beach, Florida.

km) offshore Pompano Beach, Florida (Fig. 7.39) and a cutterhead dredge pumping sand from Corpus Christi Bay to an eroded beach in Corpus Christi, Texas.

An early experiment using a hopper dredge was conducted by the Corps of Engineers (CE).[18] A hopper dredge pumps bottom material into its hopper and, when loaded, proceeds under its own power to a disposal area where the dredged material is discharged through gates in the bottom of the hopper. The New Jersey Experiment used a dredge equipped with a direct pump-out facility (Fig. 7.40). The test section was a state-owned beachfront approximately 2000 ft (610 m) offshore, with water depths sufficient for dredge flotation and a supply of suitable sand was nearby. A mooring barge was used in the experiment, and it was anchored in approximately 30 ft (9.1 m) of water with the discharge pipes connected to a 28-in- (8534 mm-) diameter submerged pipeline running ashore (Fig. 7.40). The line between the discharge pipe on the barge and the submerged line needed both flexibility and rigidity to withstand the lateral and vertical movement and the forces expected in this severe service. The possibility of using a rubber hose, which would withstand working pressures of 200 psi (1379 kN/m²), was explored, but a steel pipe was employed in the experiment. A combination of ball joints and pipe was selected, allowing movement of the barge within an elliptical area of approximately 34 by 90 ft (10.4 by 27.4 m).

This beach nourishment experiment demonstrated that a suitably equipped seagoing hopper dredge could pump sand onto an ocean beach from an offshore mooring, thereby increasing the versatility of this type of dredging plant (Fig. 7.41).

Dredged material has also been pumped out from a hopper through a pipeline discharging at Virginia Beach as part of a beach replenishment project with material dredged from a navigation channel. The connection between the pipeline and the hopper discharge line was made possible by positioning a buoy to hold the pipeline leading to the beach.

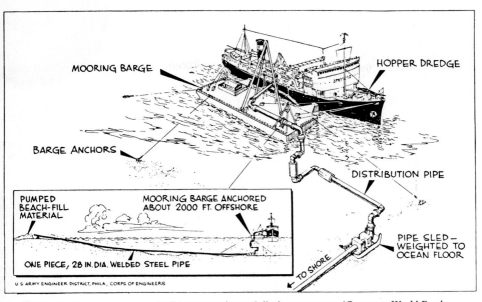

FIGURE 7.40 Arrangement of offshore mooring and discharge system. (*Courtesy, World Dredging Conference*)

The direct pump-out method of disposal is used extensively in the maintenance dredging of ship channels in the Delaware River and Norfolk Harbor. The work had been limited to the handling of light silty materials. The dredges engaged in this work were equipped with hopper jetting systems which would be suitable for pumping out medium and coarse grain sand required for beach nourishment.

Beach nourishment is routinely conducted by the CE at Rockaway Beach, New York, Miami Beach, Florida, and other locations. Figure 7.42 shows the before and after photographs of Rockaway Beach replenishment. In the 1975 contract 3,668,700 yd^3 (2,805,088 m^3) was placed at a total cost of $9,388,366, or at a unit cost of $2.56/yd^3. The contractor pumped dredged material through a floating rubber 24-in pipeline to a "spider" barge which discharged the material through articulated nozzles into scow barges having a capacity of about 4200 yd^3 (3211 m^3) but generally loaded to less than 4000 yd^3 (3068 m^3). The scow barges were then moved by tugs along the 8-mi route into Jamaica Bay to the rehandling station. Using four such barges, the contractor was able to establish an efficient queuing operation which minimized lost time between the loading and unloading of the barges. The 1976 contract provided for pumping dredged material from an offshore borrow area and placing 1,489,600 yd^3 (1,138,948 m^3) at a cost of $2,204,800 or $1.48/yd^3. The 1977 contract provided for placing 1,623,000 yd^3 (1,240,945 m^3) of sand on the beach. Figure 7.43 shows a beach rebuilding effort in Western Europe.

Beach replenishment may also be cost-effective in comparison with structural solutions if the sediment deficit is about 653,976 yd^3/yr (500,000 m^3/yr) and the length of the beach to be rebuilt about 3.1 mi (5 km) long. Figure 7.44, presented by d'Angremond et al.,[19] shows the cost per year per meter length of the beach as a function of sediment deficit per year. It should be pointed out that a recently

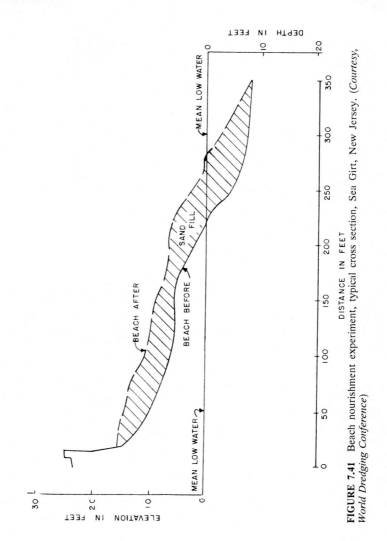

FIGURE 7.41 Beach nourishment experiment, typical cross section, Sea Girt, New Jersey. (*Courtesy, World Dredging Conference*)

7.50

FIGURE 7.42 Rockaway Beach, New York; one of many beaches routinely nourished using dredged material. (*Courtesy, U.S. Army Corps of Engineers*)

FIGURE 7.43 Beach rebuilding effort on Sylt Island, Germany. Sand is pumped to an area close to shore behind a temporary berm; it is then handled by shovels and bulldozer. The beach nourishment project was completed by Van Oord ACZ B.V. Marine & Dredging Contractors, the Netherlands. (*Courtesy, Van Oord ACZ B.V. Marine & Dredging Contractors, the Netherlands*)

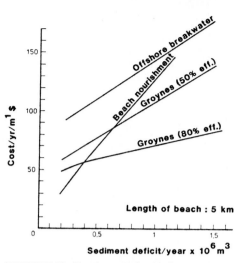

FIGURE 7.44 Beach protection works cost comparison for a 5-km beach. (*Courtesy,* Terra et Aqua)

FIGURE 7.45 The U.S. Army sidecasting dredge *Fry* operating in Barnegat Inlet, New Jersey. (*Courtesy, U.S. Army Corps of Engineers*)

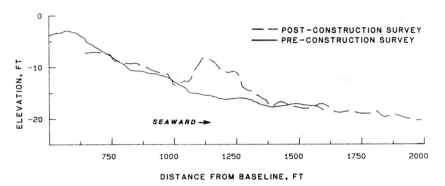

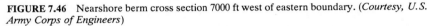

FIGURE 7.46 Nearshore berm cross section 7000 ft west of eastern boundary. (*Courtesy, U.S. Army Corps of Engineers*)

replenished beach say in the Gulf of Mexico or along the Atlantic Coast may be completely eroded in 2 to 3 days by a hurricane, while offshore (detached) breakwaters would probably not be seriously damaged by a hurricane and would provide long-term protection. A combination of offshore breakwaters and beach replenishment should also be considered for a long-term solution.

The CE conducted a program to locate and delineate offshore deposits of sand suitable for beach restoration.[20] It is important to locate new sources of sand that are close to the beaches; it is also important to find out about the total beach regime and the erosion potential of the proposed beach restoration project before embarking upon it. Large volumes dredged during maintenance and deepening of navigational channels present another source of material for beach replenishment. Not all of the dredged material is suitable for beach rebuilding; percentages would vary from place to place. However, investigations should be made to determine whether the material is suitable for that purpose and whether it could be economically placed on the eroding beaches.

The dredged material from nearshore operation can be placed directly in the surf zone and transported by waves to the beach. Such operations may be carried out by hopper dredges or by a sidecasting dredge shown in Fig. 7.45.

A new technique of nearshore placement of dredged material is being tried at various locations along U.S. coasts. Placing suitable dredged material in the nearshore zone, forming underwater berms, has a double beneficial effect (Fig. 7.46): (1) it reduces the wave energy reaching the coastline, and (2) it introduces additional sediment into the natural littoral system. McLellan et al.[21] discuss several case studies:

1. *Offshore Mobile Bay, Gulf of Mexico:* A feeder berm was successfully constructed with a relief of approximately 6 to 7 ft (1.8 to 2.1 m) in original water depths of 19 ft (5.8 m). The fine sand material was placed in the berm configuration by split-hull hopper dredges with loaded drafts of 11 to 12 ft (3.4 to 3.7 m). The district reported that no significant problems were encountered by the contractor as a result of the specific placement technique. Precision bathymetry was collected at both projects before, during, and periodically following construction (Fig. 7.47).

2. *Fire Island Inlet, New York:* The site selected for the berm demonstration project was approximately 1.5 mi (2.4 km) west of Fire Island Inlet directly off-

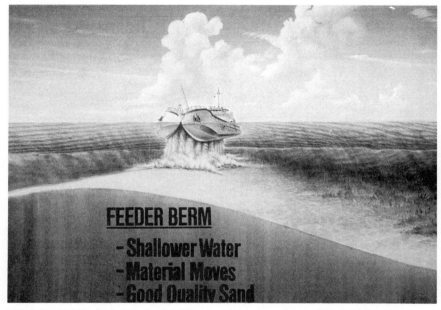

FIGURE 7.47   Feeder berm concept. (*Courtesy, U.S. Army Corps of Engineers*)

shore of Gilgo Beach. The intent of the placement plan at Gilgo Beach was to construct a shore parallel berm along the −16 ft (−4.9 m) MLW contour, 1000 to 1400 ft (305 to 427 m) offshore, with a height of approximately 6 ft (1.8 m). A 12,000-ft-long (3657 m-long) disposal site was designated and the contractor was instructed to place the material on or within the 16-ft (4.9-m) contour while providing for safe keel clearance for the disposal dredge. Horizontal positioning of ±9.8 ft (±3 m) was required on the placement vessel along with a real-time track plotting device to record position, date, and time of each placement event. Twice-weekly progress surveys were required in the placement area to assure a uniform deposit.

North American Trailing Company's (NATCO) hopper dredge *Northerly Island* conducted the dredging and placement of material for the project. The *Northerly Island* is a 2160-yd³ (1651-m³) split-hulled hopper dredge equipped with 2110 hp for propulsion and 625 hp for each pump. The 205-ft-long vessel has a loaded draft of 14.25 ft (4.3 m) and unloaded draft of 5.1 ft (1.6 m). Positioning was provided by a del Norte Transponder coupled to a Hewlett-Packard computer for data recording. Area surveys were conducted by a 42-ft (12.8-m) aluminum crew boat with twin 871-hp (650-kW) engines. A Krupp Atlas Deco 20 fathometer coupled with a Hewlett-Packard computer was used to measure and record bathymetry data.

Records kept aboard the *Northerly Island* indicate 422,500 yd³ (323,024 m³) of material was placed off Gilgo Beach. Construction was completed in approximately 1½ mo, concluding in mid-July 1987. Material was placed generally from east to west along 7500 ft (2286 m) of the designated disposal site on or within the 16-ft (4.9-m) contour. Figure 7.46 shows the results of a pre- and postconstruction

TABLE 7.3 Dredged Material Disposal Options Cost Comparison, Per Cubic Yard, Fire Island, New York

Conventional beach nourishment ($)	New York mud dump disposal site ($)	Interim designated disposal site ($)	Berm disposal ($)
5.50 (bid)	4.50 (est.)	4.00 (bid)	2.23 (act.)

Source: Murden et al., 1985.

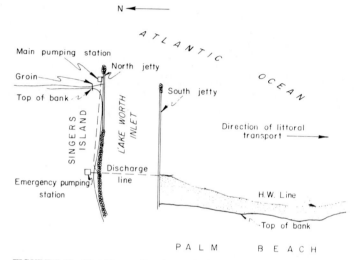

FIGURE 7.48 Fixed bypassing plant at Lake Worth Inlet, Florida. (*Courtesy, U.S. Army Corps of Engineers*)

survey. Table 7.3 indicates that the placement of dredged material in the berm was less than half the cost of placement of the material on the beach and 44 percent less expensive than placing it at the usual designated offshore disposal area.

Sand bypassing at artificial or natural inlets that are protected by breakwaters or jetties is discussed by Richardson.[22] In many cases fixed dredging plants have been installed, although floating dredged equipment is also used (e.g., Richards Bay and Durban, South Africa). Fixed dredging plants either employ conventional dredge pumps (Lake Worth Inlet, Florida, depicted in Fig. 7.48; Ennore Inlet, India,[23] Fig. 7.49; or Marina di Carrara, Italy, Fig. 7.50) or jet pumps (Nerang River Entrance, Australia, Oceanside Harbor, California, or Indian River Inlet, Delaware). Clausner et al.[24] described a bypassing operation using an eductor assembly being deployed by a crawler crane (Fig. 7.51). This "portable" system was able to bypass between 292 and 359 yd³/hr (223 and 274 m³/hr). The system's performance between February and July is shown in Fig. 7.52.

(a)

(b)

FIGURE 7.49 Fixed bypassing plant at Ennore Inlet, India. (*Herbich, 1973*)

(c)

FIGURE 7.49 (*Continued*) Fixed bypassing plant at Ennore Inlet, India. (*Herbich, 1973*)

(a)

FIGURE 7.50 Fixed bypassing plant at Marina di Carrara, Italy.

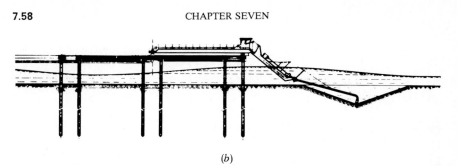

(b)

FIGURE 7.50 (*Continued*) Fixed bypassing plant at Marina di Carrara, Italy.

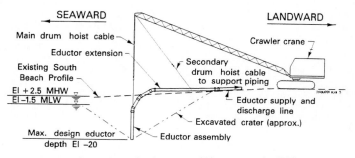

FIGURE 7.51 Crane deploying jet pump. (*Clausner et al., 1990*)

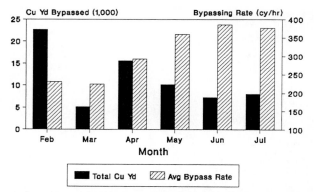

FIGURE 7.52 Bypass system performance, February to July 1990. (*Clausner et al., 1990*)

CONSTRUCTION OF ARTIFICIAL ISLANDS

Artificial islands are being constructed for airports (in Hong Kong, using dredged material, and in Japan, using land borrow material). They are also being built in the Arctic to serve as drilling platforms in the search for new oil fields to be used for production if oil reserves are discovered. In the American and Canadian Beaufort Sea, artificial island construction has become a prime method of providing a platform for drilling sites. The main reason for the construction of artificial islands is that the conventional floating drilling equipment is very difficult to use because of a very short ice-free season, and the drilling ships, jack-ups, and semi-submersibles cannot withstand the ice forces and stay on station through the long winter season.

The Beaufort Sea is covered by ice most of the time and it is only possible to construct such islands using dredges during the 3 mo of the summer. Materials such as sand and gravel required for construction of the islands have to be obtained by dredging in nearby areas or by excavation from sand and gravel pits on land. The Beaufort Sea is fairly shallow and the seafloor soil conditions are often favorable for construction. The soil composed of overconsolidated sand and gravel is available for the purpose of construction.[25] Construction of an artificial island in depths less than 65 ft (20 m) is economically advantageous.[26] However, in water depths more than 82 ft (25 m), the cost effectiveness of conventional artificial islands becomes questionable. At these depths material quantities become unmanageable in one season with any reasonable equipment capacity. The method of construction depends on three factors: (1) water depth, (2) distance from the nearest land mass, and (3) the availability of fill material. Islands have been constructed both in summer and winter. The weather has a major influence on construction methods in the Beaufort Sea.

For sites close to the shore, two efficient methods for building the islands can be used to avoid an investment in a dredging fleet. In winter months the fill can be hauled over ice roads constructed from an onshore source to the island site. In summer months the material may be loaded on barges that would then be towed out to the desired island site; there the fill can be dumped and placed with floating equipment.

If the island site is in deeper water and is further from the shore, the most common method employed in the Canadian Beaufort Sea is the use of a cutter suction dredge. Providing that suitable fill material is located at the sea bottom, the cutter suction dredge draws sand and silt from the ocean floor and transports the material through a floating pipeline to the island site. If the fill material cannot be located close enough to warrant a floating pipeline, the material can be pumped aboard a hopper dredge or a dump scow and towed to the site.

When a base for the island is developed and has risen above water level, equipment and materials are delivered by flat deck barges to begin beach development and slope protection revetments. The most widely used technique for protecting the island shores from erosion is the use of filter fabrics and sandbags. Islands built in the open water season are of two basic types, the sandbag-retained island and the sacrificial beach island. The sandbag-retained island uses a berm of sandbags to reduce the volume of fill required for construction. This method is attractive in areas where sand is scarce and must be hauled by barge from a remote source.

The sacrificial beach islands are characterized by long gradual beaches around the drilling surface. These beaches force storm waves to break and dissipate their energy before reaching the island proper. Sacrificial beach islands are restricted to sites where a sufficient quantity of suitable fill material is locally available for dredging. Such islands have been built in water depths up to 62 ft (19 m).

The key element in the construction of sacrificial beach islands is a stationary cutter suction dredge which draws sand from the seabed and delivers it to the site through a floating pipeline. A barge supporting the discharge end of the pipeline is anchored over the island site to control the placement of the discharging material. The discharge barge ceases operation when the base of the island is ready and the level of the island rises above the water level. Steel pipe is then run from the discharge barge across the island to build beaches to the required dimensions, and slope protection is placed as the beaches are completed. The sacrificial beach island usually has a long sloping foreshore with slope of 1 on 12. The slope increases to 1 on 3 above a point where the wave action is negligibly small. Although this method provides adequate protection to the beach, it requires large quantities of fill material. Typical island construction techniques are shown in Figs. 7.53 and 7.54 and island design concepts are shown in Fig. 7.55.

The second key to the economical profits is effective and efficient construction equipment. Because of the short season of less than 3 mo, the equipment must be capable of working in all types of weather conditions. Most of the present islands in the Beaufort Sea have been constructed in depths less than 65.6 ft (20 m). With the extension of Arctic exploration to deeper water, dredging systems will be required to dredge a much larger amount of soil for construction of artificial islands within their available working period. Therefore, in view of the economics of construction of artificial islands in deeper water, it is essential to extend the dredging period into the ice-infested winter season or collect a sufficient amount of fill material within the short open-water season. Extension of the dredging period to winter conditions in the Arctic is considered a major challenge in the Arctic exploration. Some of the important aspects of this problem have been studied by the Japanese and a conceptional design of hopper dredges for Arctic use has been made to clarify the problems and evaluate their feasibility in the Arctic environment.[27]

In the design of offshore gravel islands, those areas underlain by the medium stiff, fine-grained deposits are of primary concern. The shear strengths and consolidation characteristics of the sediments will have to be considered in the evaluation of the constructed island's resistance to lateral ice load forces. The medium stiff, fine-grained materials are compressible and the weight of the gravel fill will result in consolidation of the materials and contribute to settlement of the island. Shear strengths of the materials can be improved by allowing time after construction of the island for the consolidation of the materials to occur. The consolidation tests show that the time required to achieve 90 percent of the ultimate consolidation will gradually be less than 12 mo.

Different types of islands such as sacrificial beach, earth berm, and sheet-pile and concrete caisson retained fill islands have been constructed for exploration in the offshore Arctic basins. Twenty-nine fill islands have been constructed by Esso, Dome, Gulf, and others in the Canadian Beaufort Sea and by Exxon, Sun Oil Co., Sohio, and others in the American Beaufort Sea. The maximum water depth was 42.7 ft (13 m) and the volume was 39×10^6 ft^3 (12×10^6 m^3) (Isserk gravel island in the Canadian Beaufort Sea was constructed by dredge and pipeline in 80 days).[28] Tables 7.4 and 7.5 list the existing gravel islands in the Alaskan Beaufort Sea.

At present there are no dredges available anywhere in the world which can operate in the Arctic region during winter conditions with ice cover on the water. In order to get an idea of the probable size of dredge required to work during a short summer season in the Beaufort Sea in depths of 65 and 196 ft (20 and 60 m), the following hypothetical case is considered:

1. *Depth—65 ft (20 m):* Data number of working days available: 100/yr; wave climate: significant wave, 11.8 ft (3.6 m); type of material: sand, gravel, etc.; quantity: 4 million cubic meters; alternatives A, hopper dredge or B, hopper dredging with a pipeline.

Alternative A: Hopper Dredge

Daily output	65,397 yd³ (50,000 m³)
Distance to and from disposal area	60 mi (100 km)
Travel speed	16.4 ft/hr (6 m/hr)
Disposal time per trip	20 min
No. trips per day	4
Working hours per day	20

Alternative B: Hopper Dredging with Pipeline

No. working days	100
Working hours per day	20
Total hours of pumping	2000
Pumping rate to be maintained	3270 yd³/hr (2500 m³/hr), or 55 yd³/min (42 m³/min)
Pipeline size	23.6 in (600 mm)
Dredging time	1 hr
Travel time to pumping station	1/2 hr
Pumping time	1/2 hr
No. trips per day to pumping station	10

For the above two dredges, a 7848-yd³ (6000-m³) hopper capacity with over 40 percent solids content in the hopper should be adequate.

2. *Depth—60 m:* The working conditions of this site are considered to be the same as for the previous case. The total quantity to be dredged would be very high if an average beach slope of 1 on 12 is considered. Assuming that the average slope of the island beach is only 1 on 5, the total quantity worked out to about 13.1 million cubic yards (10 million cubic meters). Two dredges of 26,159-yd³ (20,000-m³) capacity would be required for alternative A and two dredges of 15,695-yd³ (12,000-m³) capacity would be adequate for alternative B.

"GLORY-HOLE" DREDGING

Glory-hole dredging is described by Brakel.[29] A glory hole is an artificial depression in the seabed in which a blowout preventer (BOP) and wellhead are placed to protect them from ice keels of pressure ridges. In the Beaufort Sea the glory-hole depth is about 32.8 ft (10 m). Since dredging depths up to 131 ft (40 m) are required, a grab dredge through the moonpool can be deployed. The material is transported to chutes and discharged into a split barge moored alongside the offshore barge (Fig. 7.56).

Phase 1   Gravel or crushed stone is discharged in dredged excavation by side-stone-dumping vessel. Vessel is
(Optional) shifted in lateral direction by means of anchors or dynamic positioning (d.p.) system. Sand could be
               released by trailing suction hopper dredge.

Phase 2   First berm of quarry-run, crushed stone or gravel is dumped. Vessel keeps on station by anchors or
               dynamic positioning system.

Phase 3   First sandfill layer is placed between berms by hopper dredge or a stationary dredge.

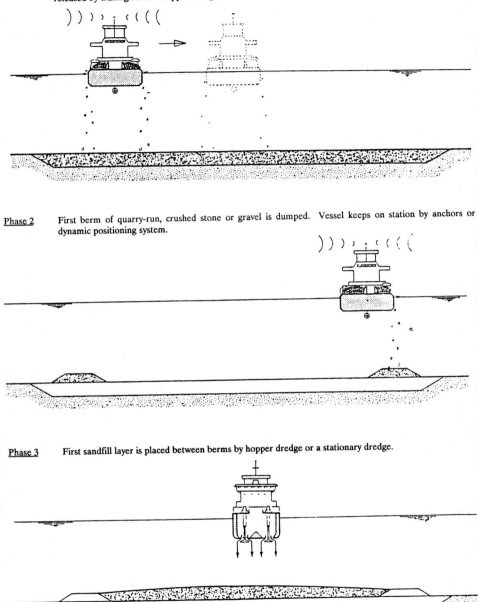

FIGURE 7.53 Artificial island construction techniques. (*Machemehl, 1985*).

Phase 4   Second berm is constructed.

Phase 5   Second sandfill layer is placed.

Final Phase The top of the island is protected either by discharging stone or gravel directly on the sandfill or by placing
(Optional)   stone or gravel in conjunction with mattress placement.

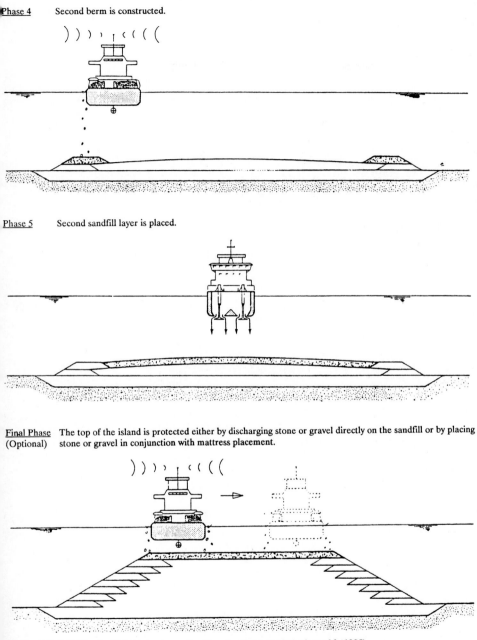

FIGURE 7.54 Artificial island construction techniques. (*Machemehl, 1985*).

SUBMERGED ISLAND CONCEPT

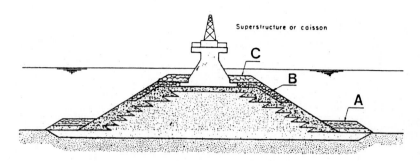

<u>Optional</u>   Protection for the toe of the island (A), slopes (B) and berms (C) can be placed by side-stone discharging
<u>Phases</u>   vessel.

EMERGENT ISLAND CONCEPT

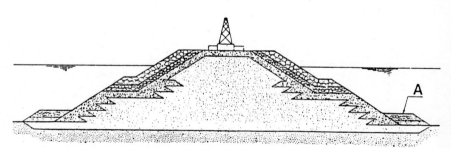

<u>Optional</u>   Extra toe protection for the island (A) can be placed by stone-discharging vessel.
<u>Phases</u>

FIGURE 7.55   Artificial island design concepts. (*Machemehl, 1985*).

TABLE 7.4 Summary of Island Fill Test Data

Year completed	Artificial island/subsea berm	Sand gradation	Fill placement method	Number of CPT soundings	Distance sounded
1981	Alerk P-23	210/6*	Pipeline spigotted	3	52
1981	Tarsiut 11-44	350/4	Berm: bottom pumped	18	190
			Core: spigotted	8	96
1982	Uviluk P-66	350/2	Berm: bottom dumped	18	368
1982	Ukalerk #1 Trial Fill	350/2	Bottom dumped	5	28
1982	Ukalerk #2 Trial Fill	350/2	Pump-out	10	62
1982	Kogyuk N-67	350/2	Bottom pumped	33	165
			Pump-out	33	132
1983	Kogyuk N-67 (prior to 1983 const.)	350/2	Bottom pumped	3	15
			Pump-out	3	15
1983	Kogyuk N-67 (CPT data collected midway through const. season)	350/2	Bottom dumped in 1982.	9	36
			Pump-out in 1982.	9	45
			Bottom pumped in 1983.	9	54
1983	Kogyuk N-67 (CPT data collected at end of const. season)	350/2	Bottom dumped in 1982.	6	24
			Pump-out in 1982.	6	30
			Bottom dumped in 1983.	6	60
1983	Kogyuk N-67 (CPT data collected with SSDC in place)	350/2	Bottom dumped in 1982.	6	24

TABLE 7.4 Summary of Island Fill Test Data (*Continued*)

Year completed	Artificial island/subsea berm	Sand gradation	Fill placement method	Number of CPT soundings	Distance sounded
1983 (*Cont.*)			Pump-out in 1982.	6	30
			Bottom dumped in 1983.	6	60
1984	Kogyuk N-67 (CPT data collected after SSDC removed)	350/2	Bottom dumped in 1983.	6	50
			Pump-out in 1982.	1	3
1983	Nerlerk B-67	320/2 Ukalerk, sand	Bottom dumped	26	—
		260/2 Nerlerk, sand	Spigotted	26	280
1984	Isserk Trial Berm	210/1	Bottom dumped	8	57
1984	Amerk P-09	290/1	Berm: bottom dumped	3	58
			Core: spigotted	3	36
1984	Tarsiut P-45	320/1	Berm: bottom dumped	18	180
			Berm: bottom dumped	7	63
1984	Tarsiut P-45	320/1	Core: center spigotted	32	784
1985			Core: spigotted	5	140
1985	Amauligak I-65	320/1	Berm: bottom dumped	9	182
1985	Amauligak I-65 (caisson in place)	320/1	Berm: bottom dumped	13	220
1985	Amauligak I-65	320/1	Core: spigotted	21	450
1986 (after April 12 event)	Amauligak I-65	320/1	Core: spigotted	9	190
1987	Amauligak F-24	320/2	Berm: bottom dumped	11	240

TABLE 7.4 (*Continued*)

Year completed	Artificial island/subsea berm	Sand gradation	Fill placement method	Number of CPT soundings	Distance sounded
1987	Amauligak F-24 (CPT data collected with mobile arctic caisson)	320/2	Berm: bottom dumped	17	—
1987	Amauligak F-24	350/1	Core: spigotted	17	440
1987	Amauligak F-24 (during densification)	350/1	Core: pump-out	8	210
				18	530
1987	Amauligak F-24 (after densification)	350/1	Core: pump-out	29	845

*210/6 represents a median grain size D_{50} of 210 μm and a silt content of 6 percent.
Source: Machemehl, J. L., 1985.

LOW-COST MAINTENANCE DREDGING

In some navigational channels advantage may be taken of tidal action to transport sediment to open water. The sediment must be resuspended before it can be moved by ebb tide. The idea is not new; an agitation dredge is said to have been operating in Holland in the fifteenth century. In recent years both ploughs and pneumatic systems have been tried.

One example refers to the removal of fine sediments at the entrance to the Zandvliet lock at the Port of Antwerp, Belgium.[30] The lock is located in the Scheldt Estuary where the saltwater wedge entering at flood tide mixes with fresh water and causes flocculation of suspended clays and organic macromolecules. The flocs settle in the dredged channel causing fairly rapid siltation (1.3 million cubic yards, 1 million cubic meters, in 2 years). The sediment is very fine with a particle size less than 0.00079 in (0.02 mm) and consists of:

1. Nonmineral clays
   a. Quartz: 45 percent
   b. Calcite: 10 percent
2. Clay minerals
   a. Montmorillonite: 25 percent
   b. Kaolinite: 10 percent
   c. Illite mica: 10 percent

The fine sediment slurries (fluid mud) exhibit nonnewtonian pseudo-plastic behavior. A plough (Fig. 7.57) is dragged along the bottom by a tug, resuspending fine sediments on a daily basis (5 days a week) during ebb tide. Another type of plough, the Lillo-type, has also been developed for agitation dredging (Fig. 7.58). A water-air mixing device has been successfully tried in the laboratory, but field

TABLE 7.5 Summary of Artificial Islands Constructed in the Canadian Beaufort Sea

Year completed	Artificial island/ subsea berm	Island type	Water depth (m)	Fill quantity (m³)	Operator
1973	Immerk B-48	Sacrificial beach island	3.0	180,000	Esso Resources
1973	Adgo F-28	Sandbag-retained island	2.1	36,000	Esso Resources
1973	Pullen E-17	Dumped gravel island	1.5	65,000	Esso Resources
1974	Unark L-24	Dumped gravel island	1.3	44,000	Sun Oil
1974	Pelly B-35	Barge enclosed in silt island	2.0	35,000	Sun Oil
1974	Netserk B-44	Sandbag-retained island	4.9	306,000	Esso Resources
1974	Adgo P-25	Sandbag-retained island	1.5	27,000	Esso Resources
1975	Adgo C-15	Dumped gravel island	1.5	70,000	Esso Resources
1975	Netserk F-40	Sandbag-retained island	7.0	291,000	Esso Resources
1975	Sarpik B-35	Dumped gravel island	4.3	118,000	Esso Resources
1975	Ikkatok J-47	Sandbag-retained island	1.5	38,000	Esso Resources
1976	Kugmallit D-49	Sandbag-retained island	5.3	280,700	Esso Resources
1976	Adgo J-27	Sandbag-retained island	1.8	69,000	Esso Resources
1976	Arnak L-30	Sacrificial beach island	8.5	1,070,000	Esso Resources
1976	Kannerk G-42	Sacrificial beach island	8.5	1,070,000	Esso Resources
1977	Isserk E-27	Sacrificial beach island	13.0	1,561,000	Esso Resources
1979	Issungnak O-61	Sacrificial beach island	19.0	4,100,000	Esso Resources
1980	Issungnak 2-061	Sacrificial beach island	19.0	1,000,000	Esso Resources
1981	Alerk P-23	Sacrificial beach island	10.5	2,360,000	Esso Resources
1981	N. Protection Island	Sacrificial beach island	4.6	2,000,000	Dome Petroleum

Year	Name	Type		Cost	Company
1981	W. Atkinson L-23	Sacrificial beach island	7.5	955,000	Esso Resources
1981	Tarsiut N-44	Caisson-retained island on subsea berm	21.0	1,800,000	Gulf Canada
1982	Uviluk P-66	Steel caisson on subsea berm	29.7	1,900,000	Dome Petroleum
1982	Itioyok I-27	Sacrificial beach island	15.0	1,940,000	Esso Resources
1983	Nerlerk B-67	Subsea berm	45.1	4,000,000	Dome Petroleum
1983	Kogyuk N-67	Steel caisson of subsea berm	28.1	1,450,000	Gulf Canada
1983	Kadluk N-67	Caisson-retained island	14.0	436,000	Esso Resources
1984	Amerk P-09	Caisson-retained island	26.0	1,162,000	Esso Resources
1984	Adgo H-29	Sacrificial beach island	3.0	75,000	Esso Resources
1984	Nipterk I-19	Sacrificial beach island	11.7	931,000	Esso Resources
1984	Tarsiut P-45	Caisson-retained island on berm	26.0	343,000	Gulf Canada
1985	Minuk I-53	Sacrificial beach island	14.7	1,986,000	Esso Resources
1985	Amauligak I-65	Caisson-retained island	31.0	1,408,000	Gulf Canada
1985	Arnak K-06	Sacrificial beach island	17.9	552,000	Esso Resources
1987	Amauligak F-24	Caisson-retained island	32.0	2,000,000	Gulf Canada

Source: Machemehl, J. L., 1985.

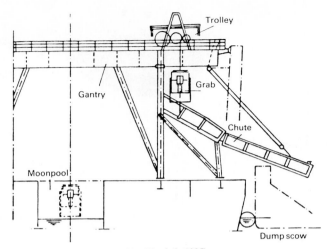

FIGURE 7.56 Grab dredge. (*Brakel, 1985*)

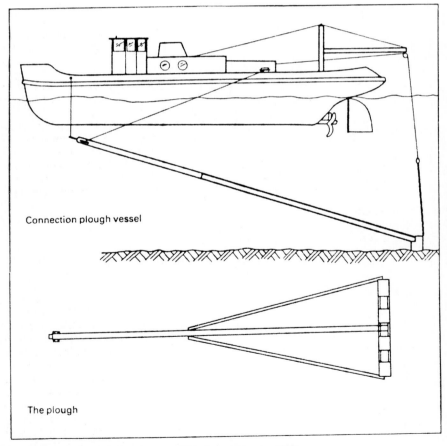

Connection plough vessel

The plough

FIGURE 7.57 Plough dragging tug. (*Meyvis and Marain, 1988*)

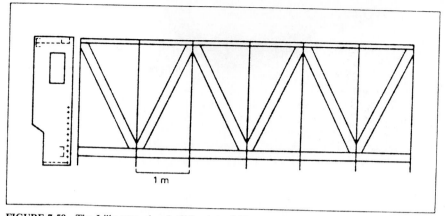

FIGURE 7.58 The Lillo-type plough. (*Meyvis and Marain, 1988*)

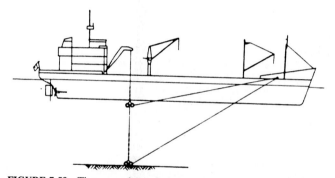

FIGURE 7.59 The sweep beam is connected to a ship with cables and winches and pulled over the bottom of the channel. (*Mayvis and Marain, 1988*)

tests have been disappointing. Research continues on several methods of agitation dredging which may reduce the cost of dredging. Environmental effects of resuspended sediments may prevent the deployment of agitation dredging in some cases.

Another device, called a *sweep beam,* has been used at the entrance to Kalo Lock at the Port of Antwerp. A sweep beam is a large steel structure equipped with winches, as shown in Fig. 7.59. Trial tests indicated that the depth of the channel was increased by 1.28 to 1.97 ft (0.39 to 0.60 m).

REFERENCES

1. Anonymous, *The Dustpan Dredge,* Report no. 90217, Ellicott Machine Corporation, Baltimore, MD, 1973.
2. Turner, T. M., *Fundamentals of Hydraulic Dredging,* Cornell Maritime Press, Centreville, MD, 215 pp., 1984.
3. Hadjidakis, A., "Optimum Utilization of Cutter Dredges," *Ports and Dredging,* no. 65, IHC Holland, Rotterdam, the Netherlands, pp. 4–8, 1970.
4. Herbich, J. B., "Operating Characteristics of Cutterhead Dredges," *9th World Dredging Conference, WODCON IX,* Vancouver, B.C., 11 pp., Oct., 1980.
5. Venezian, G., Unpublished Dredging Short Course Notes, Texas A&M University, 1983.
6. Durand, R., "Basic Relationship of the Transportation of Solids in Pipes—Experimental Research," *Proc. IAHR,* University of Minnesota, 1953.
7. Basco, D. R., "Systems Engineering and Dredging—The Feedback Problem," Texas A&M University, Sea Grant College Program, TAMU-SG-74-205, 74 pp., 1973.
8. Gibert, R., "Transport Hydraulique et Refoulement des Matières," *Annales des Ponts et Chaussées, 130,* pp. 307–373, 437–492, 1960.
9. Worster, R. C., and Denny, D. F., "Hydraulic Transport of Solid Material in Pipes," *Proc. AIME, 169,* pp. 569–586, 1955.
10. Anonymous, Report for U.S. Army Corps of Engineers, Contract DA-36-109-CIVENG-62-47, Mineraal Technologisch Instituut, Delft, the Netherlands, 1962.
11. Anonymous, "The IHC Dredging Wheel," *Ports and Dredging,* no. 102, 1979.
12. Turner, T. M., "The Compensated Cutterhead Dredge Key to Offshore Mining," *Proc. World Dredging Conference, WODCON V,* Hamburg, Germany, 1973.
13. Anonymous, "Double-Walled Dredge Pump for Longer Life," *Ports and Dredging,* no. 73, 1972.
14. Herbich, J. B., and Lou, Y. K., *Stable Catamaran Hulls for Cutterhead Dredges,* Paper no. OTC 2290, Offshore Technology Conference, 1975.
15. Marine Board, National Research Council, *Criteria for the Depths of Dredged Navigational Channels,* National Academy Press, 1983.
16. Lacasse, S., "Sizing of Containment Areas for Dredged Materials," *Proc., 13th Dredging Seminar,* Texas A&M University, pp. 146–177, 1981.
17. Murden, W. R., Herbich, J. B., Valianos, M. L., and Watts, M. G., "Dredge Equipment and Techniques of Operation for Transferring Coastal Sediments," *Proc., PIANC 26th International Congress,* Brussels, Belgium, June 16–28, 1985.
18. Mauriello, L. J., "Experimental Use of a Self-Unloading Hopper Dredge for Rehabilitation of an Ocean Beach," *Proc., World Dredging Conference, WODCON I,* New York, pp. 367–394, 1967.
19. d'Angremond, K., van Oorschot, J. H., and de Jong, A. J., "Beach Replenishment—Design Elements and Implementation," *Terra et Aqua,* no. 37, pp. 19–27, Aug. 1988.
20. Anonymous, "National Shoreline Study, Texas Coast Shores Regional Inventory Report," U.S. Army Engineer District, Galveston, TX, 1971.
21. McLellan, T. N., Truitt, C. L., and Flax, P. D., "Nearshore Placement Techniques for Dredged Material," *Proc., 23rd Dredging Seminar,* Texas A&M University, pp. 24–34, Oct. 1990.
22. Richardson, T. W., "Sand By-passing," vol. II, ch. 16, *Handbook of Coastal and Ocean Engineering,* J. B. Herbich (ed.), Gulf Publishing, Houston, TX, pp. 809–828, 1991.

23. Herbich, J. B., *Coastal & Deep Ocean Dredging,* Gulf Publishing, Houston, TX, 622 pp., 1975.

24. Clausner, J. E., Melson, K. R., Hughes, J. A., and Rambo, A. T., "Jet Pump Sand Bypassing at Indian River Inlet, Delaware," *Proc., 23d Dredging Seminar,* Texas A&M University, pp. 101–106, Oct. 1990.

25. Robertson, F. P., "Artificial Islands," *Civil Engineering,* vol. 53, no. 8, pp. 38–41, Aug. 1983.

26. Dingle, P. J., "Island Construction in the Beaufort Sea," *Proc., 16th Dredging Seminar,* Texas A&M University, College Station, TX, 1983.

27. Takekuma, T., Kawanoto, T., and Noble, P., "Feasibility of Using Hopper Dredges Under Arctic Environments," *Proc., Sixteenth Dredging Seminar,* Texas A&M University, College Station, TX, 1983.

28. Machemehl, J. L., personal communication, ARCO Resources Technology, Dallas, TX, 1985.

29. Brakel, J., "Dredging Developments in the Canadian Arctic," *Terra et Aqua,* no. 29, pp. 10–15, Apr. 1985.

30. Meyvis, L., and Marain, J., "Low Cost Tidal Inlet and Estuary Maintenance Dredging," *Terra et Aqua,* no. 37, pp. 9–14, Aug. 1988.

CHAPTER 8
DISPOSAL AND PLACEMENT OF DREDGED MATERIAL

Dredged material must be disposed of either in the ocean or on land. There are advantages and disadvantages to placing dredged material at either location. Advantages and disadvantages may be evaluated from economical and environmental impact points of view. What may be advantageous from an economic point of view (e.g., placement of material in open water close to the channel) may be disadvantageous from an environmental point of view (e.g., change of water circulation patterns, dispersion of fine sediment over a wide area under high wave energy conditions).

DISPOSAL ALTERNATIVES

Figure 8.1 shows the disposal alternatives. The disposal of dredged material has taken a variety of forms in the past. Methods which had been or are being used are listed below.

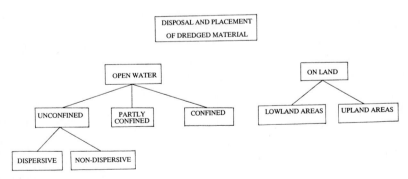

FIGURE 8.1 Disposal alternatives.

Disposal in lowland areas without the use of containment dikes: This method is seldom used today because of damage to vegetation and adjoining waters resulting from the uncontrolled flow of the dredged material. The area affected would be larger than if dikes were used. Also, the small streams and bayous that serve as major breeding grounds for aquatic life would become silted.

Disposal in lowland areas using containment dikes: This method might be used if pumping distances to upland sites are so far away as to make them impractical. In many cases, the actual laying of ‑pipelines across the lowland areas to upland disposal sites might be as detrimental to the lowland areas as the dredged material would be. With this method it is possible to direct the effluent to a single or a few well-controlled return ditches, thus largely alleviating the problem of silting up small streams and bayous. The dredged material is obviously confined in the diked area, and by proper sizing of the containment ponds within the dikes the effluent water can be ponded until it is of adequate quality to be returned to the bay. However, the disposal area may cover much of the lowland area that serves as the breeding ground and resting place for birds and aquatic life.

Disposal in open water without the use of containment dikes: This method is often used for virgin material and maintenance dredging as well. In some cases the disposal of dredged material in this manner may actually be beneficial to the environment. The release of nutrients may more than compensate for the production lost by plants and animals covered by the dredged material. The creation of shallower water after completion of dredging may also be beneficial in some areas. The material may or may not become emergent. Emergent islands are often constructed from virgin dredged material but seldom from maintenance dredged material. These islands may prove to be beneficial to the environment by providing additional nesting areas for birds. A disadvantage of this method is that the underwater slope formed with fine-grained dredged material may be very flat (thus requiring large areas for disposal) and storms in the area may completely destroy the islands before the disposal islands have had time to stabilize. Unfavorable current may cause erosion and redeposition of some material into the dredged area, thus requiring more frequent dredging.

Disposal in open water using containment dikes: The use of containment levees in open water is justified when it is desired to prevent the dredged material from spreading under the action of waves and currents. For this to be effective, the islands should be emergent, and the exterior surface of the levees should be protected to prevent erosion. To prevent the levees from forming flat slopes and covering large areas, they should be constructed of suitable material such as clay balls, coarse to medium sands, etc. Unforeseen soft foundation conditions can cause failure of the dikes, which would cause the liquid dredged material to flow through the beach back to the bay bottom. The permanent nature of the disposal area makes it suitable for bird habitats.

Disposal at sea: Hopper dredges are used in dredging navigational channels in the open sea and in sheltered waters and generally dispose of the dredged material in designated open-water disposal areas. Dredged material may also be transported in hopper barges, loaded from both cutterhead or hopper dredges, if distances from the area being dredged to the disposal site are large. Split-hull barges such as the one shown in Fig. 8.2 are suitable for that purpose. The number of barges employed will depend on the dredge production and the distance between the dredging site and the disposal area. In some cases, in congested areas, there may be significant vessel traffic problems. In shallow water areas, the draft of a seagoing hopper dredge may be greater than the water depth, and cutterhead dredges are used in combination with hopper barges to transport the dredged material after rehandling to an offshore disposal site.

Upland disposal without the use of containment dikes: This method has the same advantages as disposal in lowland areas where containment dikes are not used except when harm to vegetation in the low-lying areas will not be as great. Uncontrolled flow of sediment-laden effluent into small streams and bayous will occur, but the additional time for overland flow will tend to remove some of the sediment from the effluent before it reaches the most sensitive areas.

Upland disposal with the use of containment dikes: With this approach, the disposal area will become temporarily unsuitable as a habitat for land animals and birds, but a short time after dredging is completed, the disposal area, particularly in warm climates, should quickly revegetate and regain productivity. Dikes reduce the area required for disposal, allow the retention of effluent until it attains suitable quality before releasing it, and provide for the controlled flow of the effluent through return ditches.

FIGURE 8.2   Split hull barge. (*Courtesy, IHC Holland*)

REQUIREMENTS AND FEATURES OF CONFINED DISPOSAL AREAS

The functional requirements of confined disposal areas are to:[1]*

1. Accept dredged material during an estimated design life of the disposal area
2. Prevent solids (above the allowable limits) from escaping from a disposal area back to the water system
3. Be economical

*References for chapter sections written by John B. Herbich are at the end of the chapter.

The major features of the confined disposal area are discussed below.

Containment Dikes

Containment dikes are usually constructed from the available soil at the site by the contractor. On major projects such as the Hart-Miller disposal area in Chesapeake Bay, or disposal areas in Mobile Bay, the dikes are engineered and constructed with rip-rap or other types of revetments to keep the dikes from breaching during major storms.

An example of a large disposal area, Pleasure Island constructed in Sabine Lake, Texas, is shown in Fig. 8.3. Part of the disposal area is now used by the Corps of Engineers Area Office and as a large recreational facility.

A very large disposal area called Slufter was constructed in 1986 and 1987 in the North Sea to accommodate some of the 30 million cubic yards (23 million cubic meters) of sediments that have to be dredged from the Rotterdam and Europort area.[2] The design criteria were as follows:

1. The disposal area will be constructed partly by excavation and partly by raising the existing land level.
2. The dikes will be constructed of sand.
3. The disposal area should be located in shallow water.
4. The shape of the disposal area should conform with the existing coastline.
5. The sand for the dike construction should be excavated from inside the disposal area.

The western side of the dikes will reach an elevation of 79 ft (24 m) above MSL (the Netherlands Reference Datum—NAP). Figure 8.4 shows a cross section of this massive project; the total area of the disposal site is about 642 acres (260 ha).

FIGURE 8.3   Pleasure Island Disposal Area near Port Arthur, TX.

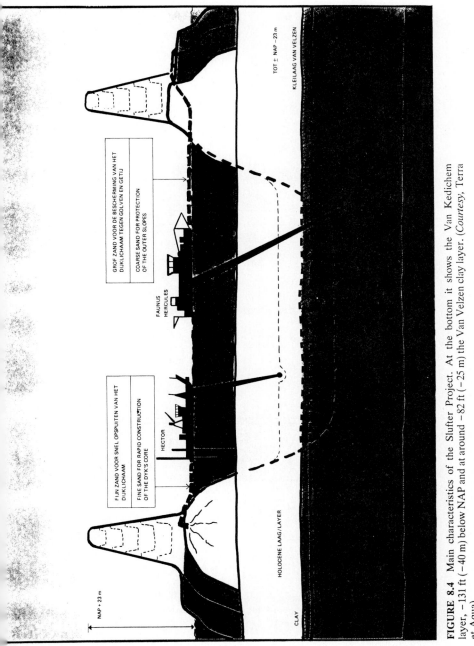

FIGURE 8.4 Main characteristics of the Slufter Project. At the bottom it shows the Van Kedichem layer, −131 ft (−40 m) below NAP and at around −82 ft (−25 m) the Van Velzen clay layer. (*Courtesy*, Terra et Aqua)

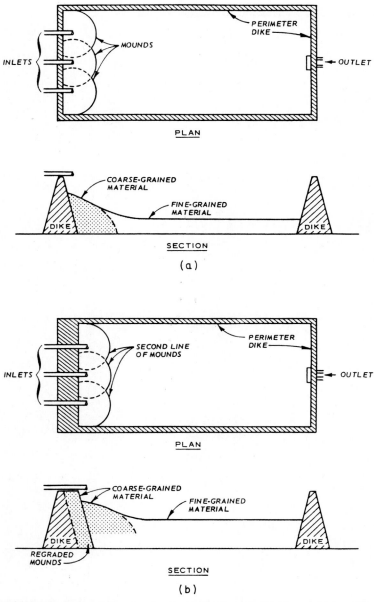

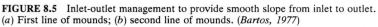

FIGURE 8.5 Inlet-outlet management to provide smooth slope from inlet to outlet. (a) First line of mounds; (b) second line of mounds. (*Bartos, 1977*)

The Inlet Structure

There are generally no inlet structures but there usually are a series of inlets to allow a more uniform distribution in the disposal area and reduce short-circuiting and dead areas. Figure 8.5 shows the inlet-outlet arrangement that provides a fairly smooth slope from inlet to outlet.[3] During a disposal operation, coarse-grained material becomes separated from fine material and settles near the inlet, forming a mound near the discharge pipe. The mound can be used:

1. To support the extended discharge pipe
2. As a supply of coarse-grained material, as a temporary surcharge to dewater the fine-grained material, or to construct drainage layers

   Figure 8.6 shows the calm areas (dead zones) in the corners of a rectangular disposal area and tendency for short-circuiting.

Outlet Structures

The rate of effluent flow is regulated by a weir structure. Proper weir design and operation can control the solids concentration in the effluent water. Because the control of solids concentration is regulated, the project may be stopped when the

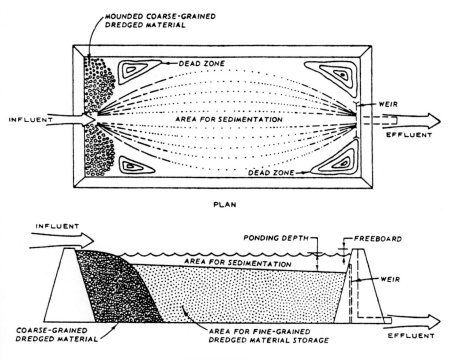

PLAN

CROSS SECTION

FIGURE 8.6 Schematic diagram of a dredged material containment area. (*EM1110-2-5025, 1983*)

TABLE 8.1 Effluent Standards Adopted by U.S. Army Engineer District (1973)

U.S. Corps of Engineers' District	Effluent standard adopted*
Galveston	8 g/l above ambient
New Orleans	None set
Mobile	None set
Jacksonville	50 JTU
Savannah	None set
Charleston	None set
Wilmington	50 JTU
Norfolk	13 g/l above ambient
Philadelphia	8 g/l above ambient
New York	8 g/l above ambient
Buffalo	50 ppm settable solids (subject to change)
Detroit	8 g/l above ambient
Chicago	None set
Sacramento	8 g/l above ambient
Portland	5 JTU
Seattle	5-10 JTU

*JTU = Jackson turbidity units.
 g/l = grams per liter.

concentration of solids is above that allowed. The effluent standards for the solids concentrations adopted by the U.S. Army Engineer Districts are shown in Table 8.1.

Determination and reporting of solids concentrations varies in practice; the terms *concentration in grams per liter, percent of solids by weight, percent of solids by volume,* and *percent of solids by apparent volume* are all being used. The methods were compared in Lacasse et al.[4] for clarification purposes and are shown in Table 8.2; the relationship between percent of solids by weight and concentration in grams per liter is illustrated in Fig. 8.7.

CONTAINMENT AREA DESIGN AND MANAGEMENT

The success of any disposal operation on land depends on proper planning, designing, operating, and managing containment areas. The goal is to provide maximum storage volume and to meet required effluent solids standards. Design procedures include consideration of dredged material sedimentation and consolidation behavior, as well as consolidation of foundation soils.

The purpose of containment area management is to promote natural dewatering of fine-grained dredged material and thus reduce the volume of containment required.[4,5] Factors to consider are:

Careful planning: Careful planning is needed because different types of soil are encountered in dredging operations. Plans should include not only a logical sequence of operations for (1) dredging, (2) disposing, and (3) dewatering of dredged material but also an estimate of the dredging requirements and an estimate of adequate storage capacity for the material dredged during that period.

TABLE 8.2 Methods of Reporting Suspended Solids

Method of reporting suspended solids	Weight-volume relationship	Method of computation	Remarks
		Preferred method	
Grams per liter or milligrams per liter	W_S, grams $V_T = 1$ litre	$S = \dfrac{W_S}{V_T}$	Common method for reporting dissolved chemical concentrations. Best method for engineering purposes.
		Other methods	
Percent by weight	W_S	$S = \dfrac{W_S}{W_T} 100$	Easy to determine by laboratory test. Does not require value for specific gravity.
Percent by volume	V_S	$S = \dfrac{V_S}{V_T} 100$	Easy to determine by laboratory test. Requires determination of percent by weight and value for specific gravity.
Percent by apparent volume	$V_T = V_S + V_I$	$S = \dfrac{V_A}{V_T} 100$	Apparent volume determined by settled solids for a bottle or flask. No standardized procedure available. Void ratio of settled solids varies with type of sediment. Can lead to errors because of nonstandard test. Not recommended. Value is meaningless in engineering calculations.

Note: W_S = oven-dry weight of solid particles V_S = volume of solid particles
       V_T = total volume V_A = apparent volume of settled solids
       W_T = total weight V_I = volume of interstitial water

Adapted from U. S. Army Engineers, EM 1110-2-5027, 1981.

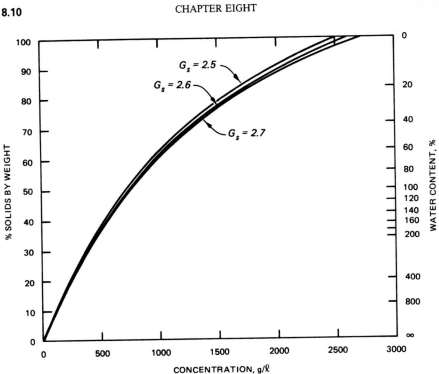

FIGURE 8.7 Relationship of concentration in percent solids by weight, percent solids by volume, concentration in grams per liter, and water content (*EM1110-2-5027, 1987*).

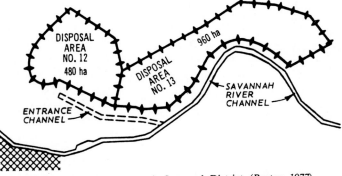

FIGURE 8.8 Containment areas in Savannah District. (*Bartos, 1977*)

Shape of disposal area: This can be any shape from square to oval to long and narrow, generally dictated by the shape of land available. It should be sloping to provide natural gravity drainage of water. There may be some advantage to long and narrow areas to increase the sedimentation (settling rates) and to facilitate the use of draglines and/or clamshells for constructing dewatering trenches. Figures 8.8 and 8.9 show some shapes employed on U.S. Army Engineers' Projects.[6]

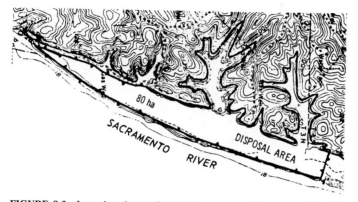

FIGURE 8.9 Irregular shape of containment area, Sacramento District. (*Bartos, 1977*)

Sizing of containment areas: The containment areas are generally designed on the basis of a bulking factor (swell factor) or on experience from previous dredging of similar materials. The factor relates the volume of material in situ (to be dredged) to the volume it is expected to occupy after being pumped into a containment area. Tables 8.3 and 8.4 summarize the bulking factors used by various agencies.

The sizing of the containment area is also a function of (1) lift thickness of the placed dredged material, (2) flow rate of dredged material, and (3) minimum effluent standards.

Dewatering, and thus reducing the volume of containment areas, can be accomplished by (1) placement of thin layers (lifts), (2) division of large areas into several compartments, and (3) cross and spur dikes.[6]

Inlets: The end of the dredge discharge pipe in a containment area is called an inlet. Inlets can be single or multiple. The multiple inlets may be uncontrolled (manifold type) or controlled by valves, as shown in Fig. 8.10. Inlets may be extended into the containment area to provide flexibility in discharging of dredged material in various parts of the containment area.

Outlets: The outlets control the flow of water from one compartment of a disposal area to another or from the containment area to a river, bay, or ocean.

TABLE 8.3 Summary of Sizing Methods Used by Selected Corps of Engineers' District Offices and Research Agencies

Source of information	Containment sizing factor to include*							Material type	Sizing factor**	Comments
	1	2	3	4	5	6	7			
Buffalo District	✓				✓	✓	✓	Sand Clay & silt	1.0 0.5–1.0	Uncertainty on volume dredged. Observed sizing factor in Cleveland, Ohio, for organic silts: 0.79.
Norfolk District	✓	✓					✓	Sand Clay & silt	1.0 2.0	Factors generally overpredict required containment size.
Mobile District	✓	✓	✓					All types	1.2	Conservative method (long term).
Detroit District	✓				✓	✓		Sand & silt	0.6–1.0	No losses during removal and transport assumed. Past volume predictions both over- and underpredicted volume. 15% swell upon bottom removal. 50 to 85% reduction in volume.
New England Division	✓							All types	1.25	
Seattle District	✓						✓	Sand Silt Clay	1.1 1.3 1.5	Sizing factors based on field observations. Use weighted average sizing factor.
Philadelphia District	✓				✓	✓	✓	Sand Silt Clay	0.56 0.73 1.0–1.12	Factors without settlement allowances are 1.0, 1.3, and 1.8–2.0 for sand, silt, clay. Settlement estimates based on field observations and column sedimentation tests in 6-cm-diameter 50-cm-high cells.

Organization	(1)	(2)	(3)	(4)	(5)	(6)	(7)	Description of material	Sizing factor**	Remarks
Galveston District	✓			✓		✓		Silt	1.35	One year after disposal, consider that settlements have reduced volume by 50%. Method does not apply to sand.
								Clay	1.65	
Jacksonville District	✓			✓				Sand	1.2–1.3	
								Clay	2.0	
J. Huston, Dredging Consultant	✓	✓					✓	Sand	1.0	Use weighted average sizing factor.
								Silt	1.5	
								Clay	2.0	
								Sandy clay	1.25	
								Rock & gravel	1.75	
Japan Dredging & Reclamation Eng. Assoc., Tokyo	✓			✓				Sand	1.0	Settlement prediction of clay very unreliable. Use laboratory tests to obtain factors.
								Silt	1.3–1.6	
								Clay	2.0	
Port & Harbour Technical Research Institute, Tokyo	✓			✓	✓	✓		Sand & silt	0.7–0.9	If swell factor only, use 1.3. Factors based on case studies. Use laboratory sedimentation tests to obtain factors.

*(1) Volume of in situ channel sediment.
(2) Overdredging.
(3) Transport efficiency.
(4) Containment area losses.
(5) Consolidation of dredged material in containment area.
(6) Containment area foundation settlement.
(7) Description of material.
**Sizing factor = ratio of volume of dredged material in containment area to volume of in situ channel sediment.

TABLE 8.4 Bulking (Design) Factors for Various Materials

Material	Dry density, γd (lb/ft^3)		Bulking factors		
	In situ	Sedimented	USAE	Huston	Others
Clay	94*	30–78	1.2–3.1	1.45	
Low plasticity clay (recent) (CL)					1.3**
Low plasticity clay (pleistocene) (CL)					2.0**
High plasticity clay (pleistocene) (CH)					2.5**
Clay balls					1.2
Silty clay					1.4–1.7
Sandy clay	94*			1.25	
Clayey sand (SC)					1.3
Silt (ML)	94*	65–82	1.1–1.4	2.00	1.3
Silty fine sand (SM)					1.1
Sand	90–110	93	1.0–1.2	1.00	
Gravel	110			1.75	
Quartz (rock)	165	93	1.8	1.75	

*Dry density of packed earth.
**Sedimented from slurry.

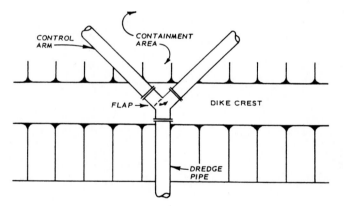

a. Y-VALVE

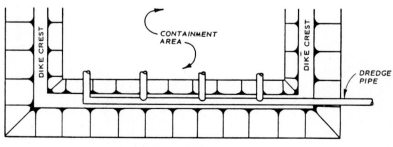

b. MANIFOLD

FIGURE 8.10 Multiple inlets. (*Bartos, 1977*)

The outlets may be (1) a weir or (2) a drop type. Outlets are located at low spots on the perimeter of the containment area close to the body of water to which effluent is to be discharged.

After the outlet locations have been specified, inlets must be located on the perimeter as far away from the outlet(s) as possible and as near as possible to the dredge to reduce pumping distance.

Compartments: The use of cross dikes to divide a containment area is common practice on many projects. The general purpose of cross dikes is to reduce the velocity of dredged material being discharged into a containment area. Lower velocities, or calm conditions, promote settlement of particles, thus improving effluent water quality.

The compartments may be used in series (Fig. 8.11) or in parallel (Fig. 8.12). When a series operation is employed, the first compartment acts as a primary sedimentation basin. As many compartments as needed can be used to meet the effluent quality standards (Fig. 8.12). An example of a disposal area in series is shown in Fig. 8.13.

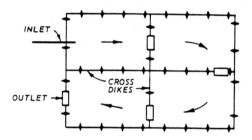

FIGURE 8.11   Series compartments, Chicago District. (*Bartos, 1977*)

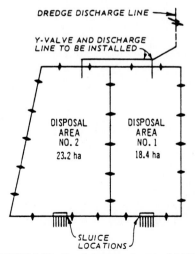

FIGURE 8.12   Parallel compartments, Mobile District. (*Bartos, 1977*)

FIGURE 8.13 An example of a disposal area arranged in series. Dredged material is pumped into basin 1, the effluent passes over to basin 2, and then passes into basin 3. The effluent satisfying the quality criteria is then released to the bay.

FIGURE 8.14 Diked disposal basin in Corpus Christi Bay (La Quinta Channel). The basin is used to confine dredged material from La Quinta Channel.

For a parallel operation, dredged material slurry is pumped into one compartment up to a desired elevation; the flow is then directed to a second compartment where uninterrupted settlement of particles can occur while the other compartment is being filled. After surface water is drained (or decanted) from the first compartment, dredged material can again be pumped into that compartment (Fig. 8.12).

A combined series-parallel compartment arrangement provides maximum flexibility in dredged-material management.

Examples of one-basin diked disposal areas are shown in Figs. 8.14, 8.15, and 8.16. In some disposal areas plastic sheeting is used to minimize the seepage through the contractor's constructed dikes (Fig. 8.17). In a drying part of the same disposal area native vegetation quickly covers the area during the growing season (Fig. 8.18). An example of open-water disposal areas forming emergent islands is shown in Fig. 8.19.

FIGURE 8.15   Diked disposal area.

FIGURE 8.16   A diked disposal basin. It appears that the effluent contains a fair amount of resuspended solids.

FIGURE 8.17 Plastic sheeting protects the confining dike constructed by a contractor. Mustang Island, Texas. Dredged material was pumped from the Corpus Christi Ship Channel.

FIGURE 8.18 Native vegetation covers the upper part of the same disposal area shown in Fig. 8.17.

SEPARATION OF COARSE MATERIAL FROM FINE-GRAINED MATERIAL

Dredged material slurry pumped into a containment area may include a large range of particle sizes and sediment characteristics. The settling and dewatering process will depend on the type of material. The materials are broadly classified as coarse-grained (sand and gravel) and fine-grained (silt and clay). The coarse-grained materials drain freely while fine-grained materials are difficult to dewater, particularly when flocculation occurs.

FIGURE 8.19 Emergent open-water disposal areas along the Port Mansfield, Texas, navigation channel.

Natural segregation occurs when the dredged material is pumped into a containment area since a mound of coarser material is formed near the discharge pipe. Finer material is carried away in suspension and deposited by a settling process some distance away from the discharge pipe, largely dependent on the topography of the area and on the presence of spur dikes and/or cross dikes. On many projects, for economical reasons, the dredged material management plan relies on natural segregation.

Selective dredging is possible and can be carried out if economically feasible. The characteristics of materials in situ must be well known prior to dredging operations so that layers of one type of material (coarse-grained) are dredged and pumped into one containment basin while the layers of another type of material (fine grained) are pumped into another compartment (Figs. 8.20 and 8.21).

There are two advantages to selective dredging:

1. Coarse-grained material will dewater quickly and the containment area will be filled quite rapidly.

2. Coarse-grained material may be used for construction of dikes or for other purposes.

Selective dredging may also be considered on a given project if suitable (coarser) material may be required for foundation fill:

1. Coarse-grained material can be used as a temporary surcharge to dewater the fine material or can be used to construct underdrainage for dewatering of subsequent lifts.

2. Overall efficiency of the disposal scheme is increased.

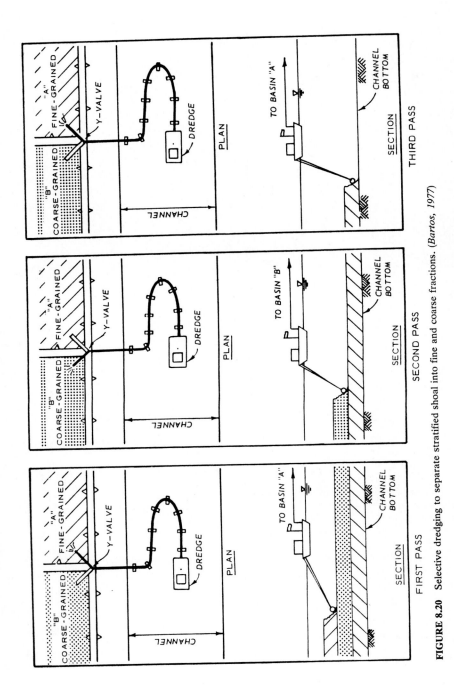

FIGURE 8.20 Selective dredging to separate stratified shoal into fine and coarse fractions. (*Bartos, 1977*)

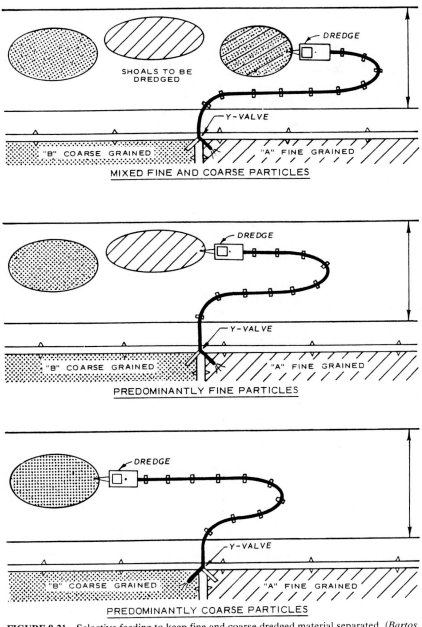

FIGURE 8.21 Selective feeding to keep fine and coarse dredged material separated. (*Bartos, 1977*)

SURFACE WATER MANAGEMENT

Since on the average, 80 percent of dredged material slurry is water, a large volume of water will accumulate in the containment area. As thickness of the dredged material deposit increases, the weir outlet elevation must be raised to provide sufficient depth for settling of solids. As soon as the upper layers of water are sufficiently clear and meet the effluent water-quality standards, the water should be discharged. The main objective here is to drain the surface water to initiate evaporative drying as soon as possible. No surface drying of material is possible as long as ponded water covers the material.

As the dredging operation begins, dredged material slurry is pumped into the disposal area and no effluent is released until the water level reaches a preset level of the outlet weir. If the disposal area(s) is properly sized, the effluent is released from the area at approximately the same rate as the slurry is pumped into the area. When pumping of slurry to the desired elevation into a given area (or compartment of an area) has been achieved, the water must be removed as quickly as the effluent water-quality standards are met. This concept is illustrated graphically in Fig. 8.22, showing—as a solid line—the elevation of ponded water increasing with time. When the ponded water reaches the preset weir outlet elevation, the effluent is discharged. The elevation of the water-solids interface also increases with time (shown as a dashed line in Fig. 8.22) until eventually there is an insufficient depth of water for the required settling to take place. This renders the effluent quality substandard and it is then necessary to increase the elevation of the outlet or to release the remaining effluent into another basin and allow evaporative drying of the dredged material.

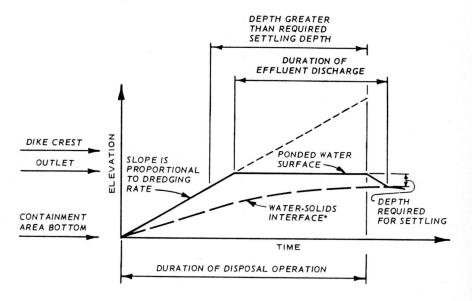

* ASSUMED APPROXIMATELY PARABOLIC (NOT LINEAR) DUE TO COMPRESSION SETTLING

FIGURE 8.22 Surface water management. (*Bartos, 1977*)

MANAGEMENT OF DREDGED MATERIAL DISPOSAL

PIANC[7] published a report recommending a strategy for selecting environmentally sound and cost-effective disposal options. A survey conducted by an ad hoc Dredging Commission in 1981[8] is shown in Table 8.5. The distribution of dredged material disposal in different categories varies among different countries, but it is clear that very large volumes of dredged material must be handled annually.

The majority of dredged material is uncontaminated and should be treated as such (e.g., it is estimated that over 95 percent of dredged material in the United States is suitable for placement in open water). The contaminated material must be handled differently. The selection of an appropriate disposal method depends on a number of variables:

1. Size of the project
2. Physical characteristics of dredged material
3. Level of contaminants present in the material to be dredged
4. Available dredging equipment
5. Site-specific conditions
6. Potential environmental impacts
7. Economical considerations
8. Social, political, and regulatory considerations

TABLE 8.5 Worldwide Method of Disposal (Volume in Thousands of Cubic Yards)

Region	Number of responses	Upland	Near wetlands*	Shore	Ocean	Other	Total
Northern Europe (21%)	26	39,196	59,502	42,936	62,044	29,412	233,091
Mediterranean (3%)	3	0	13,774	15,001	664	0	29,421
Africa (24%)	2	0	152,942	76,471	25,549	0	254,963
Southern Asia (27%)	12	62,484	11,197	121,831	89,149	0	284,661
Southeast Asia (2%)	8	0	3,078	3,698	15,190	0	21,966
East Asia (13%)	16	5,783	32,220	102,451	4,323	0	144,777
South Pacific (5%)	18	3,972	2,687	26,335	32,588	0	65,582
North America (4%)	18	6,012	9,696	8,459	16,549	159	40,875
Caribbean (1%)	5	820	646	0	2,484	0	3,950
TOTAL (100%)	108	118,267	285,742	287,202	248,522	29,571	1,079,286

*Includes estuaries, marsh, shallow water, etc.
Source: Ad Hoc Dredging Commission, 1981 (PIANC, 1990).

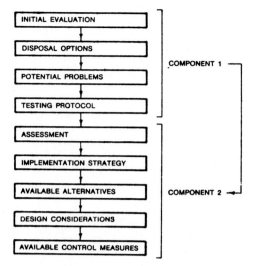

FIGURE 8.23 Technical Strategy overview flow chart. (*PIANC, 1990*)

Consideration should be principally given to:

1. Level of contamination
2. Potential environmental pathways
3. Levels of control needed
4. Identification and selection of control options

The main features of the technical strategy are the evaluation and testing, shown as component 1 in Figs. 8.23 and 8.24, and contaminate pathway assessment, selection, and implementation of an appropriate placement management strategy (component 2).

Initial evaluation should follow the pathway shown in Fig. 8.25. The three major options for placement of dredged material are dependent on whether the material may be placed in open water (aquatic or upland) or whether it may have beneficial uses.

The dredged material alternative selection strategy is summarized in Table 8.6. Figure 8.26 presents a more detailed procedure involved in the selection strategy process. In phase I the potential problems are listed as by direct contact, in the ground and surface waters, plant and animal uptakes, in the water column, and in the benthic organisms.

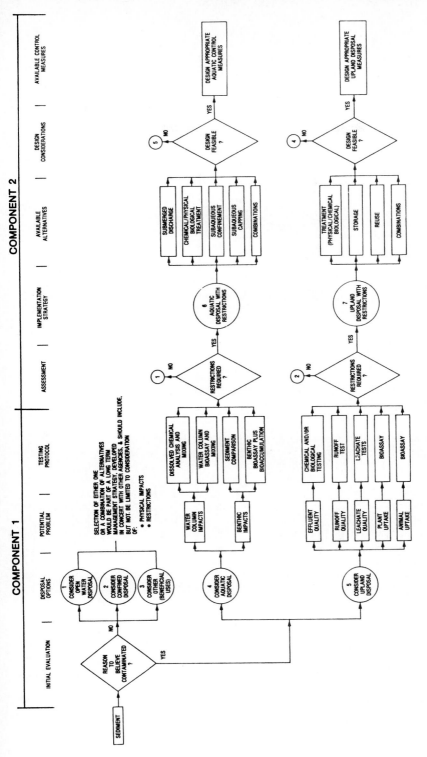

FIGURE 8.24 Strategy flow chart. *(PIANC, 1990, and Francingues et al., 1985)*

8.25

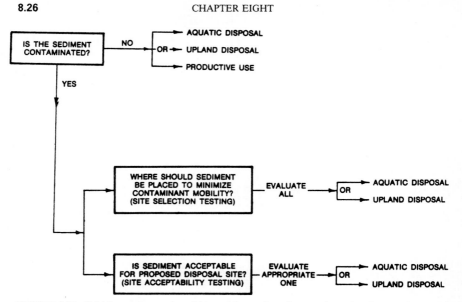

FIGURE 8.25 Initial questions to be addressed for testing of contaminated sediments. (*PIANC, 1990, and Francingues et al., 1985*)

MANAGEMENT OF LARGE CONTAINMENT AREAS

If the containment area is sufficiently large and the effluent quality standards can be met, a certain amount of ponding water may be desired to permit floating of the discharge pipeline to distribute the disposal of dredged slurry over the ponded area without interruption of the dredging operation.[9,10,11] Underwater placement of coarse-grained dredged slurry may also be beneficial to reduce the suspension of finer fractions and to provide a firm foundation within the disposal area.

Large areas do not become covered by dredged material and ponded water until after a considerable time. Division of a large area into compartments is usually required and the compartments can be filled sequentially. The slurry is pumped into one compartment while the material is permitted to dry by natural processes in other compartments. The number of compartments required will depend on:

1. Rate of slurry flow
2. Type of material
3. Size of disposal area
4. Effluent quality standards

The principal advantage of dividing an area into compartments is that drying is accomplished in some compartments during a disposal operation. This is particularly important in containment areas where disposal is almost continuous with inadequate time for dredged material drying.

TABLE 8.6 Summary of the "Dredged Material Alternative Selection Strategy" Process (DMASS)

Phase	Step	Purpose	Criteria used
I. Presumption of contamination pathway	From component 1 (Fig. 8.24)	DMF to identify contaminant type and level and pathway	DMF and related Regional Authority Decisions (RADs)
II. Confirmation of contamination pathway	1. Select potential sites	Eliminate poor or inferior sites	1. Availability 2. Distance 3. Capacity 4. Cost 5. Impact
	2. Assess site characteristics	Determine attributes of sites	
	3. Identify pathways of concern	See if pathway identified in Phase I concern is at site	
	4. Select dredge/ transport technique	For potential sites and paths, eliminate poor transport combinations	1. Impact 2. Cost 3. Compatibility
	5. Check compatibility	See if remaining site dredge/ transport options are compatible	
III. Alternative development and initial screening	1. Select potential technologies	Identify suitable combinations of technologies	1. Impact 2. Cost 3. Accepted engineering
	2. Develop alternatives	Combine technologies and sites	
	3. Screen alternatives	Eliminate poor or inferior alternatives	
IV. Detailed evaluation and ranking	1. Evaluation of alternatives	Extensive evaluation of remaining alternatives	1. Cost 2. O & M 3. Reliability 4. Safety 5. Statutory requirements 6. Implementability and availability 7. Public acceptance 8. Environmental impact 9. Technical effectiveness
	2. Ranking of alternatives	Arraying of alternatives for easy comparison	

DMF = decision-making framework.
O&M = operations and maintenance.
Source: PIANC, 1990, and Peddicord et al., 1986.

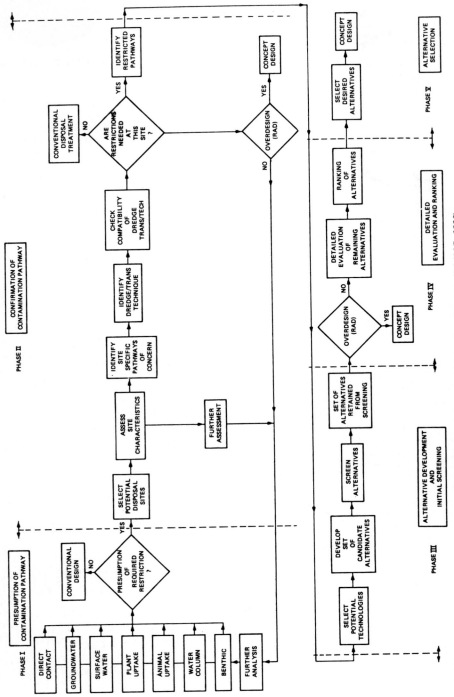

FIGURE 8.26 Detailed flow chart for the Dredged Material Alternative Selection Strategy. (*PIANC, 1990*)

Summary

1. During a disposal operation, coarse-grained material becomes separated from fine material, forming a mound near the discharge pipe (Fig. 8.5).
   a. The mound can be used to support the extended discharge pipe.
   b. The mound can be used as a supply of coarse-grained material, as a temporary surcharge to dewater fine material, or to construct drainage layers.
2. *a.* Surface water should remain ponded to a depth that will provide sufficient detention time to meet the effluent quality standards. If fine material is found to be flocculating, as is expected on the basis of laboratory testing, the top layer of clear water should be removed as quickly as possible.
   b. The surface water may also be ponded to permit the use of floating pipelines in the disposal area or to provide means of access for installation of the dewatering systems. The surface water must also be removed as soon as possible to initiate drying of dredged material.
3. There are at least three ways to optimize the natural dewatering of material dredged on a given project:
   a. Plan the dredging operations to maximize the length of time the dredged material surface is exposed to the atmosphere during the period when evaporation rates are high.
   b. Place dredged material in thin lifts which can undergo at least partial drying before being covered by subsequent lifts. This can be accomplished by compartmentalizing the diked area to allow a drying period between lifts of about 30 days (depending on the climate of a particular site).
   c. Consider the use of vegetation with a high transpiration ratio to remove moisture from dredged material. This may not be possible in land disposal areas until dredging is completed.
4. Alternative methods may include:
   a. Trenching, which can be achieved by small or large disk wheels (Figs. 8.27, 8.28, and 8.29) or with a Riverine Utility Craft (RUC) (Figs. 8.30 and 8.31). Figures 8.32 and 8.33 show the drainage trenches formed by an RUC. The trenching pattern leading to outlets is shown in Fig. 8.34. The trenches can be filled with sand and gravel to form underdrains for the next lift (Fig. 8.35).
   b. Installation of vertical drains.
   c. Placement of horizontal sand layers on soft dredged material before the surface water is drained.
   d. Bottom drainage.

Disk wheel	
Total weight dry	Diameter 2,300 mm
Total weight with water	4,000 kg
Trench depth up to	7,000 kg
Diameter	400 mm
	2,300

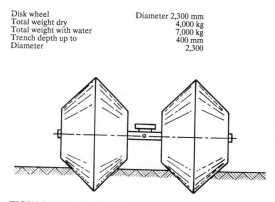

FIGURE 8.27 Small disk wheel.

Disk wheel	Diameter 3,500 mm
	Diameter 2,300 mm
Stabilizing wheel	2,300 mm
Total weight dry	9,000 kg
Total weight with water	12,000 kg
Trench depth up to	1,000 mm*

*Adjustable with stabilizing wheels

FIGURE 8.28 Large disk wheel with stabilizing wheels.

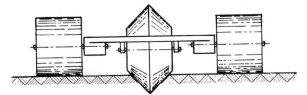

FIGURE 8.29 A disk wheel used for forming drainage trenches.

FIGURE 8.30 RUC deployed in Mobile Bay, Alabama, Confined Disposal Area.

FIGURE 8.31   A close-up of a helical screw pontoon of an RUC forming a trench in a confined disposal area.

FIGURE 8.32   Trenches formed by an RUC in a vegetated-crust confined disposal area. Note that the dredged material still has a very high water content evidenced by the volume of water almost instantaneously flowing into a drainage trench.

FIGURE 8.33 Trenches formed by an RUC to start a process of near-surface drainage in a confined disposal area.

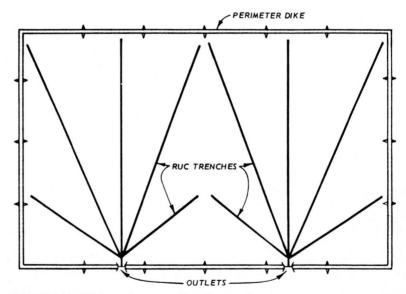

FIGURE 8.34 RUC trench pattern. Note that trenches do not cross each other because the RUC blocks trenches it crosses. (*Bartos, 1977*)

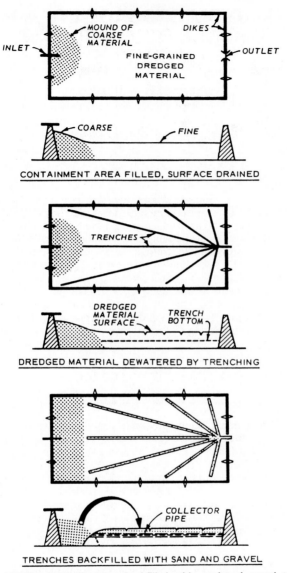

FIGURE 8.35 Trenches backfilled with sand and gravel to underdrain subsequent dredged material lift. (*Bartos, 1977*)

LONG-TERM STORAGE CAPACITY OF CONFINED DISPOSAL FACILITIES

Michael R. Palermo

Research Civil Engineer, Environmental Engineering Division, Environmental Laboratory, U.S. Army Engineer Waterways Experiment Station, Vicksburg, Miss.

Many dredging projects are located where there are excessive and often conflicting land use demands; therefore confined disposal facilities (CDFs)* for dredged material must be efficiently utilized. Furthermore, the demand for long-term management strategies to meet dredging requirements over the life of navigation projects continues to grow. Such strategies require the estimation of long-term storage capacities of CDFs for known or estimated volumes of sediment to be dredged at varying locations and times over a period of many years. Complete strategies also include plans for managing CDFs to dewater the dredged material and increase storage capacity. This chapter describes procedures for estimating long-term storage capacity of CDFs, conducting appropriate testing programs for these evaluations, and managing CDFs to increase storage capacity.

FACTORS AFFECTING LONG-TERM STORAGE CAPACITY

CDFs are diked areas constructed to contain dredged material placed using hydraulic or mechanical means. CDFs retain dredged material solids while allowing carrier water (if hydraulically filled) to be discharged from the site as effluent. The two objectives inherent in the design and operation of CDFs are to provide adequate storage capacity to meet the dredging requirements and to meet applicable effluent standards in retaining suspended solids during filling operations. These considerations are interrelated and require effective design, operation, and management of the CDF. General guidance for design and operation of CDFs to retain solids and to provide the necessary initial storage capacity for an active filling operation has been developed by the Corps of Engineers (HQUSACE, 1987).†

Storage capacity is defined as the total volume available to hold dredged material and is equal to the total unoccupied volume minus the volume associated with ponding and freeboard requirements. The total volume available is limited by the surface area of the site and the ultimate height to which dikes can be con-

---

*The terms *confined disposal facility, confined disposal area, confined disposal site, diked disposal area,* and *containment area* all appear in the literature and refer to an engineered structure for containment of dredged material.

†References for chapter section written by Michael R. Palermo are at the end of his section.

structed. If the CDF is intended for one-time use, initial storage capacity and retention of solids during filling are the only design considerations. However, if the CDF is intended for long-term use, the long-term storage capacity must also be considered.

Assuming that a CDF occupies a given surface area, the storage capacity remaining at any time will be a function of the dredged material fill height. As additional dredged material is placed in the CDF, sedimentation of the suspended solids occurs and the fill height increases. Following the completion of a filling cycle, the fill height will decrease because of three processes: continued sedimentation, consolidation, and desiccation. Sedimentation is a relatively short-term process, and the settling properties of the material will determine the requirements for ponding and initial storage during filling. However, consolidation and desiccation are long-term processes which will determine the long-term storage capacity requirements. A conceptual diagram illustrating these processes is shown in Fig. 8.36.

DREDGED MATERIAL CONSOLIDATION PROCESSES

The coarse-grained fraction of dredged material (sands and coarser material) undergoes sedimentation quickly and will occupy essentially the same volume as was occupied prior to dredging. However, the fine-grained fraction of the material (silts and clays) requires longer settling times, initially occupies considerably more volume than prior to dredging, and will undergo a considerable degree of

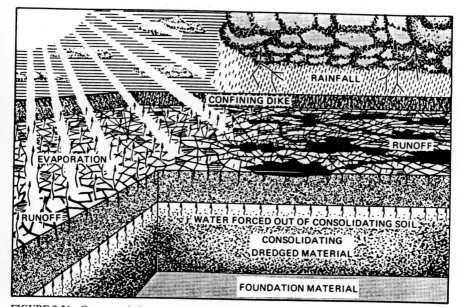

FIGURE 8.36 Conceptual diagram of dredged material consolidation and dewatering processes.

long-term volume change from consolidation if it is hydraulically placed. Such materials are essentially underconsolidated soils, and the consolidation takes place because of self-weight loading.

Dredged material placement also imposes a loading on the containment area foundation, and additional settlement may result from consolidation of compressible foundation soils. Settlement from consolidation is therefore a major factor in the estimation of long-term storage capacity. Since the consolidation process for fine-grained materials is slow, total settlement may not have taken place before the containment area is required for additional placement of dredged material. Settlement of the containing dikes may also significantly affect the available storage capacity and should be carefully considered.

Guidelines for estimating gains in long-term capacity because of settlement within the containment area are based on the fundamental principles of consolidation theory modified to consider the self-weight consolidation behavior of newly placed dredged material. Three types of consolidation may occur in dredged material containment areas: primary consolidation, secondary consolidation, and consolidation resulting from desiccation.

The Terzaghi standard theory of one-dimensional consolidation or *small strain theory* has received widespread use among geotechnical engineers and has received widespread application for consolidation problems in which the magnitude of settlement is small in comparison to the thickness of the consolidating layer. In contrast to the small strain theory, a *finite strain theory* for one-dimensional consolidation is better suited for describing the large settlements common to the primary consolidation of soft fine-grained dredged material because of self-weight (Cargill, 1983, 1985). The process of secondary consolidation, or "creep," refers to the rearrangement of soil grains under load following completion of primary consolidation. This process is not normally considered in settlement analyses and is not considered in this chapter.

Two phenomena control the amount of consolidation caused by desiccation of fine-grained dredged material. The first is the evaporation of water from the upper sections of the dredged material. The resulting reduction in its moisture content causes a reduction in void ratio or volume occupied because of the negative pore water pressure induced by the drying. This can be referred to as the dewatering process. An additional process influencing settlement involves the primary consolidation in underlying material when the free water surface is lowered. As the water surface moves downward, the unit weight acting on lower material changes from buoyant unit weight to effective unit weight. The material below the new water level is therefore subjected to an additional surcharge. Because of these factors, the consolidation and desiccation processes are interactive.

DREDGED MATERIAL DEWATERING PROCESSES

General Process Description

Once a given active filling operation ends, any ponded surface water required for settling should be decanted, exposing the dredged material surface to desiccation (evaporative drying). This process can further add to long-term storage capacity and is a time-dependent and climate-dependent process. However, active dewatering operations such as surface trenching enhance the natural dewatering process.

Desiccation of dredged material is basically the removal of water by evaporation and transpiration. Plant transpiration can also enhance dewatering but is not considered in this chapter. Evaporation potential is controlled by such variables as radiation heating from the sun, convective heating from the earth, air temperature, ground temperature, relative humidity, and wind speed. However, other factors affect actual evaporative drying rates. For instance, the evaporation efficiency is normally not a constant but some function of depth to which the layer has been desiccated and also is dependent on the amount of water available for evaporation.

Evaporative Stages

Evaporative drying of dredged material leading to the formation of a desiccated crust is a two-stage process, and the removal of water occurs at differing rates during the two stages. The first stage begins when all free water has been decanted, or drained, from the dredged material surface. The void ratio e at this point corresponds to approximately zero-effective stress as determined by laboratory sedimentation and consolidation testing. This initial void ratio can be empirically estimated as a water content of approximately 2.5 times the Atterberg liquid limit (LL) of the dredged material (Haliburton, 1978).

First-stage drying ends and second-stage drying begins at a void ratio that may be called the *decant point or saturation limit e_{SL}*. The e_{SL} of typical dredged material has been empirically determined to be at a water content of approximately 1.8 LL. Second-stage drying is an effective process until the material reaches a void ratio that may be called the *desiccation limit*, or e_{DL}. When the e_{DL} reaches a limiting depth, reduction of the water content of the dredged material from evaporation will effectively cease. However, evaporation of excess moisture from undrained rainfall and water forced out of the material as a result of consolidation of material below the crust may continue. The e_{DL} of typical dredged material may roughly correspond to a water content of 1.2 Atterberg plastic limit (PL). Also associated with the e_{DL} of a material is a particular percentage of saturation that probably varies from 100 percent to something slightly less, depending on the material (Haliburton, 1978).

ESTIMATION OF LONG-TERM STORAGE CAPACITY

Data Requirements

The data required to estimate long-term storage capacity include physical properties of the sediments and foundation soils such as specific gravity, grain size distributions, Atterberg liquid and plastic limits, and water contents; the consolidation properties of the fine-grained dredged material and foundation soils (relationships of void ratio and permeability versus effective stress); CDF site characteristics such as surface area, ultimate dike height, groundwater table elevations, average pan evaporation rates, average rainfall; and dredging data such as volumes to be dredged, rate of filling, and frequency of dredging (HQUSACE, 1987 and Stark, in preparation).

Laboratory Testing

Laboratory tests are required to determine the physical and engineering properties of fine-grained sediments and foundation soils used in estimating long-term storage capacity of CDFs. Physical properties such as water contents and Atterberg limits of the materials can be determined using conventional testing techniques (HQUSACE, 1970). Consolidation testing, especially for sediments, is more involved and should include time-consolidation data.

Consolidation tests for foundation soils should be performed using conventional procedures (HQUSACE, 1970). However, specialized procedures are necessary for consolidation testing of sediment samples because of their fluidlike consistency. Specially developed self-weight consolidation tests (Cargill, 1986; Zappi et al., 1991) can be used to determine consolidation characteristics at low effective stresses. Controlled-rate-of-strain tests (Cargill, 1986) or fixed-ring consolidometers should be used to determine characteristics at higher effective stresses. Modifications in sample preparation and the method of loading are necessary for the conventional fixed-ring procedure when testing sediments (HQUSACE, 1987; Zappi et al., 1991).

Storage Capacity-Time Relationship

Estimated time-settlements caused by dredged material consolidation and dewatering and foundation consolidation may be combined to yield a time-total settlement relationship for a single lift as shown in Fig. 8.37. These data are sufficient for estimating the remaining capacity in the short term. However, if the containment area is to be used for long-term placement of subsequent lifts, the dredged material surface height versus time should be projected. This projection

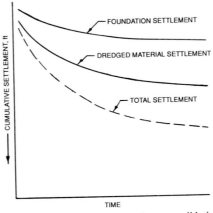

FIGURE 8.37 Illustrative time-consolidation relationships.

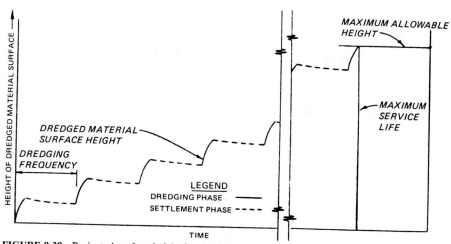

FIGURE 8.38 Projected surface height for determination of containment area service life.

can be developed using time-settlement relationships for sequential lifts, as shown in Fig. 8.38. These projections may be used for preliminary estimates of the long-term capacity of the containment area.

The maximum dike height as determined by foundation conditions or other constraints and the containment surface area dictate the maximum available storage volume. Increases in dredged material surface height during the dredging phases and decreases during settlement phases correspond to respective decreases and increases in remaining containment storage capacity, shown in Fig. 8.39. Projecting the surface height or remaining capacity to the point of maximum allowable height or exhaustion of remaining capacity, respectively, yields an estimate of the containment area service life. Gains in capacity from anticipated

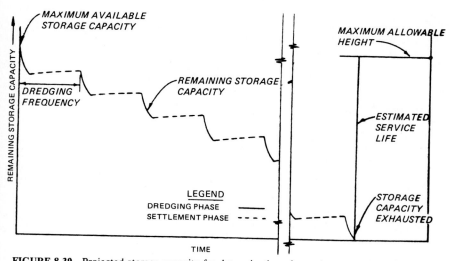

FIGURE 8.39 Projected storage capacity for determination of containment area service life.

dewatering or material removal should also be considered in making the projections.

The complex nature of the consolidation and desiccation relationships for multiple lifts of compressible dredged material and the changing nature of the resulting loads imposed on compressible foundation soils may result in errors in projections of remaining storage capacity over long time periods. Accuracy can be greatly improved by updating the estimates every few years using data from newly collected samples and laboratory tests. Observed field behavior should also be routinely recorded and used to refine the projections.

Small Strain versus Finite Strain Consolidation

The most applicable procedure for estimating consolidation in soft dredged material is the finite strain consolidation theory. The magnitude of consolidation as determined by small strain techniques is equivalent to that determined by the finite strain technique. However, the time rate of consolidation is overly conservative for small strain in that the rate of consolidation as predicted is slow when compared to field behavior (Cargill, 1983, 1985). The finite strain technique holds advantage for the estimation of dredged material consolidation settlement because it accounts for the nonlinearity of the void ratio, permeability, and coefficient of consolidation relationships that must be considered when large settlements of a layer are involved.

Empirical Methods for Estimating Desiccation Behavior

Empirical equations for estimating the settlement of a dredged material layer from desiccation and the thickness of dried crust were developed for the purpose of determining feasibility and benefits of active dewatering operations (Haliburton, 1978). The empirical relationships have been refined (Cargill, 1986) to consider the two-stage process of desiccation and the overall water balance relationships that exist within a dredged material disposal area. The interaction of the desiccation process with dredged material consolidation from self-weight has been incorporated in computer programs for estimating long-term storage capacity. The refined empirical relationships can be easily applied in determining the benefits of dewatering programs and provide increased accuracy in storage capacity evaluations.

Computer Programs for Consolidation and Desiccation

The use of computer programs can greatly facilitate the estimation of storage capacity for containment areas. Although the computations for simple cases can be done by hand, the analyses often require computations for a multiyear service life with variable disposal operations and possibly material removal or dewatering operations occurring intermittently throughout the service life. These complex computations can be done more efficiently using a computer program.

The use of computer programs holds added advantage when considering the additional settlements that occur as the result of dredged material desiccation (dewatering). The estimation of desiccation behavior can also be done by means

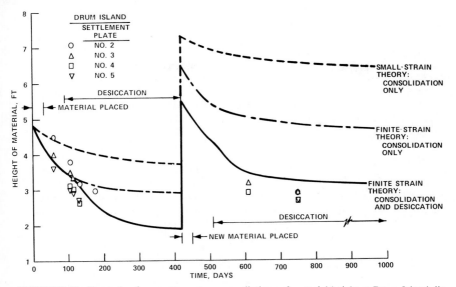

FIGURE 8.40 Example of computer program predictions of material height at Drum Island disposal area, Charleston, South Carolina.

of hand calculations; however, the interaction between desiccation and consolidation would require cumbersome iterative calculations. A computer program is well suited to handle the calculations of both consolidation and desiccation and the interaction between the two processes.

The recommended computer program for use in predicting the long-term capacity of disposal areas is entitled Primary Consolidation and Desiccation of Dredged Fill (PCDDF) and incorporates the concepts described in this chapter. This program is available as a part of the Automated Dredging and Disposal Alternatives Management System (ADDAMS) (Schroeder and Palermo, 1990). Theoretical documentation, description of solution techniques, and a user guide are available (Cargill, 1985; Poindexter-Rollings, 1989a; Stark, 1991; and Stark and O'Meara, 1991).

Example results obtained using the PCDDF program are shown in Fig. 8.40. The plot shows predicted dredged material surface elevation versus time for several cases including multiple layers deposited at varying times. Field data collected at the site are also shown for comparison. This plot also shows the relative gain in storage capacity from desiccation of the material.

DREDGED MATERIAL DEWATERING OPERATIONS

If the CDF is well managed following active filling, the excess water will be drained from the surface and natural evaporation will act to dewater the material. However, active dewatering operations should be considered to speed up the dewatering process and achieve the maximum possible volume reduction considering the site-specific conditions and operational constraints.

Dewatering results in several benefits. Shrinkage and additional consolidation of the material resulting from dewatering operations lead to creation of more volume in the CDF for additional dredged material. The drying process changes the dredged material into a more stable soil form amenable to removal and use in dike raising, other engineered construction, or other productive uses, again creating more available volume in the CDF. Dewatered material remaining in the CDF forms a more stable fast land with predictable geotechnical properties. Also, the drainage associated with dewatering helps control mosquito breeding.

A number of dewatering techniques for fine-grained dredged material have been studied (Haliburton, 1978; Haliburton et al., 1991). However, surface trenching and use of underdrains were found to be the only technically feasible and economically justifiable dewatering techniques (Haliburton, 1978). Techniques such as vacuum filtration or belt filter presses can be technically effective but are not economical for dewatering large volumes of fine-grained material.

Guidance for application of underdrains is available (Hammer, 1981), and the use of underdrains has been successfully applied in CDFs. However, use of underdrains over large surface areas is not as economical as surface drainage techniques and has not been routinely applied. Accordingly, only techniques recommended for improvement of surface drainage through trenching are described in detail here.

Dewatering by Progressive Trenching

The concept of surface trenching to dewater fine-grained dredged material was first applied by the Dutch (d'Angremond et al., 1978) and later field-verified under conditions typical of CDFs in the United States (Palermo, 1977). Surface trenching has since become a commonly used management approach for dewatering in CDFs (Poindexter, 1988; Poindexter-Rollings, 1989b).

The following considerations influence the effectiveness of dewatering through surface trenching:

1. Establishment of good surface drainage will allow evaporative forces to dry the dredged material from the surface downward, even at disposal area locations where precipitation exceeds evaporation.

2. The most practical mechanism for precipitation removal is by runoff through crust desiccation cracks to surface drainage trenches and off the site through outlet weirs.

3. To maintain effective drainage, the flow-line elevation of any surface drainage trench must always be lower than the base of crust desiccation cracks; otherwise, ponding will occur in the cracks. As drying occurs, the cracks will become progressively deeper.

4. Below the desiccation crust, the fine-grained subcrust material may be expected to exist at water contents at or above the liquid limit (LL). Thus, it will be difficult to physically construct trenches much deeper than the bottom of the adjacent desiccation crust.

5. To promote continuing surface drainage as drying occurs, it is necessary to progressively deepen site drainage trenches as the water table falls and the surface crust becomes thicker; thus, the term *progressive trenching*.

6. During conduct of a progressive trenching program, the elevation difference between the internal water table and the flow line of any drainage trench will be relatively small. When the relatively low permeability of fine-grained dredged material is combined with the small hydraulic gradient likely under these circumstances, it appears doubtful that appreciable water can be drained from the dredged material by gravity seepage. Thus, criteria for trench location and spacing should be based on site topography so that precipitation is rapidly removed and ponding is prevented, rather than to achieve marked drawdown from seepage.

Initial Dewatering (Passive Phase)

Once a filling operation is completed, dredged material usually undergoes hindered sedimentation and self-weight consolidation (called the *decant phase*), and water will be brought to the surface of the consolidating material at a faster rate than can normally be evaporated. During this phase, it is extremely important that continued drainage of decant water and/or precipitation through outlet weirs be facilitated. Weir flow-line elevations may have to be lowered periodically as the surface of the newly placed dredged material subsides. Guidelines for appropriate disposal site operation during this passive dewatering phase are available (HQUSACE, 1987).

Once the fine-grained dredge material approaches the decant point water content, or saturation limit as described previously, the rate at which water is brought to the surface will gradually drop below the climatic evaporative demand. If precipitation runoff through site outflow weirs is facilitated, a thin drying crust or skin will form on the newly deposited dredged material. The thin skin may be only several hundredths of a foot thick, but its presence may be observed by noting small desiccation cracks that begin to form at 3- to 6-ft intervals. Once the dredged material has reached this consistency, active dewatering operations may be initiated.

Perimeter Dragline Trenching Operations

Construction of trenches around the inside perimeter of confined disposal sites using draglines, as shown in Fig. 8.41, is a procedure that has been used for many years to dewater and/or reclaim fine-grained dredged material. In many instances, the purpose of dewatering has been to obtain convenient borrow material to raise perimeter dikes. Draglines and backhoes are adaptable to certain perimeter trenching activities because of their relatively long boom length and/or method of operation and control. The perimeter trenching scheme should be planned carefully so as not to interfere with operations necessary for later dewatering or other management activities.

Operations should begin at an outflow weir location, where the dragline, operating from the perimeter dike, should dig a sump around the weir extending into the disposal area. Once the sump has been completed, the dragline should operate along the perimeter dike, casting its bucket the maximum practicable distance into the disposal area, dragging material back in a wide shallow arc to be cast on the inside of the perimeter dike. A wide shallow depression will be formed and will serve as an initial drainage path.

Once appreciable desiccation drying has occurred in the dredged material ad-

FIGURE 8.41 Small dragline operation for perimeter trenching.

jacent to the perimeter trench and the material cast on the interior slope of the perimeter dike has dried, the perimeter trenches and weir sumps should be deepened. These deeper trenches will again facilitate more rapid dewatering of dredged material adjacent to their edges, with resulting shrinkage and deeper desiccation cracks providing a still steeper drainage flow gradient from the site interior to the perimeter trenches. After several cycles, trenches up to 3 to 5 ft deep may be completed. Additional guidance on the specific sequencing and timing of perimeter trenching operations is available (Haliburton, 1978).

Interior Trenching

As drying continues and perimeter trenching progresses, the construction of interior trenches spaced over the entire surface area of the CDF may be initiated. Only specialized amphibious vehicles (such as those using twin screws for propulsion and flotation) can successfully construct shallow trenches in fine-grained dredged material shortly after formation of a thin surface crust (Palermo, 1977; Haliburton, 1978). However, field experience has shown that the early stages of evaporative dewatering and crust development occur at acceptable rates considering only the natural drying processes, perhaps aided by perimeter trenching as described previously. Therefore the use of such specialized trenching equipment is not usually warranted.

Once a surface crust of 4 to 6 in has developed, use of trenching equipment with continuously operating rotary excavation devices and low-ground-pressure chassis is recommended for routine dewatering operations. This type of equipment has been used successfully in dewatering operations in numerous locations along the Atlantic and Gulf Coasts. The major features of the equipment include a low-ground-pressure chassis equipped with a mechanical excavation implement. The implement has a rotary cutting wheel or wheels used to cut a trench up to 3 ft deep. The low-ground-pressure chassis may be tracked or rubber tired, as shown in Figs. 8.42 and 8.43. The major advantage of rotary trenchers is their ability to continuously excavate while slowly moving within the containment area

FIGURE 8.42   Rubber-tired rotary trencher.

FIGURE 8.43   Track-mounted rotary trencher.

(Poindexter, 1989). This allows them to construct trenches in areas where dragline or backhoe equipment would have mobility problems. The excavating wheels can be arranged in configurations that create hemispherical or trapezoidal trench cross sections and can throw material to one or both sides of the trench, as shown in Figs. 8.44 and 8.45. The material is spread in a thin layer by the throwing action, which allows it to dry quickly and prevents the creation of a windrow which might block drainage to the trench. Based on past experience, an initial

FIGURE 8.44 Rotary trenching device in operation.

FIGURE 8.45 General view of trenches formed by rotary trencher.

crust thickness of 4 to 6 in is required for effective mobility of this low-ground-pressure equipment. This crust thickness can be easily formed within the first year of dewatering effort if surface water is effectively drained from the area, assisted by perimeter trenches constructed by draglines operating from the dikes.

Interior Trench Spacing and Pattern

Trenches should extend directly to low spots containing ponded water. However, the greater the number of trenches per unit of disposal site area, the shorter the distance that precipitation runoff will have to drain through desiccation cracks before encountering a drainage trench. Thus, closely spaced trenches should produce more rapid precipitation runoff and may slightly increase the rate of evaporative dewatering. Conversely, the greater the number of trenches constructed per unit of disposal site area, the greater the cost of dewatering operations and the greater their impact on subsequent dike raising or other borrowing operations. However, the rotary trenchers have a relatively high operational speed, and it is therefore recommended that the maximum number of drainage trenches be placed consistent with a site-specific trenching plan. Trench spacings of 100 to 200 ft have normally been used. If topographic data are available for the disposal site interior, they may be used as the basis for the trenching plan.

The most common trench pattern employs parallel trenching. A complete circuit of the disposal area with a perimeter trench is joined with parallel trenches cut back and forth across the disposal area, ending in the perimeter trench. Spacing between parallel trenches can be varied, as described above. Small disposal areas or irregularly shaped disposal areas may be well suited for a radial trenching pattern for effective drainage of water to the weir structures. The radial patterns should run parallel to the direction of the surface slopes existing within the area. Radial trenching patterns can also be used to provide drainage from localized low spots to the main drainage trench pattern. A suggested scheme for perimeter and interior trenching using a combination of draglines and a rotary trencher or other suitable equipment and incorporating both radial and parallel trenches is shown in Fig. 8.46.

CONCLUSION

Because of the shortage of suitable sites, there is a pressing need to extend the storage capacity of CDFs to meet long-term disposal requirements. The procedures discussed in this section for estimating long-term consolidation and desiccation behavior of dredged material and for implementing field dewatering programs should aid in the planning, design, and management of CDFs.

ACKNOWLEDGMENT

The information presented in this section is largely taken from publications of the U.S. Army Corps of Engineers (USACE). Permission was granted by the Chief of Engineers to publish this information. The long-term capacity of CDFs is a topic that cannot be covered in complete detail here, therefore, this section pre-

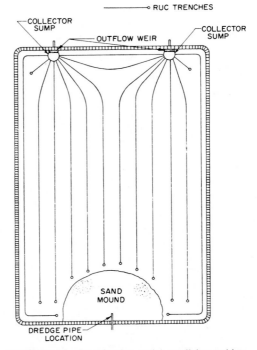

FIGURE 8.46   Combination radial-parallel trenching scheme.

sents an overview of the pertinent considerations. References cited in the text are readily available source documents containing detailed guidance. The contribution of the authors of the original publications is gratefully acknowledged.

REFERENCES

Cargill, K. W., 1983. *Procedures for Prediction of Consolidation in Soft, Fine-Grained Dredged Material,* Technical Report D-83-1, U.S. Army Engineer Waterways Experiment Station, Vicksburg, Miss.

————, 1985. *Mathematical Model of the Consolidation/Desiccation Process in Dredged Material,* Technical Report D-85-4, U.S. Army Engineer Waterways Experiment Station, Vicksburg, Miss.

————, 1986. *The Large Strain, Controlled Rate of Strain (LSCRS) Device for Consolidation Testing of Soft Fine-Grained Soils,* Technical Report GL-86-13, U.S. Army Engineer Waterways Experiment Station, Vicksburg, Miss.

d'Angremond, K. et al., 1978. *Assessment of Certain European Dredging Practices and Dredged Material Containment and Reclamation Methods,* Technical Report D-78-58, U.S. Army Engineer Waterways Experiment Station, Vicksburg, Miss.

Haliburton, T. A., 1978. *Guidelines for Dewatering/Densifying Confined Dredged Material,*

Technical Report DS-78-11, U.S. Army Engineer Waterways Experiment Station, Vicksburg, Miss.

————, et al., 1991. *Dredged Material Dewatering Field Demonstrations at Upper Polecat Bay Disposal Area, Mobile, Alabama,* Technical Report, U.S. Army Engineer Waterways Experiment Station, Vicksburg, Miss. (in print)

Hammer, D. P., 1981. *Evaluation of Underdrainage Techniques for the Densification of Fine-Grained Dredged Material,* Technical Report EL-81-3, U.S. Army Engineer Waterways Experiment Station, Vicksburg, Miss.

Headquarters, U.S. Army Corps of Engineers (HQUSACE), 1970. *Laboratory Soils Testing,* Engineer Manual 1110-2-1906, Office, Chief of Engineers, Washington, D.C.

————, 1987. *Confined Disposal of Dredged Material,* Engineer Manual 1110-2-5027, Office, Chief of Engineers, Washington, D.C.

Palermo, M. R., 1977. *An Evaluation of Progressive Trenching as a Technique for Dewatering Fine-Grained Dredged Material,* Miscellaneous Paper D-77-4, December 1977, U.S. Army Engineer Waterways Experiment Station, Vicksburg, Miss.

Poindexter, M. E., 1988. *Current District Dredged Material Dewatering Practices,* Environmental Effects of Dredging Technical Note EEDP-06-4, U.S. Army Engineer Waterways Experiment Station, Vicksburg, Miss.

————, 1989. *Equipment Mobility in Confined Dredged Material Disposal Areas; Field Evaluations,* Environmental Effects of Dredging Technical Note EEDP-09-4, U.S. Army Engineer Waterways Experiment Station, Vicksburg, Miss.

Poindexter-Rollings, M. E., 1989a. *PCDDF89—Updated Computer Model to Evaluate Consolidation/Desiccation of Soft Soils,* Environmental Effects of Dredging Technical Note EEDP-02-10, U.S. Army Engineer Waterways Experiment Station, Vicksburg, Miss.

————, 1989b. *Dredged Material Containment Area Management Practices for Increasing Storage Capacity,* Environmental Effects of Dredging Technical Note EEDP-06-6, U.S. Army Engineer Waterways Experiment Station, Vicksburg, Miss.

Schroeder, P. R., and Palermo, M. R., 1990. *The Automated Dredging and Disposal Alternatives Management System (ADDAMS),* Environmental Effects of Dredging Programs Technical Note EEDP-06-12, U.S. Army Engineer Waterways Experiment Station, Vicksburg, Miss.

Stark, T. D., 1991. *Program Documentation and User's Guide: PCDDF90, Primary Consolidation and Desiccation of Dredged Fill,* Instruction Report, U.S. Army Engineer Waterways Experiment Station, Vicksburg, Miss. (in print)

————, and O'Meara, T. J., 1991. *Database of Dredged Fill Material Properties for Use with PCDDF90,* Instruction Report, U.S. Army Engineer Waterways Experiment Station, Vicksburg, Miss. (in print)

Zappi, P. A., Schroeder, P. R., and Hayes, D. F., 1991. *Fixed Ring and Self-Weight Consolidation Test Data Reduction for Running PCDDF,* Environmental Effects of Dredging Technical Note, U.S. Army Engineer Waterways Experiment Station, Vicksburg, Miss. (in print)

*OFFSHORE AND UPLAND DISPOSAL AREAS**

Evaluation Methodology

A complete analysis of dredging and dredged material disposal for any project requires that technical, aesthetic, environmental, economical, technical, legal,

---

*The remaining material in this chapter has been written by John B. Herbich.

and political factors be considered and evaluated for alternative solutions. A systems engineering approach using multiattribute utility theory offers the engineer or planner a tool to methodically analyze and synthesize these factors to determine the relative merits of the available alternatives in a cost-effective manner. This section presents a brief overview of multiattribute utility theory and demonstrates how it can be used in evaluating dredged material disposal areas. An economic analysis of upland dredged material disposal areas and the results from an economic comparison of dredging and transport methods are presented.[12]

Multiattribute Utility Theory[12]

Multiattribute utility theory[13,14,15] provides a rational basis for evaluating alternative systems with a quantifiable unit of measure called *utility*. A system is divided into subsystems which have measurable attributes and a corresponding utility. Utility functions for each attribute must be established, preferably before identifying possible alternatives. The utility of each attribute can be added directly or weighted to give important attributes a large portion of the total system utility. Figure 8.47 shows a simplified system with two subsubsystems, each having two attributes. Representative utility functions of each attribute (A1, A2, B1, B2) are shown on the side of the figure. The quantity of A1 associated with an assumed alternative gives a utility of 0.5. This utility is multiplied by the weight factor of 0.5 and added to the weighted utility of attribute A2 to give a utility for subsystem A of 0.75. The weights of subsystems show A to be 1.5 (0.6/0.4) times as important as B. Adding the weighted utilities of each subsystem gives a total utility for this alternative of 0.71, which would be used to compare with other alternatives.

Comparison of potential dredged material disposal sites using utility theory was similarly accomplished in a recent study using the following utility mode.[16] Eight attributes, four environmental and four technical, were identified and eval-

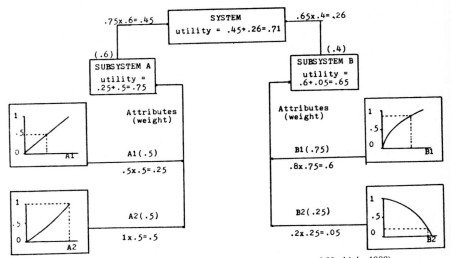

FIGURE 8.47 Multiattribute utility and system analysis. (*Glover and Herbich, 1989*)

uated for each site. The environmental attributes were wetland preservation, wildlife preservation, beneficial use of site, and existing site development. The technical attributes were storage capacity, transportation, distance, access, and other technical considerations.

Each of the eight criteria was scored from 0 to 100 and weighted by importance. Wetland preservation was considered the most important attribute of a potential site and was given a weight of 0.28, 3 times the weight of any other criteria. Disposal site area or capacity was given a weight of 0.18, 2 times that of the remaining attributes, which were given a weight of 0.09. The maximum total utility possible was 100.

In an actual study, potential upland disposal sites were identified that would minimize impact to wetland areas. To facilitate evaluating wetland areas it was assumed that the 5-ft (1.5-m) NGVD contour delineated upland and wetland area unless actual wetland determination at a site had been made and was readily available. In actuality, wetlands must be identified by a detailed site investigation required by law. The percentage of an identified area above the 5-ft (1.5-m) NGVD contours was the wetland score for that site. An area that was entirely above the 5-ft (1.5-m) NGVD contour received a utility of 100, and 0 if the entire area was a wetland area or below the 5-ft (1.5-m) NGVD contour.

Preservation of existing wildlife areas was selected so that impacts to upland areas would be included. The percentage of a potential site that was not in an identified wildlife area was the utility for this attribute. A detailed biological survey is needed to ensure that valuable upland natural resources are not destroyed. The potential for beneficial use of a site was evaluated so that a utility of 50 out of a possible 100 was given unless a potential beneficial or detrimental use was identified. The lack of development at a potential disposal site was expected to be easier to acquire and to reduce cost. A known archaeological resource on the site would reduce this attribute's utility to zero.

The utility of disposal site size or capacity was determined by the percentage of the required area that a site contained. Transportation distance was scored 100 if the potential disposal site was within 1 mi (1.6 km) of the channel, and the utility decreased 10 points for each additional mile so that a disposal site 5 mi (8.0 km) from the channel would receive a score of 50. Access to the disposal site is needed in order to monitor the site, facilitate dewatering, and remove fill material. Potential sites that have ready access to roads are given a score of 100, decreasing as access decreased or when waterborne access is required. Other technical considerations for an upland site included geotechnical or construction difficulties expected. Disposal sites that extend into marsh areas may have some special foundation requirements, lowering the score from the maximum of 100. Excess site elevation or natural slope decreased the score of this attribute.

Table 8.7 illustrates the criteria for a potential disposal site that has 90 percent of the area above the 5-ft (1.5-m) NGVD contour (score 90), is not a critical wildlife area (score 100), is neutral as to beneficial use (score 50), and has no development on the site (score 100). The site has adequate capacity for the project (score 100), is located 2 mi (3.2 km) from the channel (score 80), has good access (score 100), and has no known negative technical attributes (score 100). The overall score for this site is 90.9 out of 100 possible, which would be used to compare it to alternate sites.

In performing the analysis the three most important attributes were wetland preservation, disposal site capacity, and distance from dredge site to disposal site. The other attributes were not as important in the actual analysis because alternative sites identified fully satisfied those attributes. The use of utility theory

TABLE 8.7 Example of Potential Disposal Site Assessment Criteria. Maximum Utility Possible Is 100

	Utility	Weight	Weighted assessment
Wetland preservation	90	0.28	25.2
Wildlife preservation	100	0.09	9.0
Beneficial use	50	0.09	4.5
Not developed	100	0.09	9.0
Storage capacity	100	0.18	18.0
Transportation distance	80	0.09	7.2
Access	100	0.09	9.0
Technical	100	0.09	9.0
Total utility			90.9

Source: Glover and Herbich, 1989.

to evaluate disposal site alternative was essential in the cost-effective evaluation of the 59 potential sites identified in the study by Glover and Herbich (1989).

Upland Disposal Site Economic Analysis

This analysis was conducted to estimate the disposal site cost per cubic yard and the annual cost associated with each recommended upland disposal site in coastal Mississippi. This section describes those estimates, which are summarized in Table 8.8.

The upland cost was determined by multiplying the estimated cost per acre by the recommended disposal site acreage. Land cost was estimated based on the development in the area and proximity to the navigational channel. Land cost estimates varied from $6500 per acre ($16,062 per hectare) for a remote previously

TABLE 8.8 Economic Analysis of a Typical Upland Dredged Material Disposal Site

Area (acres)	448
Land cost ($9000/acre)	$4,032,000
Acquisition cost (5%)	$201,600
Clear/grub ($1500/acre)	$672,000
Dike height (feet)	25
Effective area (acres)	423.5
Dike volume (cubic yards)	1,365,530
Dike cost ($5/cubic yard)	$6,827,649
Weir cost	$200,000
Maintenance cost	$200,000
Dredging cycle (year)	1.5
Dewater ($1500/acre/cycle)	$4,421,024
Total capacity (cubic yard)	15,500,000
Net cost today (NPV)	$16,554,272
Unit cost (per cubic yard)	$1.07
Annual cost	$1,756,065

Source: Glover and Herbich, 1989.

used dredged material disposal area to $15,000 per acre ($37,065 per hectare) for land inside an industrial park. No actual assessment of land values was performed. The cost associated with acquiring each site was assumed to be 5 percent of the land value, including negotiating and legal fees. If litigation was required, this cost could be much higher. Clearing and rubbing of a site was based on a cost per acre, ranging from $750 to $1500 per acre ($1853 to $3706 per hectare), depending on the vegetation at the site and existing conditions.

Diking costs were estimated by assuming that the 25-ft (7.6-m) dike would be constructed at the beginning of the project and that the disposal area would be square, with side slopes of 1 vertical on 3 horizontal. An average of $5/yd^3 ($6.50/m^3) was assumed for all earth work. The weir structure was estimated to cost $50,000 for diked disposal areas less than 150 acres (60.7 hectares) and $100,000 for areas greater than 150 acres. The maintenance cost at each site was assumed to be $21,000/yr, which has an approximate net present value of $200,000. This cost was included so that the disposal areas can be minimally maintained, and to improve site aesthetics. Dewatering costs are estimated at $1500 per acre ($3706 per hectare) after each planned dredging cycle.

The initial costs and present value of future costs were added to determine the net present cost (value) of the disposal site. The net present cost was then divided by the total 30-yr dredging estimate to determine the unit cost for each site in dollars per cubic yard. The annual cost was determined by amortizing the net present cost over the 30-yr project life at 10 percent interest.

Dredging and Transportation Cost Comparison

A dredging strategy study[17] was conducted that required an economic analysis and a comparison of dredging and transportation alternatives. Figure 8.48 shows the results of this general economic analysis. Mobilization costs were not in-

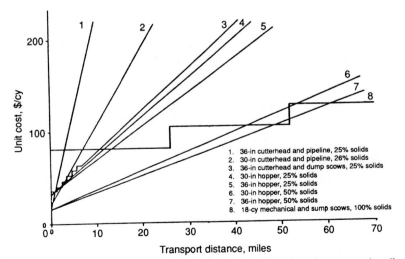

FIGURE 8.48 Unit dredging cost for alternatives as a function of transportation distance. (*Glover and Herbich, 1989*)

cluded; for the hopper dredge, no hopper overflow was assumed, but filling the hopper to capacity was assumed. For an equal slurry concentration (25 percent by volume), it can be seen that the most economical dredge type changes as the transport distance increases. A large (36-in, or 914-mm) cutter suction dredge is most economical for dredging and very short pipeline distances. A slightly smaller pipeline is more economical because lower slurry velocities can be tolerated without clogging the pipeline. Large hopper dredges were more economical because lower slurry velocities can be tolerated without clogging the pipeline. Large hopper dredges were also more economical for moderate haul distances between 3 to 10 to 15 mi (4.8 to 16.0 to 24.1 km). Increasing hopper solids concentration increases the economical haul distance to an estimated 40 mi (64 km) if solids concentration could be doubled to 50 percent. This shows the importance of the concentration of solids and the need for high solids dredging capability. The actual performance of the hopper dredge will depend on the suitability of the draghead to the bottom material and dredge operational effectiveness. At greater distances a mechanical dredge, tug, and dump scow system is most economical because of the low transportation cost and high solids concentration.

Hopper dredges used in combination with cutter suction dredges are not effective because of the estimated low solids concentration. A decrease to 50 percent from the 100 percent solids concentration of a mechanical system doubles the number of tugs and scows needed, doubling the unit ($/yd^3) ($/m^3) transportation costs. The higher production rate of hydraulic dredges also increases the required number of scows but does not affect the unit transportation cost. This cost analysis was based on equipment and operational costs presented in Table 8.9, amortized over a 50-yr life at 10 percent interest. The results presented are believed to be sufficiently accurate for the purpose of strategic comparisons, but do not necessarily reflect market prices.

This analysis compared several alternatives for navigational channel dredging.

TABLE 8.9 Alternative Initial Capital Investment and Annual Expense Estimates (in $1000)

Equipment	Capital cost ($)	Effective horsepower	Annual fuel cost ($)	Minimum personnel
36-in cutterhead	15,000	9,250	2,490	9
30-in cutterhead	10,000	6,650	1,790	9
27-in cutterhead	9,000	5,130	1,380	9
Tug/workboat	3,000	3,000	807	6
36-in hopper	40,000	13,000	3,500	16
30-in hopper	30,000	6,840	1,840	14
24-in hopper	20,000	3,040	818	11
36-in pipeline per mile	1,600	22,200	5,970	1
30-in pipeline per mile	642	6,000	1,610	1
17-in pipeline per mile	583	4,650	1,250	1
4000-cubic yards dump scow	4,000	—	—	—
18-cubic yards mechanical	15,000	3,000	818,807	9

Source: Glover and Herbich, 1989.

Hopper dredges with pump sizes of 36, 30, and 24 in (914, 762, 609 mm) were considered as well as cutter suction dredges of 36, 30, and 27 in (914, 762, 686 mm). The same size pipe as that of the dredge pump was assumed for pipelines associated with cutterhead dredges. An 18-yd^3 (13.8-m^3) mechanical dredge was used to evaluate mechanical dredging with a hopper barge and tugboat transportation system. Several assumptions were made for this economic analysis. Routine maintenance was assumed to halt operations 1 day in 2 weeks and 1 month annually for major repairs or modifications. This includes dredging 24 hr a day for 311 days, or 7464 hr/yr. No allowance was given for equipment malfunction or other problems that could stop the work. These assumptions may be compared with the experience of the Mobile District, U.S. Army Corps of Engineers, which indicated that typical dredge usage is limited to 4200 hr each year.

The maximum solids concentration by volume was estimated for 25 percent for the cutter suction dredge and 50 percent for the hopper dredge. The mechanical dredge is estimated to move 100 percent solids, or the in situ density of material. Most hopper dredges have two dragarms. One dragarm was assumed to be in service at a time, although both dragarms are usually used. Minimum flow velocities were calculated using Durand and Condolios' limiting deposit velocity for 0.0039-in (0.1-mm) particle diameter and with no net deposition of solids at the bottom of the pipe.[18]

Each alternative was evaluated over a 50-yr life, for annual fuel costs, 5-yr overhaul cycle cost, annual salaries for the crew, and annual maintenance costs. Extensive use of automation was assumed and a maximum crew size was estimated for each type of equipment. These expenses for each year were summed as total annual cost, and the present value was determined by discounting that cash flow to the present time. The present value of each annual expense was added to determine the net present value (NPV). The NPV was then amortized over the life of the project to determine the average annual cost.

The cutter suction dredge cost is based on an initial cost of $15,000,000 for a 36-in (914-mm) dredge and $6,000,000 for two work boats. The pump and pipeline costs are estimated as summarized in Table 8.10. Pipe replacement is anticipated every 10 yr because of the corrosive and abrasive properties of salt water and the dredged material. Suction and discharge pipe diameters were assumed to be of identical diameter, although in many operations the suction pipe diameter is often larger than the discharge pipeline diameter.

A hopper barge with a 4000-yd^3 (3058-m^3) capacity is estimated to have an initial cost of $4,000,000. Several of these barges and tugboats would be required to transport the material dredged to an approved ocean disposal site. The cost of a 36-in (914-mm) hopper dredge with an 8800-yd^3 (6728-m^3) hopper was estimated to have an initial cost of $40,000,000. Assumed characteristics of the hopper dredges are given in Table 8.11.

TABLE 8.10   Pipeline Cost Estimates Summary

Diameter (in)	Annual fuel cost ($)	Pipe cost per mile ($)	Pump horse-power	Pump cost ($)	Horsepower per mile
36	5,450,000	400,000	7400	400,000	22,000
30	1,610,000	282,000	5000	300,000	6,000
27	1,250,000	253,000	3875	275,000	4,650

Source: Glover and Herbich, 1989.

TABLE 8.11   Assumed Hopper Dredge Characteristics

Dredge size (in)	Hopper capacity (yd³)	Speed (knots)	Pumping capacity (yd³/hour)
36	8800	15	16,000
30	6000	12	10,500
24	4000	10	6,000

Source: Glover and Herbich, 1989.

A systems approach is recommended for navigational channel construction and maintenance planning. Each subsystem involved has an impact on the other subsystems. The optimization of a subsystem, such as an upland dredged material disposal area, may result in ineffective operation of the entire system by continuing dewatering after channel depths are reduced to less than the minimum authorized.[19] This type of analysis should provide for the continued improvement of the navigational infrastructure.

OPEN-WATER DISPOSAL SITES IN THE UNITED STATES

Specific information has been obtained and analyzed on the open-water disposal sites in the United States;[20] it includes:

1. Site location and coordinates
2. Designation status
3. Physical characteristics
4. Average distance from the channel
5. Monitoring
6. Wave and current characteristics

A 1989 survey by Herbich, De Hert, and McLellan indicated that there are about 193 open-water disposal sites, as listed in Table 8.12; Jacksonville District of the Corps of Engineers has the largest number of sites (31) and Alaska and Savannah Districts have the lowest (3). A statistical summary of site characteristics is given in Table 8.13.

Dredged Material Research Program

The U.S. Army Engineer Corps' Dredged Material Research Program (DMRP) was authorized by the River and Harbor Act of 1970 (Public Law 91-611). DMRP was administered by the U.S. Army Engineer Waterways Experiment Station between 1973 and 1978. This very extensive program dealing with all aspects of dredged material, its handling, placement, and its environmental impact resulted in numerous technical reports. The reports are indexed in Reference 21. The DMRP Synthesis Report Series is listed as follows:

TABLE 8.12 Number of Open-Water Disposal Sites per District

District	No. of sites
Alaska	3
Baltimore	7
Buffalo	10
Charleston	4
Detroit	7
Galveston	13
Honolulu	5
Jacksonville	31
Los Angeles	5
Mobile	6
New England	9
New Orleans	15
New York	16
Norfolk	4
Philadelphia	13
Portland	17
San Francisco	16
Savannah	3
Seattle	6
Wilmington	3
Total	193

TABLE 8.13 Statistical Summary of Site Characteristics of 193 Sites

Characteristics	No. of sites	Percent
Monitoring is done	26	13.5
Monitoring is not done	83	43.0
Monitoring status not given	84	43.5
Are dispersive	29	15.0
Are not dispersive	21	10.9
Dispersive status not given	143	74.1
The water depth is known	170	88.1
Any dimensions are known	155	80.3
Any mound data are known	9	4.7
Disposed volume is known	95	49.2
Any wave characteristics known	6	3.1
Any current characteristics known	43	22.3
Do experience sediment transport	10	5.2
Do not experience sediment transport	13	6.7
Sediment transport status not given	170	88.1
Disposal time periods are known	57	29.5
Are 103 designated	85	44.0
Are 404 designated	76	39.4
Are not designated or status unknown	32	16.6

Technical report no.	Title
DS-78-1	Aquatic Dredged Material Disposal Impacts
DS-78-2	Processes Affecting the Fate of Dredged Material
DS-78-3	Predicting and Monitoring Dredged Material Movement
DS-78-4	Water Quality Impacts of Aquatic Dredged Material Disposal (Laboratory Investigations)
DS-78-5	Effects of Dredging and Disposal on Aquatic Organisms
DS-78-6	Evaluation of Dredged Material Pollution Potential
DS-78-7	Confined Disposal Area Effluent and Leachate Control (Laboratory and Field Investigations)
DS-78-8	Disposal Alternatives for Contaminated Dredged Material as a Management Tool to Minimize Adverse Environmental Effects
DS-78-9	Assessment of Low-Ground-Pressure Equipment in Dredged Material Containment Area Operation and Maintenance
DS-78-10	Guidelines for Designing, Operating, and Managing Dredged Material Containment Areas
DS-78-11	Guidelines for Dewatering/Densifying Confined Dredged Material
DS-78-12	Guidelines for Dredged Material Disposal Area Reuse Management
DS-78-13	Prediction and Control of Dredged Material Dispersion Around Dredging and Open-Water Pipeline Disposal Operations
DS-78-14	Treatment of Contaminated Dredged Material
DS-78-15	Upland and Wetland Habitat Development with Dredged Material: Ecological Considerations
DS-78-16	Wetland Habitat Development with Dredged Material: Engineering and Plant Propagation
DS-78-17	Upland Habitat Development with Dredged Material: Engineering and Plant Propagation
DS-78-18	Development and Management of Avian Habitat on Dredged Material Islands
DS-78-19	An Introduction to Habitat Development on Dredged Material
DS-78-20	Productive Land Use of Dredged Material Containment Areas: Planning and Implementation Considerations
DS-78-21	Guidance for Land Improvement Using Dredged Material
DS-78-22	Executive Overview and Detailed Summary
DS-78-23	Publication Index and Retrieval System

Beneficial Uses of Dredged Material

There are many beneficial uses of dredged material.[23,25] Broad categories of use have been identified:

1. Wildlife habitat for water birds, shore birds, waterfowl, etc.
2. Aquaculture
3. Beach nourishment
4. Agriculture, forestry, and horticulture
5. Parks and recreation
5. Strip mine reclamation and solid waste management
7. Shoreline stabilization and erosion control

8. Land reclamation for industrial, airport, urban, and residential uses
9. Material transfer for fill, dikes, and roads
10. Brick manufacture
11. Multiple purpose use

REFERENCES

1. Male, R., and Basco, D. R., *A Dispersion Curve Study of Model Dredge Spoil Basins,* Sea Grant College Program, Texas A&M University, Report no. TAMU-SG-750-201, pp. 139, Sept. 1974.

2. Oosterbaan, N., "Working with Nature on the Slufter Disposal Site," *Terra et Aqua,* no. 33, pp. 7–12, Apr. 1987.

3. U.S. Army Corps of Engineers, *Guidelines for Designing, Operating and Managing Dredged Material Containment Areas,* EM 1110-2-5006, 84 pp., Sept., 1980.

4. Lacasse, S. E., Lambe, T. W., and Marr, W. A., *Sizing of Containment Areas for Dredged Material,* Technical Report D-77-21, U.S. Army Engineer Waterways Experiment Station, Oct. 1977.

5. Haliburton, T. A., *Guidelines for Dewatering/Densifying Confined Dredged Material,* Technical Report DS-78-11, U.S. Army Engineer Waterways Experiment Station, Sept. 1978.

6. Bartos, M. J., Jr., *Containment Area Management to Promote Natural Dewatering of Fine-Grained Dredged Material,* Technical Report D-77-19, U.S. Army Engineer Waterways Experiment Station, Oct. 1977.

7. Permanent International Association of Navigation Congresses, *Management of Dredged Material from Inland Waterways,* Supplement to Bulletin no. 70, 31 pp., 1990.

8. Murden, W. R., "An Overview of the Beneficial Uses of Dredged Material," *Terra et Aqua,* pp. 4–8, 1987.

9. Francingues, N. R., Jr., and Palermo, M. R., "Management Strategy for Disposal of Dredged Material, Dredging and Dredged Material Disposal, Volume 1," *Proc., Dredging '84,* American Society of Civil Engineers, Clearwater Beach, FL, Nov. 14–16, 1984.

10. Palermo, M. R., Francingues, N. R., Lee, C. R., and Peddicord, R. K., "Evaluation of Dredged Material Disposal Alternatives: Test Protocols and Contaminant Control Measures," *Proc. WODCON XI, the Eleventh World Dredging Congress,* Brighton, United Kingdom, 1986.

11. Peddicord, R. K., Lee, C. R., Palermo, M. R., and Francingues, N. R., Jr., *Decision-making Framework for Management of Dredged Material: Application to Commencement Bay, Washington,* Miscellaneous Paper D-86, U.S. Army Engineer Waterways Experiment Station, Vicksburg, MS, 1986.

12. Glover, L. B., and Herbich, J. B., "Methodology for Evaluating Offshore and Upland Disposal Areas," *Proc., XIIth World Dredging Congress (WODCON XII),* Orlando, FL, May 1989.

13. Asimov, M., *Introduction to Design,* Prentice-Hall, Englewood Cliffs, NJ, 1962.

14. Hall, A. D., *The Science of Engineering Design,* Van Nostrand, Princeton, 1962.

15. Nadler, G. (ed.), *1987 International Congress on Planning and Design Theory: Plenary and Interdisciplinary Lectures,* American Society of Mechanical Engineers, New York, 1987.

16. Baker, M., Jr., Inc., *An Identification and Assessment of Upland Dredged Material Disposal Sites in Coastal Mississippi,* Mississippi Department of Wildlife Conservation, Jackson, MS, 1987.

17. Baker, M., Jr., *North Central Gulf Dredged Material Disposal Strategy,* Mississippi Department of Wildlife Conservation, Jackson, MS, 1986.

18. Herbich, J. B., *Coastal and Deep Ocean Dredging,* Gulf Publishing Company, Houston, 1975.

19. Francingues, N. R., and Palermo, M. R., "Management Strategy for Disposal of Dredged Material," unpublished Dredging Short Course Notes, Texas A&M University, 1991.

20. Herbich, J. B., De Hert, D. O., and McLellan, T. N., "Open Water Disposal Sites in the United States," *Proc. 23d Annual Dredging Seminar,* Texas A&M University, Virginia Beach, VA, Oct. 1990.

21. U.S. Army Corps of Engineering Waterways Experiment Station, "Publication Index and Retrieval System," T:R. DS-78-23, 187 pp., April 1980.

22. Francingues, N. R., Palermo, M. R., Lee, C. R., and Peddicord, R. K., *Management Strategy for Disposal of Dredged Material: Contaminant Testing and Controls*, M.P. D-85-1, U.S. Army Engineer Waterways Experiment Station, August 1985.

23. U.S. Army Corps of Engineers, *Beneficial Uses of Dredged Material*, EM 1110-2-5026, June 1987.

24. U.S. Army Corps of engineers, *Confined Disposal of Dredged Material*, EM 1110-2-5027, September, 1987.

25. Hubbard, B. S., and Herbich, J. B., "Productive Land Use of Dredged Material Containment Areas," *Proc., 25th Annual Hydraulics Division Specialty Conference*, ASCE, August 1977.

CHAPTER 9

ENVIRONMENTAL EFFECTS
OF
DREDGING ACTIVITIES

INTRODUCTION

In the United States and other industrial nations serious environmental problems were recognized after World War II. According to Commoner,[1]* this increase in pollution in the United States and its effect on the environment are not the result of increase in population (about 43 percent) or an increase in industrial production (about 50 percent). The pollution increases, ranging from 200 to 2000 percent and higher, are mainly the result of changes in the technological processes in a number of industries. The increase in pollution has resulted in an environmental crisis affecting the whole world to varying degrees.

In the 1960s the consciousness of the environmental effects of industrialization had reached its peak and resulted in several statutes passed since 1968. There are three primary laws governing dredging and disposal in navigable waters. These are the National Environmental Policy Act of 1969, the Clean Water Act of 1977, and the Marine Protection, Research, and Sanctuaries Act of 1972. Other Acts include the Endangered Species Act, Fish and Wildlife Coordination Act, National Historic Preservation Act, Ocean Dumping Act,[2] and Oil Pollution Act of 1990.

National Environmental Policy Act

The National Environmental Policy Act of 1969 (NEPA) is the nation's basic charter for protection of the environment. The primary purpose of NEPA is to ensure that sufficient and complete environmental and alternative project information is available to both the federal agencies and the general public on issues that could have a significant impact upon the public. Each federal agency in carrying out its mandate must use NEPA as the decision-making process for all significant actions taken by that agency. This includes:

---

*References for the chapter sections written by John B. Herbich are at the end of the chapter.

1. Taking an interdisciplinary approach which will ensure the integrated use of the natural and social sciences and the environmental arts in planning and in decision-making which may have an impact on the environment

2. Identifying environmental effects and values in adequate detail so that the economic and technical analysis for each alternative can be appropriately compared

3. Developing appropriate alternatives to courses of action which involve conflicts concerning alternative uses of sources

4. Providing for cases where actions are planned by private applicants or other nonfederal entities so that the federal agencies can begin the NEPA process at the earliest possible time.

The Clean Water Act (CWA)

The purpose of the 1977 amended act, known as the Clean Water Act (CWA), is to restore and maintain the chemical, physical, and biological integrity of the waters of the United States; Section 404 of the CWA established a set of criteria for regulating the discharge of dredged or fill material into waters of the United States.

Section 404(b)(1) guidelines require a thorough review of all alternatives to the dredging and disposal operations. Dredging considerations include analyzing channel locations, the need for channel depths, and techniques for dredging. Included in the analysis of disposal site selection are the quality of materials to be disposed and the impacts on water quality, wetlands, and the benthic environment, especially related to shell fisheries.

The guidelines specify conditions which must be met for any dredging project. These include:

1. Compliance with state water quality standards
2. Compliance with EPA's toxic effluent standards
3. No adverse effect through bioaccumulation of toxic substances
4. No impact on threatened or endangered species
5. No impacts on marine sanctuaries
6. No impacts which would cause or contribute to significant degradation of the waters of the United States

Section B of the guidelines also outlines areas of special concern. These areas include sanctuaries and refuges, wetlands, mud flats, vegetated shallows, coral reefs, and riffles and pools. The guidelines define these areas and provide insight about why they should be protected.

The dredging of areas where contaminants are known or suspected to reside requires special care. Testing procedures under 404(b)(1), Subpart G, provide for categories of dredged material from clean, with no potential for harm, to very polluted requiring extensive bioassays to assess impacts.

Since wetlands had been commonly used for dredged material disposal, special consideration is given to this valuable resource. The functions of wetlands are multifold. First, they provide the primary food link for most estuarine organisms and others vital to commercial fisheries such as shrimp, salmon, oyster, menhaden, crabs, flounders, and clams. The organic production alone from a salt marsh is 2½ times greater than from a fertile hay field. As an outstanding example of wetlands values and potential loss, San Francisco Bay once had extensive wetlands and provided seven major commercial fisheries with multimillion dollar estimated values. Unfortunately, today, because of diking and filling operations only a fifth of the wetlands and none of the commercial fisheries remain.

Wetlands also act as a primary recharge for much of the nation's groundwater. They provide excellent erosion control, and they act as a pollution filtration system. Last, and of major importance, wetlands act as flood prevention buffers both by increasing sheet flow and by water storage.

The Marine Protection, Research, and Sanctuaries Act (1972) (MPRSA)

As a result of the concern for dumping of dredged spoils, sewage sludge, and industrial wastes into ocean waters, Congress in 1972 enacted the Marine Protection, Research, and Sanctuaries Act of 1972 (MPRSA). Similar to the Clean Water Act, the MPRSA provides criteria for at-sea disposal of dredged materials.

In the review of ocean dumping of dredged materials, final criteria were promulgated in 1977.[3] These criteria are similar to those under CWA, but contain additional provisions under the Convention on the Prevention of Marine Pollution by Dumping of Wastes and Other Matter. A summary of the criteria is as follows:

1. There must be justification of need and the lack of alternatives.
2. There are no unacceptable adverse effects on aesthetics, recreational, or economic values.
3. There are no unacceptable adverse effects on other uses of the ocean.
4. The material does not contain high-level radioactive wastes, radiological, chemical, or biological warfare agents, persistent inert synthetic or natural materials that float.
5. There is no greater than trace concentrations of such pollutants as organohalogens, mercury, cadmium, oil, and petroleum products and known carcinogens, mutagens, or teratogens.

Endangered Species Act (1973) (ESA)

Section 7 of the Endangered Species Act establishes a consultation process between federal agencies and secretaries of the Interior, Commerce, or Agriculture for carrying out programs for the conservation of endangered species. If endangered species or critical habitat are known to exist in the project area, a biological assessment must be conducted to evaluate the possible impact. If impact is unavoidable, reasonable alternatives for the project must be considered.

In one case an eagle's nest was found on the land disposal area during maintenance dredging operations. Even though no eagles were sighted, the dredging operations were immediately stopped and the project was delayed some 6 months.

Fish and Wildlife Coordination Act (1958) (FWCA)

The Fish and Wildlife Coordination Act provides that for any proposal for federal work affecting any stream, or other body of water, the proposing agency (such as the Corps of Engineers, or CE) must first consult with the Fish and Wildlife Agencies (federal and state) with a view to preventing damages to wildlife resources and to provide for the development and improvement of wildlife resources. The report submitted to Congress must give full consideration to recommendations provided by Fish and Wildlife Agencies. The Environmental Impact Statement must include all comments from the agencies.

National Historic Preservation Act (1966) (NHPA)

The National Historic Preservation Act directs the CE to take into account effects of the proposed project on any site, building, structure, or object that is included or is eligible for inclusion in the National Register of Historic Places. Comments from the Advisory Council on Historic Preservation (both federal and state) must be sought prior to granting a permit for construction.

Generally, magnetometer surveys must be conducted prior to the preparation of an Environmental Impact Statement to locate any possible objects of historic value underwater. In at least one case, the possible presence of a submerged Spanish galleon delayed the proposed project until it was ascertained that no Spanish galleon was located in the project area.

Assessment of Impacts

The Ocean Dumping Act of 1973 and 1977 (the United States is one of the participants in the London Dumping Convention) places emphasis on the assessment of impacts using bioassay techniques similar to those required by the CWA and upon the location of the disposal site.

The EPA and CE jointly developed a manual for implementation of MPRSA in 1977.[4] Further guidelines regarding disposal sites were published in the Federal Register in 1980.[5]

Dredged Material Research Program (DMRP)

Since many rules and criteria developed in the 1960s and early 1970s were based on limited data, there was an obvious need to conduct research on the effect of dredging on the environment and on the disposal of dredged material.[6] Congress authorized a $32.8M study which was conducted by the CE between 1973 and 1978. The study is known as the Dredged Material Research Program (DMRP) and some 250 individual studies were conducted.[7] The specific goals of DMRP were to:

1. Define water quality and biological effects on open-water upland and wetland disposal
2. Improve the effectiveness and acceptance of confined land disposal where it is a desirable alternative
3. Test and evaluate concepts of wetland and upland habitat development using dredged material
4. Develop and test concepts of using dredged material as a productive natural resource

Some 198 technical reports, 21 synthesis reports, and a summary report were published as a result of this study. A publication index and retrieval system is also available.[8]

RESUSPENSION OF SEDIMENT DURING DREDGING OPERATIONS

The bottom sediments of many navigable waterways of the United States are contaminated with potentially toxic chemicals. These contaminants are generally associated with the fine-grained fraction that is most susceptible to dispersion. Various field and laboratory studies have shown that the toxic chemical contaminants could have adverse effects if they are released in the waters through dredging. The impact could be of long and short duration. There is, therefore, concern about dredging and disposal operations.

Many factors control the dispersion of dredged material. The relative importance of these factors varies from site to site. The type of dredge used, the method of dredging, the type of sediments being dredged, environmental conditions at the site, etc., decide the nature and degree of resuspension of sediment and its subsequent dispersion.

Among the conventional dredges used in the United States are cutterhead, trailing suction hopper, dustpan, plain suction, and clamshell dredges. The major portion of dredging is done by contractors' cutterhead dredges. The CE has a number of hopper dredges which are also employed for dredging. The other types of dredges are used on a much smaller scale compared with cutterhead and hopper dredges.

The sediments to be dredged in the navigable waterways of the nation include clay, silt, sand, gravel, and organic matter of varying proportions. The contamination of these sediments by toxic chemicals varies widely from place to place. Extremely fine material such as clay and silt has a tendency to quickly go into suspension during the dredging process. Since the fall velocity of such fine particles is very small, these particles remain in suspension for a longer time compared with coarse-grained particles that settle fairly quickly. The degree of turbidity or resuspended sediments, therefore, largely depends on the size of the sediment particles.

Sediment resuspension varies from dredge to dredge and also largely depends on the dredging technique adopted. In the case of cutterhead dredges, turbidity is caused by the rotating action of the cutter and swinging action of the ladder. According to Huston and Huston,[9] the operational method of the cutterhead dredge assumes great importance, and they have suggested operating techniques to reduce turbidity at the cutterhead. Very little research effort has been made so far to properly understand the flow mechanism at the cutterhead intake that controls the pickup of suspended material. Further research is recommended to examine different aspects of the problem, including measures to reduce the turbidity at the cutterhead, such as hooded intake suggested by Apgar and Basco.[10]

In the case of hopper dredges, sediment resuspension is caused by the dragheads dragging over the soil, the overflow of hoppers, and the dumping through the bottom of hoppers. The turbidity caused at the dragheads is fairly low compared with that at the overflow. Very little attention seems to have been paid to the reduction of turbidity at the dragheads. The turbidity plume caused by overflow is observed to extend up to a distance of over 330 ft (100 m) downstream. The turbidity, however, persists for a short period only. The Japanese have conducted some research to reduce turbidity at the overflow; the antiturbidity overflow system seems to be quite promising. Some field tests carried out in Japan on existing hopper dredges with antiturbidity systems are stated to be quite successful. Data on turbidity at the bottom are very limited. The addition of chemical flocculants to increase the settling rate of sediment particles in the hopper is not recommended in view of the elaborate arrangements necessary to provide mixing and the marginal advantages it provides.

In the case of clamshell dredges, the watertight buckets appear to be quite useful since they reduce the turbidity by about 30 to 70 percent.

The plain suction dredge operates in a free-flowing sand and therefore causes less turbidity. The water jets at the bottom and the overflow at the surface can cause considerable turbidity. In general, sediment resuspension by a plain suction dredge should be much less than that by a cutterhead dredge in view of the absence of a rotating cutter in the plain suction dredge.

Among the conventional dredges, the dustpan dredge generates a considerable amount of turbidity because of high-velocity water jets, particularly in soft soil. To enhance the flow of granular material, water jets are provided along the top of the mouthpiece, and digging teeth are provided at the bottom. These features are undesirable for removal of contaminated silt (Hudson and Vann[11]) and would need to be removed. For removing the contaminated material from James River, Virginia, some modifications to the dustpan dredge were suggested. These modifications consisted of fitting the dustpan with a newly fabricated mouthpiece to present a hydraulically "clean" opening to the material without trash bars and grates, and, to overcome the entry losses of the flat rectangular mouthpiece, a rollover plate shaped like a bulldozer blade was suggested. Wing plates were also suggested on the sides and a splitter plate at the center to curb the tendency of the material to spill over the sides.

The Japanese have developed many unconventional dredging systems to remove highly contaminated fine-grained material from the bottom. The Oozer dredge is one such system. The pneumatic dredge, which was first developed in Italy, uses hydrostatic pressure and compressed air to remove contaminated sediments. By applying a vacuum to a pneumatic dredge, the Japanese were able to use the dredge in shallow water, thereby eliminating the constraint of requiring a high hydrostatic head pressure. There are specific advantages of such a system, such as continuous and uniform flow, high solids content (60 to 80 percent), and no disturbance of bed, and hence it is quite suitable for removing polluted sediments. The Japanese have also advanced other aspects of dredging technology through the development of a "cleanup" hydraulic dredge, watertight buckets, etc. Turbidity levels in the case of the Pneuma pump system, Cleanup system, and Oozer dredge are quite low. The total capacity of such systems is, however, not quite adequate to handle large volumes of highly contaminated sediments.

Other types of unconventional dredges are the Mudcat, Delta dredge, bucket wheel, etc. Very little data on sediment resuspension by these dredges are available. The capacity of these dredges is also limited.

No adequate data are available to make a comparison of dredging efficiency and dredging output between conventional and special-purpose dredges. Conventional dredges were designed to obtain high output, but little attention was paid to the environmental impact. As a result, these dredges produce more turbidity compared with those special-purpose dredges that were designed specifically to reduce sediment resuspension. The output of the special-purpose dredges is, however, low compared with that of the conventional dredges. For example, an

Oozer dredge can produce from 325 to 800 yd^3/hr (248 to 611 m^3/hr) when dredging silt with a 200 percent water content. The production is lower, however, for sandy materials. In the United States, where the quantities of contaminated bed material to be handled are large, it would be more economical to use conventional dredges with suitable modifications.

MECHANICS OF SEDIMENT RESUSPENSION

Human activities, such as dredging and filling operations and release of agricultural, industrial, and municipal effluents, are contributing to an increase in turbidity and suspended sediments.[12] A number of quantitative definitions based on optical and gravimetric principles have been employed regarding the magnitude of turbidity. The transmission of light through water (a measure of turbidity) is associated with attenuation caused by two processes, absorption and scattering. The term *turbidity* has several definitions and units. The units include the Jackson Turbidity Unit (JTU), the Formazin Turbidity Unit, and the Nephelometric Turbidity Unit. Gravimetric techniques represent a more accurate measurement of suspended solids in the water column. A multifrequency acoustic profiler to measure turbidity is also being developed.[12]

Wechsler and Cogley[13] showed that an increase in salinity levels enhances the flocculation of clay particles.

Sediment Resuspension

The nature, degree, and extent of dredged material resuspension during dredging operations depend on many factors.[14] These include:

1. Characteristics of dredged material
   a. Size distribution
   b. Solids concentration
   c. Type of sediment
2. Nature of the operation
   a. Dredge type and size
   b. Relationship between the cutter and the magnitude of hydraulic suction
   c. Type of draghead, the magnitude of hydraulic suction, and speed of the vessel
3. Waves and currents at the location of dredging operation
4. Water salinity at the site

Different types of dredges generate different rates of sediment resuspension. Specialized equipment has been developed to reduce or minimize the resuspension of sediment during dredging operations. This is particularly important while dredging contaminated sediment.

Cutterhead Dredges

A properly designed cutter, selection of the proper cutter for a given sediment, and the correct relationship between rotational speed of the cutter and the magnitude of hydraulic suction will reduce the resuspension rate of sediment. However, these conditions are rarely achieved in the field. Yagi et al.,[15] Huston and Huston,[9] and Bartos[16] all reported results of field experiments measuring the levels of suspended solids in the vicinity of the cutter. The results are summarized by Herbich and Brahme.[17]

Hopper Dredges

Resuspension of the fine-grained sediments typical of maintenance dredging is caused by the dragheads pulled through the sediment, overflow from the hoppers, propeller rotation, and dispersion during open-water disposal. Field experiments in San Francisco Bay were reported by Bartos.[16]

Plain Suction Dredge

Generally, the sediment resuspension rates with a plain suction dredge are lower than those caused by a cutter, except when water jets are used.

Dustpan Dredge

The sediment resuspension with a dustpan dredge depends on the type of sediment being dredged. For sands the resuspension rate is very small; however, for fine sands, silts, and clays the resuspension rates are significant because of water jets at the dustpan head.

Mud Cat Dredge

The sediment resuspension by a Mud Cat dredge is generally moderate and confined to within a small area around the dredge. The rates of resuspension may be reduced by properly matching the magnitude of the hydraulic suction with the rotational speed of the auger and the depth of cut.

Mechanical Dredges

With a clamshell dredge and an open bucket, the rates of resuspension rates are high since:

1. It occurs when the bucket hits the bottom and is pulled up through the water column.
2. The sediment at the surface in an open bucket is eroded as the bucket is pulled up through the water column and as it emerges through the water surface.
3. It leaks water with sediment through the openings between the jaws.

A watertight bucket with a clamshell dredge significantly reduces resuspension of the sediment, as shown in Figs. 9.1 and 9.2.[18]

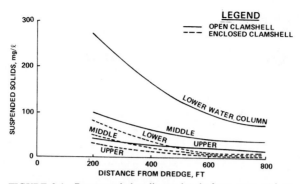

FIGURE 9.1   Resuspended sediment levels from open and enclosed clamshell dredged operations in the St. John's River. (*Hayes, 1986*)

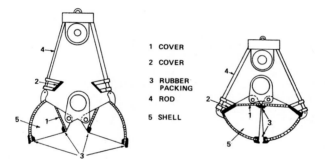

FIGURE 9.2   Open and closed positions of the watertight clamshell bucket.

Special-Purpose Dredges

Special-purpose dredges have been either adapted from existing design or specially developed to handle contaminated material. The main reason for such dredges is to minimize resuspension of the sediment. The production of such dredges is generally low. They are described below:

> *"Clean-up" dredge:* A Clean-up dredge head consists of a shielded auger that collects sediment as the dredge swings back and forth and guides the sediment toward the suction pipe of a submerged centrifugal pump.[19] Clean-up dredges have been successfully used in Japan for removal of contaminated material.
> *"Refresher" dredge:* A refresher dredge was developed for removal of contaminated materials.[20] A flexible enclosure completely covers the cutter, preventing escape of sediments to the outside of the immediate dredging area (Figs. 9.3 and 9.4).
> *"Matchbox" suction head:* A matchbox suction head was developed to replace the conventional cutterhead[21] (Fig. 9.5). The main design points are as follows:

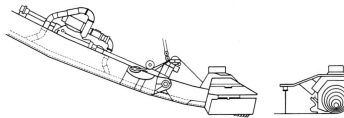

SIDEVIEW OF LADDER   FRONT VIEW

FIGURE 9.3   Refresher dredge. (*Shinsha, 1988*)

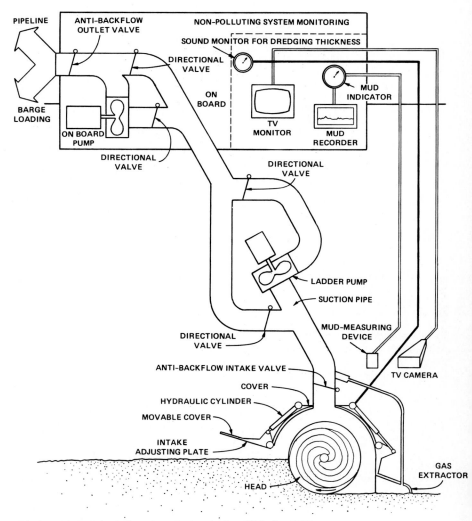

FIGURE 9.4   Description of a refresher dredge. (*Shinsha, 1988*)

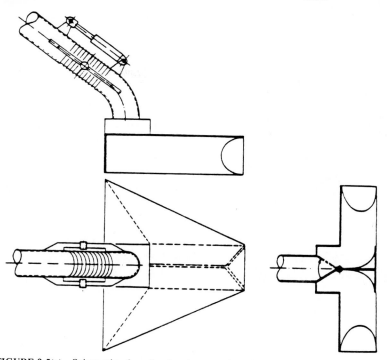

FIGURE 9.5(a)   Schematic of suction head mounted on dredge *Otter*.

FIGURE 9.5(b)   Matchbox suction head employed in a demonstration project at Calumet Harbor, IL. (*Courtesy, U.S. Army Engineer District, Environmental Engineering Section, Chicago*)

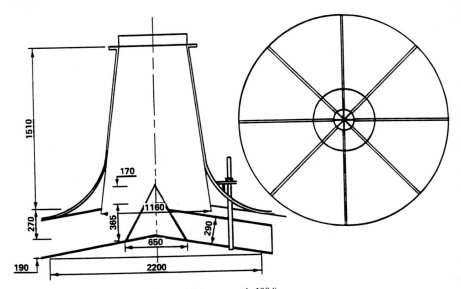

FIGURE 9.6 Schematic of diffuser. (*d'Angremond, 1984*)

1. A large plate covers the top of the dredge head to avoid inflow of water and escape of gas bubbles.

2. There is an adjustable angle between the dredge head and the ladder to create an optimum position of the dredge head independent of the dredging depth.

3. There are openings on both sides of the dredge head to improve dredging efficiency. During swinging action the leeward side is closed to prevent water inflow.

4. Dimensions of the head must be carefully designed for the average flow rate and swing rate.

To minimize dispersion of suspended sediment at the discharge point, a diffuser may be attached at the end of the discharge pipe.[21] Discharge velocities are reduced to 0.2 to 0.3 m/s (0.7–1.0 ft/s) (Fig. 9.6). Experiments conducted in the Calumet, Illinois, harbor[22] indicated that the clamshell dredge generated the largest suspended sediment plume, whereas a properly operated cutterhead dredge or a matchbox-head-equipped dredge was able to limit the sediment resuspension to the lower portion of the water column, as shown in Table 9.1.

"Wide sweeper" dredge: A wide sweeper hydraulic suction dredge does not have a cutterhead and was also designed for removal of contaminated sediments with minimum resuspension of sediments.[20]

"Pneuma" pump: The Pneuma pump is a compressed-air-driven displacement pump. According to the literature published by the manufacturer the pump can handle high solids contents with little generation of turbidity. The pump itself was evaluated by Richardson.[23]

Oozer dredge: The pump on the oozer dredge is similar to the Pneuma pump. It is mounted at the end of a ladder and is equipped with special suction heads and cutter units, depending on the type of sediments being dredged.

TABLE 9.1 Plume Area for 10-mg/liter Contour for the Cutterhead, Clamshell, and Matchbox Dredges[22]

Depth percent	Cutterhead acres	Clamshell acres	Matchbox acres
5	0	1.7	0
50	0	1.8	0
80	0	—	0.40
95	1.2	3.5	2.95

Evaluation of Model and Field Experiments

The possible correlation of the operating parameters, as well as dredge and sediment characteristics with varying levels of suspended sediments, is best investigated by means of dimensional analysis. The following list of variables includes all of the pertinent characteristics thought to be important in influencing the generation of suspended sediment.[24,25]

C Concentration of suspended sediment at or near the cutterhead
ρ Water density
ρ_s Sediment density
μ Water dynamic viscosity
d_s Median sediment diameter
d Inside diameter of the suction pipe at the cutterhead
D Diameter of the cutterhead
x Angle between ladder axis and vertical
t_c Thickness of cut
V_T Tangential speed of the cutterhead at its perimeter
V Flow velocity at the mouth of the suction intake pipe
S Swing velocity
w Settling velocity of the sediment particles
g Acceleration caused by gravity

Systematic application of the dimensional analysis procedure results in a functional relationship which may be written in dimensionless form as:

$$C/\rho = f\left(\frac{\rho_s}{\rho} \frac{d_s}{D} \frac{d}{D} x \frac{t_c}{D} \frac{V_T}{V} \frac{S}{V} \frac{w}{V} \frac{\mu}{VD\rho} \frac{gD}{V^2}\right)$$

The parameter t_c/D is equivalent to the percentage of a full cut, denoted %FC, where D is the limiting dredging thickness.

The model study[25] provided the following parameters for evaluation: V_T, S, percentage of full cut (%FC), x, W, D, d, d_s, and ρ_s. The plot of $V_T(x/90)d_s/S(\%FC)^{1/3}D$ as a function of the log C/ρ show a fairly good correlation (Fig. 9.7). The best correlation (coefficient = 0.956) was obtained for a plot of $V_T D/V d_s(\rho_s/\rho)$ as a function of C/ρ, as shown in Fig. 9.8.

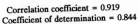

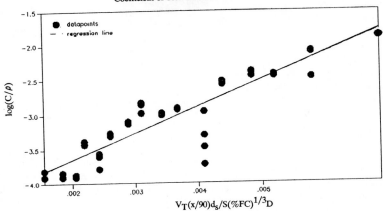

FIGURE 9.7 C/ρ versus $V_T(x/90)d_s/S(\%FC)^{1/3}D$. (*Andrassy and Herbich, 1988*)

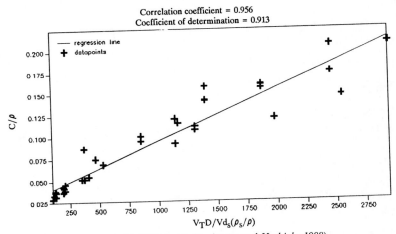

FIGURE 9.8 C/ρ versus $V_TD/Vd_s(\rho_s/\rho)$. (*Andrassy and Herbich, 1988*)

Good correlation was obtained in a model study by Brahme[26] between a dimensionless parameter V_TD_d/Vd_s and C/ρ.

Resuspended Sediment Concentrations

Special-purpose dredges may be used to reduce resuspension of contaminated sediments. Reported concentration levels of suspended sediments are shown in Table 9.2.

Conventional dredging equipment may be operated in a "modified" procedure to handle contaminated sediments. Since the volumes of contaminated sediments are relatively low, the decision whether to use conventional or special-purpose dredges should be made on economical grounds.

TABLE 9.2 Resuspended Sediments by Special Purpose Dredges[17]

Type of dredge	Reported suspended sediment concentrations*
Pneuma pump	48 mg/liter 3 ft (1 m) above bottom
	4 mg/liter 23 ft (7 m) above bottom [16 ft (4.9 m) in front of pump]
Clean-up system	1.1 to 7.0 mg/liter at 10 ft (3 m) above suction
	1.7 to 3.5 mg/liter at surface
Oozer pump	Background level (6 mg/liter) 10 ft (3.0 m) from head
Refresher system	4 to 23 mg/liter 10 ft (3.0 m) from head

*Suspended solids concentrations were adjusted for background concentrations.

Several special-purpose dredges were developed, principally overseas, and have been successfully employed in the removal of contaminated sediments.

Capabilities of mechanical, mechanical-hydraulic, and hydraulic suction dredges are shown in Tables 9.3, 9.4, and 9.5. Capabilities of pneumatic dredges are shown in Table 9.6.

TABLE 9.3 Summary Table—Mechanical Dredges

Type	Production	Depth limitation	Resuspension of sediment	Comments
Open clamshell, water tight	Low	30–40 ft (9.1–12.2 m)	High	
Water tight clamshell bucket	Low	30–40 ft (9.1–12.2 m)	Low	Experiments conducted in St. John's River

TABLE 9.4 Summary Table—Mechanical-Hydraulic Dredges

Type	Production	Depth limitation	Resuspension of sediment	Comments
Mud Cat	Moderate	15 ft (4.6 m)	Low to moderate	Extensively used
Remotely controlled Mud Cat	Low	15 ft (4.6 m)	Low to moderate	New development
Clean-up system	Moderate	70 ft (21.3 m)	Low to moderate	Extensively used in Japan

TABLE 9.5 Summary Table—Hydraulic Suction Dredges

Type	Production	Depth limitation	Resuspension of sediment	Comments
Refresher	Moderate to high	60–115 ft (18.2–35.0 m)	Low	Extensively used in Japan
Waterless	Moderate		Low	Limited experience
Matchbox	Moderate to high	85 ft (25.9 m)	Low	Experiments conducted at Calumet Harbor
Wide sweeper	Moderate	100 ft (30.5 m)	Low	Used in Japan

TABLE 9.6   Summary Table—Pneumatic Pumps (Dredges)

Type	Production	Depth limitation	Resuspension of sediment	Comments
Pneumatic	Low to moderate	+100 ft (+30.5 m)	Low	Evaluated by USAE Waterways Experiment Station
Oozer	Moderate to high	59 ft (18.0 m)	Low	Used extensively in Japan

EFFECTS OF DREDGED MATERIAL RESUSPENSION ON THE AQUATIC ENVIRONMENT

Pequegnat et al.[27] discussed in detail the various impacts of dredged material disposal on the ocean. These impacts are of short- and long-term nature and can be grouped into the following three broad categories: (1) physical impacts, (2) chemical impacts, and (3) biological impacts. Increase in turbidity is one of the most important of the physical impacts. The cloudiness associated with turbidity causes considerable unfavorable public response to some dredging projects. The increase in turbidity attenuates light to some extent. Pequegnat et al.[27] found that turbidity is one of the important factors controlling horizontal and vertical distributions of bacteria and fungi in the ocean. A significant increase in turbidity is often accompanied by an increase in bacteria counts, while a decrease in turbidity generally causes decreasing numbers of bacteria. Other important physical impacts of turbidity are the aesthetically displeasing nature of turbidity, decreasing availability of food, migration of mobile organisms out of the environment, topographic modifications, and moderate modifications of bottom currents.

Pequegnat et al.[27] examined the chemical impacts of dredged material on the disposal environment and stated that they are difficult to predict and even more difficult to control. The dredging and disposal practices are likely to adversely affect the water quality parameters such as oxygen, nutrients, trace metals, and pesticides that are known to affect marine life. In order to evaluate these effects, the EPA[28,29] recommends the elutriate test, which is designed to simulate open-water disposal of dredged material.

Keeley and Engler[30] discussed the rationale behind the elutriate test development as follows:

> Regulatory agencies faced with the legislative requirements of establishing dredged material criteria must strive to establish meaningful criteria based on the best possible knowledge, and avoid the tendency to set forth criteria that preceded the current technical state of the art. Furthermore, regulatory criteria should be based on laboratory procedures that can be performed satisfactorily in routine testing laboratories as opposed to complicated procedures that can only be conducted in sophisticated research-level laboratories.

Figure 9.9 shows the standard elutriate test. This test involves mixing sediment to be dredged with water from the dredging or disposal site, separating the two, and analyzing the water, especially for nutrients and known contaminants.

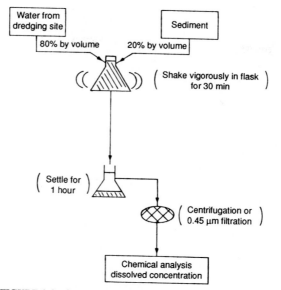

FIGURE 9.9 Standard elutriate test.

The elutriate test has added greatly to the understanding of contaminant releases into the water column. Among the important chemical impacts are the changes in oxygen concentration; the uptake and release of nutrients; and the uptake and release of toxins such as trace metals, halogenated hydrocarbons, petroleum hydrocarbons, and unknown toxins as detected by bioassay. It is generally recognized that some oxygen loss will occur when any sediment is exposed to oxygenated water, but the magnitude of the loss will depend on the particular sediment and the chemical and physical factors in the disposal environment.[27] One measure of the potential loss is the chemical oxygen demand (COD) of the material. The Federal Water Quality Administration (FWQA) criteria for the suitability of the sediment for dredging and disposal gave a maximum value of 5 percent (i.e., 5 percent of the sediment, by dry weight, requires oxygen to stabilize). The uptake and release of nutrients can have potentially significant effects on both pelagic and benthic organisms. The extent of biological effects, however, is largely dependent on the rate of dilution of added nutrients and the rate of renewal of water. Studies conducted earlier by Ketchum[31] have revealed that human activities are drastically increasing the input of many metals to the ocean.[27] Some metals such as mercury, cadmium, arsenic, chromium, copper, and lead can act as powerful toxins, not only to marine organisms but also to people who consume seafood contaminated with them. The magnitude of impact from these sources would depend on the mobility of these materials and their availability to the biota. Lee[32] found that the effects of chemical contaminants on the water column are primarily short term and are manifested either as toxicity to water column organisms or stimulation of noxious aquatic plants. He further stated that the real concern over chemical contaminants is from the potential long-term chronic toxicity effect and the transfer of contaminants from sediments to fish and other organisms.

Stern and Stickle[33] concluded that turbidity and suspended material can play both a beneficial and a detrimental role in aquatic environments. Suspended material absorbs and removes contaminants from the water column and stimulates photosynthesis through the introduction of inorganic nutrients. There is also a possibility that the nutrients might stimulate excessive biological growth and that turbidity might reduce photosynthetic activities because of its interference with light penetration. Pequegnat et al.[27] made a detailed study of the biological impacts of contaminated dredged material. Their analysis indicated in some detail the types of biological effects likely to occur within the areas most heavily affected by disposal materials. Among the important biological impacts listed are destruction of spawning areas, smothering and suffocation of organisms, and absorption of toxic materials. Both Pequegnat et al.[27] and Stern and Stickle[33] have concluded from their studies that, although some temporary and local damage may occur to the benthic species, the temporary increases in turbidity and suspended material will not cause significant or long-lasting effects to the benthic species of the marine ecosystems.

References 34 and 35 discuss the evaluation of dredged material pollution potential. This is a difficult and much-researched topic. It is generally agreed that the properties of a dredged sediment affect the fate of contaminants, and the short- and long-term physical and chemical environment of the dredged material at the disposal site influences the environmental consequences of contaminants. These factors should be considered in evaluating the environmental risk of a proposed disposal method for contaminated sediment. The process involved with release or immobilization of most sediment-associated contaminants is regulated to a large extent by the physical-chemical environment and the related bacteriological activity associated with the dredged material at the disposal site. Important physical-chemical parameters include pH, oxidation-reduction conditions, and salinity. Where the physical-chemical environment of a contaminated sediment is altered by disposal, chemical and biological processes important in determining environmental consequences of potentially toxic materials may be affected.

The major sediment properties that will influence the reaction of dredged material with contaminants are the amount and type of clay; organic matter content; amount and type of cations and anions associated with the sediment; the amount of potentially reactive iron and manganese; and the oxidation-reduction, pH, and salinity conditions of the sediment. Although each of these sediments properties is important, much concerning the release of contaminants from sediments can be inferred from the clay and organic matter content's initial and final pH and oxidation-reduction conditions. Much of the dredged material removed during harbor and channel maintenance dredging is high in organic matter and clay and is both biologically and chemically active. It is usually devoid of oxygen and may contain appreciable sulfides. These sediment conditions favor effective retention of many contaminants, provided the dredged materials are not subject to mixing, resuspension, and transport. Sandy sediments low in organic matter content are much less effective in retaining metal and organic contaminants. These materials tend not to accumulate contaminants unless a contamination source is nearby. Should contamination of these sediments occur, potentially toxic substances may be readily released upon mixing in a water column or by leaching and possibly by plant uptake under intertidal or upland disposal conditions.

Many contaminated sediments are reducing and near neutral in pH, initially. Disposal into quiescent waters will generally maintain these conditions and favor contaminant retention. Certain sediments (noncalcareous and containing appreciable reactive iron and particularly reduced sulfur compounds) may become moderately to strongly acid upon gradual conditions. This altered disposal environment greatly increases the potential for releasing potentially toxic metals. In addition to the effects of pH changes, the release of most potentially toxic metals is influenced to some extent by oxidation-reduction conditions, and certain of the metals can be strongly affected by oxidation-reduction conditions. Thus, contaminated, sandy, low organic-matter-content sediments pose the greatest potential for release of contaminants under all conditions of disposal. Sediments which tend to become strongly acid upon drainage and long-term oxidation also pose a high environmental risk under some disposal conditions. The implications of the influence of disposal conditions on contaminant mobility are discussed below.

METHODS OF CHARACTERIZING POLLUTION POTENTIAL

Bioassay

Bioassay tests are used to determine the effects of a contaminant on biological organisms of concern. They involve exposure of the test organisms to dredged material (or some fraction such as the elutriate) for a specific period of time, followed by determination of the response of the organisms. The most common response of interest is death of an organism. Often the tissues of organisms exposed to dredged material are analyzed chemically to determine whether they have incorporated (or bioaccumulated) any contaminants from the dredged material. Bioassays provide a direct indication of the overall biological effects of dredged material. They reflect the cumulative influence of all contaminants present, including any possible interactions of contaminants. Thus, they provide an integrated measurement of potential biological effects of a dredged material discharge. For precisely these reasons, however, a bioassay cannot be used to identify the causative agent(s) of impact in a dredged material. This is of interest but is seldom of importance since usually the dredged material cannot be treated to remove the adverse components even if they could be identified. Dredged material bioassay techniques for aquatic animals have been implemented in the ocean-dumping regulatory program for several years and are easily adapted for use in fresh water. Dredged material bioassays for wetland and terrestrial plants have also been developed and are coming into ever-wider use.

Water Column Chemistry

Chemical constituents contained in or associated with sediments are unequally distributed among different chemical forms depending on the physical-chemical conditions in the sediment and the overlying water. When contaminants introduced into the water column become fixed into the underlying sediments, they rarely if ever become part of the geological mineral structure of the sediment. Instead, these contaminants remain dissolved in the sediment interstitial water, or pore water, become absorbed or adsorbed to the sediment ion exchange portion as ionized constituents, form organic complexes, and/or become involved in complex sediment oxidation-reduction reactions and precipitations. The fraction of a chemical constituent that is potentially available for release to the water column when sediments are disturbed is approximated by the interstitial water concentrations and the loosely bound (easily exchangeable) fraction in the sediment. The elutriate test is a simplified simulation of the dredging and disposal process wherein predetermined amounts of dredging site water and sediment are mixed together to approximate a dredged material slurry. The elutriate is analyzed for major dissolved chemical constituents deemed critical for the proposed dredging and disposal site after taking into account known sources of discharges in the area and known characteristics of the dredging and disposal site. Results of the analysis of the elutriate approximate the dissolved constituent concentration for a proposed dredged material disposal operation at the moment of discharge. These concentrations can be compared to water quality standards and mixing zone considerations to evaluate the potential environmental impact of the proposed discharge activity in the discharge area.

Total or Bulk Sediment Chemistry

The results of these analyses provide some indication of the general chemical similarity between the sediments to be dredged and the sediments at the proposed disposal site. The total composition of sediments, when compared with natural background levels at the site, will also, to some extent, reflect the inputs to the waterway from which they were taken and may sometimes be used to identify and locate point source discharges. Since chemical constituents are partitioned among various sediment fractions, each with its own mobility and biological availability, a total sediment analysis is not a useful index of the degree to which dredged material disposal will affect water quality or aquatic organisms. Total sediment analysis results are further limited because they cannot be compared to any established water quality criteria in order to assess the potential environmental impact of discharge operations. This is because the water quality criteria are based on water-soluble chemical species, while chemical constituents associated with dredged material suspensions are generally in particulate and solid-phase forms or mineralogical forms that have markedly lower toxicities, mobilities, and chemical reactivities than the solution-phase constituents. Consequently, little information about the biological effects of solid-phase and mineral constituents that make up the largest fraction of dredged material can be gained from total or bulk sediment analysis.

ENVIRONMENTAL IMPACTS

Environmental impacts of dredging on the water column and the bottom sediments are discussed in Ref. 35 and are reproduced here.

Impacts in the Water Column

Contaminants. Although the majority of heavy metals, nutrients, and petroleum and chlorinated hydrocarbons are usually associated with the fine-grained and organic components of the sediment,[36] there is no biologically significant release of these chemical constituents from typical dredged material to the water column during or after dredging or disposal operations. Levels of manganese, iron, ammonium, nitrogen, orthophosphate, and reactive silica in the water column may be increased somewhat for a matter of minutes over background conditions during open-water disposal operations; however, there are no persistent well-defined plumes of dissolved metals or nutrients at levels significantly greater than background concentrations.

Turbidity. There are now ample research results indicating that the traditional fears of water quality degradation resulting from the resuspension of dredged material during dredging and disposal operations are for the most part unfounded. The possible impact of depressed levels of dissolved oxygen has also been of some concern because of the very high oxygen demand associated with fine-grained dredged material slurry. However, even at open-water pipeline disposal operations where the dissolved oxygen decrease should theoretically be greatest, near-surface dissolved oxygen levels of 8 to 9 ppm will be depressed during the operation by only 2 to 3 ppm at distances of 75 to 150 ft (23 to 46 m) from the discharge point. The degree of oxygen depletion generally increases with depth and increasing concentration of total suspended solids; near-bottom levels may be less than 2 ppm. However, dissolved oxygen levels usually increase with increasing distance from the discharge point because of dilution and settling of the suspended material.

It has been demonstrated that elevated suspended solids concentrations are generally confined to the immediate vicinity of the dredge or discharge point and dissipate rapidly at the completion of the operation. If turbidity is used as a basis for evaluating the environmental impact of a dredging or disposal operation, it is essential that the predicted turbidity levels are evaluated in light of background conditions. Average turbidity levels, as well as the occasional relatively high levels that are often associated with naturally occurring storms, high wave conditions, and floods, should be considered.

Other human activities may also be responsible for generating as much or more turbidity than dredging and disposal operations. For example, each year shrimp trawlers in Corpus Christi Bay, Texas, suspended 16 to 131 times the amount of sediment that is dredged annually from the main ship channel. In addition, suspended solids levels of 0.1 to 0.5 ppt generated behind the trawlers are comparable to those levels measured in the turbidity plumes around open-water pipeline disposal operations. Resuspension of bottom sediment in the wake of large ships, tugboats, and tows can also be considerable. In fact, where bottom clearance is 3 ft (0.9 m) or less, there may be scour to a depth of 3 ft (0.9 m), and the fine sediment is easily resuspended.

Impacts on the Benthos

Physical. Whereas the impact associated with water column turbidity around dredging and disposal operations is for the most part insignificant, the dispersal of fluid mud dredged material appears to have a relatively significant short-term impact on the benthic organisms within open-water disposal areas. Open-water pipeline disposal of fine-grained dredged material slurry may result in a substantial reduction in the average abundance of organisms and a decrease in the community diversity in the area covered by fluid mud. Despite this immediate impact, recovery of the community apparently begins soon after the disposal operation ceases.

Disposal operations will blanket established bottom communities at the site with dredged material which may or may not resemble bottom sediments at the disposal site. Recolonization of animals on the new substrate and the vertical migration of benthic organisms in newly deposited sediments can be important recovery mechanics. The first organisms to recolonize dredged material usually are not the same as those which had originally occupied the site; they consist of opportunistic species whose environmental requirements are flexible enough to allow them to occupy the disturbed areas. Trends toward reestablishment of the original community are often noted within several months of disturbance, and complete recovery is approached within a year or two. The general recolonization pattern is often dependent upon the nature of the adjacent undisturbed community, which provides a pool of replacement organisms capable of recolonizing the site by adult migration or larval recruitment.

Organisms have various capabilities for moving upward through newly deposited sediments, such as dredged material, to reoccupy positions relative to the sediment-water interface similar to those maintained prior to burial by the disposal activity. Vertical migration ability is greatest in dredged material similar to that in which the animals normally occur and is minimal in sediments of dissimilar particle-size distribution. Bottom-dwelling organisms having morphological and physiological adaptations for crawling through sediments are able to migrate vertically through several inches of overlying sediment. However, physiological status and environmental variables are of great importance to vertical migration ability. Organisms of similar life style and morphology react similarly when covered with an overburden. For example, most surface-dwelling forms are generally killed if trapped under dredged material overburdens, while subsurface dwellers migrate to varying degrees. Laboratory studies suggest vertical migration may very well occur at disposal sites, although field evidence is not available. Literature review indicates the vertical migration phenomenon is highly variable among species.

Dredging and disposal operations have immediate localized effects on the bottom life. The recovery of the affected sites occurs over periods of weeks, months, or years, depending on the type of environment and the biology of the animals and plants affected. The more naturally variable the physical environment, especially in relation to shifting substrate because of waves or currents, the less effect dredging and disposal will have. Animals and plants common to such areas of unstable sediments are adapted to physically stressful conditions and have life cycles which allow them to withstand the stresses imposed by dredging and disposal. Exotic sediments (those in or on which the species in question do not normally live) are likely to have more severe effects when organisms are buried than sediments similar to those of the disposal site. Generally, the physical impacts are minimized when sand is placed on a sandy bottom and are maximized when mud is deposited over a sand bottom. When disposed sediments are dissimilar to bottom sediments at the sites, recolonization of the dredged material will probably be slow and will be carried out by organisms whose life habits are adapted to the new sediment. The new community may be different from that originally occurring at the site.

Dredged material discharged at disposal sites which have a naturally unstable or shifting substrate because of wave or current action is rather quickly dispersed and does not cover the area to substantial depths. This natural dispersion, which usually occurs most rapidly and effectively during storms and hurricanes, can be assisted by conducting the disposal operation so as to maximize the spread of dredged material, producing the thinnest possible overburden. The thinner the layer of overburden, the easier it is for mobile organisms to survive burial by vertical migration through dredged material. The desirability of minimizing physical impacts by dispersion can be overridden by other considerations, however. For example, dredged material shown by biological or chemical testing to have a potential for adverse environmental impacts might best be placed in a retention area rather than dispersion. This would maximize habitat disruption in a restricted area but would confine potentially more important chemical impacts to the same small area.

Since larval recruitment and migration of adults are primary mechanisms of recolonization, recovery from physical impacts will generally be most rapid if disposal operations are completed shortly before the seasonal increase in biological activity and larval abundance in the area. The possibility of impacts can also be reduced by locating disposal sites in the least sensitive or critical habitats. This can sometimes be done on a seasonal basis. Known fish migratory routes and spawning beds should be avoided just before and during use but might be acceptable for disposal during other periods of the year. However, care must be taken to ensure that the area returns to an acceptable condition before the next intensive use by the fish. Clam or oyster beds, municipal or industrial water intakes, highly productive backwater areas, etc., should be avoided in selecting disposal sites.

All the above factors should be evaluated in selecting a disposal site, method, and season in order to minimize the habitat disruption of disposal operations. All require evaluations on a case-by-case basis by persons familiar with the ecological principles involved, as well as the characteristics of the proposed disposal operations and the local environment.

WATER QUALITY ASPECTS OF DREDGING AND DREDGED SEDIMENT DISPOSAL*

G. Fred Lee, Ph.D., P.E.
President, G. Fred Lee & Associates
El Macero, Calif.

R. Anne Jones, Ph.D.
Vice-President, G. Fred Lee & Associates
El Macero, Calif.

BACKGROUND

The dredging of U.S. waterways and harbors is recognized by Congress to be highly beneficial to the country as a whole. It is further recognized that dredging and dredged sediment disposal practices as part of waterway and harbor navigation depth maintenance will have some impact on beneficial uses and water qual-

---

*References for this chapter section are at the end of the section.

ity of the waters at the dredging and dredged sediment disposal sites. Congress, in the 1972 amendments to the Federal Water Pollution Control Act (PL 92-500), specified in Section 404 that the disposal of dredged sediments in U.S. waters may take place, provided that there is an avoidance of "unacceptable effects." It further stated that the disposal of dredged sediments should not result in violation of applicable water quality standards after considering dispersion and dilution, toxic effluent standards, and marine sanctuary requirements and should not jeopardize the existence of endangered species. While Congress indicated that some adverse impacts associated with dredging and dredged sediment disposal are to be expected, these impacts are to be minimized. Further, water quality criteria and state water quality standards are not to be violated as part of dredged sediment disposal operations.

It is important to understand how U.S. EPA (EPA) water quality criteria were developed and how they are conventionally used in water pollution control programs in order to understand problems with applying these criteria and state water quality standards based on these criteria to evaluate the potential water quality impacts of dredging and dredged sediment disposal.

DEVELOPMENT OF EPA CRITERIA AND THEIR USE IN WATER QUALITY CONTROL PROGRAMS

The 1972 amendments to the Federal Water Pollution Control Act (PL 92-500) required that the EPA develop water quality criteria for determining excessive concentrations of contaminants in ambient waters for various types of beneficial uses, such as fish and aquatic life, domestic water supplies, agricultural use, etc. The first of these criteria were released in July 1976 as the "Red Book" criteria (U.S. EPA, 1976). Congress also ordered the EPA to develop a set of priority (most important) pollutants and water quality criteria for the priority pollutants. The EPA did not comply with this congressional mandate by the specified date, with the result that environmental groups filed suit against the agency. This led to a court order decree in which approximately 130 chemicals or groups of chemicals were designated as priority pollutants by the court.

It is recognized by many professionals in the water quality management field that this list of chemicals was not necessarily an appropriately developed list. However, these are the chemicals that receive primary attention in almost all of current water pollution control programs. This list of chemicals is largely composed of compounds which are known or suspected animal and/or human carcinogens. Many of the chemicals on this list are chlorinated or brominated hydrocarbons. In November 1980, the EPA promulgated water quality criteria for the Priority Pollutants (U.S. EPA, 1980).

In July 1985, the EPA promulgated revised water quality criteria for a small group of chemicals, principally heavy metals and ammonia, which can cause significant adverse impacts on aquatic life. These criteria supersede the Red Book criteria. All of these criteria have been reissued by the EPA in what is called the "Gold Book," released in 1986 (U.S. EPA, 1987).

Congress in 1972 specified that the EPA water quality criteria are to serve as a guide to state pollution control agencies in developing the state water quality standards. While typically EPA water quality criteria are not enforceable, state standards based on these criteria are enforceable and are the primary basis upon which the water pollution control programs in the United States are now formulated. Since the EPA must approve state water quality standards, typically the state standards are at least as protective of beneficial uses of receiving waters as EPA water quality criteria upon which they are based. States have the option of making their standards more protective than the EPA criteria. They, however, have great difficulties obtaining approval for less protective criteria.

In order to keep the state standards somewhat up to date with the recent information developed on the impact of contaminants on aquatic life and other beneficial uses of water, the Federal Water Pollution Control Act specifies that the states must review their standards every 3 yr and bring them up to date in accord with EPA requirements. The EPA must approve these standards every 3 yr.

It is important to point out that federal and typical state water pollution control regulations specify that the objective of the regulations is the protection of designated beneficial uses of a waterbody. Frequently this is translated into attainment of numeric water quality criteria and standards. However, as discussed below, there are situations, especially associated with dredging and dredged sediment disposal, where the attainment of state water quality standards or EPA criteria at the edge of a mixing zone associated with a dredging or dredged sediment disposal operation is unnecessarily overprotective of aquatic life. While congressionally mandated requirements associated with dredging and dredged sediment disposal operations specify that water quality criteria and standards shall not be violated, violation of standards would not necessarily be adverse to aquatic life at the dredged sediment disposal site. It should be noted that the water quality impacts at the dredging site are considered to be *de minimis* by federal regulations.

The basic approach used today in water pollution control programs for regulation of wastewater discharges is to require that any discharger of contaminants, such as a city, industry, etc., must obtain a National Pollutant Discharge Elimination System (NPDES) permit for such a discharge. This permit typically specifies that the concentrations of various contaminants that are likely to be in the wastewater discharge shall be no greater than the state water quality standard at the edge of a mixing zone for the wastewater effluent and the receiving waters. Each state has the ability to specify to some extent the size of the mixing zones that are allowed. Within the mixing zone, the concentrations of contaminants can exceed the water quality standard.

Many states do not define a physically sized mixing zone but instead define a minimum flow in the receiving water (river), such as a Q_{7-10}, which is the low flow in the river that occurs for 7 consecutive days once in 10 years. For this flow, the states will specify a load of various contaminants in the wastewater discharge that will not exceed the water quality standard when the wastewater discharge is mixed into the receiving waters. Using this approach, the states do not have to define a mixing zone, nor do they have to measure the concentrations of various contaminants in the receiving waters. All that needs to be done is to measure the concentrations of contaminants in the effluent and determine the effluent flow. This provides sufficient information to determine whether the concentrations of contaminants in the receiving water would exceed the water quality standard at the point where they are mixed into these waters.

In contrast to the mixing allowed under PL 92-500 wherein the size and nature of the mixing zone are left up to the individual states, very specific constraints regarding mixing are required for ocean disposal under the Marine Protection, Research, and Sanctuaries Act (PL 92-532). In that instance the mixing zone is defined as that volume of water which overlies the disposal site. Federal water quality criteria are not to be exceeded outside the boundary of the mixing zone nor anywhere in the marine environment 4 hr after disposal has taken place. Determination of whether the criteria are met is accomplished through the use of a mixing model provided with the ocean disposal implementation manual. Where criteria do not exist for all contaminants of concern or synergism is suspected, the results of acute toxicity tests conducted with water column organisms are used to determine compliance with the criteria. Results of acute toxicity tests with water column organisms may also be used to determine compliance with state water quality standards under PL 92-500.

EPA water quality criteria are, in general, based on worst-case or near-worst-case assumptions in which aquatic organisms have been exposed to essentially 100 percent available (toxic) forms of contaminants for chronic or near chronic (extended) periods of time conditions [see Lee et al. (1982a, 1982b)]. In some cases acute toxicity tests are conducted, and chronic values are mathematically derived through an application factor (i.e., 10 or 1 percent of acute value). These criteria are generally highly protective of aquatic life since, in many instances, chemical contaminants exist in aquatic systems in a variety of chemical forms, only some of which are toxic to aquatic life. Further, there are many situations, especially near the point of discharge of contaminants, where the duration of exposure of aquatic organisms to the contaminants of concern is less than the duration of exposure applicable to the criterion values.

Figure 9.10 shows the typical relationship between the concentration of available forms of contaminants and the duration of exposure that organisms may experience without adverse effects. As shown, there is a stippled area where aquatic organisms can be adversely affected by available forms of contaminants for a certain period of exposure. The EPA criteria are typically developed by extrapolating the horizontal line under the stippled area to the ordinate value. These

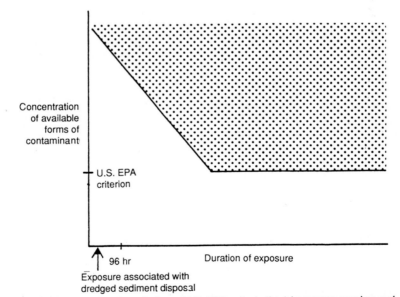

FIGURE 9.10 Aquatic toxicology. U.S. EPA criteria list 1-hr-average maxima and 4-day-average maxima. Not valid for assessing potential impacts of dredged sediment disposal.

criteria, therefore, are protective of aquatic life under all durations of exposure. However, it is well known that organisms can experience short durations of exposure in waters that contain concentrations of available forms of contaminants well above the EPA criterion value without adverse impact on the organisms. Typically, it has been found that there is a factor of at least 10, and more commonly 50 to 100, between the concentrations of contaminants that will kill half of the aquatic organisms in a 4-day exposure and the chronic safe concentration (EPA criterion value) for a chemical.

APPLICATION OF EPA WATER QUALITY CRITERIA TO DREDGED SEDIMENT DISPOSAL PROJECTS

Figure 9.10 diagrammatically shows a typical water column duration of exposure for aquatic organisms associated with dredged sediment disposal operations. The typical dredged sediment disposal project is such that organisms in the water column for which EPA criteria or state water quality standards based on these criteria are based can receive exposures of a few minutes to a few hours duration. There are very few situations where the organisms would encounter a 4-day exposure, much less chronic exposure. Therefore, the application of EPA criteria or state water quality standards based on these criteria is unnecessarily overprotective of aquatic life in the water column associated with dredging and dredged sediment disposal operations.

Another significant problem with applying the EPA-based water quality criteria and standards to dredged sediment disposal projects is that the contaminants in dredged sediments are largely associated with particulate matter where they occur as precipitates or are sorbed (attached) onto sediment particles. It has been known for over 20 years that contaminants associated with particulates are largely unavailable (nontoxic) to aquatic life. This situation makes EPA criteria and state water quality standards based on these criteria overprotective of aquatic life in a water column. It is therefore evident that applying EPA water quality criteria and state water quality standards based on these criteria to dredging and dredged sediment disposal projects is not a valid approach to determine potential adverse impacts of "excessive" concentrations of contaminants. Concentrations of many contaminants, such as heavy metals, can be present in the water column near dredging and dredged sediment disposal activities in concentrations many orders of magnitude above EPA criteria without adverse effects on aquatic life.

It is extremely important to understand that the EPA criteria and state standards are not applicable to the protection of aquatic life in sediments, including their interstitial waters, where the total concentrations of contaminants in the sediments are compared to the criterion values. While some regulatory agencies are proposing to use this approach, it is technically invalid and should not be adopted. This issue is discussed further below.

In addition to concern about the direct toxic effects of chemical contaminants in aquatic systems, there is also concern about the bioaccumulation of contaminants in aquatic organisms that would occur to a sufficient extent to cause these organisms to become unsuitable for use as food by humans or other fish-eating animals. In the 1960s and 1970s, there was widespread concern about DDT, PCBs, mercury, and some other persistent chemicals' bioaccumulation within aquatic organism tissue to the extent that the edible tissue was judged unsafe for use for human food based on U.S. Food and Drug Administration (FDA) action levels. The FDA has promulgated acceptable concentrations (action levels) of certain chemicals in food. It should be noted, however, that these action levels are also based to some extent on economic considerations. These action levels are widely used as critical tissue concentrations for excessive bioaccumulation of contaminants in aquatic organisms. The EPA has developed a number of their water quality criteria, such as for mercury, DDT, PCBs, dioxin, etc., based on worst-case assumptions about the potential for bioaccumulation of contaminants present in water within the fish or other aquatic organisms' edible tissue.

The worst-case nature of these criteria is based primarily on the assumption that the bioaccumulation that occurs in relatively clean aquatic systems with little or no particulate matter present will occur in other situations where particulates are present. It is well known, however, that in the presence of particulate matter, the bioaccumulation factors that the EPA uses in developing their criteria are overly protective. Far less bioaccumulation occurs from contaminants associated with particulates than for many dissolved contaminants. Since dredging and dredged sediment disposal projects involve large amounts of particulate matter, EPA criteria and state water quality standards based on these criteria are overly protective of beneficial uses of water where the concern is the bioaccumulation of chemicals within aquatic life that can be used as food and are potentially affected by the dredging or dredged sediment disposal operations.

One of the areas of frequent concern associated with bioaccumulation of chemicals in aquatic life is what constitutes an excessive amount of accumulation for chemicals for which the FDA has not established an action level. Also of concern is the allowable amount of bioaccumulation in nonedible organisms (i.e., worms) or within specific organs of an organism, such as the liver. Some regulatory agencies will apply FDA limits to these organisms or organs as being appropriate for determining excessive bioaccumulation. This approach is technically invalid. What constitutes an excessive concentration of a chemical in the human diet does not necessarily translate to an excessive body burden in worms or other forms of aquatic life, nor is this a valid basis to determine an excessive concentration of a chemical in an aquatic organism's liver or some other organ unless that organ is in fact used as food in accord with the assumptions that were used by the FDA in developing the action levels.

It is evident from the above discussion that there are considerable problems with using EPA criteria or state standards based on these criteria as a basis for determining "unacceptable impacts" of contaminants associated with a dredging or dredged sediment disposal operation. It can be generally if not universally assumed that associated with a dredging or dredged sediment disposal operation, if the EPA criteria and state standards are not violated, there is little or no likelihood that the contaminants present in the dredged sediments will have an adverse effect on aquatic life in the water column near the dredging site or at the disposal site. If, however, the criteria or standards are violated, it is highly likely that such violations do not constitute an adverse impact on the beneficial uses of the water. As noted above, the objective of the U.S. water pollution control program is protection of beneficial uses, not attainment of standards. The EPA's criteria, while having some applicability to regulating wastewater discharges, have little or no technical applicability to regulating dredging or dredged sediment disposal operations. Certainly, it would be highly inappropriate to cause a particular dredging project to alter the proposed approach that is to be used for dredged sediment disposal based on the finding that the EPA criteria and/or state water quality standards would be violated.

In the early 1970s, the predecessor to the EPA and some state pollution control agencies adopted regulatory approaches for dredged sediments that significantly increased the costs of dredging projects by 20 to as much as 50 percent, based on "excessive" concentrations of contaminants in the sediments. This approach was based on an inaccurate analysis of the potential impacts of the contaminants in the sediments by the regulatory agency personnel and resulted in a significant unnecessary expenditure of public funds in the name of pollution control associated with dredging projects. Further, in some instances, the alternative, more expensive approaches adopted, some of which are still being used today, were more harmful to aquatic life than the less expensive methods.

THE DEVELOPMENT AND USE OF HAZARD ASSESSMENT APPROACHES FOR DREDGED SEDIMENT PROJECTS

Lee et al. (1982a) discuss the importance of properly evaluating the water quality impacts of contaminants on aquatic life in formulating technically valid, cost-effective water pollution control programs. They recommended that EPA water quality criteria and state standards based on these criteria be used as an indicator of potential water quality problems. With few exceptions, because of the highly overprotective character of these criteria and standards, if the concentrations of contaminants in a water are less than the criteria or standards, there is little or no likelihood that the contaminants will have an adverse effect on aquatic life. If, however, the concentrations exceed the criteria or standards, there is the potential for adverse impact. However, as discussed above, it should not be assumed that violation of the criteria or standards represents an adverse impact on beneficial uses. Lee et al. (1982b) recommended that in most instances, the discharger of contaminants which cause apparent violations of criteria or standards should be given the opportunity to conduct site-specific studies to evaluate whether the violations of the criteria or standards represent adverse impacts on beneficial uses of the water. These site-specific studies represent an aquatic life hazard assessment evaluation in which the real hazard that a chemical contaminant or group of contaminants represents to aquatic life-related beneficial uses of a water are evaluated.

One of the first steps in developing a hazard assessment scheme for use as an environmental quality management tool is to define the potential adverse impacts that should be considered. That is, the beneficial uses of the water or area in question desired by the public must be defined. They can include aesthetic enjoyment, fishery, boating, swimming, navigation, benthic worm farm, drinking water source, habitat for migratory waterfowl—anything that the public wants for a water and can convince others through the regulatory process that that is an appropriate use of that water. The hazard assessment is then designed to evaluate the impact of the activity on the uses desired.

The hazard assessment approach involves a sequential, tiered evaluation of the expected concentrations of available forms of contaminants and aquatic organisms' duration of exposure to these available forms compared to the critical concentration-duration of exposure relationships which are known to have an adverse impact on a particular form of aquatic life. The first tier is typically a "back of the envelope" evaluation which is based largely on a preliminary analysis of the situation utilizing existing information. Higher-level tiers require laboratory and/or field studies to develop the necessary information to determine whether adverse impacts are likely, compared to the critical concentration-duration of exposure situation. As discussed below, Lee and Jones (1981a,b) have applied hazard assessment techniques that they developed to several dredging projects with considerable success. While the hazard assessment approach typically involves the expenditure of more funds in field studies than are normally expended associated with a particular dredging or dredged sediment disposal operation, when taken in the context of the cost of alternative dredged sediment disposal operations, often a few dollars spent in hazard assessment studies may save many tens of thousands to millions of dollars in dredging project costs.

FACTORS CONTROLLING RELEASE OF CONTAMINANTS FROM SEDIMENTS

Figure 9.11 presents a diagrammatic representation of a sediment-water interface for deposited sediments associated with dredging projects. The sediment-water column consists of four regions. The uppermost region is the water column in which there is typically a small amount of suspended particles. The next region is just above the sediment-water interface where there is typically a significant increase in the number of suspended sediment particles. At times this can be a slurry of suspended sediments. This is the area where contaminants released from the bedded sediments are mixed into the overlying waters. This is also the area where those organisms that are present in the sediment that depend on oxygen obtain their oxygen as well as exposure to contaminants present either in the water column or released from the sediments.

The sediment-water interface can be fairly diffuse. This depends upon the nature of the sediments and the turbulence of the water in this area. Below the sediment-water interface is the zone of active deposition in which there is active mixing of the sediments and their associated contaminants within the sediments.

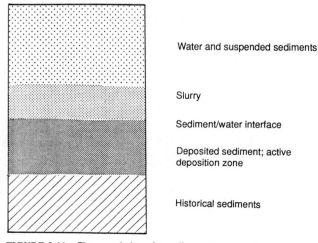

FIGURE 9.11   Characteristics of a sediment-water system.

This mixing can arise from a variety of causes such as wind, tidal, or flow-induced currents; ship traffic; and organism activity. The latter includes biogenic formation of gases such as methane, which would stir the sediments as the gas bubbles rise through them.

Below the active deposition zone are the historical sediments. Any contaminants found in these sediments would not be expected to be brought to the surface through normal mixing processes. New work dredging, however, frequently involves removal of historical sediments. While sometimes these sediments are contaminated by human activity, in many instances, these sediments are relatively clean and their impacts on beneficial uses of water are largely restricted to physical impacts. In some instances the characteristics of new work dredging project sediments are such that they have appreciable adsorption capacity that can remove contaminants from the water column by this process upon dredged sediment disposal.

As discussed by Lee (1970), the primary factor governing the release of contaminants from sediments to the overlying waters is the mixing of the sediments into these waters. This mixing causes a change in liquid-solid ratio which would tend to promote release of contaminants from a sediment surface to water. Further, the stirring enables the contaminants in the interstitial water to be brought to the surface at a significantly greater rate than typically occurs by diffusion controlled processes. Since in most instances the concentrations of contaminants in interstitial waters are greater than in the overlying waters, the primary factor controlling release of contaminants from sediments is the mixing or stirring process. Therefore, concern should be focused on dredging activities and especially hydraulic dredging where the sediments are slurried in an approximate one part sediment to four parts water (20 percent by volume). This slurrying would tend to result in rapid release of contaminants from sediments caused by the change in liquid-solid ratio and from the mixing of the interstitial waters with the waters used to slurry the sediments.

Lee et al. (1978) and Jones and Lee (1978) as part of the Corps of Engineers Dredged Materials Research Program (DMRP) conducted extensive laboratory and field studies at dredging and dredged sediment disposal sites located in many parts of the United States. They found that as long as the sediment water slurry was oxic (contained dissolved oxygen) that of the over 30 chemical parameters they measured, including heavy metals, a variety of organics, and other contaminants, only ammonia and manganese were released from the sediments. However, if the slurrying of the sediments with dredging site water took place in the absence of dissolved oxygen (anoxic), a large number of contaminants were released to the water. This pattern of release under anoxic conditions and no release under oxic conditions is strongly supportive of the role of ferric hydroxide in acting as a highly efficient scavenger for contaminants released from the sediments.

Iron in most sediments exists in a ferrous form. Upon contact with waters containing dissolved oxygen, it is rapidly oxidized by the dissolved oxygen to ferric iron, which precipitates as ferric hydroxide. Freshly precipitated ferric hydroxide has a large surface which can sorb significant amounts of a wide variety of contaminants. It is for this reason that few contaminants present in interstitial water or which are desorbed during slurrying remain in the water column. While they are released, they are rapidly taken back to the sediments by the ferric hydroxide scavenging.

Lee (1975) pointed out that it is important to distinguish between the scavenging ability of freshly precipitated ferric hydroxide and aged ferric hydroxide precipitate. While freshly precipitated ferric hydroxide has a high sorption capacity, aged ferric hydroxide has limited sorption capacity and will release contaminants sorbed at the time of formation of the ferric hydroxide, especially if it has become dry. This is an important phenomenon that occurs in some confined dredged sediment disposal projects that leads to the potential for water quality problems associated with this method of dredged sediment disposal.

Organism-Induced Release and Uptake

In addition to a solubilization of precipitated contaminants and desorption of contaminants associated with sediment particles typically associated with stirring of the sediments into the overlying waters, sediment contaminant release can occur in the intestinal tract of aquatic organisms that consume sediment. The pH, organic content, and other characteristics of the intestinal tract of some aquatic organisms are significantly different from many ambient waters with the result that contaminants that may not be released upon slurrying could be released within organisms. The possibility of this mode of release of contaminants should be evaluated. While in some instances the release of contaminants from aquatic organisms results in their being made available to the water column, most of the time this mode of release has been reflected in tissue residues which are best assessed through bioaccumulation studies.

The trophic-level buildup of contaminants in aquatic organisms where higher-level trophic state organisms have greater concentrations of contaminants in their tissue than lower-level organisms is of concern. Unlike terrestrial organisms, contaminants in aquatic organisms tend to reach an equilibrium with the external environment because the organisms are permeable. Differences in contaminant concentrations in organisms at different trophic levels are generally understandable in the context of lipid content, age, and other factors rather than trophic considerations. However, as noted by Kay (1984), there are a few contaminants which have the potential for trophic magnification in aquatic environments. These include chlorinated hydrocarbons, pesticides, methylmercury, PCBs, and PAHs, and possibly kepone and mirex. The bioaccumulation of contaminants within organisms occurs from both direct uptake of contaminants from the water and the consumption of particles that have contaminants associated with them as well as the consumption of other organisms. It is often very difficult if not impossible to distinguish between these various modes of uptake. It is for this reason that water concentrations of contaminants are often very poor predictors of the amount of contaminants that will accumulate within higher-trophic-level organisms associated with a dredging or dredged sediment disposal project.

Another potential problem that has occurred with the interpretation of bioaccumulation data is the assumption that the concentrations in lower forms of aquatic life such as polychaetes (worms) are representative of what would be present in higher-trophic-level organisms such as fish. This is certainly not the case. Polychaetes could have very high concentrations of a contaminant without adverse effect to them or higher trophic levels which may feed on them. The significance of body burdens of contaminants within lower-trophic-level organisms is largely unknown at this time. The only valid basis by which the significance of the concentration of a contaminant in an organism's tissue can be evaluated is by the FDA action level. This level is based on edible tissue that is used for human food. It is technically invalid to use the FDA action levels for critical concentrations of contaminants in polychaetes or other organisms not used directly as food for people.

SIGNIFICANCE OF THE CONCENTRATIONS OF CONTAMINANTS IN SEDIMENTS

By far, the greatest cause of regulatory agencies' development of inappropriate dredged sediment management approaches is the failure of those who develop those approaches to understand that there is no relationship between the concentrations of contaminants in sediments and the potential impact that contaminants could have on the beneficial uses of the water associated with the sediments. Typically, the concentration of a contaminant in water shows an increasing impact with increasing concentration. A concentration of contaminants in sediments, however, does not always show this pattern. The difference is that concentrations in water are on a mass per unit volume of water basis. The water matrix (typically 1 liter) is of constant composition (H_2O). However, the concentrations of contaminants in sediments is on a mass per unit mass of sediment basis. The unit mass of sediment is of variable composition. In some cases, it can be an oily waste or residue. In others, it is a quartz sand or a calcium carbonate precipitate. In other cases, it can be a clay or combination of types of clay. Each of these and other sediment matrices affects the impact that contaminants present in them have on aquatic organisms and many other beneficial uses of the water. Since the impact that the contaminant associated with the sediment has on aquatic organisms is related to the release of contaminants from the sediments, it is not surprising that there is no general relationship between the concentration of contaminants in sediments of various types and the impacts these contaminants in sediments have on water quality.

The authors have found that some sediments of a region with lower concentrations of contaminants had greater impacts on water quality than those with the higher concentrations. This arose out of the fact that the basic sediment matrix for the two types of sediment were different. The sediments with the higher concentrations of contaminants tended to hold the contaminants much more strongly and therefore did not release them upon slurrying with water.

However, there may be situations where a region will have all the same types of basic sediment matrix. Under these conditions, since the binding power of contaminants to the sediments is a constant, a relationship may be developed between the concentrations present in this sediment and their impact on water quality. It would indeed be very rare, however, that such a situation would exist and it should not be assumed, as it is often done, that there is a relationship between the concentration of contaminants in sediments and their impacts on water quality.

Since the late 1960s, various regulatory agencies have been trying to develop dredged sediment regulations using the bulk composition of sediments as a regulatory parameter. Many millions of dollars of public funds have been unnecessarily expended because of this approach as a result of regulatory agencies requiring that more expensive methods of dredged sediment disposal be used because the concentrations of contaminants in the sediments exceeded some arbitrary "critical" concentration. While it has been known for over 20 years that this approach is technically invalid and wasteful of public funds as well as possibly doing more harm to the environment than would occur otherwise, some regulatory agencies are still regulating dredging practices using bulk sediment chemical composition in developing dredged sediment disposal criteria. At this time the EPA and some state regulatory agencies are actively pursuing numerical chemical sediment-based quality criteria for the purpose of regulating the discharge of contaminants to surface waters and for the purpose of assessing the water quality significance of contaminants associated with sediments. A number of approaches have been adopted and/or proposed which involve the use of numeric sediment quality criteria as regulatory requirements that must be met. As discussed below, these approaches could ultimately cause significant unnecessary expenditure of public funds.

POTENTIAL WATER QUALITY PROBLEMS ASSOCIATED WITH DREDGING PROJECTS

Congress mandates that the U.S. Army Corps of Engineers (CE) maintain the navigation depth for commercial and public transport of approximately 25,000 mi (40,000 km) of U.S. waterways. To accomplish this, the CE, either directly or indirectly through contractors, dredges or issues permits to dredge about 500 million cubic meters per year of sediment from these waterways. Generally, the least expensive and most expedient method to dispose of dredged sediments is to discharge them in open water off-channel or dump them in open water at a location removed from the dredging site. In the early 1970s, considerable concern was expressed about the potential water quality impacts of open-water disposal of chemically contaminated dredged sediments.

There are several distinct aspects of water quality problems associated with dredging and dredged sediment disposal projects. It is important to distinguish among them to properly evaluate whether contaminants in a sediment that is to be dredged in a certain manner could have a significant impact on the beneficial uses of water at the dredging or the dredged sediment disposal sites.

In evaluating the potential adverse impacts of disposal of dredged sediments, it is important to distinguish between physical and chemical impacts. Within the potential chemical impacts, it is important to distinguish between water column and redeposited sediment impacts. For the chemical impacts of redeposited sediments, consideration should be given to not only toxicity to aquatic organisms but also to the potential for bioaccumulation of contaminants present in the dredged sediments that would be adverse to using edible aquatic organisms as a source of human food. A summary of the information available in each of these topic areas is presented in this section.

Physical Impacts of Dredging and Dredged Sediments

At both the dredging and disposal sites, consideration has to be given to the physical impacts of dredging and dredged sediment disposal. These physical aspects range from burial of organisms because of the settling of the suspended or dumped sediment to physical abrasion or clogging of gills or other organs by suspended sediment. Dredging and especially dredged sediment disposal can have adverse impacts on certain organism populations such as a coral reef. Great care must be exercised in dredging in the vicinity of special organism habitats, such as coral reefs, to be certain that the suspension and deposition of sediment are not adverse to the organisms. Problems of this type are primarily caused by the presence of sediment and can be largely independent of its chemical characteristics.

While redeposited dredged sediments especially at the disposal site can result in the burial of aquatic organisms, leading to their death, the DMRP studies showed that many organisms that live in sediments are able to migrate through appreciable depths of sediments dumped on them. The key factor seems to be the hydrogen sulfide concentration of the sediments since this chemical is fairly toxic to aquatic life. It also removes all oxygen from the sediment column and therefore could deprive oxygen from organisms that are attempting to migrate to the surface of the sediment for a sufficient period to cause their death. Phenomena of this type are to be expected at designated dredged disposal sites where large amounts of anoxic sediments are dumped onto mobile aquatic organisms. There is little that can be done about this problem except to be certain that the designated disposal areas do not represent ecologically important areas for the region.

The physical impacts of grain size and texture of the sediments must be considered in evaluating the impact of a dredged sediment disposal operation. In the studies conducted by the authors and their associates (Lee et al., 1978) near the Galveston, Texas, Bay Entrance Channel in the Gulf of Mexico, it was found that dumping of large amounts of highly contaminated sediments resulted in little long-term impact on the aquatic organism populations in and upon the sediments that could be explained by other than physical grain size changes. This site was a geologically high-energy disposal site where the dumped sediments were rapidly dispersed throughout the region. This dispersion quickly diluted the contaminants to where they had little or no impact.

While some regulatory agencies assert that increased suspended sediment concentrations are significantly adverse to aquatic life, as discussed by Jones and Lee (1978), work that was done under the sponsorship of the Corps of Engineers Dredged Material Research Program showed that very high concentrations of suspended sediment, in some cases approaching grams per liter, can be present without adversely affecting the organisms either by abrasion or by organ blockage. The relative insensitivity of many organisms to high concentrations of suspended sediment is not surprising considering that these organisms have evolved in an environment where they have been frequently exposed to high concentrations of suspended sediment from natural causes such as storms, flow-induced currents, tides, etc.

It is also sometimes asserted that dredging activities increase the water column turbidity to the point where the amount of photosynthesis that can occur in the water is drastically curtailed, with the result that there is an overall impairment of the function of the ecosystem as a result of limiting algal growth. While dredging projects can and usually do increase water column turbidity for a short period of time, it would be very rare that the decreased photosynthesis associated with this situation would have a significant adverse impact on the overall functioning of an ecosystem. It would also be rare that the suspended sediment associated with dredging and dredged sediment projects would have a significant adverse effect on aquatic-life-related beneficial uses of a waterbody.

Impacts on Water Movement, Bay Circulation, and Waste Assimilative Capacity

One of the potential consequences of the dredging of channels in harbors and other waterways is a change in the water transport and circulation patterns in the area. In an estuarine situation, the dredged channel can allow saltwater to move further into the estuary than would occur otherwise. The dredged channels can also allow a change in the location of impacts of municipal and industrial wastewater discharges caused by a more rapid transport of pollutants down the channel than would occur in the shallow or nondredged area.

The dredging of a channel can also affect the flow of groundwater into a bay or river as well as the flow of saltwater out of the estuary to the groundwaters of the area. The dredging of deep channels in areas in which the groundwaters of an area have been protected from saltwater by a confining (low permeability clay) layer within the sediments can result in a greater potential for saltwater intrusion into coastal freshwater aquifers. Obviously, before a channel is dredged or significantly deepened near an area that serves or could serve as a freshwater source, fairly detailed hydrogeological investigations should be conducted to be relatively certain that problems of this type will not occur.

While there are known physical impacts of dredging and dredged sediment disposal of the types described above, evaluation of the potential significance of them has to be done carefully on a site-specific basis. Further, it is not always found that the physical effects of dredged channel impacts are adverse to water quality.

Impacts of Chemical Contaminants in Dredged Sediments on Water Quality

In the late 1960s, it became generally known that many U.S. waterway sediments, especially those in urban and industrial areas, were highly contaminated with a wide variety of chemicals that, if released from the sediments, could have significant adverse impacts on the beneficial uses of the water. This finding caused pollution control agencies at the federal and state levels to attempt to develop criteria by which the water quality significance of dredged sediment-associated contaminants could be judged. At the federal level, the Federal Water Quality Administration (a predecessor of the EPA) developed what became known as the *Jensen criteria* (Boyd et al., 1972) which were based on bulk sediment composition. These criteria specified the total concentrations of a few water pollution indicator parameters or contaminants, such as COD, volatile solids, zinc, etc., which, if exceeded, indicated that the sediments were unsuitable for open-water disposal and would have to be disposed of by alternative methods. At the time of the adoption of these criteria, it was known by many who were working on the water quality aspects of sediment-associated contaminants that the total concentration of a contaminant in a sediment is rarely a reliable indicator of the amount of the contaminant that is available to affect water quality. Thus, the fundamental assumption of the Jensen criteria was invalid.

Nonetheless, the Jensen criteria have had a pronounced impact on the CE's dredging operations since they were used by many EPA regions as the principal basis for evaluating potential dredged sediment disposal impacts. As a result, sediments that for many years had been dredged and disposed of in nearby open water were judged by these criteria to be polluted and alternative methods of disposal had to be undertaken. In the U.S.-Canadian Great Lakes region where there was considerable concern about "polluted" sediments, Congress authorized a $250 million construction program to undertake so-called "confined" disposal of dredged sediment as an alternative to open-water disposal. Because many of the CE districts faced significant increases in the cost of dredging operations associated with the implementation of the Jensen criteria, including the CWA amendments of 1972, Congress authorized the CE to conduct a $30-million, 5-year DMRP designed to evaluate, among other things, the water quality significance of contaminants associated with dredged sediments. Through this research program, which was conducted by the CE Waterways Experiment Station in

Vicksburg, Mississippi, and completed in 1978, several hundred reports were generated and many professional papers were published based on the results of these studies. In addition to the DMRP study, a number of the CE districts conducted studies specifically designed to address problems of concern to them that were not being addressed by the DMRP. It is estimated that the total funding devoted to various aspects of dredged sediment disposal investigations and research in the 1970s was in excess of $40 million. About 25 to 30 percent of this amount was specifically directed toward evaluating the water quality significance of contaminants associated with dredged sediments.

The authors and their associates were involved in several laboratory and field studies, both as part of the DMRP and with several Corps districts, devoted to evaluating the water quality significance of contaminants associated with dredged sediments, with particular emphasis on open-water disposal of these sediments (Lee et al., 1978; Jones and Lee, 1978; Lee and Jones, 1977; Lee and Jones 1981a). In this work, the reliability of the elutriate test, developed by the CE and the EPA as an alternative to bulk sediment criteria as a measure of release of contaminants to the water column during open-water disposal of dredged sediment was evaluated. An aquatic organism bioassay-toxicity test procedure was also developed and evaluated for its suitability for assessing the potential toxicity of sediment-associated contaminants to organisms living on or within the redeposited sediments. Part of this work was devoted to evaluating the potential for bioaccumulation within edible aquatic organisms of sediment-associated contaminants that could render these organisms unsuitable for use as human food or be adverse to higher-trophic-level organisms, such as fish-eating birds.

Elutriate Test

The results of the DMRP studies conducted by the authors (Lee et al., 1978; Jones and Lee, 1978) confirmed what was known at the time of the start of the DMRP (Lee and Plumb, 1974), namely that the total concentration of a contaminant or indicator parameter in a dredged sediment is not related to the release of the contaminant in available forms to the water column during open-water disposal or to the toxicity of the sediment to aquatic life. This is because a variety of physical, chemical, and biological factors, principally sorption, reactions with the iron system, and hydrodynamics, control contaminant uptake and release in sediment-water systems (Lee and Jones, 1987). Thus, the concentration of a contaminant in a sediment cannot be used directly to estimate the potential impact of the sediment on aquatic life or other beneficial uses of a disposal site water.

Because of the importance of physical and chemical factors in controlling the release of sediment-associated contaminants during disposal, the elutriate test was developed by the EPA and the CE to imitate in the laboratory conditions that could exist during a hopper dredging, open-water disposal operation. The procedure involves mixing dredged sediment with site water (20 percent sediment by volume) for 30 min with compressed air, allowing the mixture to settle under quiescent conditions for 1 hr, and filtering and analyzing the filtrate for the chemicals of interest. The factors influencing the release of contaminants during the elutriate test and the ability of this procedure to mimic the release of the chemical contaminants during actual dredged sediment disposal operations were evaluated by Lee et al. (1978) by conducting more than 300 elutriate tests on a variety of waterway sediments from across the United States and monitoring about 20 dredged sediment disposal operations. It was found that when conducted under oxic conditions as prescribed, the elutriate test generally predicted both the direction and the approximate magnitude of contaminant release upon open-water disposal of dredged sediments; failure to maintain oxic conditions during the test, however, renders the results of the test uninterpretable.

Ammonia was consistently released during elutriation of the dredged sediments. Concentrations in the elutriates were often higher than existing EPA water quality criteria and many state water quality standards. However, as discussed by Jones and Lee (1978), this does not necessarily mean that water quality problems would result from the open-water disposal of these sediments. It does indicate that site-specific hazard assessments of the potential water quality problems associated with ammonia should be made when "high" concentrations are found in elutriates to determine the likelihood that sufficiently high concentrations would occur in association with the particular disposal operation to cause an impairment of the beneficial uses of the water at the disposal site.

Generally, the studies by Lee et al. (1978) showed that zinc, iron, nitrate, copper, lead, cadmium, and phosphate, as well as a variety of chlorinated hydrocarbon pesticides, were not released during an oxic elutriate test. Some relatively clean sediments did show small amounts of PCB releases, however. Highly contaminated sediments with large amounts of organics did not show PCB releases.

No relationship was found between the bulk chemical content of the sediments and the release of contaminants in the elutriate test. Field studies showed that the elutriate test results predict the direction and approximate magnitude of contaminant release during open-water disposal of dredged sediments. It is important, however, that the elutriate tests be conducted under oxic conditions (with air mixing) in order to reliably assess the potential for contaminants released from the sediments to have an adverse impact on aquatic organisms at the dredged sediment disposal site water column. The elutriate test was found to be a valuable tool for testing the potential water column impact of the contaminants present in dredged sediments during open-water disposal operations.

The EPA and CE have recently released a testing manual for *Evaluation of Dredged Material Proposed for Ocean Disposal* (U.S. EPA, CE, 1991). This guidance manual is designed to implement Section 103 of Public Law 92-532 (Marine Protection, Research, and Sanctuaries Act of 1972). In accord with this section, the EPA is to develop regulations governing ocean disposal of dredged sediments which are to be used as a basis for evaluating the suitability of such disposal. The EPA regulations governing ocean disposal of dredged sediments, Title 40 of the Code of Federal Regulations, Parts 220-228 (40 CFR 220-228), contained a number of significant technical errors when they were first promulgated in the 1970s. Unfortunately, these regulations have not been brought up to date. As a result, the recently released EPA-CE guidance manual presenting testing procedures for ocean dumping of dredged sediments continues to use a highly inappropriate approach of applying EPA water quality criteria at the edge of a mixing zone or anywhere 4 hr postdisposal to determine if the criteria have been exceeded. These criteria are not appropriate for such an assessment. As discussed above, exceedence of the EPA criteria at the edge of a mixing zone associated with a dredging project rarely, if ever, would represent impairment of the beneficial uses of the waters of the region where the dredged sediment disposal would take place.

Water column toxicity tests on the elutriates are used when there are not enough criteria for all contaminants or synergism is expected. They are used to evaluate whether contaminants present in dredged sediment elutriates could cause acute toxicity at the dredged sediment disposal site. It is important to emphasize that such an approach must properly consider the duration of exposure that organisms could receive at a particular dredged sediment disposal site. As conducted now, the duration of exposure of the acute toxicity test significantly exceeds the duration of exposure that organisms could receive at a dredged sediment disposal site involving open-water dumping of the sediments. The exposures that are possible under these conditions are on the order of minutes. The proposed acute toxicity testing approach of elutriates is highly overprotective of what is needed to protect aquatic organisms at the dredged sediment disposal site water column from toxicity that would be found in the elutriates. It is indeed rare, if ever, that ocean or inland water disposal of dredged sediments involving dumping could cause toxicity to water column aquatic life. The inappropriateness of the approach in this section again arises out of the outdated approaches adopted by the EPA in the mid-1970s regulating the ocean disposal of dredged sediments.

Bioassay-Toxicity Sediment Testing

In the DMRP studies conducted by the authors, about 30 chemical parameters were measured in the dredged sediment elutriates. These tests showed that except for ammonia, there was little or no likelihood of contaminant release from dredged sediments that would be adverse to aquatic life at the disposal site water column based on examining the relationship between the concentrations of contaminants released and the concentration-duration of exposure relationships that exist for aquatic organisms at a dredged sediment disposal site involving dumping of dredged sediments. In order to assess whether other unmeasured chemicals could have an adverse impact on water column organisms as well as organisms that might colonize the redeposited dredged sediments shortly after deposition and whether it was possible for synergistic adverse impacts of various chemicals to occur, Lee et al. (1978) developed a dredged sediment elutriate bioassay screening toxicity test.

The dredged sediment bioassay-toxicity tests developed by Lee et al. (1978) were generally conducted by proceeding through a standard oxic elutriate test procedure and introducing the test organisms into the settled elutriate. Grass shrimp (*P. pugio*) were used for marine conditions and daphnids were used for freshwater. The survival of the organisms in the test systems and controls were recorded over a 96-hr (or longer) period. Thus, the organisms were exposed not only to the contaminants released from the sediment during the elutriate test but also to any contaminants released over the 4-day test period and obtained by the organisms directly from the sediment particles present in the bottom of the test system. These test conditions represent worst-case conditions, more harsh than would likely occur at a dredged sediment disposal site.

The work of Lee et al. (1978) on the potential toxicity of U.S. waterway sediments to aquatic life showed that many sediments, especially those near urban and industrial areas, contained contaminants which caused them to be toxic to aquatic life under the conditions of the laboratory test. Typically, from 10 to 50 percent of the test organisms were killed in the 96-hr test period. Some tests were carried out over a period of 21 days; they did not, in general, show significantly greater toxicity than that observed over the 4-day test period. The toxicity found in sediments obtained near urban-industrial centers was, however, much less than what would be predicted based on the concentrations of contaminants that were present in the sediments. Further, it was found that there was no relationship between the bulk chemical content of contaminants in sediments, either individually or collectively, and the toxicity of the sediments to aquatic organisms. While many sediments had very high concentrations of heavy metals and chlorinated hydrocarbons of various types, these contaminants were present in the sediments in nonavailable, nontoxic forms. In the case of New York Harbor sediments, Jones and Lee (1988) found that the toxicity was caused by ammonia. Other investigators are also finding that ammonia is one of the principal causes of sediment toxicity to aquatic life.

Recent work by the EPA and others directed toward developing numeric sediment quality criteria is developing information that is helpful in explaining why many chemical contaminants in sediments are not toxic to aquatic life. While this has been known to some extent for some time, it is now very clear that particulate forms of contaminants are typically not available-toxic to aquatic life, including benthic and epibenthic forms. When the chemical characteristics of sediments are examined in light of possible chemical reactions that could cause chemical contaminants of concern to be converted to particulate nontoxic forms, it is found that aquatic sediments typically contain a wide variety of constituents which would convert some of the most hazardous chemicals, such as heavy metals, chlorinated hydrocarbons, pesticides, and PCBs, into particulate, nontoxic forms. For example, it has been known for many years that many aquatic sediments contain high concentrations of sulfides and polysulfides primarily in the form of iron sulfides. Almost all of the heavy metals, such as Cu, Zn, Cd, Ni, Pb, etc., tend to form highly insoluble sulfides. The solubility of many of these sulfides is less than iron sulfide and therefore would replace iron as a precipitating metal for the sulfide species. The recent studies by the EPA and others have shown that heavy metal sulfides are nontoxic to aquatic life. Since the concentrations of noniron heavy metals in sediments would rarely exceed the concentrations of sulfides in sediments on a molar basis, it would be rare that a sediment would not contain sufficient concentrations of sulfides to detoxify all of the heavy metals present in them.

It should be noted, however, that there are a variety of other mechanisms for detoxification of heavy metals in sediments that must also be considered. These include carbonates, hydrous metal oxides, solid-phase organic complexing agents, clays, etc. Therefore it would not be surprising to find that even if the total noniron heavy metals in a sediment exceeded the sulfides and polysulfides measured as acid volatile sulfides (convertible to H_2S under acid conditions) that the sediments were nontoxic due to heavy metals.

The work of the EPA and others has included examining the role of particulate organic matter present in sediments measured as total organic carbon (TOC) in binding potentially toxic, nonpolar organic chemicals to sediment particles. The chemicals of greatest concern are the chlorinated hydrocarbon pesticides (DDT, aldrin, dieldrin, chlordane, etc.), PCBs, and the polynuclear aromatics (PNAs-PAHs). All of these chemicals can be highly toxic to aquatic life under water column conditions. However, in sediments they would all tend to bind to the organic matter through sorption reactions. The EPA studies have shown that high TOC-containing sediments tend to be much less toxic for given concentrations of these chemicals than low TOC-containing sediments. It is reasonable to explain this observation by understanding that the high TOC sediments, which have a higher amount of organic matter present, can bind the organic toxicant in such a way as to detoxify it. It is therefore evident that one of the primary reasons why the chlorinated hydrocarbon pesticides and other organic chemicals present in sediments are not toxic to aquatic life is because they are tightly bound to particulate organic matter.

In the 1970s, the EPA and CE developed an unnecessarily complex approach for sediment bioassay testing (U.S. EPA, CE, 1977). This approach was and still is required by 40 CFR Parts 220–228. They required individual tests on various component parts of the elutriate, using three different types of test organisms. Not only did this requirement increase the cost from a few hundred dollars to several thousand dollars per sample, but also the results were not readily interpretable. Essentially the same amount of useful information can be obtained from a simple screening toxicity test using a moderately sensitive, benthic or epibenthic organism, such as grass shrimp for estuarine and marine waters and daphnids for freshwater, as is obtained using three different kinds of organisms (Lee and Jones, 1977). Although the regulations have not been revised, current guidance (U.S. EPA, CE, 1991) recommends using a combined liquid and suspended phase (rather than using them separately) but, in accord with regulations, three organisms continue to be used.

If the screening toxicity tests were to show potentially excessive toxicity, site-specific follow-up testing could be undertaken on an array of organisms, various sediment to water ratios, and under other conditions that may influence the results, in a hazard assessment format to assess the potential for adverse impacts under the particular conditions being considered; thus the mechanical conduct of an expensive array of tests may be reduced. Similarly, if the screening toxicity tests indicated that there was essentially no toxicity under worst-case conditions, testing could be terminated and considerable time and money saved.

One of the controversial issues that still has not been appropriately resolved for either marine or freshwater disposal of dredged sediments is what is an allowable percentage of mortality in a sediment toxicity test. It is certainly inappropriate to assume, as has been done in some parts of the country, that if there is a statistically significant difference in organism deaths in the test systems and the control systems, open-water disposal of the sediments would result in a significant adverse effect on aquatic life at the disposal site. It is important to understand that there may be little or no relationship between a 10, 20, or some other percent mortality in the sediment toxicity test and what will actually occur at the dredged sediment disposal site.

It is important that the test protocol be designed in such a way as to properly consider the "noise" in the test results. In this regard, there seems to be considerable confusion about the difference between *statistical* significance and *water quality* significance. As discussed by Jones and Lee (1978), it is important to clearly distinguish between these two concepts. The former is simply a comparison of numbers and is a function of sample size, degree of confidence, etc. While of importance in ensuring the integrity of sampling and analytical programs and in describing the variability of a system, the statistical significance has essentially no relationship to water quality significance. It can and does readily occur that the concentration of a chemical increases by a statistically significant (at some prescribed confidence level) amount without there being an environmentally significant impact. The former indicates that there are sufficient data to indicate that an increase in organism impacts has occurred at the prescribed confidence level. If the confidence level is changed, the increase may no longer be significant statistically. If the increase has water quality significance, it means that the increase is sufficient to cause an adverse impact on beneficial uses of the water.

Jones and Lee (1978) discussed the approach that should be used in interpreting the dredged sediment toxicity test data. As conducted, this test estimates the impact on fairly sensitive aquatic organisms that could occur if the organisms spent 4 days in settle water that is the same as that in the dredge discharge pipe.

Not considered in this assessment are the dilution of the released contaminants with area water and incoming water, the dilution and mixing of the dredged sediments with other nontoxic sediments of the disposal region, the mobility and avoidance behavior of many types of organisms, or the myriad physical and chemical reactions that typically take place at a disposal site. The physical and chemical reactions, principally the scavenging of chemicals by ferric hydroxide, tend to render many chemicals less toxic. As a result, the toxicity observed during the worst-case elutriate toxicity tests is likely to be different from (i.e., less than) that exhibited by a dredged sediment dumped within a disposal area.

Laboratory-based toxicity tests can represent well beyond worst-case situations that can occur under field conditions (Jones and Lee, 1988). Under field conditions, contaminants released to the sediment-water interface have the opportunity for significant dilution. Organisms that derive their contaminants through uptake of water at or near the sediment-water interface would typically be exposed to significantly lower concentrations under field conditions than in the laboratory test. Therefore, significantly lower toxicities would probably be observed in the field than in the laboratory.

During the past few years, increasing use of partial chronic toxicity tests has been made to assess whether a particular water has the potential for toxicity to aquatic life. Some tests of this type involve the use of recently hatched fish embryos which are exposed to the water and contaminant for about a 1-week duration. The survival and growth of the larvae are assessed during this week-long exposure. The results of such tests are found to provide similar levels of toxicity to those that are typically found with full-scale chronic aquatic organism testing involving several years of exposure. The EPA has published two guidance manuals for conducting these tests (Peltier et al., 1985; Weber et al., 1988). These tests are relatively simple to conduct and are relatively inexpensive. When testing an ambient water, they provide significant information on whether there is toxicity in the water that could be of significance to aquatic life in the area. It is likely that these tests will soon receive widespread use in dredging and dredged sediment disposal projects. It is important that their use, however, be conducted in such a way as to properly mimic the exposure that aquatic organisms could receive associated with the dredging project. It is virtually impossible for any dredged sediment disposal operation to provide an exposure of aquatic organisms to contaminants in a water column for a week's duration. Therefore, the use of these tests should involve shorter exposure periods to the dredged sediments or dredged sediment elutriates than the typical test period.

Bioaccumulation

By far the greatest difficulty with the EPA-CE dredged sediment testing manual (1991) is its prescribed test for bioaccumulation. The EPA and CE (1991) use a laboratory-based, 28-day bioaccumulation test for organic contaminants. The lengthening of the time from the previously used 10 days to 28 days for bioaccumulation is appropriate for some chemicals, such as dioxins, since it appears that 10 days is too short an exposure to allow full accumulation. The lengthening of the test period does not, however, address the major problem with this test, that of not being able to extrapolate the laboratory results to field conditions.

All that can be said from the results of a bioaccumulation test is that under the test conditions bioaccumulation to a certain degree occurred in the test organisms. This provides little or no useful information on the bioaccumulation that will actually occur in the field at a dredged sediment disposal site. With few exceptions, although it could occur, the accumulation that will occur under laboratory conditions will greatly exceed what will occur in the field. Therefore the bioaccumulation test, whether conducted for 14 or 28 days, is not a valid test to determine whether excessive concentrations of contaminants present in a dredged sediment will accumulate in edible tissues of organisms of concern with respect to their use as human food in excess of Food and Drug Administration guidelines. Since the relationships between actual bioaccumulation under field conditions and the bioaccumulation found under various sediment:water: organism biomass couplings are not defined, it is difficult, if not impossible, to extrapolate the results of such laboratory tests to a field situation. On the other hand, this test is environmentally conservative since field exposure is typically far less than that achieved in the laboratory under the conditions used.

Lee and Jones (1977) and Jones and Lee (1978) have described an approach to evaluate bioaccumulation with open-water disposal of dredged sediments. It involves collecting organisms from a disposal site which has received some of the sediments that are in question. These organisms should be analyzed for the various contaminants of potential concern and the results compared to FDA action levels for these contaminants. The organisms should be collected at least twice a year (e.g., spring and fall) to detect major seasonal differences. Also, care must be exercised in selecting the types of organisms for analysis. Distinction should be made between those organisms which are transitory (i.e., migrating through the area) and those that tend to stay near the disposal area for a sufficient period of time to accumulate the contaminants of concern. While it is of potential interest to see if organisms in a particular area of concern contain higher concentrations of contaminants not on the FDA list than those in other areas, at this time, the interpretation of the significance of body burden data for these chemicals to the organism or to organisms that use these organisms as food is not reliably possible.

Hazard Assessment Approaches for Dredged Sediment Disposal

There is a need to use a combination of elutriate tests, dredged sediment bioassays, and knowledge of the physical, chemical, and biological characteristics of the disposal area to determine the potential for adverse impacts of contaminants present in a particular dredged sediment on the beneficial uses of a disposal site water. Lee and Jones (1981a) developed a hazard assessment approach to provide a framework which can be used to identify information needs, integrate the information as it is generated, and provide guidance in the integration and interpretation of elutriate test, dredged sediment bioassay, and disposal site characterization results in making dredged sediment management decisions.

Figure 9.12 presents a hazard assessment scheme that was developed by the authors for assessing the potential impact of dredging and disposal in the Upper Mississippi River (Lee and Jones, 1981a). It shows how elutriate tests, dredged sediment bioassays, and the physical, chemical, and biological characteristics of the disposal site can be used in a tiered hazard assessment scheme to evaluate dredging and disposal alternatives.

With reference to Fig. 9.12, the first feasible methods should be identified and evaluated in order of cost; if the least expensive method does not have unacceptable hazard associated with it, it would probably be the method of choice. Generally, open-water disposal is the least expensive disposal alternative. The pro-

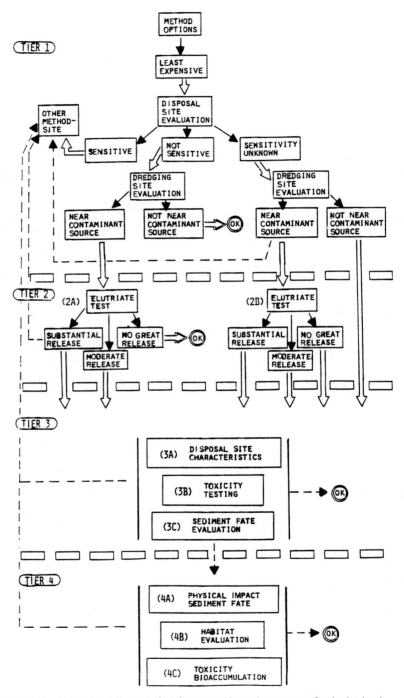

FIGURE 9.12 Abbreviated diagram of environmental hazard assessment for dredged sediment disposal.

posed disposal site should be examined to be certain that it is not an ecologically sensitive area, such as a fish breeding ground, oyster bed, etc. This is generally of greatest concern if the area has not received dredged sediments in the recent past. If it is a sensitive area, another site should be selected and evaluated. If it is known to not be a sensitive area, other aspects of potential concern with respect to dredged sediment disposal can begin to be evaluated. If the sensitivity of the proposed disposal site is not known and cannot be readily determined, potential impacts may be evaluated by examining the proposed dredged sediments. Under the latter two conditions, the testing continues in the first tier by evaluating the dredging site conditions. If the disposal site is not a sensitive area and the sediments are not being derived from an area near a source of contaminants, the technical recommendation to the public would probably be that there should not be significant impacts on beneficial uses of the disposal site water.

If the sediments were identified in the Tier 1 evaluation as being from near a source of contaminants, the potential for the release of chemicals from them should be examined in a more advanced, Tier 2, testing level. That is, insufficient information was obtained from the Tier 1 investigation to make a recommendation regarding the potential adverse impact of dredged sediment disposal. Tier 2 testing, in the scheme outlined in Fig. 9.12, consists of elutriate tests run under oxic conditions. Depending on the types and amounts of chemicals released during elutriation (a worst-case procedure), either additional testing should be undertaken, other methods of disposal considered, or a recommendation should be given that the expected hazard should be considered acceptable.

Where there are still questions about the potential impacts of the particular dredging or disposal scenario being considered after the Tier 2 tests, it may be desirable to undertake the next level of testing. In this and subsequent levels, the tests become much more site-specific, involved, and expensive. Additional information is collected to fill gaps in understanding of the disposal site characteristics such as dilution and dispersion, the acute toxicity of chemicals that were released in the elutriate tests and in a screening bioassay, and the fate and persistence of the sediments at the disposal site. Depending on the results of these tests, a decision could be made that another method of dredging or disposal should be considered, that a recommendation should be made that the risk should be considered acceptable, or that additional information is needed in order to make the technical assessment of potential impact.

Under the last condition, Tier 4 testing and the cost of conducting those types of assessments compared to the cost of evaluating other methods would be considered. Tier 4 testing would focus on the more subtle potential impacts of dredged sediment disposal at the particular area and would be largely in the form of experimental dredged sediment disposal operations using the site and method evaluated through Tier 3, but with detailed monitoring. In undertaking Tier 4, those responsible must carefully evaluate whether irreparable damage could be done to the ecosystem of the regions as a result of deposition of dredged sediment in the area. Based on past studies, it appears that there will be very few situations where irreparable damage will be done to an aquatic ecosystem by the disposal of a single or several sediment loads at a site; normally such potential situations would have been identified in earlier tiers of testing.

This hazard assessment scheme is an illustration of an approach that should be developed for a particular area; it is not intended to be an all-inclusive statement. The particular tests prescribed depend on the information known and lacking, etc. Before it is used, the back-up papers and reports (Lee and Jones, 1981a, 1981b) should be referred to and the subtleties and implications of the approach understood.

It is important to distinguish between a tiered hazard assessment of the type described above and a tiered assessment that has been adopted by regulatory agencies such as the EPA and the CE for evaluating potential water quality impacts of contaminants in dredged sediments. Typically, the first tier of the EPA-CE focuses on evaluation of existing information to determine if there is reason to believe that the sediments have been contaminated. Part of Tier 2 focuses on bulk sediment inventory in which an evaluation is made as to whether the dredged sediments are more contaminated than the disposal site sediments. While this is rarely the case, if this is found, caution should be exercised in making any judgments about the potential impacts of dredged sediment disposal based on bulk sediment chemical analyses. The relative levels of contamination between the dredging site and disposal site sediments or for that matter any other sediments, such as a reference sediment, provide no reliable, useful information on the potential impacts of contaminants in these sediments on beneficial uses of water at the disposal site.

Another part of Tier 2 involves the use of the elutriate test in which a comparison is made between the amounts of contaminants released in the test with water quality criteria and standards that would be applied at the edge of a mixing zone. Again, as discussed above, it is technically invalid to determine potential impacts of contaminants released in the elutriate test on water quality at an open-water dredged sediment disposal site, by comparison to water quality criteria. Violation of such criteria-standards would rarely, if ever, represent an adverse effect on water quality at the dredged sediment disposal site.

Tier 3 focuses on acute aquatic organism bioassays using the EPA-CE multiple species multiple test approach. This is an overly complex, overly expensive approach that for most dredging projects unnecessarily and significantly increases the cost of testing. Tier 3 also focuses on bioaccumulation assessment involving laboratory testing of sediments and various organisms. This testing approach, however, does not provide reliable information on the bioaccumulation that will occur at a dredged sediment disposal site.

As part of developing updated guidance for implementing regulations governing ocean disposal of dredged sediments, the EPA and CE (1991) have used a tiered assessment approach for determining whether dredged sediments may be disposed of in the ocean. This approach suffers from some of the same deficiencies described above.

The EPA-CE tiered assessment approach is not a properly developed hazard assessment approach of the type described by Lee and Jones (1981a). Considerable testing funds could be saved at many sites through the use of the Lee and Jones approach. An example of such savings is provided by the work performed for the Norfolk District of the Corps of Engineers (Lee and Jones, 1981b), where through a true selective tiered hazard assessment, it was possible with limited additional testing to demonstrate to regulatory agencies that the ban that had been imposed on dredging of part of the Intercoastal Waterway because of the presence of certain contaminants in the sediments was technically invalid.

IMPACT OF EQUIPMENT USED FOR DREDGING AND DREDGED SEDIMENT DISPOSAL

There are basically two types of equipment used to dredge U.S. waterway sediments. These are characterized as mechanical and hydraulic dredging. Mechanical dredging typically involves use of a clamshell or dragline to remove sediments in a bucket, which are typically deposited in a barge for disposal. Normally, mechanical dredging and dredged sediment disposal have the least potential for water quality impacts because of contaminants associated with the sediments. This is because there is very little mixing of the sediments with water and therefore there is limited opportunity for the release of contaminants from the sediments to the water.

Hydraulic dredging, on the other hand, involves slurrying the sediments with water in a one-part sediment to four-parts water mixture where this mixture is typically then pumped as a slurry to either open water or to confined upland disposal. The slurrying of sediments results in mixing of the sediments with water, which tends to promote the release of contaminants from the sediments and in the interstitial water into the slurry water (see Fig. 9.13). However, it is well documented that such release does not occur for most elements once the slurried waters are in contact with dissolved oxygen because of the scavenging of the released contaminants by ferric hydroxide.

There are basically two types of disposal operations associated with hydraulic dredging. One of these involves pumping the slurry into a hold of a ship (hopper dredge) where the dredged sediments are transported to the disposal site. Typically hopper dredging involves open-water disposal of the sediments, although it is possible to pump from the hoppers to on-land disposal. Figure 9.14 presents a diagrammatic representation of the principal components of a hopper dredge dumping of sediments. Basically, when the dredge is at the appropriate location, the doors on the hoppers, which are on the underside of the ship, open and the contaminated sediments are allowed to fall out. Most of these sediments rapidly descend to the bottom as a cohesive mass. While the sediments were placed into the hoppers as a 20% slurry, typically the excess water is allowed to drain off at the dredging site or in transport to the disposal site with the result that the slurry characteristics of the sediment are rapidly lost after being placed in the hopper.

As indicated in Fig. 9.14, small amounts of water column turbidity are associated with the descent of the dredged sediments from the hopper to the bottom of the water column. This forms a turbid cloud in the waters of the region. Based on the studies by Lee et al. (1978), while this turbid cloud does show some release of chemical contaminants, its size and the magnitude of the release are

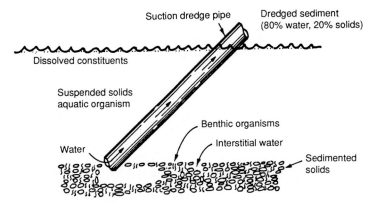

FIGURE 9.13 Hydraulic dredging.

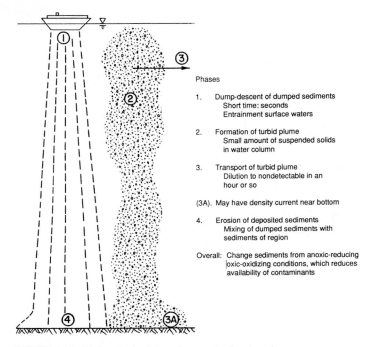

FIGURE 9.14   Hydraulic dredging—hopper dredge dumping.

such that no water quality problems have been found nor are they expected from contaminant release associated with the disposal operations.

In the studies conducted by Lee et al. (1978), the only contaminant released that was of potential concern was ammonia. Figure 9.15 shows the typical passage of the turbid plume associated with the open-water disposal of dredged sediments. As indicated in this figure, near the surface (2 m depth) the turbidity persisted at a location a few tens of meters down current from the dump for about 2 min. Near the bottom at 14 m, the turbid plume turbidity persisted for about 7 min. The dissolved oxygen (DO) depletion for this dumping operation near the surface is shown in Fig. 9.16. The DO depletion caused by the oxygen demand in the sediments, while measurable, would not have an adverse effect on aquatic life.

The release of ammonia from two hopper dredged dumps of contaminated sediments in open waters that took place about 2 hr apart is shown in Fig. 9.17. It is evident from Fig. 9.17 that the release of ammonia would present no adverse impact on aquatic life unless the organisms were able to stay in the rapidly moving turbid plume for a considerable period of time and the plume persisted for a long period of time. Studies by Lee et al. (1978) at several locations showed that typically the turbid plume was no longer identifiable after about 1 hr following the dump. The concentration of released ammonia and the duration of exposure that organisms encounter associated with release are sufficiently short (minutes) so that no water quality problems caused by ammonia release would be expected.

It may be concluded that the hopper dredged disposal of even highly contam-

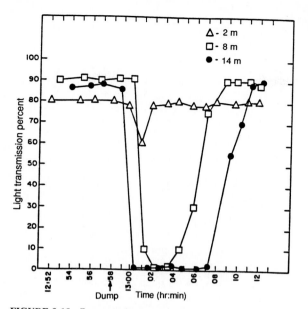

FIGURE 9.15 Passage of the turbid plume.

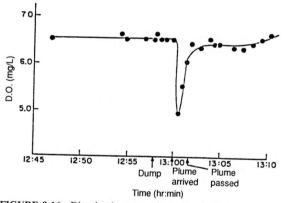

FIGURE 9.16 Dissolved oxygen depletion during passage of turbid plume.

inated sediments would not be expected to cause any water column water quality problems because of the very short exposures that aquatic organisms could experience from such releases. A similar situation exists for the dumping of mechanically dredged sediments in open waters. The magnitude of release of contaminants to the water column and the duration of exposure that water column organisms could experience in the turbid plume that forms during the descent of the dredged sediments are such that no water column water quality problems would be expected.

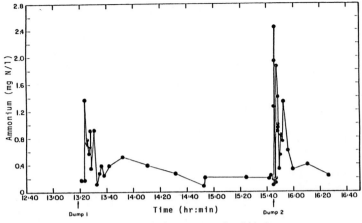

FIGURE 9.17 Ammonia release during passage of turbid plumes.

The open-water disposal of hydraulically dredged sediments involving pipeline transport to the disposal site represents a significantly different situation than those found for dumping of mechanically or hydraulically dredged sediments in open waters. This situation is pictured in Fig. 9.18. The approximately 20% sediment slurry discharge from a hydraulic pipeline transport operation quickly forms a density current which moves along the bottom in about a 1-m-thick layer down ambient current from the discharge. This density current has been found to persist for thousands of meters from the point of discharge. It is typically devoid of dissolved oxygen and could therefore represent a significant adverse impact on aquatic organisms residing on the bottom in the path of the current caused by low DO and/or the release of contaminants such as ammonia. This would be especially true if the density current were to persist in one location for a period of many hours to a day or so. Ordinarily, however, such a situation does not occur

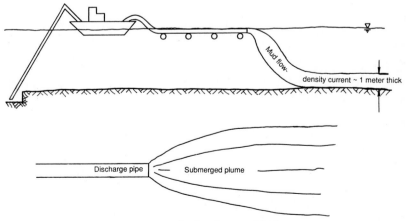

FIGURE 9.18 Hydraulic dredging—open-water pipeline disposal.

because of mechanical problems which cause the dredging operations to be shut down. Further, because of the movement of the dredge as well as accumulations of sediments near the point of discharge, there is frequent need to change the point of discharge, thereby shortening the exposure time that organisms located on the bottom could experience from hydraulic dredging and pipeline transport operations.

From an overall point of view, mechanical dredging of sediments would tend to have the least environmental impact at the dredging and dredged sediment disposal sites. Hydraulic dredging with open-water dumping of the sediments would also be expected to have little or no impact on water quality in the water column at either the dredging or dredged sediment disposal site. Hydraulic dredging with pipeline transport of the dredged sediment slurry could have an adverse impact on aquatic organisms located at the sediment-water interface because of the extended exposure that such organisms could receive associated with the density current arising from this type of dredged sediment disposal operation.

There are a number of other types of dredging equipment which are modifications of hydraulic dredging which attempt to minimize the stirring of sediments in the water column at the dredging site. There is also equipment that involves a vacuum sucking of sediments into a vessel which is then raised to the surface and dumped into a barge for transport. These various types of equipment do, or would probably, reduce the amount of suspension of sediments in the water column at the dredging site and thereby minimize the problems of adverse impacts of contaminants associated with the dredging operation; however, there has been no demonstrated case where using conventional hydraulic dredging equipment has ever caused a water quality problem at the dredging site. Therefore there is justifiable reason to question the need to spend additional funds to use modified dredging equipment because of alleged potential environmental impacts.

UPLAND CONFINED DISPOSAL OF DREDGED SEDIMENTS

While it has been commonly assumed that confined disposal of dredged sediments involving placing them on land or behind dikes to form an island in water is less environmentally damaging, these assumptions are not necessarily correct. In fact, it is now beginning to be more widely recognized that at some locations upland disposal of contaminated dredged sediments has a greater potential for adverse environmental impact than open-water disposal of dredged sediments. As generally practiced today, confined disposal does not truly confine deposited sediment but rather provides a settling area where the larger, more dense particles are removed and the water associated with the hydraulic dredging operation is allowed to enter the watercourse. With few exceptions, confined disposal operations have overflows to the nearby watercourse where any contaminants released from the sediments during the dredging operation, as well as those contaminants associated with the fine materials that are not removed in the disposal area, enter the water. Since it is the fine materials that often have the greatest affinity for contaminants, if the contaminants are or become available from the fines, more ecological and/or water quality damage could result from so-called confined disposal than with open-water disposal, since the area where the confined disposal overflow occurs (i.e., nearshore) is generally the most ecologically sensitive area of the waterbody.

Studies conducted at the CE Waterways Experiment Station (Palermo, 1986) have shown that the sediments in a confined disposal area, which when wet retain the contaminants such as heavy metals, upon drying tend to release these heavy metals. This appears to be related to the oxidation of the amorphous sulfides present in the sediments which are keeping the heavy metals in particulate form, and the development of acidic conditions. It may also be related to the aging of the ferric hydroxide precipitate (hydrous oxides). While freshly precipitated ferric hydroxide has a large holding power for contaminants, aged precipitates, especially those that dry out, lose this holding power. It is therefore not surprising to find that heavy metals associated with dredged sediments that are placed in a confined disposal area and allowed to dry out will show appreciable release of heavy metals to the overflow waters the next time that dredged sediments are introduced into the area. This situation can also occur with atmospheric precipitation drainage and runoff from these areas. It is very important that those who advocate upland disposal of contaminated dredged sediments conduct a proper, critical review of the adverse impacts of the contaminants associated with the sediments which leave the confined disposal area during overflow, during dredging operations, or via drainage from the disposal site.

NEW AND PENDING REGULATIONS

The current EPA regulations governing disposal of dredged sediments in freshwater, estuarine, and marine waters are badly out of date and require a number of technically invalid approaches to be used to assess the potential water quality significance of chemical contaminants in the sediments. There is a need for the EPA to update the regulations governing both the ocean and inland water disposal of contaminated dredged sediments to reflect the large amount of information that has been developed over the years on the impact of such sediments and associated contaminants on water quality as well as the approaches that should be followed to evaluate this impact on a site-specific basis. There are also significant inconsistencies between the ocean disposal of dredged sediments and disposal in U.S. waters.

The current regulations governing the ocean disposal of dredged sediments prohibit any disposal of contaminated sediments that could have a long-term adverse impact on the numbers and types of organisms present at the disposal site. Short-term impacts from physical effects, such as sediment grain size, etc., are allowable. Long-term impacts, such as those that might be associated with persistent chemical contaminants, are prohibited. There is no significant justification for such an approach. Ocean disposal of dredged sediments occurs at designated disposal areas. Such areas are well defined and, contrary to the current EPA regulations, there should be no requirement that the dredged sediment associated contaminants have no impact on the numbers and types of aquatic organisms present within these areas. It has been well established that changing the grain size of the uncontaminated sediments can have a dramatic impact on the numbers and types of benthic organisms present in an area. The impacts on the numbers and types of organisms within the designated disposal area due to chemical contaminants should also be considered acceptable. Such impacts can readily occur without having a significant adverse impact on the nearshore marine water quality and aquatic life resources of the area.

One of the areas of greatest concern in the regulations governing ocean disposal of dredged sediment is the limitation on the toxicity of these sediments to aquatic life. The approach that should be followed in sediment toxicity tests is to evaluate the real or potential impacts at the edge of the designated disposal area, where the dredged sediments will be mixed with the natural sediments of the region. Dilution of contaminated sediments with small amounts of noncontaminated sediments typically greatly reduces the toxicity found for the contaminated sediments.

If large amounts of toxicity are found under laboratory test conditions which simulate field conditions, additional studies should be done to determine whether the simulated conditions of the laboratory tests appropriately mimic the field conditions with respect to dilution of the sediments and the dilution of any contaminants released from the sediments with the waters of the type that would occur at the edge of the designated disposal area. It is only with testing of this type that a proper evaluation can be made of what is an allowable percentage of mortality of the test organism in a sediment toxicity test which is needed to protect the designated beneficial uses of the waters outside of a designated disposal area.

Reference Sediment

The EPA and the CE (1991) in their testing manual governing ocean disposal of dredged sediments make use of a reference sediment for testing purposes. This reference sediment approach is part of the EPA's regulations governing the ocean disposal of dredged sediments. The same tests are performed on the reference sediments and on the sediments scheduled to be dredged. It is not clear, however, how the behavior of a contaminant in a reference sediment as well as the toxicity to or bioaccumulation in an organism exposed to a reference sediment can in any meaningful way be used to evaluate the water quality impacts that a particular dredging or dredged sediment disposal operation will have on water quality.

Sediment Quality Criteria

The EPA is pursuing an attempt to develop numeric sediment quality criteria as a means of regulating the impacts of chemical contaminants that become associated with aquatic sediments. Basically, this effort is an outgrowth of the Jensen criteria discussed above in which the bulk chemical composition or some modification of this approach is used to determine "excessive" concentrations of contaminants in sediments. While it is well known that numeric criteria based on the bulk chemical composition of sediments are technically invalid, the EPA is attempting to find some way to normalize sediment chemical data so that a single numeric value can be developed which can be used to determine excessive concentrations of contaminants in sediments. One of the normalization procedures that at this time is being aggressively pursued by the EPA is the Equilibrium Partitioning (EqP) approach in which the amount of organic carbon present in the sediments is used to estimate the detoxification of the sediments for nonpolar organic chemicals such as chlorinated hydrocarbons and PAHs. This approach involves the use of EPA water column criteria to determine excessive concentrations of contaminants in the waters in equilibrium with the sediments based on a sediment-water partition coefficient. As discussed above, this approach has considerable merit in explaining why chemicals of this type are not toxic in sediments. It has very limited utility, however, in predicting whether a sediment will be toxic or not.

The other approach that is being used to normalize contaminant concentrations in sediments in an attempt to try to develop numeric criteria is through the use of acid volatile sulfides (AVS). Di Toro et al. (1990) have reported that when heavy metals such as cadmium, zinc, copper, and lead are present in sediments that have on a molar basis more sulfides in the sediments than noniron heavy metals, the heavy metals are nontoxic to aquatic organisms. As discussed above, this is to be expected since these heavy metals form highly insoluble sulfides which should be nontoxic. While sulfide precipitation can be used to explain why heavy metals present in sediments are nontoxic, there are a variety of other chemical reactions that occur with heavy metals that cause them to also become nontoxic. Therefore trying to establish numeric criteria for heavy metals in sediments based on their concentrations relative to the sulfide concentrations of the sediments as the EPA is proposing to do can readily lead to significant overestimates of toxicity of the heavy metals to aquatic life.

The complexity of aquatic sediments as chemical systems requires that an effects-based approach be used to assess whether chemical contaminants in the sediment could have an adverse impact on aquatic life and other beneficial uses of the nearby waters. As discussed by Wright in the next section, the EPA-CE evaluation procedures for determination of the suitability of open-water disposal of contaminated sediments are to a considerable extent based on effects-based approaches such as sediment toxicity testing and bioaccumulation. Rather than trying to use arbitrarily developed chemically based numeric criteria where exceeding a particular numeric value is related in some way to the concentration of contaminants in sediments, the EPA-CE chose to follow directly measuring potential impacts such as measuring toxicity, bioaccumulation, etc. This is the most technically valid approach and the one that should be followed. It provides a much more appropriate estimate of the potential impacts of contaminants in sediments than chemically based sediment criteria.

As discussed by Lee and Jones (1992), the use of total organic carbons and/or sulfides as parameters for normalization of chemical concentrations can both under- and overestimate the toxicity that a sediment can have to aquatic life. The studies by Jones and Lee (1978) have provided a database on the chemical characteristics of sediments from many different U.S. waterways. A review of this database shows that typically there are sufficient sulfides in the sediments to precipitate all the noniron heavy metals and there is sufficient organic matter in the sediments to bind all the nonpolar organic chemicals so that neither of these groups of chemicals should be toxic. However, there is also sufficient ammonia in many sediments to cause toxicity to aquatic life. It is very clear that the EPA's current approach of trying to develop sediment quality criteria based only on acid volatile sulfides and organic matter can lead to highly erroneous conclusions on the potential impacts of the contaminants in the sediments to aquatic life. For many sediments it will underestimate the detoxification of sediment constituents for heavy metals and nonpolar organic chemicals. Further, it does not consider the role of ammonia as a toxicant in the sediments to aquatic life. It is clear that the only reliable way to proceed is through an effects-based testing program using sediment bioassays to directly assess potential toxicity.

It is concluded that numeric sediment criteria of the type that the EPA is developing, including the equilibrium partitioning and acid volatile sulfide approaches, have limited reliable applicability to assessing the real impact that a dredging or dredged sediment disposal will have on the beneficial uses of the waters in the vicinity of where the dredging and disposal activities take place.

Empirical Sediment Criteria

At several locations in the United States, especially in the Puget Sound, Washington, area, regulatory agencies have been attempting for a number of years to use empirically based sediment criteria to regulate dredging and dredged sediment disposal activities. It is likely that the EPA's current efforts in developing sediment quality criteria will lead to attempts to use these criteria to regulate dredging programs. Such an effort could be of great significance to many dredging projects for waterways near urban-industrial centers where, because of the use of inappropriately based sediment quality criteria, the potential impacts of contaminants associated with dredged sediments will be overestimated, causing alternative, usually more expensive, methods of dredged sediment management to be adopted. Wright has discussed the problems with attempting to use numeric chemical criteria for regulating dredging projects. His review on this topic (see next chapter section) should be consulted for further information on it.

Some of the approaches being used to attempt to judge sediment quality and therefore regulate dredging projects include the Triad and the Apparent Effects Threshold (AET) approaches. These approaches are similar in that they make use of a combination of bulk sediment chemical characteristics, benthic organism community structure (numbers and types of organisms present), and sediment bioassays to develop numeric values which are purported to be an index of sediment quality. Basically, both of these approaches are formulated on technically invalid assessments of the potential impacts of sediment associated chemical contaminants on aquatic organisms. They make use of bulk chemical analyses as a foundation for assigning a numeric value to a sediment. Further, the AET numeric values used in the Puget Sound, Washington, area are based on the use of the Microtox testing of sediment extracts using the effect of the extracts on photoluminescent bacteria as a measure of toxicity to aquatic life. This procedure has been known to be unreliable for evaluating aquatic life toxicity for many different aquatic systems. It is highly inappropriate to assume, as has been done in the AET approach, that the Microtox measured toxicity is in any way related to the toxicity that would be present for aquatic organisms of interest in fresh or marine systems.

Another problem with the AET approach is that it is assumed that the numbers and types of organisms present in a sediment are controlled by the contaminants present. It is well known that a variety of factors, such as grain size and organic matter content, can significantly affect the numbers and types of organisms present in a sediment.

The AET and Triad numeric values which are supposed to reflect sediment quality do not provide a proper evaluation of the potential significance of contaminants in sediments in affecting water quality and aquatic life resources at a dredging or dredged sediment disposal site. Therefore, these approaches have little technical validity in regulating the impacts that chemical contaminants associated with sediments can have on the water quality and other beneficial uses of the waters of the area where the sediments are found or are deposited in a dredging project.

The State of California Water Resources Control Board staff have recently proposed to develop sediment quality criteria based on the geometric mean of the AET, sediment spiking toxicity test, and equilibrium partitioning approach. This numeric value is to be multiplied by some yet to be determined "uncertainty factor" to develop an overall numeric value that would be used to characterize the so-called sediment quality. As discussed by Lee and Jones (1991), this is not a technically valid approach for properly characterizing the water quality significance of contaminants in sediments. It can readily lead to a misclassification of sediment quality since it has in it a number of components which are known to be invalid approaches, such as bulk sediment chemical analyses, spiked sediment bioassays, etc., which are not related to the impact of the contaminants in the sediments on the beneficial uses of the waters in the vicinity of the sediments. Lee and Jones recommend rather than using a highly empirical approach for assessing sediment quality that the State of California should adopt an effects-based assessment of the potential water quality significance of contaminants in sediments using the sediment toxicity test, aquatic organism assemblages, and bioaccumulation of contaminants from the sediments.

Another approach that is being investigated for possible use to judge sediment quality based on chemical analyses is through the use of the chemical characteristics of sediment pore (interstitial) waters. The aquatic chemistry of interstitial pore water is significantly different from the water from which benthic aquatic organisms typically obtain exposure to chemical contaminants. Almost all interstitial water associated with chemical contaminants is anoxic (i.e., oxygen-free) and therefore is not a suitable environment for aquatic organisms which depend on oxygen for respiratory purposes. Further, many of the aquatic organisms that burrow into the sediments develop protective tubes which tend to isolate them from the sediments and interstitial water-associated contaminants.

There is no relationship between the concentration of interstitial water-associated contaminants and the concentrations that would be present at the sediment-water interface. Typically, the concentrations in interstitial waters are much higher than that present at the sediment-water interface, where most organisms would be exposed to the contaminants of potential concern. This situation is the result of several factors, one of which is dilution that occurs at the sediment-water interface. Another factor is that anoxic interstitial pore waters typically have elevated concentrations of iron in the ferrous form which upon contact with oxygen in the overlying waters is oxidized to ferric iron and precipitates as ferric hydroxide. The ferric hydroxide is a known, highly efficient scavenger for most contaminants, making them less available if not totally unavailable for organism uptake.

CONCLUSIONS

U.S. waterway sediments, especially near urban-industrial centers, were found in the late 1960s to contain large amounts of chemical contaminants. If these contaminants are released in the waterway or during dredging and/or disposal, they could have significant adverse effects on the beneficial uses of the water in the waterway, near open-water disposal sites, as well as in areas where confined disposal overflow occurs. Studies conducted during the 1970s under the auspices of the CE Dredged Material Research Program reaffirmed that bulk sediment criteria based on total contaminant concentration are not technically valid for judging the toxicity of sediment-associated contaminants to aquatic life. Elutriate tests and dredged sediment bioassay-toxicity tests provide an approach for evaluating the potential water quality significance of contaminants associated with dredged sediments. Significant problems have developed with the use of these procedures in making dredged sediment disposal management decisions because of problems in the interpretation of the results of the tests. A hazard assessment approach, such as described in this chapter, should be used for this purpose, in which the hazard that the contaminants present in the dredged sediment represent to the beneficial uses of the water at the disposal site is evaluated.

There is a need to update the EPA regulations governing dredged sediment disposal in fresh and marine water systems to properly consider the vast amount of information that was generated in the 1970s and 1980s on the water quality significance of contaminants associated with sediments.

ACKNOWLEDGMENTS

Much of the authors' work upon which this chapter is based was supported by the Corps of Engineers Dredged Material Research Program, as well as the New York, Norfolk, and Rock Island Districts of the Corps of Engineers. The assistance of a number of the senior author's former students and associates, P. Bandyopadhyay, J. Butler, D. Homer, R. Jones, J. Lopez, G. Mariani, C. McDonald, M. Nicar, M. Piwoni, R. Plumb, and F. Saleh, in data collection and report preparation is acknowledged. The authors wish to acknowledge the significant assistance provided to them over the years by personnel at the U.S. Army Corps of Engineers Waterways Experiment Station in Vicksburg, Mississippi, especially the assistance of Dr. R. Engler and Dr. T. Wright.

REFERENCES

Boyd, M. B., Saucier, P. T., Keeley, J. W., Montgomery, R. L., Brown, R. D., Mathis, D. B., and Guice, C. J., *Disposal of Dredge Spoil. Problems Identification and Assessment and Research Program Development,* Technical Report H-72-8, U.S. Army Corps of Engineers Waterways Experiment Station, Vicksburg, MS, 1972.

Di Toro, D. M., Mahony, J. D., Hansen, D. J., Scott, K. J., Hicks, M. B., Mayr, S. M., and Redmond, M. S., "Toxicity of Cadmium in Sediments: The Role of Acid Volatile Sulfide," *Environmental Toxicology and Chemistry,* 9:1487–1502, 1990.

Francingues, N. R., Palermo, M. R., Lee, C. R., and Peddicord, R. K., *Management Strategy for Disposal of Dredged Material: Contaminant Testing and Controls,* Miscellaneous Paper D-85-1, Department of the Army, Waterways Experiment Station, Corps of Engineers, Vicksburg, MS, Aug. 1985.

Jones, R. A., and Lee, G. F., *Evaluation of the Elutriate Test as a Method of Predicting Contaminant Release during Open Water Disposal of Dredged Sediment and Environmental Impact of Open Water Dredged Material Disposal, Vol. I: Discussion,* Technical Report D-78-45, U.S. Army Engineer Waterways Experiment Station, Vicksburg, MS, 1978.

——— and ———, "Toxicity of U.S. Waterway Sediments with Particular Reference to the New York Harbor Area," In: *Chemical and Biological Characterization of Sludges,*

Sediments, Dredge Spoils and Drilling Muds, ASTM STP 976, American Society for Testing and Materials, Philadelphia, PA, pp. 403–417, 1988.

Kay, S. H., *Potential for Biomagnification of Contaminants within Marine and Freshwater Food Webs,* Technical Report D-84-7, U.S. Army Waterways Experiment Station, Vicksburg, MS, 1984.

Lee, G. F., *Factors Affecting the Transfer of Materials between Water and Sediments,* University of Wisconsin Eutrophication Information Program, Literature Review no. 1, 50 pp., 1970.

——, "Role of Hydrous Metal Oxides in the Transport of Heavy Metals in the Environment," *Proc., Symposium on Transport of Heavy Metals in the Environment, Progress in Water Technology,* 17:137–147, 1975.

—— and Jones, R. A., *An Assessment of the Environmental Significance of Chemical Contaminants Present in Dredged Sediments Dumped in the New York Bight,* Occasional Paper no. 28, Department of Civil and Environmental Engineering, New Jersey Institute of Technology, Newark, NJ, 1977.

—— and ——, *A Hazard Assessment Approach for Assessing the Environmental Impact of Dredging and Dredged Sediment Disposal for the Upper Mississippi River,* Report to Rock Island District, Corps of Engineers, Occasional Paper no. 68, Department of Civil and Environmental Engineering, New Jersey Institute of Technology, Newark, NJ, 1981a.

—— and ——, "Application of Hazard Assessment Approach for Evaluation of Potential Environmental Significance of Contaminants Present in North Landing River Sediments upon Open Water Disposal of Dredged Sediment," *Proc., Old Dominion University/Norfolk District Corps of Engineers Symposium, Dredging Technology: A Vital Role in Port Development,* Aug. 1981b.

—— and ——, comments on Draft document, "Workplan for the Development of Sediment Quality Objectives for Enclosed Bays and Estuaries of California," submitted to D. Maughan, Chairman Water Resources Control Board, Sacramento, CA, May 1991.

—— and ——, "Water Quality Significance of Contaminants Associated with Sediments: An Overview," In: *Fate and Effects of Sediment-Bound Chemicals in Aquatic Sediments,* Pergamon Press, Elmsford, NY, pp. 1–34, 1987.

——, ——, and Newbry, B. W., "Water Quality Standards and Water Quality," *J. Water Pollut. Control Fed.* 54:1131–1138, 1982a.

——, ——, and ——, "Alternative Approach to Assessing Water Quality Impact of Wastewater Effluents," *J. Water Pollut. Control Fed.* 54:165–174, 1982b.

—— and ——, "Sediment Quality Criteria Development: Problems with Current Approaches," lecture notes, *1992 National R&D Conference on the Control of Hazardous Materials,* Hazardous Materials Control Research Institutes, Greenbelt, MD, 1992.

——, ——, Saleh, F. Y., Mariani, G. M., Homer, D. H., Butler, J. S., and Bandyopadhyay, P., *Evaluation of the Elutriate Test as a Method of Predicting Contaminant Release during Open Water Disposal of Dredged Sediment and Environmental Impact of Open Water Dredged Materials Disposal, Vol. II: Data Report,* Technical Report D-78-45, U.S. Army Engineers Waterways Experiment Station, Vicksburg, MS, 1978.

—— and Plumb, R. H., *Literature Review on Research Study for the Development of Dredged Material Disposal Criteria,* U.S. Army Corps of Engineers Dredged Material Research Program, Vicksburg, MS, 1974.

Palermo, M. R., *Development of a Modified Elutriate Test for Estimating the Quality of Effluent from Confined Dredged Material Disposal Areas,* Technical Report D-86-4, Department of the Army, U.S. Army Corps of Engineers, Washington, DC, Aug. 1986.

Peltier, W., and Weber, C. I. (eds.), *Methods for Measuring the Acute Toxicity of Effluents to Freshwater and Marine Organisms,* Environmental Monitoring and Support Laboratory, 3d ed., EPA/600/4-85/013, U.S. EPA, Cincinnati, OH, 230 pp., 1985.

U.S. Environmental Protection Agency (U.S. EPA) and U.S. Army Corps of Engineers (CE), *Ecological Evaluation of Proposed Discharge of Dredged Materials into Ocean Wa-*

ters, Implementation Manual for Section 103 of PL 92-532, U.S. Army Engineer Waterways Experiment Station, Vicksburg, MS, July 1977.

U.S. EPA, "Notice of Quality Criteria Documents: Availability," *Federal Register,* 45, 79318, Nov. 1980.

————, *Quality Criteria for Water,* EPA-440/9-76-023, U.S. EPA, Washington, DC, July 1976.

————, *Quality Criteria for Water 1986,* EPA 440/5-86-001, Office of Water, Washington, D.C., May 1987.

———— and U.S. Army Corps of Engineers, *Evaluation of Dredged Material Proposed for Ocean Disposal Testing Manual,* U.S. EPA Office of Water (WH-556F), EPA-503-8-91/001, Feb. 1991.

Weber, C. I., Horning, W. B., II, Klemm, D. J., Neiheisel, T. W., Lewis, P. A., Robinson, E. L., Menkedick, J., and Kessler, F. (eds.), *Short-Term Methods for Estimating the Chronic Toxicity of Effluents and Receiving Waters to Marine and Estuarine Organisms,* EPA-600/4-87/028, U.S. EPA, Cincinnati, OH, May 1988.

EVALUATION OF DREDGED MATERIAL FOR OPEN-WATER DISPOSAL: NUMERICAL CRITERIA OR EFFECTS-BASED?*

Thomas D. Wright

Environmental Laboratory
U.S. Army Engineer Waterways Experiment Station
Vicksburg, Miss.

INTRODUCTION

Approximately 500 million cubic yards of material are dredged each year from navigable waterways. Where open-water disposal is proposed for the material, the Corps of Engineers (CE) evaluates the material for suitability under the Clean Water Act (CWA, P.L. 92-500, as amended) or the Marine Protection, Research, and Sanctuaries Act (MPRSA, P.L. 92-532, as amended). If the material does not meet the CWA guidelines or the MPRSA criteria the CE cannot dispose of the material in open water nor will it issue a permit for a private applicant to utilize such disposal. The CWA guidelines and MPRSA criteria are promulgated by the EPA and it exercises oversight on CE decisions regarding disposal. Further, CWA disposal requires state certification that it will not violate state water quality standards.

The CWA guidelines (40 CFR, Part 230) for the evaluation of dredged material were issued in 1975. These guidelines allow a comparison of contaminants in the dredged material with those at the disposal site and allow open-water disposal

---

*References for this chapter section are at the end of the section.

where contaminants at the two sites are "substantially similar" or where it can be shown that unacceptable concentrations of contaminants will not be transported beyond the boundaries of the disposal site. In addition, the guidelines provide that where there is such a large number of contaminants as to preclude identification of all of them by chemical analyses, or where chemical-biological interactive effects may occur, bioassays may be used in lieu of chemical tests. In response to these guidelines, the CE issued an implementation manual (CE, 1976) which described the bioassay procedures. This manual is currently being revised.

The MPRSA criteria (40 CFR Parts 220–228) for the evaluation of dredged material were issued in 1977. These criteria are clearly effects-based. At 40 CFR 227.6 certain constituents (organohalogen compounds, mercury and mercury compounds, cadmium and cadmium compounds, and oil of any kind or in any form) are prohibited from disposal other than as "trace contaminants." No numerical limits are given for these contaminants. Rather, the results of biological tests are to be used to determine whether or not the prohibited constituents are present in greater than trace amounts. In response to the 1977 criteria, the EPA and the CE issued a joint implementation manual (EPA/CE, 1977) which described the bioassay procedures. A revision of this manual was issued in 1991 (EPA/CE, 1991). In general, the revision focused on refinements of the 1977 procedures and retained the effects-based approach.

DISCUSSION

It is important to understand that dredged material is a highly complex substance and is not comparable to other materials, such as sewage sludge or industrial waste, which may be discharged into open water. Both the MPRSA and the CWA make this distinction and provide evaluatory procedures for dredged material that are different from those used for other materials. In the case of new projects, the excavated material is usually "virgin," that is, it is sediment which has been exposed to few, if any, anthropogenic contaminants. Material excavated as a maintenance operation may come from a variety of sources, such as littoral drift, riverine input, and sheet erosion adjacent to the project. Such material may have been contaminated at its source or may become contaminated during transport or deposition at the project. Because the initial source of the material is soil or existing sediments, it will contain all of the elements in the periodic table as well as both natural and anthropogenic compounds. Insofar as many of these are classified as "contaminants," virtually all dredged material could be considered to be contaminated. In actual practice, the mere presence of a contaminant or its concentration in dredged material can rarely be used to predict whether or not it will have adverse effects upon biota (Engler, 1980), and the effects-based approach described below appears to be environmentally conservative (Jones and Lee, 1988; Lee and Jones, 1987).

The bioavailability of contaminants in sediments, including dredged material, is governed by a variety of factors. In the case of metals, sulfides appear to be a major controlling factor whereas total organic carbon is involved in the bioavailability of nonpolar organic contaminants, such as PCBs, PAHs, chlorinated hydrocarbon pesticides, and dioxins (McFarland and Clarke, 1987). Clay minerals, such as the kaolinites, smectites, and the hydrous micas, humic and fulvic acids, the cation exchange capacity of the sediment, pH, Eh, and organic complexes also play a role in determining bioavailability (Pequegnat, Gallaway, and Wright, 1990). In addition, synergistic and antagonistic interactions between and among contaminants are frequent occurrences.

All of these factors and interactions take place in a highly complex and poorly understood manner. This is not to say that limited predictive capabilities do not exist. Di Toro et al. (1989) have suggested a method, based upon the relationship between acid-volatile sulfides and metals, which may be of utility to predict the toxicity of metals such as nickel, zinc, cadmium, lead, copper, and mercury, either singly or in combination. McFarland and Clarke (1987) have utilized the relationship between total organic carbon, nonpolar organic contaminants, and lipids in organisms to predict the potential bioaccumulation of these compounds from sediments. Wright (1977), Kraft (1979), and Malueg et al. (1984) demonstrated a clear relationship between toxicity, benthic community structure, and contaminant concentration in an ecosystem dominated by a *single sediment contaminant* (copper from mine tailings) where other contaminant contributions or sources were either absent or minimal.

With the above exceptions, attempts to establish cause-and-effect relationships between the concentration of a particular contaminant and a biological effect in sediments have proved futile other than under laboratory situations where a "clean" sediment was "dosed" with a single contaminant. Even in those cases, when the "dose" was delivered to different sediments, the relationship in one sediment did not hold true for other sediments. Results from regulatory testing of sediments proposed for open-water disposal and broad field studies during the past decade which have yielded vast databases, such as the Status and Trends Program, have failed to demonstrate clear relationships between sediment contaminants and biological effects (O'Connor, 1990).

Despite the lack of cause-and-effect relationships, sediment quality criteria have been developed and applied. Among the first were the so-called Jensen criteria promulgated by the Federal Water Quality Administration (predecessor of EPA) in 1971 for use in the Great Lakes. These appear to have had little, if any, technical validity and, in some cases, the criteria were well below the average crustal abundance for several contaminants (Engler, 1980) and did not take into account natural background concentrations (Wright, 1974). More recently, criteria were developed for use in Puget Sound (CE/Washington State, 1988). These were developed using an approach known as the apparent effects threshold. Although originally applied to exclude or allow open-water disposal (sediments which were not clearly excluded or allowed would be biologically tested to determine their status for disposal), the current use of these criteria is as a screening tool. In essence, the criteria provide a "trigger" to conduct biological tests, and decisions on disposal of the material are made on the basis of the biological tests rather than the criteria. Reviews of the various approaches used to derive sediment quality are found in Brannon et al. (1990) and Marcus (1991).

In the development of sediment quality criteria it is important that the criteria take into account the activity to which they will be applied. In the case of navigation dredging, it is a given that the material will be removed, and the question to be addressed concerns potential contaminant effects at the disposal site. For remediation, dredging concerns are the effects of in-place sediments, the benefits of removal, and potential effects at the disposal site. Several of the approaches proposed for the development of criteria, specifically the apparent effects threshold (PTI, 1988) and the sediment quality triad (Chapman, 1986, 1989) have failed to make this distinction. The threshold and the triad incorporate benthic community structure at the excavation site as a component, thereby raising serious questions regarding their applicability to navigation dredging. The benthic community structure at the excavation site is not a particularly useful indicator of sediment effects since the community is subject to a variety of influences other than the sediment. These include dredging, navigation traffic, degradation of water quality from outfalls, thermal discharges, surface runoff, the effects of droughts and floods, and other perturbations. The threshold and the triad may be useful tools in evaluating the overall health of an aquatic environment but should not be used in the determination of the suitability of dredged material for open-water disposal. Unfortunately, this seems to have been overlooked in a recent controversy over the applicability of the threshold and triad (Spies, 1989; Chapman et al., 1991).

In contrast to the current situation with sediment quality criteria, water quality criteria have been available for many years. The development of water quality criteria is quite straightforward. Organisms are exposed to a range of concentrations of a particular contaminant and the observed effects are used to establish criteria for a given level of protection. In the experimental procedure it is ensured that the contaminant of interest is bioavailable, and this is accomplished by ensuring that it is in solution so that cause-and-effect relationships can be established. These criteria do not take into account synergistic or antagonistic effects. As discussed above, many factors control the bioavailability of contaminants in sediments, and attempts to develop sediment quality criteria following an approach similar to that used to develop water quality criteria have been unsuccessful.

There is an inherent desire on the part of regulatory agencies to use, whenever possible, numerical criteria in their regulatory activities. Clearly, this greatly simplifies decision-making because an activity is either in compliance or is it not. Numerical criteria are also easily understood by the general public, environmental groups, the regulated community, the courts, and others, whereas more subjective approaches tend not to be. The current approach used in determining the suitability of dredged material for open-water disposal uses biological toxicity and/or bioaccumulation as criteria, but the endpoints are not absolute values. Rather, the potential for effects is measured by comparing the response of the organism in dredged material to its response to sediment from a reference area (MPRSA) or the disposal site (CWA). This holistic method does not distinguish which contaminant or combination of contaminants is responsible for an observed effect. It does, however, take into account possible synergistic or antagonistic effects and is a direct measure of the bioavailability of all of the contaminants present (Wright and Saunders, 1990).

It is important to understand the intended purpose of criteria. In regulatory usage, there are three levels of protection: objectives, criteria, and standards. *Objectives* are aims or goals toward which to strive and which may or may not represent an ideal condition. They are frequently very broad statements which are not legally enforceable nor are they intended to be. They may or may not have any technical or scientific basis and tend to be somewhat philosophical in nature. *Criteria* are means by which something is evaluated in forming a correct judgment about it. They are developed through the application of widely accepted technical and scientific procedures. They are not intended to be legally enforceable because they must be sufficiently broad to encompass a variety of circumstances and, hence, require a certain degree of interpretation in their application. *Standards* are usually (but not always) developed from criteria but may be entirely arbitrary, especially when the technical and scientific basis of the criteria are thought to be incomplete and where a safety factor is deemed necessary. Standards are usually much more restrictive and narrower than criteria and are legally enforceable.

In the water quality arena, the establishment of standards has generally been delegated to the states, with the federal government providing criteria upon which to base the standards. This is reasonable because criteria cannot take into account all local considerations and circumstances. Under the MPRSA, the discharge of dredged material must comply with federal water quality criteria whereas under the CWA compliance with state water quality standards is required.

The desire for numerical limits, be they criteria or standards, is not only found in regulatory agencies. Senate Bills S.1178 and S. 1179, introduced but not passed by the 101st Congress, would require the establishment of federal sediment quality criteria for a large number of contaminants. These criteria would apply to all contaminated sediments. As noted above, concerns with contaminated sediments outside of navigation projects are quite different from those which must be dredged to establish or maintain navigation. The criteria would be established over a period of several years by the EPA and the various coastal states would be required to establish sediment quality standards based upon the criteria. Failing action by a state, the EPA would establish sediment quality standards for that state. This political solution to the perceived problem that the current procedures do not provide adequate environmental protection may create a set of problems of its own.

Under current procedures water quality criteria (MPRSA) or standards (CWA) must be met. Further, unless it can be shown that water quality criteria or standards exist for all contaminants of concern and that synergistic effects will not occur, *biological testing is mandatory.* This creates a potential situation under the MPRSA where one or more criteria might not be met but that the subsequent biological testing indicates no potential effects. This does not indicate that there is a defect in the criteria but, rather, that criteria are general measures and are not to be construed as standards. Such cases will be considered on a case-by-case basis by the CE as the permitting authority and by the EPA under its oversight authority. Under the CWA, if the proposed discharge will not meet state water quality standards, the state may decline to issue the Section 401 certification or may waive the standards which are not met. For sediment, inasmuch as there are no standards or criteria (except for 50 ppm PCBs under TSCA, the origin and technical rationale for which is not known), the determination of suitability for open-water disposal is evaluated on the basis of biological testing (although comparison of the contaminants at the extraction and disposal site may also be used under the CWA).

At best, because of the underlying technical deficiencies in many of the approaches being used to develop sediment quality criteria, such criteria should be used only as a screening tool and should absolutely not be arbitrarily converted into standards. An analogy can be drawn with water quality criteria or standards. If a sediment does not meet the numerical criteria or standards, open-water disposal could be prohibited. At the present time, it is not known whether or not this would be considered on a case-by-case basis when material not meeting numerical criteria or standards meets the biological criteria. At the time that the governing regulations (MPRSA and CWA) were promulgated there were no sediment quality criteria or standards. Hence, the regulations are silent on this, but a consistent approach would be to treat sediment criteria or standards in a manner similar to water quality criteria or standards. Whether or not this will be done remains to be seen.

The fact that for many years federal, state, and academic agencies have strived to develop sediment quality criteria at great public cost and have succeeded in developing none should convey a clear message. There are hundreds, if not thousands, of potential contaminants in sediments, many of which are biologically innocuous despite their concentration whereas others may be biologically active at concentrations which cannot be measured with current analytical chemistry techniques other than those found in sophisticated research and development facilities. In contrast, the current effects-based procedures have been in use since the mid-1970s and evidence of their effectiveness in environmental protection is provided by the observation that in spite of intensive monitoring of disposal sites, there is no documentation of adverse effects as the result of materials which were evaluated under these procedures. Further, the contaminants listed in S.1179 for which criteria are to be established number at least 500, with the criteria to be established within 3 yr of enactment of the bill. This would require the development of criteria at the rate or over three per week and, after development, the states would have only 2 yr within which to convert these criteria into standards. Considering that years of effort have yielded no criteria, this is a truly stupendous task.

Between 1973 and 1978 the CE conducted a major $33 million program on dredged material disposal. This program consisted of over 250 individual studies and, in contrast to previous largely site-specific project investigations, the studies were generic in nature so as to have the widest applicability. A specific goal was to define the biological and water quality effects of open-water, wetland, and upland disposal. A major finding was that no single disposal option is presumptively suitable for a geographic region or group of projects. What may be desirable for one project may be completely unsuitable for another; consequently, each project must be evaluated on a case-by-case basis (Saucier et al., 1978). An additional finding was that open-water disposal resulted only in physical, rather than contaminant, effects on biota at the disposal site and that biotal recovery was rapid following the cessation of disposal (Wright, 1978).

A further effort was initiated as a cooperative program between the CE and the EPA. This $7 million program was designed to compare new evaluatory techniques with those in use and to investigate the effects of the disposal of material from a single site in three different environments (open-water, wetland, and upland). Of the various new biological techniques examined to determine the suitability of material for open-water disposal, only a few showed significant potential as evaluatory tools and these were not suitable for regulatory application without additional research and development. None appeared to predict the effects of open-water disposal better than the acute toxicity and bioaccumulation techniques which are still in use; field investigations following the laboratory tests verified the predictive ability of the tests (Gentile et al., 1988). Upland disposal produced the greatest and most persistent effects, including the release of metals and extreme toxicity whereas open-water disposal showed relatively minor and nonpersistent effects; effects from wetland disposal were intermediate between upland and open-water disposal (Peddicord, 1988).

Based upon these and other studies, it appears that numerical sediment quality criteria or standards may be environmentally underprotective in that they cannot take into account the many and complex interactions in sediment which control contaminant bioavailability and, hence, potential biological effects. Alternately, they may be overprotective because there is currently no widely accepted method with which to establish the cause-and-effect relationship which is crucial to the development of technically valid regulatory criteria or standards for sediments. Because of the imprecision of the methods which are presently being investigated, a conservative "safety factor" will almost surely need to be incorporated into whatever criteria or standards are developed.

Overprotection will have significant economic and environmental impacts. In those instances where open-water disposal is the only available alternative, navigational dredging will not be allowed. Where there are other alternatives, such as upland or wetland disposal, these may be more costly because of transportation costs and other factors, such as the construction of facilities to contain the material. Wetland, and especially upland, disposal has been shown to frequently have environmental effects that are more severe than those commonly associated with open-water disposal. Following disposal in upland or wetland sites, major costs may be subsequently incurred in management of the material and maintenance of the sites which, in some cases, could even require removal of the material and transport to a more suitable site.

This is not to say that all material is suitable for open-water disposal and should be so disposed. Some may be unsuitable for physical reasons, some for economic reasons, and some, although suitable, can be better used for beneficial purposes which do not involve open-water disposal. For a given project, all disposal alternatives should be thoroughly explored prior to any disposal. Where open-water disposal is a serious consideration, the most common reason for abandoning that consideration concerns contaminants. The current effects-based approach has been shown to be environmentally protective and is technically sound. Even with the development of numerical sediment quality criteria or standards, the effects-based approach should remain the primary determinant in evaluating the potential effects of contaminants in dredged material for open-water disposal.

CONCLUSION

Current regulations require an effects-based approach for the evaluation of contaminants in dredged material that is proposed for open-water disposal. Experience with this approach over the past 15 yr by the CE and the EPA has shown it to be environmentally protective. Although numerical sediment quality criteria have been under development for many years, they still do not exist and technically valid criteria cannot be promulgated by legislative mandate. The primary reasons for their nonexistence are the nature of dredged material and technical flaws in the techniques that have been used in attempts to develop them. If, on whatever basis they are developed, sediment quality criteria supplant the current approach, they may be environmentally over- or underprotective. This will carry not only environmental costs but direct economic costs through a loss of navigation by making dredging unfeasible or where disposal costs increase through the unnecessary use of more expensive alternatives than open-water disposal.

ACKNOWLEDGMENTS

This section summarizes investigations conducted under the Dredged Material Research Program, Long-Term Effects of Dredging Program, Field Verification Program, Dredging Operations Technical Support Program, and field reimbursable work funded by the U.S. Army Corps of Engineers. Permission to publish this material was granted by the Chief of Engineers.

REFERENCES

Brannon, J. M., McFarland, V. A., Wright, T. D., and Engler, R. M., 1990. "Utility of Sediment Quality Criteria (SQC) for the Environmental Assessment and Evaluation of Dredging and Disposal of Contaminated Sediments." *Coastal and Inland Water Quality Seminar Proceedings no. 22,* U.S. Army Corps of Engineers Committee on Water Quality, Washington, DC, pp. 7–19.

CE, 1976. *Ecological Evaluation of Proposed Discharge of Dredged or Fill Material into Navigable Waters: Interim Guidance for Implementation of Section 404(b)(1) of Public Law 92-500 (Federal Water Pollution Control Act Amendments of 1972).* Miscellaneous Paper D-76-17, U.S. Army Engineer Waterways Experiment Station, Vicksburg, MS.

CE/State of Washington Dept. of Natural Resources. 1988. *Final Environmental Impact Statement—Unconfined Open-Water Disposal Sites for Dredged Material, Phase 1 (Central Puget Sound).* U.S. Army Engineer District, Seattle, WA.

Chapman, P. M., Long, E. R., Swartz, R. C., DeWitt, T. H., and Pastorok, R., 1991. "Sediment Toxicity Tests, Sediment Chemistry and Benthic Ecology *Do* Provide New Insights into the Significance and Management of Contaminated Sediments—A Reply to Robert Spies." *Environ. Toxicol. Chem.,* vol. 10, pp. 1–4.

———, 1989. "Current Approaches to Developing Sediment Quality Criteria." *Environ. Toxicol. Chem.,* vol. 8, pp. 589–599.

———, 1986. "Sediment Quality from the Sediment Quality Triad—An Example." *Environ. Toxicol. Chem.,* vol. 5, pp. 957–964.

Di Toro, D. M., Mahony, J. D., Hansen, D. J., Scott, K. J., Hinks, M. B., Mayr, S. M., and Redmond, M. S., 1989. "Toxicity of Cadmium in Sediments: The Role of Acid Volatile Sulfide." *Environ. Toxicol. Chem.*, vol. 9, pp. 1487–1502.

Engler, R. M., 1980. "Prediction of Pollution Potential Through Geochemical and Biological Procedures: Development of Regulation Guidelines and Criteria for the Discharge of Dredged and Fill Material." *Contaminants and Sediments,* vol. 1, R. A. Baker (ed.), Ann Arbor Science Publishers, Inc., Ann Arbor, MI, pp. 143–169.

EPA/CE, 1977. *Ecological Evaluation of Proposed Discharge of Dredged Material into Ocean Waters: Implementation Manual for Section 103 of Public Law 92-532 (Marine Protection, Research, and Sanctuaries Act of 1972).* U.S. Army Engineer Waterways Experiment Station, Vicksburg, MS.

———, 1991. *Evaluation of Dredged Material Proposed for Ocean Disposal (Testing Manual).* U.S. Army Engineer Waterways Experiment Station, Vicksburg, MS.

Gentile, J. H., Pesch, G. G., Lake, J., Yevich, P. P., Zaroogian, G., Rogerson, P., Paul, J., Galloway, W., Scott, K., Nelson, W., Johns, D., and Munns, W., 1988. *Synthesis of Research Results: Applicability and Field Verification of Predictive Methodologies for Aquatic Dredged Material Disposal.* Technical Report D-88-5, U.S. Army Engineer Waterways Experiment Station, Vicksburg, MS.

Jones, R. A., and Lee, G. F., 1988. "Toxicity of U.S. Waterways with Particular Reference to the New York Harbor Area." *Chemical and Biological Characterization of Sludges, Sediments, Dredge Spoils, and Drilling Muds.* ASTM STP 976, J. J. Lichtenberg, F. A. Winter, C. I. Weber, and L. Franklin (eds.), Amer. Soc. for Test. and Mat., Philadelphia, PA. pp. 403–417.

Kraft, K. J., 1979. "*Pontoporeia* Distribution Along the Keweenaw Shore of Lake Superior Affected by Copper Tailings." *Internat. Assoc. Great Lakes Res.,* vol. 5(1), pp. 28–35.

Lee, G. F., and Jones, R. A., 1987. "Water Quality Significance of Contaminants Associated with Sediments: An Overview." *Fate and Effects of Sediment-Bound Chemicals in Aquatic Systems.* Pergamon Press, New York, pp. 3–34.

McFarland, V. A., and Clarke, J. U., 1987. *Simplified Approach for Evaluating Bioavailability of Neutral Organic Chemicals in Sediment.* Environmental Effects of Dredging Tech. Note EEDP-01-08, U.S. Army Engineer Waterways Experiment Station, Vicksburg, MS.

Malueg, K. W., Schuytema, G. S., Krawczyk, D. F., and Gakstatter, J. H., 1984. "Laboratory Sediment Toxicity Tests, Sediment Chemistry and Distribution of Benthic Macroinvertebrates in Sediments from the Keweenaw Waterway, Michigan." *Environ. Toxicol. Chem.,* vol. 3, pp. 233–242.

Marcus, W. A., 1991. "Managing Contaminated Sediments in Aquatic Environments: Identification, Regulation, and Remediation." *Env. Law Reporter,* 1-91, pp. 10020–10032.

Peddicord, R. K., 1988. *Summary of the U.S. Army Corps of Engineers/U.S. Environmental Protection Agency Field Verification Program.* Technical Report D-88-6, U.S. Army Engineer Waterways Experiment Station, Vicksburg, MS.

Pequegnat, W. E., Gallaway, B. J., and Wright, T. D., 1990. *Revised Procedural Guide for Designation Surveys of Ocean Dredged Material Sites.* Technical Report D-90-8, U.S. Army Engineer Waterways Experiment Station, Vicksburg, MS.

PTI, 1988. *The Apparent Effects Threshold.* Briefing Report to the U.S. Environmental Protection Agency Science Advisory Board. PTI Environmental Services, Bellevue, WA.

O'Connor, T. P., 1990. *Coastal Environmental Quality in the United States, 1990, Chemical Contamination in Sediment and Tissues.* National Oceanic and Atmospheric Administration, Rockville, MD.

S.1178, 1989. *A Bill to Improve and Expand Programs for the Protection of Marine and Coastal Waters.* 101st Congress, First Session.

S.1179, 1989. *A Bill to Establish a Comprehensive Marine Pollution Restoration Program, to Amend the Federal Water Pollution Control Act and the Marine Protection, Research and Sanctuaries Act, and for Other Purposes.* 101st Congress, First Session.

Saucier, R. T., Calhoun, C. C., Engler, R. M., Patin, T. P., and Smith, H. K., 1978. *Executive Overview and Detailed Summary.* Technical Report DS-78-22, U.S. Army Engineer Waterways Experiment Station, Vicksburg, MS.

Spies, R. B., 1989. "Sediment Bioassays, Chemical Contaminants and Benthic Ecology: New Insights or Just Muddy Water?" *Mar. Env. Res.,* vol. 27, pp. 73–75.

Wright, T. D., and Saunders, L. H., 1990. "U.S. Army Corps of Engineers Dredged Material Testing Procedures." *The Environmental Professional,* vol. 12, pp. 13–17.

————, 1978. *Aquatic Dredged Material Disposal Impacts: Synthesis Report.* Technical Report DS-78-1, U.S. Army Engineer Waterways Experiment Station, Vicksburg, MS.

————, 1977. *Study Completed on the Effects of Dredging the Keweenaw Waterway.* U.S. Army Corps of Engineers Dredged Material Research Information Exchange Bulletin D-77-2, U.S. Army Engineer Waterways Experiment Station, Vicksburg, MS, pp. 3–7.

————, 1974. "Is Dredge Spoil Confinement Always Justified?" *Great Lakes Basin Communicator,* vol. 4(12), pp. 5–8.

CONTAMINATED SEDIMENTS*

Dredging and disposal do not introduce new contaminants to the aquatic environment but simply redistribute the sediments which are the natural depository of contaminants introduced from other sources. The potential for accumulation of a metal in the tissues of an organism (bioaccumulation) may be affected by several factors such as duration of exposure, salinity, water hardness, exposure concentration, temperature, the chemical form of the metal, and the particular organism under study. The relative importance of these factors varies from metal to metal, but there is a trend toward greater uptake at lower salinities. Elevated concentrations of heavy metals in tissues of benthic invertebrates are not always indicative of high levels of metals in the ambient medium or associated sediments. Although a few instances of uptake of possible ecological significance have been

---

*This section was written by John B. Herbich. References for it are at the end of the chapter.

shown, the diversity of results among species, different metals, types of exposure, and salinity regimes strongly argues that bulk analysis of sediments for metal content cannot be used as a reliable index of metal availability and potential ecological impact of dredged material but only is an indicator of total metal content. Bioaccumulation of most metals from sediments is generally minor. Levels often vary from one sample period to another and are quantitatively marginal, usually being less than one order of magnitude greater than levels in the control organisms, even after 1 mo of exposure. Animals in undisturbed environments may naturally have high and fluctuating metal levels. Therefore, in order to evaluate bioaccumulation, comparisons should be made between control and experimental organisms at the same time.

Organochlorine compounds such as DDT, dieldrin, and polychlorinated biphenyls (PCBs) are environmental contaminants of worldwide significance which are artificial and, therefore, do not exist naturally in the earth's crust. Organochlorine compounds are generally not soluble in surface waters at concentrations higher than approximately 20 ppb, and most of the amount present in waterways is associated with either biological organisms or suspended solids. Organochlorine compounds are released from sediment until some equilibrium concentration is achieved between the aqueous and the solid phases and then reabsorbed by other suspended solids or biological organisms in the water column. The concentration of organochlorines in the water column is reduced to background levels within a matter of hours as the organochlorine compounds not taken up by aquatic organisms eventually settle with the particular matter and become incorporated into the bottom deposits in aquatic ecosystems. Most of these compounds are stable and may accumulate to relatively high concentrations in the sediments. The manufacture and/or disposal of most of these compounds is now severely limited; however, sediments that have already been contaminated with organochlorine compounds will probably continue to have elevated levels of these compounds for several decades. The low concentrations of chlorinated hydrocarbons in sediment interstitial water indicate that during dredging operations, the release of the interstitial water and contaminants to the surrounding environment would not create environmental problems. Bioaccumulation of chlorinated hydrocarbons from deposited sediments does occur. However, the sediments greatly reduce the bioavailability of these contaminants, and tissue concentrations may range from less than one to several times the sediment concentration. Unreasonable degradation of the aquatic environment caused by the routine maintenance dredging and disposal of sediment contaminated with chlorinated hydrocarbons has never been demonstrated.

The terms *oil* and *grease* are used collectively to describe all components of sediments of natural and contaminant origin which are primarily fat soluble. There is a broad variety of possible oil and grease components in sediment, the analytical quantification of which is dependent on the type of solvent and method used to extract these residues. Trace contaminants, such as PCBs and chlorinated hydrocarbons, often occur in the oil and grease. Large amounts of contaminant oil and grease find their way into the sediments of the nation's waterways either by spillage or as chronic inputs in municipal and industrial effluents, particularly near urban areas with major waste outfalls. The literature suggests long-term retention of oil and grease residues in sediments, with minor biodegradation occurring. Where oily residues of known toxicity become associated with sediments, these sediments retained toxic properties over periods of years, affecting local biota. Spilled oils are known to readily become adsorbed to naturally occurring suspended particulates, and oil residues in municipal and industrial effluents are commonly found adsorbed to particles. These particulates are deposited in sediments and are subject to suspension during disposal. Even so, there is only slight desorption, and the amount of oil released during the elutriate test is less than 0.01 percent of the sediment-associated hydrocarbons under worst-case conditions. Selected estuarine and freshwater organisms exposed for periods up to 30 days to dredged material that is contaminated with thousands of parts per million of oil and grease experience minor mortality. Uptake of hydrocarbons from heavily contaminated sediments appears minor when compared with the hydrocarbon content of the test sediments.

Ammonia is one of the potentially toxic materials known to be released from sediments during disposal; it is routinely found in evaluations of sediments using the elutriate test and in the water near a disposal area where concentrations rapidly return to baseline levels. Similar temporary increases in ammonia at marine, estuarine, and fresh water disposal sites have been documented in several DMRP field studies, but concentrations and durations are usually well below levels causing concern.

The potential environmental impact of contaminants associated with sediments must be evaluated in light of chemical and biological data describing the availability of contaminants to organisms. Information must then be gained about the effects of specific substances on organism survival and function. Many contaminants are not readily released from sediment attachment and are thus less toxic than contaminants in the free or soluble state on which most toxicity data are based.

There are now cogent reasons for rejecting many of the conceptualized impacts of disposed dredged material based on classical bulk analysis determinations. It is invalid to use total sediment concentration to estimate contaminant levels in organisms since only a variable and undetermined amount of sediment-associated contaminant is biologically available. Although a few instances of toxicity and bioaccumulation of possible ecological consequence have been seen, the fact that the degree of effect depends on species, contaminants, salinity, sediment type, etc., argues strongly that bulk analysis does not provide a reliable index of contaminant availability and potential ecological impact of dredged material.

FIELD STUDIES TO EVALUATE OPEN-WATER DISPOSAL

The impact of dredging and disposal operations on the aquatic environment was studied under DMRP.[37] Five field studies were conducted at the following locations, which represent different types of physical environment:

1. Eatons Neck, N.Y.
2. Columbia River, Or.
3. Ashtabula, Oh.
4. Galveston, Tex.
5. Duwamish Waterway, Wash.

The results of the study indicated that

1. It appears that open-water disposal had a negligible impact upon physical, chemical, and biological variables. The impacts observed were site-specific

and it was suggested that the results of field studies cannot be universally applied.

2. The release of manganese and ammonia during and after disposal may pose a problem, and there was limited evidence this may also apply to iron, mercury, and PCBs.

3. Overall, most impacts seemed to be short-term. The condition of the water column generally returned to ambient within minutes to hours after a disposal operation.

EXTENT OF CONTAMINATED MARINE SEDIMENTS

The problem of contaminated marine sediments has emerged as an environmental issue of national importance. Harbor areas in particular have been found to contain high levels of contaminants in bottom sediments from wastes from municipal, industrial, and riverine sources. Herbich[38] summarized the Marine Board Committee Study on the extent of Contaminated Marine Sediments. The study examined the extent and significance of marine sediment contamination in the United States, reviewed the state of the art of contaminated sediment cleanup, and identified research and development needs.

Contamination of marine sediment in all areas of the world, particularly in the shallow water areas, poses a potential threat to marine resources and human health. Improving the capability to assess, manage, and remediate these contaminated sediments is critical not only to the well-being of the marine environment but also to its use for navigation, commerce, fishing, and recreation.

The nature of the problem has resulted from using coastal waters, intentionally or unintentionally, for waste disposal for many decades. Confined or partly confined areas where low wave and current energies are present (such as harbors) contain high levels of contaminants in bottom sediments from wastes from urban, industrial, and riverine sources. Such areas where flushing action is unlikely (except during hurricanes) have accumulated contaminants which may now be buried by fine sediment deposits of recent years that contain no, or low levels of contaminants.

Legislative authority for the management of contaminated marine sediments falls largely under three statutes: the Comprehensive Environmental Response, Compensation, and Liability Act of 1980 (CERCLA), the Marine Protection, Research, and Sanctuaries Act (MPRSA), and the Clean Water Act (CWA). The CERCLA, as amended by the Superfund Amendments and Reauthorization Act (SARA) of 1986, is aimed at the cleanup and remediation of inactive or abandoned hazardous waste sites, regardless of location. Superfund sites are currently ranked by the U.S. EPA based on the hazard they may pose to human health and the environment via releases to groundwater, surface water, and air. Underwater accumulations of hazardous wastes in marine environments are unlikely to threaten human health except by way of food chain exposure, which is not currently addressed in the EPA's hazard-ranking process. Under the 1986 Superfund amendments, however, the EPA was required to modify its Hazard Ranking System to address "the damage to natural resources which may affect the human food chain and which is associated with any release (of a hazardous substance)" [Section 105(a)(2)].

Meanwhile, the CWA, as amended by the Water Quality Act of 1987, gives the

EPA lead responsibility for safeguarding the quality of U.S. coastal and inland waters. This includes regulating the disposal of dredged and fill materials (shared with the U.S. Army Corps of Engineers, under Section 404) and removing in-place toxic pollutants in harbors and navigable waterways (under Section 115). The 1987 amendments added new authorities requiring the EPA to study and conduct projects relating to the removal of toxic pollutants from Great Lakes bottom sediments [Section 118(c)(3)] and to identify and implement individual control strategies to reduce toxic pollutant inputs into contaminated waterway segments [Section 304(1)].

In response to Title II of the MPRSA (PL 92-532) and the National Ocean Pollution Planning Act, the National Oceanic and Atmospheric Administration (NOAA) Office of Marine Pollution Assessment conducts comprehensive inter-disciplinary assessments of the effects of human activities on estuarine and coastal environments. Among these assessment activities is the National Status and Trends Program (NST), which attempts to create, maintain, and assess a long-term record of contaminant concentrations and biological responses to contamination in the coastal and estuarine waters of the United States. This assessment provides some insight into the extent of contamination nationally.

As a result of legislative responsibility and programmatic interests, a wide variety of federal agencies have shown active interest in this subject. The EPA's responsibilities under Superfund and the CWA are the source of its interests in water quality concerns and remediation of uncontrolled hazardous waste sites. The U.S. Army Corps of Engineers (CE) is involved because of its responsibility to dredge and maintain navigable rivers and harbors. The CE also assists in the design and implementation of remedial cleanup actions under Superfund. NOAA has responsibility for assessing the potential threat of Superfund sites to coastal marine resources as a natural resource trustee as well as under its NS&T program. The U.S. Fish and Wildlife Service has legal authority for various endangered coastal species, food chain relationships, and habitat considerations, all of which are potentially affected by contaminated sediments. The Navy has had experience in assessing contaminated sediments and now must grapple with such problems in locating and maintaining homeports for Navy vessels.

Procedure

There are many definitions of contamination of sediments. The Committee defined the contaminated sediments for the purpose of the study as follows:

> Contaminated sediments are those that contain chemical substances at concentrations which pose a known or suspected environmental or human health threat.

A symposium and a workshop were organized to examine the extent of contamination nationwide, the methods for classification of sediment contamination, risks to human health and the ecosystem, sediment resuspension and contaminant mobilization, remedial strategies, and technologies for handling contaminated sediments. In addition, five case studies of the different ways in which a variety of sediment contamination problems are being handled were examined. They include the PCB problem in New Bedford Harbor, Mass.; PCBs in the upper Hudson River, N.Y.; kepone contamination of the James River, Va.; the variety of chemicals contaminating Commencement Bay, Wash.; and the Navy Homeport Project in Everett Bay, Wash.

The main purposes of the study included (1) examination of the extent and significance of marine sediment contamination, (2) a review of the state of the art of contaminated sediment removal and remediation technology, (3) identification and appraisal of alternative sediment management strategies, and (4) identification of research and development needs and issues for future technical assessments.[39]

Extent of Contamination

Many contaminated marine sediments are found along all coasts of the contiguous United States, both in local "hot spots" and distributed over large areas. There is a wide variety of contaminants including: heavy metals, polychlorinated biphenyls (PCBs), DDT, polynuclear aromatic hydrocarbons (PAHs), mononuclear aromatic hydrocarbons, phthalate esters, pesticides, etc. At present, there are no generally accepted sampling techniques or testing protocols.

Classification Methodologies for Determining Sediment Contamination

There have been some research efforts in classifying the extent of contamination and some states have collected data for special purposes. No uniform methods have been adopted by various states or federal agencies. A variety of classification methods are available:

1. Sediment bioassays: Sediment toxicity on a crustacean, infaunal bivalve, and infaunal polychaetes. Essentially, marine life is subjected to various levels of toxicity and their survival noted.
2. Sediment quality triad:
    a. Contamination quantified by chemical analysis
    b. Toxicity determined by laboratory bioassays
    c. Benthos community structure determined by taxonomic analysis of biofauna
3. Apparent effects threshold technique equilibrium partitioning (AET): A tool for deriving sediment quality values for a range of biological indicators to assess contaminated sediments. The AET is the contaminant concentration in sediment above which adverse effects are always expected for a particular biological indicator.

Recommendations.   To ensure that decision making is informed and scientifically based, continued research and use of assessment methodologies should provide information to determine:

• A range of concentrations of chemicals in sediments that will result in biological effects
• Whether in-place sediments are causing biological impacts

A tiered approach to the assessment of contaminated sediments should be used. The approach would progress from relatively easy and less expensive (but perhaps less definitive) tests to more sensitive methods as needed.

Contaminated Sediment Management Strategies

Although the dredged material management strategy developed by the CE may be relevant to severely contaminated sediments, it is important from a management standpoint to differentiate them from less contaminated sediments. In particular, most highly sophisticated remedial technologies (i.e., those involving treatment or destruction of associated contaminants) are likely to be cost effective only in small areas and for sediments with relatively high contamination levels. Sediment contamination problems often involve large volumes of sediment with relatively low contamination levels. As a result, some highly sophisticated technologies may be inapplicable or inefficient for remediating contaminated sediments.

"No action" may be the preferred alternative in cases in which the remedy may be worse than the disease (e.g., where dredging or stabilizing contaminated sediments results in more biological damage than leaving the material in place). Contaminants generally accumulate in depositional zones, and, if the source is controlled, new clean sediments will deposit and cap the contaminated material over time. In effect, no action alternatives in such cases may result in natural capping.

1. No action may be an acceptable option if the contamination degrades or is buried by natural deposition of clean sediment in a short period of time.

2. In-place capping may be a useful option if the sediments are not in a navigation channel or if groundwater is not flowing through the site.

3. Removal and subaqueous burial offsite may be a viable option, although the experience with this technique is limited to relatively shallow water (less than 100 ft).

4. Incineration seems to be viable only for sites with relatively small amounts of sediments containing high concentrations of combustible contaminants.

5. Other techniques to assist in remediation of contaminated sediment may be appropriate in special cases. Examples include a variety of sediment stabilization or solidification techniques and biological and/or chemical treatment.

Recommendations

- Additional evaluation should be conducted to determine the applicability of the CE's dredged material management strategy to more severely contaminated sediments.

- No action should always be considered as an alternative strategy for minimizing biological damage. In using the no-action strategy as a form of natural capping of contaminated material, consideration should be given to the length of time it takes for contaminants to be isolated from the food chain.

Remedial Technologies

From a remediation standpoint, the most important factors are likely to be a definition of the cleanup target, technical and cost feasibility, natural recovery estimates, and ability to distinguish and/or control continuing sources of contaminants.

Recommendations

- Source control measures must be considered in all cases, including no action. Federal and state regulatory agencies requiring remedial action should implement source control measures as a component of remedial action when applicable and appropriate. Use of financial incentives through strict liability for assessment cost, remedial actions, and damages also may play an important role in source control, provided that trustees make aggressive efforts to hold responsible parties liable for releases into the environment.

- Aggressive technology and information transfer mechanisms are needed to ensure that knowledge gained and lessons learned from all remedial actions are available and accessible to managers confronting new remediation problems at federal, regional, and local levels. Knowledge gained should be systematically compiled in guidance documents. Lessons learned regarding the feasibility of sophisticated remedial technologies under varying conditions of contamination severity and extent should be documented and made widely available to facilitate future decision making. Lastly, experience gained through the use of screening procedures at large sites should be distilled and generalized into routine methodologies for economically assessing smaller sites.

- When possible, remediation projects should be designed to take advantage of existing navigational dredging activities that may already be authorized in conjunction with the Clean Water Act, Section 115 or Section 10/404.

- Research and development should be encouraged by the federal government to develop technology and equipment for efficiently removing contaminated sediments and to make that technology and equipment available in the United States. Foreign technologies should continue to be examined relative to their appropriateness in this country. Efforts to conduct and fund research and development as a partnership between government and industry should be encouraged.

- Although capping might not, in the strictest terms, be considered a remedial technology, it should not be ignored because it can play a valuable role in remediating contaminated sites.

- Monitoring programs should be well focused on testing forecasts made during design of the remediation plan. To the extent possible, monitoring should be extended to remove uncertainties in the basic understanding of contaminated sediment behavior. For example, monitoring of capped areas might focus on changes of cap thickness, erosion around boundaries, and leakage of contaminant through the cap.

Remediation and Source Control: Economic Considerations

Remedial actions are costly and become more expensive as additional levels of cleanup or treatment are pursued. The role of tradeoffs between possible technologies at and among sites must be considered, given the scarcity of funds to clean up contaminated sites and the potentially great number of sites.

The use of a benefit-cost analysis as part of the remedial action decision process would provide perspective on the issues involved. It would place investments in this area on the same footing as other public investments. However, difficulty in quantifying benefits from remedial actions in monetary terms makes reliance on benefit-cost analysis infeasible in a number of cases. Nonetheless, in light of the high cost of remedial actions, it is important that implicit (if not explicit) consideration be given to potential benefits before remedial actions are undertaken.

Removal of contaminated sediments can be very expensive, varying widely from several hundred thousand dollars to tens of millions of dollars. Data on 15 cleanup sites indicate that total cleanup costs can reach \$500,000 to \$1,000,000 per acre* (\$1,235,521 to \$2,471,043 per hectare; U.S. Congress Office of Technology Assessment, 1988). This compares with an average unit cost of navigation dredging of \$1 to \$2/yd^3 (\$2.18/m^3) of sediment dredged. The average unit cost of all dredging, both government and private, is estimated in 1988 at \$1.67/yd^3 (\$2.18/m^3) of material dredged. On-site incineration, one of the remedial measures proposed at various sites, is also very expensive. The estimates quoted are from \$186 to \$750/yd^3 (\$243 to \$980/m^3).

Recommendations

- In view of the high cost of remedial actions in most cases, greater use should be made of benefit-cost comparisons over ecologically relevant time periods in order to place investments in this area on the same economic footing as investments in other public projects.

- Cost-effectiveness analysis of alternative remedial actions should consider both short- and long-term costs. Comparisons at and among sites should be based on costs estimated using a consistent approach.

- In evaluating the degree of remediation to be conducted at a site, it should be recognized that incremental costs typically will increase rapidly as additional levels of cleanup are sought.

- The decision as to whether or not remedial actions are undertaken should be based on a balanced comparison of the anticipated environmental and public health benefits of actions with their costs, including possible environmental and health risks.

- Clearly infeasible options should be eliminated at the outset, before alternative remedial actions are considered in depth.

CAPPING OF CONTAMINATED SEDIMENTS

Contaminated dredged material placed in an open-water disposal site may be capped with clean dredged material to isolate the contaminants from the water column. Field studies were conducted for many years to evaluate the effectiveness of capping by Bokuniewicz,[40] Truitt,[41,42] Palermo,[43,44] and Sumeri.[45] An example of a completed project in the Duwamish Waterways is shown in Fig. 9.19.

A Dredging Research Technical Note[46] reviews design requirements for capping. Some of the information described in the Technical Note is reproduced here.

When dredged material is placed in open-water sites, there is potential for

---

*For purposes of comparison, assume that a 1-acre cleanup involved removing overburden to a depth of 1 yd, or a total of 43,560 yd^3 (33,304 m^3) of contaminated material. In that event, total cleanup costs would range from \$11.50 to \$23.00/yd^3 (\$15.04 to \$30.08/m^3).

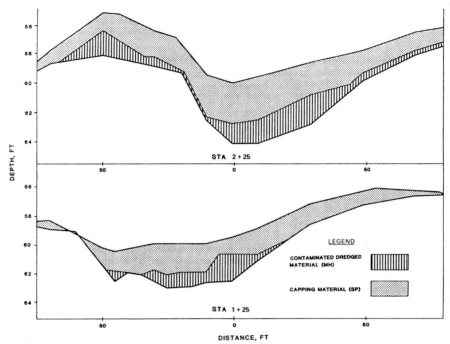

FIGURE 9.19 Typical composite profiles through completed disposal mound. (*Truitt, 1987*)

both water column and benthic effects. The release of contaminants into the water column is not generally viewed as a significant problem for dredged material from most navigation projects. The acceptability of a given material for unrestricted open-water disposal is therefore mostly dependent on an evaluation of potential benthic effects. Capping is considered an appropriate contaminant con-

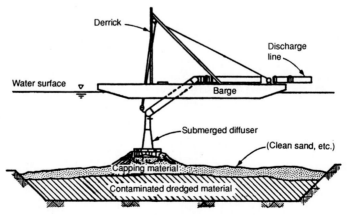

FIGURE 9.20 Contained aquatic disposal. (*Adapted from Truitt, 1987*)

trol measure for benthic effects in the CE's dredging regulations (33 CFR 335-338) and supporting technical guidelines (Francingues et al.[47]) and is recognized by the London Dumping Convention as a management technique to rapidly render harmless otherwise unsuitable materials.

Guidelines are available for planning capping projects and for selection of placement techniques for capping projects (Truitt[41,42]). An artist's conception of contained aquatic disposal capped with clean material is shown in Fig. 9.20.

Design requirements for capping are shown on a flowchart in Fig. 9.21. There

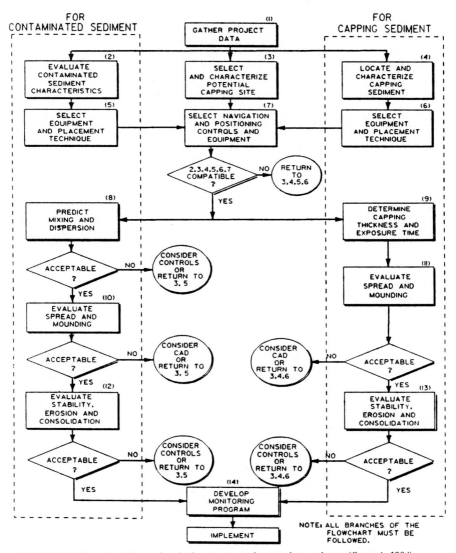

FIGURE 9.21 Flowchart illustrating design sequence for capping projects. (*Sumeri, 1984*)

is a strong interdependence among all components of design for a capping project. For example, the initial consideration of a capping site and placement techniques for both the contaminated and capping materials will strongly influence all subsequent evaluations, and these initial choices must also be compatible for a successful project (Shields and Montgomery[48]). When an efficient sequence of activities for design of a capping project is followed, unnecessary data collection and evaluation can be avoided.

Capping should not be viewed merely as a form of unrestricted open-water placement. A capping operation should be treated as an engineered project with carefully considered design, construction, and monitoring to ensure that the design is adequate. The basic criterion for a successful capping operation is simply that the cap thickness required to isolate the contaminated material from the environment be successfully placed and maintained.

Contaminated sediments should not be placed in a high-energy environment (i.e., high waves and currents). Capping in such cases will not be effective since the clean-sediment cap may erode, exposing contaminated sediments. Vyas and Herbich[49] have evaluated erosion of dredged-material islands caused by waves and currents.

LONDON DUMPING CONVENTION

An international treaty was signed by many countries, which is implemented through the "Convention of the Prevention of Marine Pollution by Dumping of Wastes and Other Matter," commonly called the London Dumping Convention (LDC).[50] United States is a signatory to this treaty which requires that signatory countries shall take all practicable steps to prevent the pollution of the sea which is liable to create hazards to human health, to harm living resources and marine life, and to damage amenities or to interfere with other legitimate uses of the sea (Article 1 of LDC).

Engler[51] discusses the evolution of the "Guidelines for the Application of the Annexes to the Disposal of Dredged Material" and their meaning to the dredging industry. Engler states:

> In the special case of dredged materials, sea disposal is often an acceptable disposal option, though opportunities should be taken to encourage productive use of dredged material for, for example, marsh creation, beach nourishment, land reclamation or use in aggregates. For contaminated dredged materials, consideration should be given to the use of special methods to mitigate their impact, in particular with respect to contaminated inputs. In extreme cases of pollution, containment methods (including land-based disposal) may be required but very careful consideration should be given to the comparative assessment of the factors listed above* before this option is pursued.

PRODUCTIVE USE OF DREDGED MATERIAL

Hubbard and Herbich[52] reviewed information available on locations where dredged material was put to productive use. About 150 sites were documented and classified. The categories of uses were

---

*The factors listed in Ref. 51 are human health risks; environmental costs; hazards (including accidents) associated with treatment, packaging, transport, and disposal; economics (including energy costs); exclusion of future uses of disposal areas, and they are for both sea disposal and alternatives.

1. Commercial (25 percent)
2. Industrial (21 percent)
3. Recreational (16 percent)
4. Wildlife habitats (10 percent) (Fig. 9.22)
5. Hydraulic control (9 percent)
6. Agricultural (6 percent)
7. Transportation (4 percent)
8. Research (4 percent)
9. Multipurpose (5 percent)

The owners of the sites were usually local, state, or national governments.

Landin[53] discusses habitat development from dredged material for migratory and nesting use by waterbirds, shorebirds, and waterfowl.

The CE[54] developed a manual for beneficial uses of dredged material. The following beneficial uses are identified:

1. Habitat development
2. Beach nourishment
3. Aquaculture
4. Parks and recreation
5. Agriculture, forestry, and horticulture
6. Strip mine reclamation and solids waste management
7. Shore stabilization and erosion control
8. Construction and industrial use
9. Material transfer
10. Multiple purpose

FIGURE 9.22 Aerial view of Drake Wilson Island in Apalachicola Bay, FL, before habitat development.

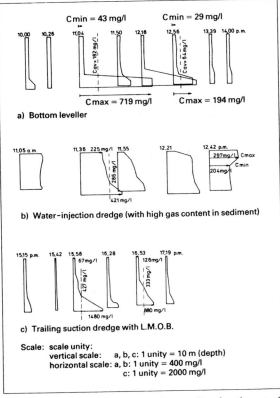

FIGURE 9.23 Vertical concentration profiles for three techniques. (*Dijkman et al., 1989*)

EVALUATION OF CONTAMINATED SEDIMENT IMPACTS ON AQUATIC ENVIRONMENT IN THE NETHERLANDS

Dijkman et al.[55] discuss the problems with the quality of surface water in the Rhine and Meuse rivers, which originate in other countries and pass through the Netherlands on the way to the North Sea. The rivers are already polluted entering the Netherlands by discharging salt effluent from salt mines, industrial effluents, and accidental discharges of chemicals and oil. In addition contamination enters the river system from the bioindustry and industrial areas near Rotterdam. Thus much of the dredging in the Netherlands is faced with handling contaminated sediments. Government agencies are setting up safe and economically justifiable methods for the control of environmentally sensitive dredging projects. This will be achieved by a system of classification standards and by control measures.

Blokland[56] reviews the studies to evaluate the extent of resuspension of sediment during maintenance dredging operations in Dutch harbors. Three types of dredging equipment were evaluated:

1. Bottom leveler
2. Water-injection dredge
3. Hopper dredge

Figure 9.23 shows the magnitudes of resuspended sediment for three types of dredging equipment. The resuspended sediment by the bottom leveler is contained mostly in a sharply defined layer near the bottom. The water-injection dredge resuspends sediments throughout the water column while the resuspended sediment by a hopper dredge is mostly near the bottom. This is similar to the results obtained by Herbich and de Vries[25] for the sediment resuspension by a cutterhead (i.e., the greatest concentration of sediment is near the bottom).

Velinga[57] discusses the environmental effects of dredging and disposal operations, particularly relating to the port of Rotterdam.

NEED, CONSTRUCTION, AND MANAGEMENT OF DREDGED MATERIAL ISLANDS FOR WILDLIFE*

Dr. Mary C. Landin
Research Biologist
U.S. Army Engineer Waterways Experiment Station
Vicksburg, Miss.

As a result of dredging operations over the past 100 years, the U.S. Army Corps of Engineers (CE) has built over 2000 islands, primarily while constructing the Intracoastal Waterway System and maintaining navigation channels and harbors. Until the 1970s, the CE did not build these islands with wildlife and fish habitats as objectives. Rather, material was sidecast or mounded up in adjacent shallow water areas until islands were formed, usually incidental to the primary goal of clearing and maintaining navigation channels. A number of the older islands have eroded and are no longer in existence or have sustained heavy damage from wind and wave action. Others are replenished with fresh quantities of dredged material through maintenance dredging operations.[1,2] Increasingly, however, wherever the CE finds that dredging needs and islands coincide, it will try to minimize environmental impacts as well as enhance wildlife and fish habitats using the dredged material resource available from the dredging project.[3]

The older islands were constructed for the purpose of disposing of large quantities of material, and at the time of dredging, wildlife was not a consideration. Because of their very nature—isolated, sometimes barren, and considered waste sites—they were avoided by most development enterprises and had little human use. At the same time, natural wildlife habitats were disap-

---

*References for this chapter section appear at the end of the section.

FIGURE 9.24 On Goat Island, Bolivar Peninsula, Texas, sandy dredged material was used to build a salt marsh that was protected by a temporary sandbag breakwater. This breakwater, mud flat, and salt marsh are densely colonized with oysters.

pearing from America's coastlines at a rapid rate. Culliton et al.[4] note that the U.S. population has shifted to the coasts, and they project that there will be an average 60 percent increase in all U.S. coastal populations by 2010, with 210 million people in coastal counties.

Dredged material islands and their surrounding shallows are home or stopover points for numerous species of wildlife and fish. These include relatively few mammals, although raccoons, white-tailed deer, foxes, harbor seals, river otters, nutria, muskrats, beavers, coyotes, opossums, armadillos, cottontails, small rodents, and goats and other domestic or feral livestock and animals may visit or live year-round on larger islands. Island use by alligators and other reptilian and amphibious animals is also relatively common. Shrimp, blue crabs, and numerous species of commercially important and sport fishes use the shallows in and around dredged material islands at various stages in their life cycles (Fig. 9.24).

The primary use of dredged material islands is by numerous species of birds. These include a variety of songbirds, especially on islands close to the mainland and in a major migratory flyway. Migratory use by waterfowl, waterbirds, and raptors with water-related feeding habits is a very important use of such islands. Perhaps more importantly and certainly more conspicuously, these dredged material islands provide habitats for about 1 million waterbirds each year from 37 different species[1,2] (Fig. 9.25). The CE considers waterbird nesting on these islands as a highly desirable beneficial use of dredged material and encourages such use whenever possible.[5]

FIGURE 9.25 A million sea and wading birds nest on dredged material islands in North America each year. These Caspian terns are nesting on dewatered silty dredged material in Mobile Bay, Alabama.

TYPES OF DREDGED MATERIAL ISLANDS AND SITES

There are generally four types of CE dredged material islands and sites used by colonially nesting waterbird species throughout North America. The first type, mainland disposal sites (diked and undiked), is less frequently used because the sites allow access to ground predators such as raccoons and coyotes. However, under isolated conditions, nesting colonies will occur on mainland sites. The second type, older undiked islands that were built prior to the 1970s and a few undiked islands built since that time (Figs. 9.26 and 9.27), and the third type, diked islands (both new and modified), are the most commonly occurring colony islands, especially along the Gulf and Atlantic coasts.[2]

The CE is moving more and more to construction of the fourth type of island, very large confined disposal facilities (CDFs) that can be used for decades for placement of material from more than one dredging project (Fig. 9.28). There are 47 CDFs in the Great Lakes; all receive wildlife use to some extent. Three CDFs are located in the northern Gulf Coast region, and all three are used extensively by nesting seabirds (terns, gulls, skimmers, and pelicans). Others occur in the Mid-Atlantic Coast region and are currently receiving limited wildlife use due to high human use of the sites.

In the U.S. northern Gulf Coast region, which includes the coastal area from Bradenton, Fla., to the Mexican border, there are a total of 645 remaining dredged material islands and sites. Table 9.7 shows numbers of islands by states, and those islands with nesting colonies. The states of Florida (north of Bradenton) and Texas have the largest numbers of islands and sites, with 120 and

FIGURE 9.26 Bird and Sunken Islands in Tampa Bay were built of dredged material in 1930 and 1951, respectively. The islands are managed by the National Audubon Society, and an estimated 30,000 waterbirds nest there each year.

FIGURE 9.27 Older undiked dredged material islands often resemble this one, located in North Carolina. It is providing a combination of four nesting habitats: bare nesting areas, sparse grasses, dense shrubs, and maritime forest.

FIGURE 9.28 CDFs such as this one at Pointe Mouillee, western Lake Erie, Michigan, provide multiple-purpose use, including waterbird habitats. Pointe Mouillee was designed with wildlife and fisheries habitat as part of the project objectives.

414, respectively. However, Florida still has a number of natural islands available for nesting waterbirds, and therefore only 28 (23 percent) of the state's dredged material sites have nesting colonies.

By contrast, there is much less available natural habitat for nesting waterbirds in Mississippi, Louisiana, and Texas, where 68, 56, and 58 percent of all dredged material islands and sites have nesting colonies. The state of Alabama, with only 26 possible dredged material nesting sites, has only 7 with colonies. However, the Gaillard Island CDF in lower Mobile Bay, Ala., has approximately 30,000 waterbirds from more than 20 species nesting on it and is an extremely important nesting site (Fig. 9.29).

Along parts of the Atlantic Coast, natural island habitats for nesting are so scarce that waterbirds are nesting almost exclusively on dredged material islands. This is especially true in North Carolina, where up to 99 percent of seabirds nest

TABLE 9.7 Dredged Material Islands and Sites in the Northern Gulf Coast Region

	Total sites	With colonies	Use (%)
Florida (north of Bradenton)	120	28	23
Alabama	26	7	27
Mississippi	19	13	68
Louisiana	66	37	56
Texas	414	242	58
Totals	645	327	51

FIGURE 9.29 Brown pelicans began nesting in 1983 on Gaillard Island CDF in Mobile Bay, Alabama, when the site was only 2 yr old. In 1988, there are over 500 pairs of adult pelicans nesting there and about 2000 brown pelicans living at the CDF.

on estuarine islands. In New Jersey and further north, nearly every dredged material mound in the marsh, island, and isolated beach is a nesting site.

In addition to the CDFs in the Great Lakes, waterbirds will be found using virtually every isolated or semiisolated dredged material island and site. However, these species still use natural islands in the lakes in large numbers.

On the Pacific Coast, waterbirds and other wildlife can be found on every dredged material island and site where disturbance is at a minimum. This is true of islands in Coos Bay, islands in the lower Columbia, and Jetty Island in Puget Sound. Where protective fencing is provided to prevent disturbance, dredged material islands and sites are also used for nesting in southern California by the California least tern.

NESTING SPECIES

While 37 colonial species and a number of other wildlife species (primarily birds) use dredged material island habitats (Tables 9.8 and 9.9), this use varies according to location, island availability, availability of natural habitat, and human disturbance factors. Twelve of the 37 species are listed on federal or state endangered, rare, or threatened lists and are considered to be in trouble and not maintaining their population levels.

TABLE 9.8 Colonial Waterbird Species Found Nesting on Dredged Material Islands in Seven Regions of the Corps-Maintained Waterways*

Species	Regions†					
	TX	FL	NC	NJ	GL	PNW
White pelican	X					
Brown pelican	X	X	X			
Double-crested cormorant		X	X			
Olivaceous cormorant	X					
Anhinga		X				
Great blue heron	X	X	X	X		X
Green heron	X	X	X	X		
Little blue heron	X	X	X	X		
Cattle egret	X	X	X	X	X	
Reddish egret	X	X				
Great egret	X	X	X	X		
Snowy egret	X	X	X	X		
Louisiana heron	X	X	X	X		
Black-crowned night heron	X	X	X	X	X	
Yellow-crowned night heron	X	X	X	X		
White-faced ibis	X					
Glossy ibis	X	X	X	X		
White ibis	X	X	X			
Roseate spoonbill	X	X				
Glaucous-winged gull						X
Great black-backed gull			X	X		
Herring gull			X	X	X	
Western gull					X	
Ring-billed gull					X	X
Laughing gull	X	X	X	X		
Gull-billed tern	X	X	X	X		
Forster's tern	X		X	X	X	
Common tern		X	X	X	X	X
Roseate tern		X	X	X		
Least tern	X	X	X	X		
Royal tern	X	X	X			
Sandwich tern	X	X	X			
Caspian tern	X	X	X		X	X
Black tern					X	X
Black skimmer	X	X	X	X		

*The Upper Mississippi River study is not listed since none of the nesting colonies found were located on dredged material.

†TX = Texas; FL = Florida; NC = North Carolina; NJ = New Jersey; GL = Great Lakes; PNW = Pacific Northwest.

Northern Gulf Coast

As an example of the importance of dredged material islands in the United States, 27 colonially nesting waterbird species and several other noncolonial species are nesting on dredged material islands and sites in the northern Gulf Coast. Some species, such as least terns, brown pelicans, and laughing gulls, seem to prefer dredged material sites since most of the nesting colonies of these species are on dredged material.

TABLE 9.9 Noncolonial Species Nesting on Dredged
Material Islands in CE-Maintained Waterways

Canada goose	Yellow-billed cuckoo
Mallard	Grove-billed ani
Black duck	Short-eared owl
Mottled duck	Common nighthawk
Gadwall	Scissor-tail flycatcher
Marsh hawk	Long-billed marsh wren
Osprey	Short-billed marsh wren
Kestrel	Fish crow
Bobwhite quail	Mockingbird
American bittern	Brown thrasher
Least bittern	Ruby-crowned kinglet
Sora	Loggerhead shrike
Black rail	Yellow warbler
Clapper rail	Chestnut-sided warbler
King rail	Prairie warbler
Common gallinule	Louisiana waterthrush
American oystercatcher	Yellowthroat
American avocet	Eastern meadowlark
Black-necked stilt	Red-winged blackbird
Piping plover	Boat-tailed grackle
Snowy plover	Great-tailed grackle
Wilson's plover	Common grackle
Kildeer	Painted bunting
Spotted sandpiper	Savannah sparrow
Willet	Grasshopper sparrow
Sooty tern	Seaside sparrow
Mourning dove	Field sparrow
Ground dove	Song sparrow

Black Skimmers. While black skimmers nest on isolated barrier islands and beaches in large numbers along the northern Gulf Coast, they also use dredged material islands extensively. For example, the largest black skimmer colony on the Gulf Coast (over 4000 birds) is located on Gaillard Island CDF.

Tern Species. Seven species of terns—least, common, royal, Caspian, Sandwich, Forster's, and gull-billed—nest on Northern Gulf Coast dredged material islands. In the case of the least tern, hundreds of pairs nest on Gaillard Island CDF in Alabama and on protected dredged material beaches in Mississippi, as well as numerous other dredged material and natural sites along the coast.

In most cases, royal terns will nest alone or with Sandwich terns. Caspian and gull-billed terns may also be found on the same island (but not in the same colony). Forster's terns seek herbaceous vegetation for nesting sites and are becoming more and more common on older gulf dredged material islands. All seven species nest on Gaillard Island CDF.

Brown Pelicans. The only brown pelican colonies in Texas and Alabama are located on dredged material, on Pelican Island near Corpus Christi, and on Gaillard Island CDF near Mobile. Other brown pelican colonies occur on dredged material in Florida and in Louisiana (and have expanded as far north on the Atlantic

Coast as Virginia). No nesting colonies occur in Mississippi. The first nesting on Gaillard Island occurred in 1983 with one successful nest.[6] By 1987, there were 331 nests and over 1000 brown pelicans on the island. By 1988, there are over 500 nests and about 2000 brown pelicans using the island.[7]

In Florida and Louisiana, brown pelicans are nesting in mangroves and other low-growing coastal trees. In Alabama and Texas, they are nesting directly on the dredged material in nests built from 6 in to 2 ft high of twigs and driftwood. In 1988, a few brown pelicans on Gaillard Island were nesting in low-growing shrubs that have reached sufficient height and strength to support nests.

American White Pelicans.   There is only one American white pelican nesting colony in the northern Gulf Coast region, and it is located in the Laguna Madre between Corpus Christi and Brownsville, Tex. This dredged material island has been used by nesting white pelicans for decades[8] but has not expanded into other parts of the coast with one temporary exception. In 1979, white pelicans were found establishing nests on a dredged material island in Galveston Bay, but the colony did not persist.

The only other year-round occurrence of white pelicans in the northern Gulf is on Gaillard Island, where between 600 and 800 immature white pelicans live. They feed in the CDF containment pond, and it is assumed that as they reach breeding age, they are migrating to the large white pelican nesting colonies in the western United States. Preliminary breeding behavior has been observed on Gaillard Island CDF, but no nesting has occurred there.[6]

Gull Species.   Laughing gulls by far make up the largest numbers of individual waterbirds nesting on both dredged material and natural islands along the northern Gulf Coast. They are apparently highly successful nesters and have abundant food sources since colonies with 10,000 to 20,000 birds are not unusual in Florida, Texas, and Alabama.[6,8,9] All of these largest colonies are located on big dredged material islands or CDFs.

A very low number of herring gulls nest in scattered locations in the northern Gulf Coast region. These nesting occurrences can hardly be considered colonies since usually only one to five pairs are found nesting. Herring gulls' primary breeding range is the northern United States, along the northeast coast and in the Great Lakes. Herring, ring-billed, and other northern-nesting gulls will overwinter in the northern Gulf Coast region by the thousands. They are frequently observed on dredged material islands and sites from October to March.

Heron and Egret Species.   Five heron and three egret species nest on dredged material islands and sites in the northern Gulf Coast region. These include great blue herons, little blue herons, tricolored herons, yellow-crowned night-herons, black-crowned night-herons, great egrets, reddish egrets, and cattle egrets.[2,10] With the exception of the tricolored heron and the addition of the green heron, these same species will nest in fresh water sites throughout the Mississippi River basin and other fresh water wetland sites. Because newer dredged material sites are usually not well vegetated, the species tend to congregate on older islands where successional stages have progressed to provide woody vegetation large enough to support nests. However, in south Texas, herons have been observed nesting in clumps of cacti as well as in small shrubs.[8] Most of these species also tend to nest in mixed colonies with other species.

With the exception of Florida and Texas dredged material islands, mixed species heronries are not generally as large as those that have occurred for many years on natural sites and islands. However, large heronries have occurred for 30 to 50 years on such dredged material sites as Bird and Sunken Islands in Tampa Bay, Big Pelican and North Deer Islands in Galveston Bay, and on other larger dredged material islands in the Texas waterway system.[8,11]

Ibis Species. Three species of ibis nest on dredged material islands, primarily in Texas and Florida. In Florida, thousands of white ibises nest in Tampa Bay and other coastal waterway sites. In Texas, white ibises, glossy ibises, and white-faced ibises nest, the latter nesting only in Texas.[8] Although white ibises have been found nesting in mixing heronries 300 mi inland in Mississippi as far upriver as Yazoo City on isolated artificial sites,[12] colonies in Mississippi, Alabama, and Louisiana tend to be located on natural islands and sites and only those in Louisiana have large nesting numbers of ibises.

Cormorants. The double-crested cormorant overwinters and feeds during migration in all five states of the region. However, they primarily nest on dredged material in Texas and Florida and on natural sites in the other states. Their colonies are not large but consist of up to 50 nests. Since they will nest in both fresh and salt water, they have nesting colonies in the Mississippi River as well. The olivaceous cormorant is found in Texas only, and nests in small numbers on artificial structures and other sites, including a few dredged material islands.

Roseate Spoonbill. An increasing number of roseate spoonbills nest in Florida, and large numbers nest along the Texas Intercoastal Waterway. Many of these nest on dredged material islands in mixed heronries with herons, egrets, and ibises. This species may occasionally wander into Mississippi, Alabama, and Louisiana but does not occur in large numbers. Roseate spoonbills are contact feeders (compared to terns, herons, and egrets that are visual feeders). They are often observed feeding in the shallow water and soupy mud flats of dredged material placement sites. It is a distinct possibility that the occurrence of large dredged material placement sites which provide this type of feeding habitat is a boon to the spoonbill population, especially in Texas where most such feeding observances have been made.

Other Nesting Species. Four other species nest in large enough numbers on dredged material islands and sites in the Northern Gulf Coast region to be worthy of mention. They do not, however, tend to nest in large congregations but nest a few pairs together or in individual pairs. The species of most concern is the black-necked stilt, which is quite common on such sites as Gaillard Island CDF and on dredged material sites such as Big Pelican Island in Galveston Bay. They tend to nest in low vegetation around borrow pits and shallow swales formed by dredging and construction operations. The stilt population at Gaillard Island has increased dramatically since the island was built in 1981, with over 30 nesting pairs in 1987.[13]

American oystercatchers and willets are more solitary nesters, and in general, there is usually one or more nests on dredged material islands in low-growing herbaceous vegetation each year. This is a more common occurrence in Florida. Clapper rails are also very common nesters in both planted and naturally colonizing salt marshes on Gulf and Atlantic dredged material islands. Up to three nesting pairs have been found nesting in the relatively small planted marshes on Wilson Island in Apalachicola Bay, Fla.,[7] and on Gaillard Island CDF in Mobile

Bay.[6] In addition to the above, snowy plovers are rare nesters and killdeers are frequent nesters on dredged material islands in the region.

Other Regions

As with the northern Gulf Coast region, certain species are attracted to and use dredged material islands for nesting, resting, and feeding. More information and data on the use of Pacific Northwest islands are available.[1,3,7,10] More information is available on dredged material islands on the Atlantic Coasts.[1,9,14,15] More detailed information has been published on dredged material island use in the Great Lakes.[5,7,16]

WHAT MAKES DREDGED MATERIAL SO ATTRACTIVE TO NESTING WATERBIRDS?

There are some generalities about characteristics of dredged material islands and sites that can be stated. All five characteristics make these sites very attractive to nesting waterbirds, and depending upon the state of island development at a given time, regulate the species or group of species that will be found there.

First, dredged material sites in U.S. waterways tend to provide isolation from ground predators and human disturbance. In the past, dredged material has often been viewed by citizen and developer alike as "spoil" and not considered useful for construction or for recreation. Therefore, these islands and sites were relatively undisturbed, a feature that is of great importance to nesting waterbirds or any other wildlife raising young.

Since waterbirds began to use these artificial islands for nesting in large numbers as early as the 1930s, some of the larger ones have come under the protection of such organizations as the National Audubon Society, county Boards of Supervisors with natural resource interests, State Departments of Natural Resources, and the National Park Service. Newer ones such as Gaillard Island CDF are posted to trespassers during the breeding season by the CE.

Second, dredged material sites generally provide a wide range of habitats and diversity to accommodate nesting waterbirds. Four successional stages can exist on a given dredged material island or on one large dredged material island, depending upon placement schedules, climatic conditions, and construction factors. These are (1) bare ground, which is usually the habitat available immediately after the completion of a placement operation; (2) sparse herbaceous cover, the stage that occurs about 1 to 2 yr after placement of dredged material; (3) denser herbaceous cover with some sparse, low shrubs that occurs about 3 to 10 yr after placement; and (4) tall shrubs and trees, which is the climax stage of vegetation on an island that has not had a new deposit of dredged material for about 10 to 20 yr. Times of successional stages vary within areas of the region. For example, south Texas is so hot and arid that trees seldom attain a height of over 20 ft and are of different species from those occurring in Galveston Bay.[8] Likewise, mangroves occur in dense stands on dredged material in the Tampa Bay area,[17] while willows, cottonwoods, cypresses, and other freshwater trees are more likely to occur above the winter kill zone that limits mangrove survival.

Third, ongoing dredging operations usually keep early successional stage habitats available without a great deal of expense or difficulty. Using Gaillard Island CDF as an example, this 1300-acre island built in 1980 to 1981 already has three of the four stages of habitat available for nesting waterbirds.[6] The CDF has been used for disposal of dredged material every year since its construction, and at the same time dike upgrading and repair has taken place. Ongoing construction activities have maintained large expanses of bare ground habitat ideal for black skimmers and tern species. The CDF's south dike has remained relatively undisturbed and has become densely vegetated. In its earlier plant growth stages, there were thousands of black skimmers and laughing gulls nesting on the south dike. The habitat is changing to support larger shrubs and small trees, and in 1988 herons and egrets have also moved to the south dike to nest, along with fewer skimmers and gulls. The larger gull and skimmer colonies are tending to move to less vegetated parts of the CDF. This rapidly evolving CDF already supports 30,000 nesting birds, and the limitations of available nesting habitat have not been reached.[7]

Fourth, dredged material islands are usually located close to shallow water areas which provide feeding habitat. Over the years, dredged material has been placed in shallow water habitats, displacing that habitat type with islands. However, the surrounding sloping dredged material from unconfined disposal leading into the island beaches have generally been colonized rapidly with marine and aquatic organisms and seagrasses. These provide feeding areas for nesting waterbirds and enhance the possibility that they will select a certain island for nesting purposes.

Fifth, larger dredged material islands, especially CDFs, provide shallow water feeding habitats within the island complexes. Shifting sediment from current and wave action on newly placed dredged material islands, especially in Florida, has caused some islands to form shallow enclosed or semienclosed ponds. In addition, all CDFs have containment ponds that are usually quite large (Gaillard Island's is 700 to 800 acres of shallow water). These containment ponds are not only protected from bay wave action but are provided with tidal interchange through weirs. Abundant shrimp, crabs, and fishes are found in containment ponds, and they are prime feeding habitat for the nesting birds. On Gaillard Island, for example, both brown and white pelicans congregate and feed by the hundreds at the weir in the containment pond where water exchange takes place.

HABITAT NEEDS AND PLANNING

Studies conducted by the CE and the U.S. Fish and Wildlife Service have brought attention to the needs of sea and wading birds which use dredged material islands. They have also pointed out means by which islands could be altered, constructed, and/or managed for wildlife.

At the present time in many U.S. waterways, there is a scarcity of undisturbed bare sand habitat for those species such as terns and skimmers who require these bare habitats for nesting. The lack of undisturbed bare ground habitat is especially critical in the Great Lakes, the Pacific Coast, the interior river systems, and along the Gulf Coast. Forested islands are needed along the Jersey coast where wading birds nest in the salt marsh for lack of tree and shrub habitat. These needs increase as more and more people find older islands attractive for recreational purposes, thus preventing much use by wildlife. Many other needs are region-specific but are known to local biologists who can be used as experts to make recommendations for ways to develop these islands beneficially.

Needs can be fulfilled using one or a combination of three techniques: habitat manipulation, habitat establishment, and habitat protection. Manipulation of island habitats is by far the most likely technique to be used. For an engineer, this would include proper placement and/or shaping of dredged material to maintain or reestablish habitats, increase the size of existing islands, and/or change configuration, elevation, vegetation, and other features for more desirable habitats (Fig. 9.30). Biologists would use habitat manipulation to establish new vegetation and/or manage existing vegetation on islands through various agronomic or horticultural techniques, including bioengineering and vegetation removal or control.

Establishment of new habitats is generally only needed when a nesting habitat is lacking and new islands must be created, with the resulting need for vegetation establishment, or when a nesting habitat is expanded by island additions. New habitats can also be established by removal of unwanted vegetation and the introduction of most desirable vegetation in order to attract certain wildlife species to the sites.

There are many ways to offer habitat protection, but the most effective of these is posting and fencing of islands or nesting sites to obtain isolation, and the hiring of wardens to patrol these islands, as the National Audubon Society does. Although all of these wildlife species are protected by migratory bird laws, their habitats are not protected except when the birds are actually using them for nesting. Year-round protection to prevent destruction of habitats from year to year and seasonal protection to prevent nesting colony disruption by humans and

FIGURE 9.30 Dredged material, especially sand, may require contouring after placement on an island to achieve correct elevations, slope, and topographic relief desired for target wildlife species.

predators are necessary. Some of these species are very site-tenacious, and will return to the same island to nest until the habitat has been so totally disturbed that they fail as a colony.

Management of dredged material islands already in existence has been shown for a number of years to be an effective disposal technique and wildlife management practice.[3] Disposal of large quantities of dredged material offers considerable potential for the alteration and improvement of critical habitat. For example, sandy dredged material can be placed over existing undesirable vegetation so that the site becomes habitat. Management of existing islands receives considerable attention in the United States because the potential environmental impacts of disposal material on an existing site are less than those of developing new islands which would cover water bottoms or wetlands. This is a matter of great concern to such agencies as the National Marine Fisheries Service and the U.S. Fish and Wildlife Service and has already been a point of contention between them and the CE.

ISLAND CONSTRUCTION

Once the need and development feasibility for a new island or an addition to an island has been determined, island design must be considered. There are three basic aspects important to consider for wildlife use: site location, timing of development, and physical design. Long-term manageability must also be considered as an island is being designed.

Location

Site location of an island should be worked out with knowledgeable wildlife biologists and concerned agencies to establish the best locations for optimum wildlife use. Building or adding to an island in an area that does not conform to the biological and engineering specifications known to be required for target wildlife species will fail to produce the desired wildlife habitat. The islands and additions must be placed where the birds will be isolated from predators and human disturbances unless the islands are going to be actively protected. With active protection, colonies of sea and wading birds have been successfully close to human activities and have provided tourist attractions that could be observed from outside the colonies.

Timing

Timing of island development is important. Ideally, an island built for nesting should be built during the fall or winter preceding the spring breeding season. Sea birds generally do not use a dredged material site until after the initial sorting of fine materials by wind and water. If it is built in the spring, this sorting will not have had time to take place, and a colony of birds trying to nest there may not be successful. Frequently, their eggs will be covered by drifting fine material. In addition, they cannot use a site of silty dredged material until it has had adequate time to dewater.

Physical Design

The physical design of an island is important for wildlife use. In general, islands must be permanently emergent at high water levels. Birds have been found nesting on all sizes and configurations of islands as long as they met this one crucial breeding requirement. However, observations of hundreds of bird colonies on dredged material islands and the kinds of islands they select have led to the following four categories of recommendations concerning size, configuration, substrate, and elevation.

There are numerous problems that may be encountered in island construction. One important aspect to remember is that during early planning stages, coordination, cooperation, and communication with federal, state, and local agencies and concerned private and public organizations should be sought to prevent the erection of obstacles to project success. Public awareness of the positive effects of disposal of dredged material to build wildlife habitats should be part of every dredging operation where wildlife habitat creation is a goal.

The development of specifications for dredged material disposal to build islands for habitat and at the same time satisfy the need to dispose of a given amount of dredged material requires considerable care. Exact locations, timing of disposal, size of deposit, elevation of deposit, and movement of dredging pipe to assure that habitat plans are carried out should be in the specifications before the project gets under way. On-site monitoring is highly desirable, and dredging inspectors should be well informed as to the habitat goals of the dredging project. Monitoring is absolutely necessary when disposal is onto an island with an existing bird colony or populations of vulnerable wildlife. Dredging windows to prevent dredged material placement during critical nesting or breeding seasons are now often part of the contract specifications, but if it is necessary to dredge when nesting is taking place, it must be done with extreme care to avoid the nesting area and to prevent disturbance of the critical wildlife areas.

Protective Structures

Temporary dikes or breakwaters may be required in island construction and habitat development, including bioengineering (a combination of physical structure and plant material). If a dike is built on an existing island and filled to create an addition or to raise the elevation of the existing island, the dike should be at least partially removed or breached to allow ground access to water by young sea birds. This will require returning to a site with earth-moving equipment. For best use by wildlife, dikes do not need to be erected until just prior to disposal operations. Periodic monitoring to determine any impacts of disposal will provide useful information for future disposal efforts.[1,2,7]

Size

Nesting islands generally should be no smaller than 5 acres and no larger than 50 acres. However, birds have been found nesting on both smaller and larger islands. Islands larger than 50 acres are more difficult to manage and would be more likely to support predator populations such as coyotes, snakes, foxes, raccoons, and feral cats and dogs. Smaller islands tend to erode and overtop during storms more frequently than larger ones. Islands between the two extremes can be easily managed and considerable habitat diversity can be achieved on them. However, larger isolated islands have been found to be very successful (Gaillard Island in Mobile Bay, Miller Sands Island in the Columbia River, Pointe Mouillee in Lake Erie, Michigan, and others). Generally, the greater the amount of habitat diversity to be maintained for wildlife populations, the larger the island should be. For example, it is often possible to include upland (woody, herbaceous, bare), wetland (fresh/brackish and salt marsh or mangroves), and aquatic (shallow and deep water) habitats in an island design. This has already been done at several larger CE dredged material island sites.[5,6,7]

Configuration

The configuration of an island will depend upon the target wildlife species. Steep slopes such as those found on high dikes should be avoided if possible for all sea and wading bird species. It is also very difficult to establish and maintain vegetation on such steep slopes before erosion affects the dike and surrounding wetland habitats. A slope with no greater than a 3-ft rise per 100 linear feet has been recommended.[8] Substrate configurations for ground nesters must have gentle slopes to prevent their eggs from rolling from nest scrapes. There is also evidence that the formation of a swale or pond on an island or between two islands or additions makes the nesting site more attractive to nesting birds because they can feed in these shallow water habitats.

Nesting Substrate

The general nesting substrate requirements of colonial bird species are quite varied and can be obtained in detail.[3,11] In general, the coarser material such as sand or cobble make better nesting substrates because of their greater stability. Fine material such as silt and clay are subject to wind and rain erosion and usually have desiccation cracks and settle and pond. A mixture of sand, small gravel, and/or crushed shell material makes good nesting substrate for most ground nesting birds which prefer bare areas. They traditionally nested on such sites before being forced off by human use. Fine-grained unstable dredged material may be stabilized to form suitable nesting substrate by adding coarse materials such as shells, pebble gravel, or a layer of coarse sand over its surface or by planting a ground cover on the material to provide vegetation for those species which prefer that kind of habitat, usually Forster's terns and gull species. Wading bird species prefer woody habitats which often grow best on silty, more fertile dredged material substrates.

Elevation

Elevations of constructed islands should be high enough to prevent flooding of the areas that could be used by wildlife, especially for nesting. However, elevations do not need to be so high that the substrate will not become stabilized because of wind erosion. Generally, the optimal elevation for an island is 3 to 10 ft above mean high water. The desirable elevation to be achieved will depend upon texture of the exposed dredged material, wind exposure, and the habitat objectives or target species. Coarser materials will stabilize at higher elevations than finer materials. If islands or additions could be constructed of coarser material, it would be acceptable in some cases to exceed the recommended elevation. In general, the higher the elevation, the more slowly the island will be colonized by plants. Therefore, lower final elevations are desired to achieve plant cover for some ground nesting species and all tree nesting species, where those are the target wildlife species, and where substrates are of fine-textured material. It should be remembered that, given the proper substrates and vegetation for nesting, none of the species using dredged material islands for nesting choose one elevation over another elevation, as long as they are above the tide or flood lines and not eroding severely.

Island Additions

Construction of additions to existing islands is a viable disposal alternative and is one that may be useful in situations where valuable nesting sites are altered by erosion until they eventually are abandoned by nesting birds. Such additions will prolong the life of an island and its usefulness as a wildlife habitat. Additions to islands which are already covered with vegetation will increase habitat diversity by providing some bare ground habitat, at least temporarily, for those forms of wildlife requiring bare ground. Such additions in south Florida, for example, placed in such a manner as to encourage the growth of mangroves, have provided successfully used habitat for tree nesting birds.[17] Many colonies along the Gulf and Atlantic Coasts have responded favorably to island additions, especially bare ground nesters.

THE POTENTIAL FOR ISLAND MANAGEMENT

Tremendous, relatively inexpensive potential exists for dredged material island and site management using creative dredging work and cooperative, far-sighted management techniques. There are three ways in which the CE can manage dredged material islands and sites for wildlife and fish.[2,3] The first is through the placement of dredged material on a timely, rotational basis to stabilize existing islands and to provide habitat diversity. The second is to work with the National Audubon Society and other interested public and private organizations on a long-term basis to find ways dredged material can be used beneficially where waterbird nesting colonies or marine habitats are concerned. The third way the CE can and is managing dredged material islands is to post and to provide habitat protection at known nesting colonies on active CE projects.

The responsibility for management does not rest solely with the CE. Other federal, state, and local governmental agencies must play active roles in nesting colony protection (Fig. 9.31) and in cooperative efforts with the CE to actively manage and maintain waterbird nesting sites on public lands. This includes allowing the CE to place dredged material that will keep diverse habitats available for the various wildlife species who use dredged material islands and sites.

The National Audubon Society and other resource-oriented private organizations also have a responsibility for management. They too should be involved in active nesting colony protection on private lands, in acquiring such sites so that they can remain valuable habitats, and in cooperative efforts with the CE where

FIGURE 9.31 It is the responsibility of federal and state agencies and interested private groups to protect dredged material islands and their wildlife inhabitants. Miller Sands Island in the Columbia River has two nesting colonies and a great diversity of wildlife use. It is posted by the U.S. Fish and Wildlife Service.

sites under their organizational control can be nourished through beneficial applications of dredged material.

It is the responsibility of all of the above agencies and organizations to actively work with U.S. citizens and to educate them concerning nesting colonies, critical habitats, and migratory bird protection. It is also the individual responsibility of trained scientists and engineers to explain why the United States has to dredge and what can be done with this dredged material resource that is mutually beneficial to us all.

SUMMARY

This brief overview into a very large subject at least gives an indication of the magnitude of use and the importance of 2000 dredged material islands and sites to wildlife and fish in the United States; the northern Gulf Coast region was made a focus area because of the importance of its dredged material islands to wildlife. There are 645 Gulf Coast dredged material islands and sites offering wildlife habitat development, natural resource recreation, marine and wetland enhancement or creation, and a myriad of other potential beneficial uses. Broad baseline data are available to give us insight on biological requirements of species using these sites and to allow us to develop intelligent strategies for dredged material wildlife and fish management. This is especially true for nesting habitats for the thousands of waterbirds using dredged material placement sites.

ACKNOWLEDGMENTS

The overview presented here was based on the results of studies conducted since the mid-1970s of waterbird and other wildlife and fish use of dredged material islands and sites begun under the Dredged Material Research Program and continued under the Environmental Effects of Dredging Program, Dredging Operations Technical Support Program, and Wetlands Research Program, based at the U.S. Army Engineer Waterways Experiment Station, Vicksburg, Mississippi. Funding was provided by the Office of Chief of Engineers, Washington, D.C.

REFERENCES

1. Landin, M. C., 1980. "Building and Management of Dredged Material Islands for North American Wildlife." In *Proc., 9th World Dredging Conference,* Vancouver, British Columbia, Canada. pp. 527–538.

2. Landin, M. C., 1990. "Use of Dredged Material Islands by Colonial Nesting Waterbirds in the Northern Gulf Coast." In *Proc., Beneficial Uses of Dredged Material in the Northern Gulf Coast,* April 26–28, 1988, Galveston, TX. Technical Report D-90-3. U.S. Army Engineer Waterways Experiment Station, Vicksburg, MS. pp. 160–173.

3. U.S. Army Corps of Engineers, 1986. *Dredged Material Beneficial Uses.* Engineer Manual 1110-2-5026. Office, Chief of Engineers, Washington, DC. 297 pp. (author: Landin).

4. Culliton, T. J., et al., 1990. *Fifty Years of Population Change along the Nation's Coasts, 1960–2010.* NOAA Strategic Assessment Branch, Rockville, MD. 41 pp.

5. Landin, M. C., Patin, T. R., and Allen, H. H., 1989a. "Dredged Material Beneficial Uses in North America." In *Proc., 12th World Dredging Conference,* May 1989, Orlando, FL. pp. 821–832.

6. Landin, M. C., 1986. "The Success Story of Gaillard Island: A Corps Confined Disposal Facility." In *Proc., 19th Dredging Seminar and WEDA,* Baltimore, MD. pp. 41–54.

7. Landin, M. C., Webb, J. W., and Knutson, P. L., 1989b. *Long-Term Monitoring of Eleven Corps of Engineers Habitat Development Field Sites Built of Dredged Material, 1974–1987.* Technical Report D-89-1. U.S. Army Engineer Waterways Experiment Station, Vicksburg, MS. 192 pp. + app.

8. Chaney, A. H., Chapman, B. R., Kargas, J. P., Nelson, D. A., Schmidt, R. R., and Thebeau, L. C., 1978. *The Use of Dredged Material Islands by Colonial Seabirds and Wading Birds in Texas.* TR D-78-08. U.S. Army Engineer Waterways Experiment Station, Vicksburg, MS.

9. Schrieber, R. W., and Schrieber, E. A., 1978. *Colonial Bird Use and Vegetation Succession of Dredged Material Islands in Florida,* vol. I (wildlife). TR D-78-14. U.S. Army Engineer Waterways Experiment Station, Vicksburg, MS.

10. Soots, R. F., Jr., and Landin, M. C., 1978. *Development and Management of Avian Habitat on Dredged Material Islands.* TR DS-78-18. U.S. Army Engineer Waterways Experiment Station, Vicksburg, MS.

11. Dunstan, F. M., 1978. *Forty Years' Colonial Waterbird Data for Bird and Sunken Islands, Tampa Bay, Florida.* Subcontract report to TR D-78-14 by the National Audubon Society. U.S. Army Engineer Waterways Experiment Station, Vicksburg, MS.

12. Landin, M. C., 1985. "Bird and Mammal Use of Selected Lower Mississippi River Borrow Pits." Dissertation, Mississippi State University, Miss. State, MS. 405 pp.

13. Landin, M. C., and Miller, A. C., 1988. "Beneficial Uses of Dredged Material: A Strategic Dimension of Water Resource Management." In *Trans., 53d North American Wildlife and Natural Resources Conference*, Louisville, KY. pp. 325–334.

14. Parnell, J. F., Dumond, D. M., and Needham, R. N., 1978. *A Comparison of Plant Succession and Bird Utilization on Diked and Undiked Dredged Material Islands in the North Carolina Estuaries*. Technical Report D-78-9. U.S. Army Engineer Waterways Experiment Station, Vicksburg, MS.

15. Parnell, J. F., Dumond, D. M., and McCrimmon, D. A., 1986. *Colonial Waterbird Habitats and Nesting Populations in North Carolina Estuaries: 1983 Survey*. Technical Report D-86-3. U.S. Army Engineer Waterways Experiment Station, Vicksburg, MS.

16. Scharf, W. C., et al., 1978. *Colonial Birds Nesting on Manmade and Natural Sites in the U.S. Great Lakes*. Technical Report D-78-10. U.S. Army Engineer Waterways Experiment Station, Vicksburg, MS.

17. Lewis, R. R., III, and Lewis, C. S., 1978. *Colonial Bird Use and Vegetation Succession of Dredged Material Islands in Florida*, vol. II (vegetation). TR D-78-14. U.S. Army Engineer Waterways Experiment Station, Vicksburg, MS.

DREDGING CONTRACTS

William H. (Bill) Sanderson
Dredging Consultant
The Sand Hen Corporation
Wilmington, N.C.

Previous chapters of this handbook deal with the technical aspects of the mechanics of dredging. An understanding of the theory and the mastery of the operating techniques of the dredging process should improve production and lead to greater profitability. These are some of the essential elements of a successful project.

Profit is not always guaranteed. The best equipment and the most efficient operators are sometimes squandered on projects that suffer from poor contract decisions, inadequate investigation, inept bidding, or poor administration. Profit is often left on the table or wasted in futile attempts to recover from overlooked obscure requirements or obligations not fully understood. Enormous legal costs are often incurred when disputes arise.

From the owner's viewpoint, administrative mistakes, faulty contracts, and inept specifications can be just as devastating. Contracts and the precontracting process are almost always given less attention than they deserve. Poor performance, construction mistakes, and delays with high legal costs often result. Careful review of contract clauses and contract plans and specifications are seldom made by either party until the damage is done. The ounce of prevention is too often exchanged for the pound of cure when contracts are being written and bids are being made.

The practical meaning of contract language and the interpretations given in previous court decisions are seldom seriously considered during the contract formulation stage. The risks assumed, the potential for disagreement, and subsequent protracted litigation are given little consideration in the excitement of the bidding process.

This section attempts to alert the reader to some of the common mistakes made in formulating and administering dredging contracts. The contractor's and the owner's prospectives are explored with a view to encouraging a practical evaluation of both parties' obligations under a variety of contract circumstances.

CONTRACTS, SPECIFICATIONS, AND CLAIMS

Contracts Generally

A good and simple definition to remember is that:

A contract is a promise, or a set of promises, the breach of which the law provides remedy for, or the performance of which the law in some way recognizes as a duty.

Contracts are the subject of considerable federal and state law and are the root of much litigation over disagreements as to terms thereof. Contracts may be enforced under American contract law if the terms are reasonably definite, the subject matter is lawful, the parties are legally competent, and some considerations (either financial or mutual promises) are exchanged.

Contracts may be oral, the so-called handshake agreement, and still be subject to legal recognition. To enter into a dredging contract in such a fashion would be insane. Some vaguely written and ill-conceived contracts also may be equally dangerous. The author has been involved in some contract disputes that would have been easier to settle equitably had the contract not been written. The obligate oneself, either as contractor or owner, by the terms of a vague, ambiguous, or impracticable contract can easily end in economic disaster.

The contract in lay terms usually means the contract document which bears the signature of the principals and also the specifications and the voluminous laws, regulations, publications, and rules that may be incorporated by reference.

A recognized contract authority, Max E. Greenberg, wrote in the *ASCE Magazine* of May 1975 his assessment of construction contracts as follows:

A contract is a dangerous instrument. It should always be approached with trepidation and caution. Theoretically, the aim of a written contract is to achieve certainty of obligation of each party, the avoidance of ambiguities, and such definiteness of understanding as to preclude ultimate controversy. In practice, construction contracts are generally found not to definitely fix obligations, but to avoid obligations.

In this statement, Greenberg identified the root problem of most contract disputes, namely, an inappropriate contract whether by oversight or by deliberate design. A contract that contains ambiguous and exculpatory language intended as an escape from obligation and risk assignment is a poor contract indeed.

Contracts take many forms, all of which should attempt to be specific and to address the scope of the work to be done. Certain parts of most written contracts are "standard" clauses or forms that are used in a variety of procurement documents. Site-specific and work-specific requirements must be tailored to the job at hand. Commercial construction contracts most often follow those published by and available from the American Institute of Architects (AIA). Federal contracts are specified by federal law and regulation that have evolved over time from government experience with procurement and its problems.

Contract Regulations

Federal contract regulations are changed from time to time. Formats and nomenclature may change dramatically but the content and substance of contract clauses and provisions change very slowly. Changes of significance occur as a result of legal decisions and interpretations by the federal courts. Most changes over the years are minor and relatively obscure. The federal regulations that specify the duties of and restrictions on all Corps of Engineer (CE) personnel involved in the contracting process during recent years have been known by many titles such as: Engineer Contract Instructions (ECI), Defense Acquisition Regulations (DAR), Engineer Regulations (ERs), as well as local division (DIVRs), and district regulations (DRs).

Currently, all contract instructions are included in the format known as Federal Acquisition Regulations (FAR). There are certain deviations permitted, which in the case of federal construction contracts administered by or in accordance with CE authority, are known as Engineer, Federal Acquisition Regulations (EFAR). Special permission is required to deviate from the published FAR.

American dredging contracts, which are the subject of this text, are most often prepared by the CE using the FAR or by others such as port authorities, states, and municipalities who almost always elect to use CE's format and practices. In addition to instructions and other provisions, the FAR contains many standard clauses that for federal contracts must be used verbatim.

The contracting agency, such as the CE, using a matrix system selects the contract clause applicable to the work being done. Special permission is required to deviate from the standard language. Such deviations usually appear in the technical provisions section of a construction contract. The FAR is divided into sections pertaining to (1) construction, (2) supply, and (3) service contracts. All dredging contracts use *construction* FAR clauses and EFAR deviations.

Contract Form

Dredging contracts, both federal and private, may take several forms. The most usual contract type is known as the unit price contract. This format requires the measurement of the dredging done and payment based upon a unit price established by the bid document.

An acceptable form of contract, which in some cases is preferable to the unit price contract, is the format known as a *rental contract*. This practice avoids detailed measurement of work done but provides for payment based upon the time the dredge plant is operating. Inspection and record keeping become more important in this instance if the owner is to be protected. A rental contract is more detailed in its description of an acceptable plant and must contain a payment schedule that compensates for operating, idle, standby, and/or moving time. Some question whether the average contractor will produce as much per unit of time on rental as on lump sum.

A seldom used contract form, especially in public work, is a lump-sum contract that contemplates a turn-key performance without benefit of detailed description of quantities and other subsurface conditions. Such a contract would be entered into to restore a section of channel or to excavate an area to agreed-upon dimensions for a fixed amount, using whatever device the contractor may choose. The CE does not permit this form of contracting except in rare instances. It may be expedient to resort to this procedure in situations when time is the most important consideration.

Other forms of construction contracts, such as the different versions of cost-plus agreements, are not considered appropriate by the federal regulations but may be beneficial to some private owners. Such agreements must be made between parties that are prepared to rely heavily upon mutual trust and business friendship. The federal agencies are not prone to be good candidates for this arrangement.

Unless otherwise described, the comments and examples to follow will contemplate dredging contracts which follow the FAR and are based upon the unit price format of bidding and contract administration.

It is convenient for discussion to separate the contract documents into three parts: (1) the contract agreement, (2) the contract clauses, and (3) the plans and specifications.

The Contract Agreement

The contract agreement is the form used to make the offer to perform the work (contractor bid form) and the acceptance of the offer (award). These documents are legalistic in nature and contain information used by the contract administrator to establish the price and to identify the legal parties to the agreement. The bid form contains many certifications and acknowledgments that are often very important. Rules concerning surety, small business administration, and bid opening procedures are described.

Standard Contract Clauses

The contract clauses are those provisions that describe a wide variety of agreed-upon procedures and policies that must be followed by all parties. These provisions are often not fully appreciated by either party until contract administration has failed and disagreements have escalated to litigation. Both parties are usually surprised to learn the extent of the promises they made to each other when the bid was submitted and the award given. These clauses are the proverbial "fine print" that is often overlooked and misunderstood.

Among the contract clauses are the requirements included by reference, which are words that make applicable such restrictions as "all federal, state and local laws and ordinances" that may appertain to the work to be done. The requirement for permits, and the regulations of other federal agencies such as OSHA, Federal Fish and Wildlife, EPA, U.S. Coast Guard, etc., may be a part of the contract. Policy and procedure are laid out in the contract clauses that determine the process to be followed in cases of protests, disputes, modification, termination, default, etc. These clauses cover general requirements such as the necessity to adhere to laws concerning equal opportunity, affirmative action, a body of labor law, clean air and water acts, and the list is extensive. Currently there are more than 85 specified contract clauses that are found in most federal contracts. These clauses are applicable to all construction contracts, including dredging.

Special Clauses

Usually bound with the bid form and the construction contract clauses are the requirements, known currently as *special clauses*. A simple maintenance dredging contract may contain 35 to 50 such clauses, some of which are deviations from the FAR standard clause that have specific application to the work under the contract. These clauses are more job- and site-oriented than the standard contract clauses; many pertain to dredging as opposed to general construction.

Some of the more important special clauses treat such subjects as the time for commencement and completion, liquidated damages, physical data, wage rates, time extensions, variations in estimated quantities, survey procedures, plant requirements, quality control, and final examination and acceptance.

Technical Provisions

The smallest portion of the contract is usually the most important from the standpoint of performance and inspection of the work. The clauses, known as the *general requirements* or *technical provisions,* follow the other clauses and usually number from 15 to 20. As the terms imply, these clauses describe in detail the manner in which the work must be accomplished, how the work will be measured, the details of the survey procedure that will be used, more specific details of reporting and record keeping, inspections, compliance with navigation rules and regulations in the specific area of work, and most importantly, the character of the material that will be encountered.

Most contracts provide the contractor with sample report forms that are applicable to the work. These are usually appendices to the specifications. Some or all of these may be mandatory submissions required of the contractor.

Dredging Contract Problems

Since the FAR clauses are standard and appear in most construction contracts, it is beneficial for a student of the contract administration process to first obtain a set of dredging contract documents and read and try to understand the meaning of each clause. A complete FAR may be found in federal offices that have contracting authority, or it can be purchased from the Superintendent of Documents, U.S. Government Printing Office.

Seminars and training courses are available from a variety of sponsors that concentrate on specific aspects of contracts and contract administration problems. The subject is extremely broad, and contract situations that may arise are infinitely large in number. Unfortunately, the only way to become reasonably proficient is to work diligently with the real problems for a long time. For example, no one can appreciate the complexities of the litigation of a construction claim until he or she has been personally involved as a principal witness. A few such experiences will develop sensitivities to potential contract problems that will last a lifetime.

While it is impossible to cite solutions to every problem encountered in dredging contract administration, there are a few recurring situations that should be fully understood so that trouble can be recognized immediately.

The contracting and contract administration process must be understood. The use of standard clauses makes this task easier. The clauses are written in a straightforward manner and the words are not hard to understand. The meaning of the clauses and the impact of one upon the other as they relate to accepted principles of contract law become more complicated. We find that the clauses mean what the courts have said they mean. Board and court decisions, therefore, become the textbook of what one might expect as an outcome of a particular set of circumstances. Dredging contracts are not as complicated as some brick and mortar contracts are because they deal with fewer separate parts and the process is relatively uncomplicated.

Authority

The first important question to ask may be "Who is in charge of the contract?" In government work there is one person designated as the contracting officer (CO) and one person designated as the contractor. All others act for these principal parties and have little, or sometimes no authority, on their own. It is important to understand that certain actions can only be done by people with the proper legal designation. Inspectors have no authority to make any contract decision; they only report what they see and hear to the contracting officers. In matters that affect the contractor's interest, therefore, an inspector must be certain that the contracting officer is a party to any agreement or compromise. In private work the same principle is true. The person signatory to the contract must personally act or someone to whom specific authority has been given must act in his or her name if the rights given by the contract are to be protected. The author has been employed to assist in contract dispute settlements where many of the standard procedures set out in the contract clauses concerning notification, submission of timely complaints, etc., were not followed. Such circumstances allow the other party to plead that his or her rights were damaged thereby, and the judge may agree.

The contractor must understand and must rigorously follow the procedure and the chain of command established and the owner must do likewise. The authority of the participants should always be stated in writing for the record.

Disputes, Claims, and Appeals

Owners and contractors are perpetual optimists. Everyone hopes that each dredging contract will begin on time, proceed smoothly, and be completed ahead of schedule. To make these events a perfect contract, the work would pass final inspection and the contractor would be promptly paid. There would be no contract modifications and no disputes.

Some contracts turn sour and disagreements become disputes, disputes lead to lawsuits, and both parties usually lose money and sometimes more. When the parties resort to litigation, it is evidence of a complete failure of contract administration.

The Disputes Article

In federal contracts the disputes article (FAR 52.233) gives the rule concerning the procedure of appeal from an adverse ruling of the owner or contracting officer. The clause makes clear that all the provisions of the Contract Dispute Act of 1978, *et seq.,* apply and states the procedures that must be followed. The article also requires the contractor to proceed diligently with performance of the work while the appeal is being processed and perhaps a board hearing or court trial is occurring. The author pursued a termination for default of a major dredging contractor because the contractor failed to prosecute the work and insisted upon delaying the job until his appeal was handled to his satisfaction. The ensuing litigation lasted over 10 yr. The termination was upheld by the court and the work was completed by another contractor. The only winner was the contractor's lawyer.

All private contracts should have a disputes article, but some do not. In the absence of a specified set of rules to settle disputes that arise, the aggrieved party must resort to civil court. This is a protracted and costly procedure, and for business disputes, equity in the courtroom is not assured. The more enlightened private contracts now contain provision for arbitration of disputes.

The author recalls a dispute arising from a contract performed in a foreign country under a very poor contract that did not provide for disputes. As a matter of fact, the hapless contractor accepted a contract that actually said that no dispute would be allowed. The contractor fell behind schedule; the owner ceased making payments and invited the contractor to leave the job. Needless to say, the American contractor was at a great disadvantage and did not pursue a claim in the foreign court. Because of this, and a few other mistakes, this company is out of business.

In federal contracts an appeal can be taken from CO decision either to the agency board (the CE's Board is the Engineer Board of Contract Appeals [Eng. BCA]) or to the U.S. Court of Claims (since 1982 known as the U.S. Claims Court). Most contract appeals within the CE's system are taken, at the contractor's election, to the Eng. BCA.

If a board decision is unacceptable to either party, an appeal can be taken to the U.S. Court of Appeals for the Federal Circuit. Needless to say, few CE decisions are appealed by the government but contractors often resort to the court to attempt to have an adverse board decision overruled. The Eng. BCA is very sensitive to being overruled by the court and for that reason, among others, the board decisions are often felt to be slanted in favor of the contractor.

Differing Site Conditions

A differing site condition is the subject of most dredging claims. A claim of differing site conditions presents some very difficult problems for both sides. Contracts never perfectly describe a subsurface condition. Either too much or too little is said. The character of materials paragraphs are often complicated by highly technical language and descriptions including the specification writer's opinions concerning the meaning of the geologic terms used and his or her attempts to discuss dredgeability. Board decisions are often influenced by contractor's experts who dissect the words of the character of materials paragraph on a technical level far above that understood by the contractor, his or her estimator, or his or her operating people. Such experts have the privilege of examining the words and the physical evidence after the fact and rely upon subtle and often obscure details to show defects in the specifications. I am convinced that the best specification for describing the subsurface condition is to describe the testing and investigatory methods used, make the raw data available for review by the bidders, and refrain from writing multiple pages of descriptions, opinions, and interpretations of the data.

There are two recognized types of differing site conditions, they are:

Type I, subsurface or latent physical conditions at the site differing materially from those indicated in the contract

Type II, unknown physical conditions at the site, of an unusual nature, differing materially from those ordinarily encountered and generally recognized as inhering in the work of the character provided for in the contract.

In the first instance, the contractor shows that the specification was defective in that it misdefined, misclassified, or misled the bidder by its words and inferences.

In the second instance, the contractor shows that he or she ran into unexpected obstructions that were not mentioned in the specification. To qualify as a Type II, the unexpected obstructions are assumed to have been unknown to the owner as well as to the bidder.

The Right to Make Changes

The changes article is the vehicle by which the contract may be modified after award. The contract reserves the right to the owner to make changes within the scope of the contract and provides that equitable adjustment be made (in time and money) for any change made. To make a change that is not within the general scope of the work requires a supplemental agreement. Such an agreement must be by mutual consent of the parties. Claims often arise from changes made under the changes article and usually these claims involve the calculation of cost to the contractor and the effect of sequential damage—the ripple effect.

The Right to Terminate

The contract provides for termination of a contractor's right to proceed with the work at any stage of the construction process. There are two types of terminations, as follows:

Termination for convenience of the owner is provided for when it is viewed to be in the best interest of the owner to cease work on a contract. This is a business decision made by the owner and does not often result in a claim. The contract provides that the contractor shall be paid for all work done and for costs incurred in work in process, including inventory, etc. This process is usually costly for the owner and requires considerable work by the contract administrators and auditors to reach a fair and equitable adjustment in the contract amount.

Termination for default is a last resort in the process of trying to accomplish a dredging project. Termination for default almost always leads to a claim. As a matter of fact, most cases that involve termination for default already are fraught with disagreement. Defending a default action against a viable contractor is the ultimate contract administration challenge.

Suspension of Work

Related to termination, but usually not as severe, is the right of the owner to suspend work on a contract for a variety of reasons. A suspension is temporary and is used to mitigate the cost of prolonged delays without resorting to cancellation altogether. This action is a great exercise for project personnel who must inventory the work in progress, the equipment status, and the demobilization expenses and submit many reports and records.

Board and Court Decisions

When we examine board decisions over an extended period of time, we find certain board concepts that are axiomatic in that they usually affect the decision in a very predictable way. Some of these often seen in dredging claims follow.

Ambiguities are statements that are subject to more than one interpretation by reasonable people. Such statements are the root of most disagreements and the rule followed by the board is that since the government wrote the contract, it generated the ambiguity and therefore must suffer the consequences of any *reasonable* interpretation of it.

Exculpatory clauses are statements that are intended to negate an obligation that other parts of the specification provide for. Such clauses are very often used to avoid risk or to transfer risk to the contractor. For example, a contract includes a two-page description of the nature of the material to be dredged by giving geologic names, laboratory test data, and physical descriptions. The specification writer then often adds a caveatory and exculpatory paragraph such as:

> The information and data furnished are not intended as representations or warranties but are furnished for information only. It is expressly understood that the government will not be responsible for the accuracy thereof or for any deduction, interpretation or conclusion drawn therefrom by the contractor. (Quoted from a CE contract.)

The overriding rule in this case is that the contractor is entitled to rely upon all the data and information given in the contract if it appears reasonable and is without obvious error.

The courts and the board are hostile toward disclaimer clauses generally. A frequent attempt is made by the government in dredging contract litigation to plead that the contract tells the contractor to make site investigations and to satisfy him- or herself as to the nature of the material, the site conditions, etc. This defense does not usually provide a shelter from a claim of differing site conditions.

Warranty of Plans and Specifications.   Dredging contracts are supposed to contain the information necessary to allow a contractor to make an appropriate bid. Engineering is performed by the owner by use of hired labor, or by AE contract, and the owner is assumed to warrant the information presented as being correct and complete. Contractors are not expected to add contingency amounts to ensure against inaccurate information or faulty engineering. The theory is that in accepting the liability for its own possible mistakes, the owner secures the lowest possible bid for performing the work. This theory admittedly is idealistic, and there are endless circumstances that complicate the application of the principle. The standard FAR clauses provide remedy for claims based upon poor or defective contract information.

Equity. The theory used by the Eng. BCA and the ASBCA, differing somewhat from the civil courts, is derived from the idea that contract disputes should be settled so as to provide *equity*. Equity means that the board should listen to all the arguments and base its decision primarily upon correcting an injustice that has been inflicted upon the damaged party. The federal rules of evidence do not necessarily apply in board hearings.

Impossibility of Performance. The doctrine of impossibility has recently been liberalized to include the theory of impracticability, which means that the performance should not only be possible but economically realistic. A test of impracticability may be stated as whether the cost of performance is so much greater than the anticipated cost as to render performance commercially senseless. This remains a controversial legal issue and one that has many variations.

Technicalities. In civil court actions we see technicalities argued much more successfully than before the Board of Contract Appeals. It is a rare case indeed that is decided by the board solely upon the technicality of such arguments as improper notice, failure to notify the contracting officer directly (as the contract may require), or for taking instructions from an owner's employee not authorized to give them.

The board usually examines the contractor's view of a disputed fact in light of whether his or her conclusions were reasonable and, if so, whether these reasonable conclusions were the root cause of the injury he or she subsequently suffered.

Documentation of the Record

The most important single lesson to be learned by both the contractor's project personnel and the owner is the value of good documentation of the facts. Facts are evidence. Evidence is the essence of the negotiation and the litigation or administration process.

The job record must be accurate, complete, and clear to those who must rely upon it during attempts to settle disagreements at the job level and later in the courts. Good documentation is direct and contains no emotional statements; if opinions are stated, they must be justified. Attempts to mislead or to falsify the record usually fail to achieve the purpose intended; this is a very poor practice on either side. Good documentation is made as the events occur. Recollection of facts after the event takes place is never as good as a contemporaneous record. A lawyer cannot be expected to win a favorable decision at the bargaining table, or before the judge, if the facts were not properly documented.

The Effect of Government Practice on Private Contracts

The comments and examples previously given indicate that the author uses the terms *owner* and *government* interchangeably in most cases. Private industry A & E firms that prepare dredging specifications very often adopt the principal FAR clauses for their contracts. Some so-called private contracts are in essence copied from CE dredging contracts, at least in the most important parts. Some private contracts are partially federally funded and are thus required to follow certain parts of the FAR.

Whether the actual government contract is used or not, disputes between dredging contractors and private owners that are tried in court or settled by arbitration almost always are affected by expert testimony. Most expert testimony will relate to industry practice, which will usually have strong ties to the CE practice and procedure since the CE is universally recognized as the premier authority on dredging matters in the United States. The author has been involved in many cases that did not involve the government in any way, but the judge was always interested in what the CE would do or would say concerning the matter before the court.

Since the CE has such influence on dredging and dredging contract administration in all segments of the industry, it is wise for all so-called private contracts to adhere whenever possible to the CE's principles in writing and in interpreting the meaning of the contract terms.

Defensive Contract Administration

A summary of what must be done to prevent unsatisfactory results is perhaps oversimplified in the following statements:

1. The contract documents must be drafted and reviewed by persons who know what the project requirements are or what they should be. All levels of administration must make certain that the standards of performance are clearly set out in the contract.
2. Contract administrators must know the contract requirements. Those individuals with most authority must be sure that those with the least authority (usually the ones most involved with the work) understand the contract and communicate with those who must make the decisions.
3. The record should be thoroughly, accurately, and faithfully documented, and those at the upper levels of authority must constantly review and understand the significance of the daily documentation.
4. Problems must be anticipated. Experience and practice are the only teachers that will allow the project manager to be a good anticipator of problems. Those at higher levels of authority must understand the requirements and the level of performance that is being observed by inspection and must concentrate supervision and inspection where problems are most likely to occur.
5. All levels of the administration network must approach their work with a professional attitude, being certain that they are knowledgeable of the requirements and of the daily work performance, that personalities are not allowed to be a factor in decisions, and that each event is viewed with the proper amount of skepticism so that a worst-case scenario will not be a surprise.

Disagreements, disputes, claims, and appeals will always occur so long as the contract is an imperfect instrument. There has never been a perfect contract. The better the contract promises are stated and the better the understanding between the parties is established, the less likely we will be forced to resort to lawsuits for settlements of differences. The English language encourages ambiguous statements and obscure meanings; human nature encourages self-serving interpretations and selfish motives.

ALTERNATIVE DISPUTE RESOLUTION

Litigation the Conventional Way

The disputes that arise from contracts with the government and between private sector parties have historically been settled by litigation. The CE and some other government agencies that do considerable business through contracts with outside entities established their own agency boards that adjudicate claims and appeals. The FAR provides standard contract language which establishes the policy and procedure that must be followed in processing a dispute through the system. The system is litigation. The ultimate resolution of litigated disputes is often found in the U.S. Claims Court system.

The system operated well for many years and most results were equitable to both parties. Changes have occurred in the philosophy of business transactions and in the attitudes of contracting parties that have rendered the present system of litigating disputes in some respects inadequate. The court system is overburdened with cases, and the administrative boards which function as the courts do cannot be expanded fast enough to provide room for the cases to be heard. The approaching crisis has been apparent for at least two decades. All parties agree that different methods must be adopted to overcome the slow, tedious, and expensive process that can no longer guarantee equity. The very fact that disputes that proceed through the system require in excess of 3 to 5 yr to reach final settlement is generically inequitable, regardless of the correctness of the decision. Time erodes the value of otherwise equitable settlements and the expense of litigation has become overwhelming.

New Thinking

The alternatives to litigating business disputes are not yet universally accepted. As a matter of fact, controversy still prevails even though everyone agrees that the court system is burdened beyond its ability to function. The reasons are many, not the least of which is the reluctance of some lawyers to agree to vacate the court system and the inability of the client to operate without legal counsel. Alternative dispute resolution (ADR) techniques are, however, proliferating and are sure to emerge as accepted standard procedures in the not too distant future. Necessity is, after all, the mother of invention.

The Alternatives

Mediation-based techniques are forms of voluntary procedure that are gaining acceptance for a variety of disputes and can be beneficial in the contract (business) disputes in which we are interested. Mediation is a process that depends on a third party to help the disputing parties reach a mutually satisfactory agreement. The mediator encourages the parties to identify the problem and may suggest solutions. The mediator should not coerce or pressure but should work toward an agreement the parties make themselves. This works best when the parties wish to continue a business relationship and wish to settle the issues between them. Successful mediation always requires a skillful person to act as the third-party mediator.

Arbitration-based techniques are more formally structured and involve the submission of the dispute to a third party or a panel who renders a decision after hearing the case argued and reviewing evidence. The procedure is more informal than litigation but more structured than mediation. Arbitration as a technique is not new and is often (in private contracts) an accepted alternative dispute resolution technique. There are a number of private arbitration activities affiliated with state and federal agencies as in-house arbitrators of specific classes of disputes. There are arbitration centers in universities which offer a variety of services to the public concerning binding and nonbinding arbitration; the insurance industry often provides and encourages the use of arbitration to settle insurance-related disputes, as does the Department of Labor. The courts in some jurisdictions have ordered arbitration at the federal and state levels for some types of disputes.

The American Arbitration Association (AAA) is a nonprofit, private corporation engaged in a variety of arbitration variations often tailored to fit a business situation. The federal and state courts recognize the procedures and encourage their use. All arbitration is voluntary dispute settlement. If the parties agree to binding arbitration, the courts will enforce the arbitrator's decision. The best results are obtained when the contract document includes a requirement that stipulates dispute settlement by means of arbitration and references the rules under which arbitration will proceed. The justification for this method of dispute settlement is to provide a simple, economical, and understandable system for obtaining practical solutions to business disputes. Arbitration seems to be well suited to the settlement of dredging contract claims and other fact-oriented disputes.

Advantages of Arbitration

Arbitration is a legally recognized dispute settlement technique that, if managed correctly and with the proper contract provisions, is final and binding on both parties. The process is almost always less expensive and much faster than litigation. The outcome should be equally as equitable as litigation. Since the parties select the arbitrator and the rules of civil procedure do not apply, the parties to the dispute are less restrained at the hearing. The proceedings are private and the matter, including the settlement, is confidential.

What Is Better than Arbitration or Litigation?

Settling the disagreement at the earliest possible point is always preferable to any alternative. All construction contract claims and appeals usually begin with a disagreement in the field or in the interpretation of a contract provision. The disagreement escalates into a dispute, the dispute into a claim, and the claim results in a contracting officer's decision. From the decision an appeal is usually made and litigation follows.

The dispute settlement alternatives escalate similarly. The dispute settlement continuum takes the following path (*Source*: American Arbitration Association):

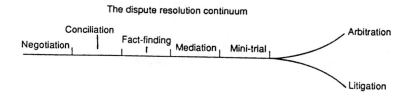

The dispute resolution continuum

As the process moves from left to right on the diagram of options, the cost increases. With each escalation of confrontation, each party spends more time in preparation and presentation; as the process is moved toward the ultimate conclusion, the parties lose more and more control.

The Corps of Engineers' Position

At the present time the CE must use the FAR and the contract procedures that are consistent with those regulations. The use of agency boards and the U.S. Claims Court system are the options in the contract. Arbitration is not specified in the government construction contract and an agreement to use that technique must presently be a voluntary choice agreeable to both parties. Binding arbitration, absent a contract provision, is not enforceable by the courts under the present status of the federal contract clauses, even if both parties agree to enter into such arbitration.

The Mini-Trial

A March 1989 article written by Lester Edelman, Esq., the Chief Counsel of the Corps of Engineers, Office of Chief of Engineers, Washington, D.C., is summarized as follows:

> The U.S. Army Corps of Engineers, concerned about the increasing time and expense to settle government contract claims, examined alternatives to the traditional method of resolving disputes before boards of contract appeals. The option that was chosen was the mini-trial, a voluntary, expedited, and non-judicial process whereby the top management officials of each party meet to resolve a dispute. The Corps of Engineers adapted the mini-trial to best suit its own organizational needs.

When the "adapted" mini-trial is examined, we see that it is a blend of characteristics from several dispute resolution sources—negotiation, arbitration, and mediation.

The mini-trial is not a trial at all; the misnomer is unfortunate. It is better described as an elaborate form of mediation. There is no judge or lengthy procedures. Decisions are made by the managers who begin by determining that they desire a settlement without litigation. A mini-trial agreement is developed which defines what the parties want to happen before and during the mini-trial. The agreement is a guide which specifies roles, time limits, schedules, procedures, and limits on the discovery process. Attorneys are not allowed to present lengthy briefs and arguments; the process is deliberately kept short and to the point. The focus in preparing the case must be on the best argument that will convince the other side, not a judge or jury.

Both sides must be represented by the top management people who are assisted by a neutral advisor (optional). The advisor may actually preside over the conference. Having top management present is the unique feature that makes the process work. These people usually have heard only their organization's position, not the other side, and they have the authority to make decisions and settlements. Mini-trials are not appropriate for all issues. To be a candidate for this alternative dispute settlement method, the issue should be of such importance as to justify the expense of the case preparation and the conferences that will be required. Also, the case should involve disagreements concerning facts, not purely matters of law, and the case should be subject to negotiation as opposed to the necessity to declare one side a complete winner or loser.

The CE is committed to employ any means that is likely to result in settlements that are satisfactory to both sides without resorting to litigation.

Some day soon the FAR will be revised to include an arbitration clause or perhaps a clause that may make a variety of procedures, if selected, to be binding on the parties.

The common contract clause used in other than government contracts which is referred to as the Arbitration Clause is:

> Any controversy or claim arising out of or relating to this contract, or the breach thereof, shall be settled by arbitration in accordance with the Commercial Arbitration Rules of the American Arbitration Association, and judgment upon the award rendered by the arbitrator(s) may be entered in any court having jurisdiction thereof.

BIBLIOGRAPHY

Arbitration Journal, a Dispute Resolution magazine, American Arbitration Association, 140 West 51st St., New York, NY.

Colson, R., *Business Arbitration, What You Need to Know*, American Arbitration Association, New York, Oct. 1987.

———, *Business Mediation, What You Need to Know*, American Arbitration Association, New York, May 1988.

Construction Industry Arbitration Rules, American Arbitration Association, New York, Jan. 1, 1990.

Edelman, L., Carr, F., and Creighton, J. L., *The Mini Trial,* IWR Pamphlet 89-ADR-P-I, U.S. Army Corps of Engineers, April 1989.

Federal Acquisition Regulation System, Title 48 Code of Federal Regulations. U.S. Government Printing Office.

Greenberg, M. E., "The Disadvantages of Arbitration," *McGraw-Hill Construction Business Handbook,* McGraw-Hill, New York, 1978.

House, L. P., Coleman, A. L., and Smith, G. A., *Construction Contract Litigation,* Federal Publications, Course manual, 1988.

Poulin, T. A., "Avoiding Contract Disputes," *Symposium Construction Division American Society of Civil Engineers,* Oct. 21–22, 1985.

Public Law 95-563, "Contract Disputes Act of 1978," 41 USC 601, Nov. 1978.

*REFERENCES**

1. Commoner, B., *The Closing Cycle,* Knopf, New York, 1972.

2. Herbich, J. B., "Environmental Effects of Dredging—The United States Experience," *Dock and Harbour Authority,* pp. 55–57, July/Aug. 1985.

3. "Ocean Dumping, Final Revision of Regulations and Criteria," Environmental Protection Agency, *Federal Register,* Jan. 11, 1977.

4. *Biological Evaluation of Proposed Discharge of Dredged Material into Ocean Waters,* Environmental Protection Agency/Corps of Engineers Technical Committee on Criteria for Dredged and Fill Material, U.S. Army Engineer Waterways Experiment Station, Vicksburg, MS, July 1977.

5. "Guidelines for Specification of Disposal Sites for Dredged and Fill Material," Environmental Protection Agency, *Federal Register,* Dec. 24, 1980.

6. Herbich, J. B., *Coastal & Deep Ocean Dredging,* Gulf Publishing Company, Houston, TX, 1975.

7. *Dredged Material Research Program, Executive Overview and Detailed Summary,* T.R. DS-78-22, U.S. Army Engineer Waterways Experiment Station, Vicksburg, MS, Dec. 1978.

8. *Dredged Material Research Program, Publication Index and Retrieval System,* T.R. DS-78-23, U.S. Army Engineer Waterways Experiment Station, Vicksburg, MS, Apr. 1980.

9. Huston, J., and Huston, W. C., *Techniques for Reducing Turbidity with Present Dredging Procedures and Operations,* C.R. D-76-4, U.S. Army Engineer Waterways Experiment Station, Vicksburg, MS, May 1976.

10. Apgar, W. J., and Basco, D. R., *An Experimental and Theoretical Study of the Flow Field Surrounding a Suction Pipe Inlet,* Report no. CDS-172, TAMU-SG-74-203, Texas A&M University, Oct. 1973.

11. Hudson, R. E., and Vann, R. G., "An Overview of a Dredging Demonstration in Contaminated Materials, James River, VA," *Proc., U.S./Dutch Memorandum of Understanding of Dredging Technology Meeting,* New Orleans, LA, Sept. 1982.

12. Herbich, J. B., Brahme, S. B., and Andrassy, C., "Generation of Re-Suspended Sediments by Dredges," *Proc., XIIth World Dredging Congress,* Orlando, FL, May 1989.

13. Wechsler, B. A., and Cogley, D. R., *A Laboratory Study of the Turbidity Generation Potential of Sediments to be Dredged,* T.R. D-77-14, U.S. Army Engineer Waterways Experiment Station, Vicksburg, MS, Nov. 1977.

14. Barnard, W. D., *Prediction and Control of Dredged Material Dispersion Around Dredging and Open-Water Pipeline Disposal Operations,* T.R. DS-78-13, U.S. Army Engineer Waterways Experiment Station, Vicksburg, MS.

15. Yagi, T., et al., "Turbidity Caused by Dredging," *Proc., Seventh World Dredging Conference, WODCON VII,* San Francisco, CA, June 1976.

16. Bartos, M. J., *Classification and Engineering Properties of Dredged Material,* T.R. D-77-18, U.S. Army Engineer Waterways Experiment Station, Vicksburg, MS, Sept. 1977.

17. Herbich, J. B., and Brahme, S. B., *Literature Review and Technical Evaluation of Sediment Re-Suspension During Dredging,* C.R. HL-91-1, U.S. Army Engineer Waterways Experiment Station, Vicksburg, MS, 1991.

18. Hayes, D. R., *Guide to Selecting a Dredge for Minimizing Re-suspension of Sediment,*

---

*These are the references for the material written by John B. Herbich.

Environmental Effects of Dredging, EEDP-09-1, U.S. Army Engineer Waterways Experiment Station, Vicksburg, MS, Dec. 1986.

19. Sato, E., "Bottom Sediment Dredge *Clean Up,* Principles and Results, Management of Bottom Sediments Containing Toxic Substances," *Proc., 8th U.S.-Japan Experts Meeting,* T. R. Patin (ed.), pp. 403–418, July 1984.

20. Shinsha, H. (personal communication), *Refresher* Dredge, Engineering Research Institute, Penta-Ocean Construction Company, Ltd., Japan, Feb. 1988.

21. D'Angremond, K., de Jong, A. J., and de Waard, C. P., "Dredging of Polluted Sediment in the First Petroleum Harbor, Rotterdam," *Proc., 3d U.S.-the Netherlands Meeting on Dredging and Related Technology,* U.S. Army Engineer Water Resources Support Center, Fort Belvoir, VA, 1984.

22. Hayes, D. F., McLellan, T. N., and Truitt, C. L., "Demonstration of Innovative and Conventional Dredging Equipment at Calumet Harbor, Illinois," MP EL-88-1, U.S. Army Engineer Waterways Experiment Station, Vicksburg, MS, 1988.

23. Richardson, T. W., Hite, J. E., Shafer, T. R., and Etheridge, J. D., *Pumping Performance and Turbidity Generation of Model 600.100 Pneuma Pump,* T.R. HL-82-8, U.S. Army Engineer Waterways Experiment Station, Vicksburg, MS, 1982.

24. Andrassy, C., and Herbich, J. B., "Generation of Suspended Sediment at the Cutterhead," *The Dock and Harbour Authority,* vol. 68, no. 797, pp. 207–216, Jan. 1988.

25. Herbich, J. B., and De Vries, J., *An Evaluation of the Effects of Operational Parameters on Sediment Re-suspension During Cutterhead Dredging Using a Laboratory Model Dredging System,* Report no. CDS 286, Center for Dredging Studies, Texas A&M University, College Station, TX, 1986.

26. Brahme, S., "Environmental Aspects of Suction Cutterheads," Dissertation presented to Texas A&M University in partial fulfillment of the requirements for the degree of Doctor of Philosophy, 1983.

27. Pequegnat, W. E., Smith, D. D., Darnell, R. M., Presley, B. J., and Reid, R. O., *An Assessment of the Potential Impact of Dredged Material Disposal in the Open Ocean,* Technical Report D-78-2, U.S. Army Engineer Waterways Experiment Station, Vicksburg, MS, Jan. 1978.

28. U.S. Environmental Protection Agency, "Ocean Dumping Final Criteria," *Federal Register,* vol. 38, pp. 12872–12877, 1973*a.*

29. U.S. Environmental Protection Agency, "Ocean Dumping Final Criteria," *Federal Register,* vol. 38, pp. 28610–28621, 1973*b.*

30. Keeley, J. W., and Engler, R. M., *Discussion of Regulatory Criteria for Ocean Disposal of Dredged Materials: Elutriate Test Rationale and Implementation Guidelines,* Miscellaneous Paper D-74-14, U.S. Army Engineer Waterways Experiment Station, Vicksburg, MS, Mar. 1974.

31. Ketchum, B. K., *The Water Edge: Critical Problems of the Coastal Zone,* MIT Press, Cambridge, MA, 1972.

32. Lee, G. F., "Recent Advances in Assessing the Environmental Impact of Dredged Material Disposal," Dredging, Environmental Effects and Technology: *Proc., Seventh World Dredging Conference,* WODCON VII, San Francisco, CA, pp. 551–578, July 1976.

33. Stern, E. M., and Stickle, W. B., *Effects of Turbidity and Suspended Material in Aquatic Environments, Literature Review,* Technical Report D-78-21, U.S. Army Engineer Waterways Experiment Station, Vicksburg, MS, June 1978.

34. Dredged Material Research Program, *Evaluation of Dredged Material Pollution Potential,* T.R. DS-78-6, U.S. Army Engineer Waterways Experiment Station, Vicksburg, MS, 39 pp., Aug. 1978.

35. U.S. Army Corps of Engineers, *Dredging and Dredged Material Disposal,* EM 1110-2-5025, 43 pp., Mar. 1983.

36. Dredged Material Research Program, *Water Quality Impacts of Aquatic Dredged Material Disposal (Laboratory Investigations),* T.R. DS-78-4, U.S. Army Engineer Waterways Experiment Station, Vicksburg, MS, Aug. 1978.

37. Wright, T. D., *Aquatic Dredged Material Disposal Impacts,* T.R. DS-78-1, U.S. Army Engineer Waterways Experiment Station, Vicksburg, MS, 57 pp., 1978.

38. Herbich, J. B., "Extent of Contaminated Marine Sediments and Cleanup Methodology," *Proc., 22d International Conference on Coastal Engineering,* ASCE, Delft, the Netherlands, 1990.

39. Marine Board, National Research Council, *Contaminated Marine Sediments—Assessment and Remediation,* National Academy Press, 493 pp., 1990.

40. Bokuniewicz, H. J., et al., *Field Study of the Mechanics of the Placement of Dredged Material at Open-Water Disposal Sites,* Technical Report D-78-7, U.S. Army Engineer Waterways Experiment Station, Vicksburg, MS, 1978.

41. Truitt, C., *Engineering Considerations for Subaqueous Dredged Material Capping,* Background and Preliminary Planning, Environmental Effects of Dredging Technical Note EEDP-01-3, U.S. Army Engineer Waterways Experiment Station, Vicksburg, MS, 1987a.

42. Truitt, C., *Engineering Considerations for Capping Subaqueous Dredged Material Deposits Design Concepts and Placement Techniques,* Environmental Effects of Dredging Technical Note EEDP-01-4, U.S. Army Engineer Waterways Experiment Station, Vicksburg, MS, 1987b.

43. Palermo, M. R., "Capping Contaminated Dredged Material in Deep Water," *Proc., Specialty Conference Ports '89,* American Society of Civil Engineers, Boston, MA, 1989.

44. Palermo, M. R., *Design Requirements for Capping,* Dredging Research Technical Note DRP-5-03, U.S. Army Engineer Waterways Experiment Station, Vicksburg, MS, Feb. 1991.

45. Sumeri, A., "Capped In-Water Disposal of Contaminated Dredged Material," *Proc., Conference Dredging '84,* American Society of Civil Engineers, vol. 2, pp. 644–653, Nov. 1984.

46. U.S. Army Engineers, *Design Requirements for Capping,* Dredging Research Technical Note DRP-5-03, U.S. Army Engineer Waterways Experiment Station, Vicksburg, MS, Feb. 1991.

47. Francingues, N. R., Palermo, M. R., Lee, C. R., and Peddicord, R. K., *Management Strategy for Disposal of Dredged Material: Contaminant Testing and Controls,* Miscellaneous Paper D-85-1, U.S. Army Engineer Waterways Experiment Station, Vicksburg, MS, 1985.

48. Shields, F. D., and Montgomery, R. L., "Fundamentals of Capping Contaminated Dredged Material," *Proc., Specialty Conference Dredging '84,* American Society of Civil Engineers, Clearwater, FL, 1984.

49. Vyas, Y. K., and Herbich, J. B., "Erosion of Dredged-Material Islands Due to Waves and Currents," Paper H2, *Proc., 2d International Symposium on Dredging Technology,* BHRA and Texas A&M University, College Station, TX, Nov. 1977.

50. Edgar III, C. E., and Engler, R. M., "An Update on the London Dumping Convention and Its Application to Dredged Materials," *Terra et Aqua,* no. 29, pp. 16–23, Apr. 1985.

51. Engler, R. M., "The London Dumping Convention (LDC). Its Role in Regulating Dredged Material," *Proc., XII World Dredging Congress,* Orlando, FL, pp. 67–86, May 1989.

52. Hubbard, B. S., and Herbich, J. B., *Productive Land Use of Dredged Material Containment Areas: International Literature Review,* Center for Dredging Studies, Texas A&M University, CDS Report no. 199, 118 pp., Jan. 1977.

53. Landin, M. C., "Habitat Development Using Dredged Material," *Proc., Dredging and Dredged Material Disposal,* ASCE Conference, Tampa, FL, pp. 907–918, 1984.

54. U.S. Army Corps of Engineers, *Dredged Material Beneficial Uses,* EM 1110-2-5026, Mar. 1988.

55. Dijkman, F., Haas, A. W., Moll, R., de Nekker, J., and van Raalte, G. H., *Impact of Contaminated Sediments and River Regulation on the Aquatic Environment,* PIANC, 1989.

56. Blokland, T., "Determination of Dredging-Induced Turbidity," *Terra et Aqua,* no. 38, Dec. 1988.

57. Velinga, T., "Environmental Effects of Dredging and Disposal Operations," *Proc., XIIIth World Dredging Congress,* Orlando, FL, pp. 235–252, May 1989.

CHAPTER 10
INSTRUMENTATION AND AUTOMATION*

FLOW AND VELOCITY METERS

The most desired information in a dredging operation is the accurate determination of the amount of solid material which passes through the dredge pump. If this information is known, many other parameters for the dredge system can be accurately determined and controlled. To know the amount of material pumped requires knowledge of the density of the water mixture passing through the pump and the mixture flow velocity, or rate of flow. Numerous methods have been developed and employed to determine these values.

Magnetic Flowmeters

The magnetic flowmeter has been found to be capable of accurately measuring the flow velocity "within 1% or a percentage of full velocity reading, depending on adjunct instruments."[1] The meter determines the average velocity in the fluid passing through it, which, when multiplied by the area of the pipe, results in the fluid flowrate.

The magnetic flowmeter, basically an electromagnetic generator, operates according to Faraday's law of induction, which says that the voltage induced in a conductor moving at right angles through a magnetic field will be proportional to the velocity of the conductor through the field (Fig. 10.1). In a magnetic flowmeter the liquid serves as a conductor. As the liquid flows through the magnetic lines of force produced by the electromagnets of the flowmeter, a voltage directly proportional to its velocity is induced in the liquid. This induced voltage, which is perpendicular to the liquid and the magnetic lines of force, is drawn off through the electrodes of the flowmeter and transmitted to the electronic receiver instruments (Fig. 10.2).

A flow velocity V through the region of the magnetic field causes a voltage E to appear between the electrodes, or

$$E(V) = K(H)V \qquad (10.1)$$

*References for the chapter section written by John D. Herbich appear at the end of the chapter.

10.1

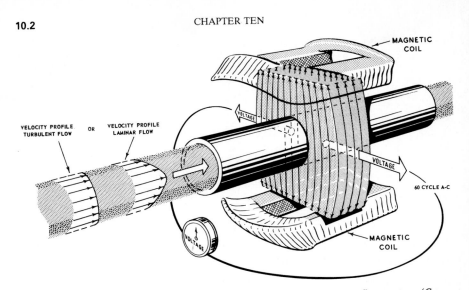

FIGURE 10.1   Basic operating principle of magnetic flowmeter typical mass flow system. (*Courtesy, Fischer & Porter Co.*)

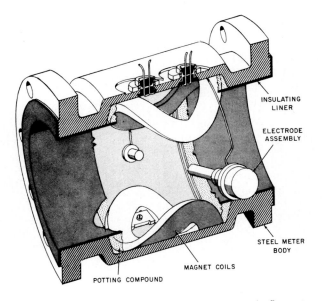

FIGURE 10.2   Section through a short-form magnetic flowmeter with characterized coil showing the construction used for 6-in pipe size and larger. (*Courtesy, Fischer & Porter Co.*)

where H = magnetic field strength
K = factor of proportionality

Thus, the voltage appearing between the electrodes is a direct measure of velocity.[2] The low voltage (a few millivolts) has to be amplified to provide a useful signal.

The main advantages of the magnetic flowmeter for dredging applications are the following:

1. There is no obstruction to flow in the pipeline.
2. There is no additional head loss due to obstruction, because the friction head loss is only due to length of pipe housing the meter.
3. The signal output is linear with respect to liquid flowrate.
4. The electric signal generated may be analyzed on digital computers.
5. The induced voltage is not affected by temperature, viscosity, turbulence, or conductivity (as long as conductivity is above a minimum level).

The magnetic flowmeter is usually mounted in the suction or discharge line of the dredge system since it does not provide an obstruction inside the pipe. As the dredge mixture passes through the meter, it generates a voltage proportional to the velocity of the mixture. The solid-water mixture must have a conductivity of at least 5 μmhos.

The liner of a magnetic flowmeter should be at least 1 in (2.5 cm) thick and preferably made of polyurethane. The meter should be periodically rotated to provide a more uniform wear. Magnetic flowmeters have proved to be accurate and reliable.

Sonic Flowmeters

Sonic (ultrasonic) flowmeters are of two types, pulse-type units and doppler meters. Pulse-type units typically employ a pair of transducers mounted on opposite sides of the pipe with one transducer upstream from the other. These units operate by emitting a burst of sonic vibrations from one crystal (Fig. 10.3). The

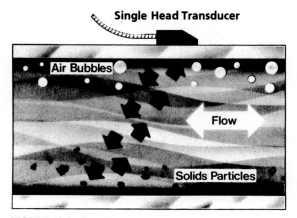

FIGURE 10.3 Doppler velocity meter.

burst of energy travels upstream to the other crystal. After several of these signals have been emitted upstream, the receiving crystal now becomes the trans-
mitting crystal, sending a series of bursts downstream. The sonic flowmeter measures the time required for these bursts to travel upstream and downstream.
Knowing the time difference, the flow velocity can be determined. Attempts to
employ this type of meter on dredges have not been successful.

Doppler Sonic Flowmeter

This type of meter is similar to doppler radar used in speed traps or airplanes.
The doppler meter emits a continuous wave and operates with a very narrow
band, which allows the majority of the noise to be filtered out of the system,
achieving a high degree of sensitivity. The doppler meter is basically a reflection-
type measurement and operates on the basis of reflections from the solids in the
fluid mixture near the pipe wall. Doppler meters are not very accurate, as recent
measurements show,[3] but are considerably cheaper than the magnetic
flowmeters.

DENSITY METERS

The nuclear density gauge is used to accurately determine the density of the material passing through the pipe. No information is presently available on the accuracy of these gauges, particularly when used for large-scale applications, such
as are necessary in the hopper dredge. The nuclear density gauges emit a stream
of gamma ray energy (the usual source is cesium 137) which is absorbed by the
material passing through the dredge pump (Fig. 10.4). An ionization chamber, located on the opposite side of the pipe from the emitter (across the pipe diameter),
measures the amount of energy reaching it and converts the radioactive energy
into electrical energy. These systems are calibrated for known density, and the
resulting ionization (when in use) is proportional to the density of the material
being pumped.

The flow velocity signal from the magnetic flowmeter and the density signal
from the nuclear density gauge are then multiplied electronically to produce the
mass flowrate. When this signal is integrated with time limits, the amount of solid
material pumped can be determined.

Since the density meter is a nucleonic device and contains radioactive material, licenses from the regulatory agencies are required. Before a manufacturer
can ship a radiation source, the buyer must obtain a license to own and to make
use of the density meter.

According to Erb[2] the basic safety requirements which have been designed
into the device are:

1. The radioactive material itself shall be encapsulated in such a way that it cannot be dispersed even in the event of accidental destruction of the head.

2. The head itself shall provide enough shielding to assure that radiation levels in
   the vicinity of the head do not exceed prescribed limits.

3. The head shall have a lockable shutter, which when closed, will reduce radiation in the region normally exposed to the beam to safe levels.

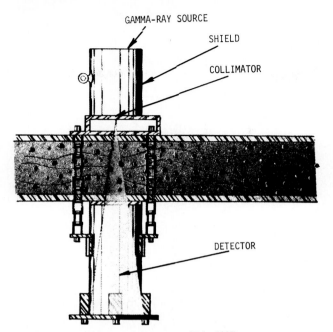

FIGURE 10.4   Nuclear density gauge. (*Erb, 1985*)

Depending on the pipe size and density range, the source head for a nuclear density gauge may be fitted with an electromechanical actuator which will close the shutter automatically when the pump is turned off. This would assure that excess of radiation levels, at the detector side of the pipe and within it, are safe when the discharge pipe is empty.

Almost always, the user's license will require that the source head be installed and checked out by duly licensed vendor personnel. The user will be forbidden to disassemble, demount, or otherwise tamper with the source head.

The magnetic flowmeter and nuclear density gauge represent the most accurate means of determining the dredged material density and flow velocity. Because it is expensive and requires trained personnel to maintain it, this type of equipment has not been fully adopted by the dredging industry. It is believed, however, that in order to precisely control a truly automatic dredged system, it will be necessary to use the output signals from measuring devices such as these.

PRODUCTION METERS

A real-time production meter provides the dragtender, or leverman, with production rates during the actual dredging. The dredge production system consists of a velocity meter, a density meter, and a display which indicates dredge production in units of weight (or volume) per unit time (usually tons/h, or yd^3/h or ft^3/h or m^3/h). The outputs from the flowmeter and the density meter are supplied to the production meter (Fig. 10.5) and a microprocessor calculates the production.

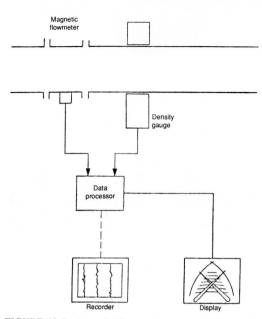

FIGURE 10.5 A production meter system with a nucleonic density gauge and a cross-pointer display. (*Erb, 1980*)

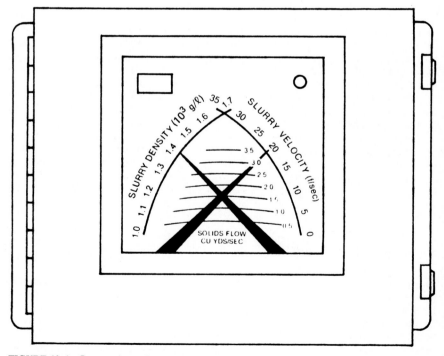

FIGURE 10.6 Cross-pointer display. (*Erb, 1980*)

FIGURE 10.7 Leverman control room on cutterhead dredge *Dave Blackburn* (note the cross-pointer display). (*Courtesy, Bean Dredging Corporation*)

A very useful visual display is provided by a cross-pointer (Figs. 10.6 and 10.7). The cross-pointer display provides instantaneous readings of solid-water mixture density (on the left scale) and mixture velocity (on the right scale), and where the pointers cross, the scale indicates solids flow per unit of time. A totalizer is usually supplied with a production meter which will provide the dredge operator with a total production in an hour, per shift, or per day. More than 90 percent of dredge operators who have measuring equipment agree (according to a recent survey[4]) that instrumentation is useful in improving production rates and in evaluating total solids production.

DREDGING AUTOMATION

Decades ago dredges had large crews, and dredging efficiency depended on the skill of a dragtender. As the cost of equipment and labor have increased, the government and industry have been looking for ways of reducing unit cost of dredging which can be achieved through more efficient equipment and better training for personnel operating dredges. The previous section discussed the instrumentation which permits the operator to know, in real time, the density of material being pumped and production being achieved. There are limits to the efficiency that can be obtained unless some of the functions are automated. The overall efficiency can be improved by the selective use of instruments and automatic controllers. These devices not only assist the operator but also allow a more accurate analysis of the whole dredging process. Many types of instruments installed for land operations are not suitable for operations on dredges because of vibration,

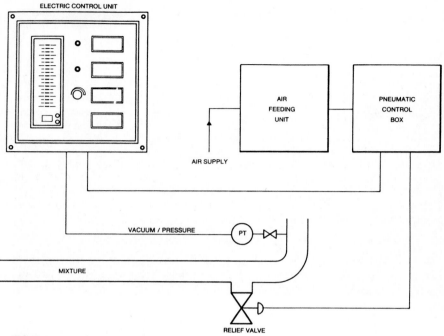

FIGURE 10.8 Vacuum relief valve. (*Courtesy, Meconaut*)

high humidity, or high or low temperatures. Consequently much of the instrumentation had to be developed especially for dredging and in many cases the accuracy had to be sacrificed in favor of ruggedness of instruments and reliability. Efforts have been made by many instrument manufacturers to provide a more reliable and accurate system. The purpose of automation is to obtain maximum efficiency by relieving the operator of repetitive manual tasks which may be better completed by automatic controls. In an advanced automated system, the aim is to control and monitor the dredging process to achieve optimum efficiency. Operation of a modern dredge is quite complex, and it would be impossible for a person to observe all the gauges and at the same time operate the dredge. The dragtender and captain's tasks are thus changed from moving levers and pressing buttons to supervisory functions, making sure that all the systems are operating correctly.

The various measuring instruments have three basic components:

1. Transmitter
2. Signal amplifier
3. Visual display

The transmitter is the most important component and the most vulnerable to damage, since it must be located on parts of the dragarm, or vessel, underwater.

In addition to production meters discussed earlier, the following equipment is employed:

1. *Vacuum relief valve:* This valve lets the water enter the suction pipe when the suction pressure (vacuum) becomes excessive. Both manual and automatic versions are available (Fig. 10.8).

2. *Bypass valve:* If the discharge pressure falls below a specified minimum value, the valve is opened to let water enter the suction line. This is an effective protection against plugging the discharge line. The bypass valve is automatically activated by a pressure sensor.

3. *Dredge profile monitor (DPM):* This monitor indicates the position of the cutter or draghead and prevents over- or underdredging (Figs. 10.9 and 10.10). A visual display is also available to the dragtender on a hopper dredge, as shown in Fig. 10.11. An air-feeding panel for the pneumatic measuring system is shown in Fig. 10.12. System layout is shown in Fig. 10.13. It may also be equipped with a device to automatically adjust to variations in the draft of the vessel and to tidal variations. The DPM system may be equipped with automatic dredging profile controls, acting as an autonomous unit as well as an input to the fully automatic dredging control.

4. *Automatic cutter or draghead controller:* This device controls the swing winches, the ladder winch, the spud carriage (if available), and the cutter drive. The controller automatically actuates the draghead winch to compensate for variations in the dredging depth, the vessel's draft, and the effects of tides and waves. The swell compensator is equipped with a positional transmitter which actuates the ladder winch.

5. *Load and draft indicator:* This operates on hydrostatic principles. The sensors are placed either amidship or one fore and one aft, as indicated in Fig. 10.14. The system may be combined with a load and draft indicator. A visual display shows the mean draft, displacement, and draft fore and aft (Fig. 10.15). The slope of the line indicates the extent of trim.

6. *Loading rate indicator:* As the hoppers are filled, the dredge settles deeper in the water. After the overflow is reached, any increase in draft results from the sediment settled in the hopper. The dragtender can visually determine the

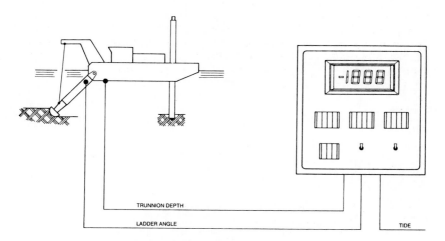

FIGURE 10.9 Cutterhead depth indicator. (*Courtesy, Meconaut*)

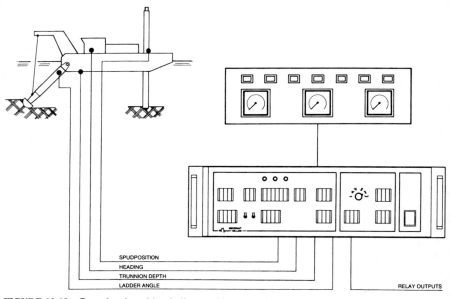

SPUDPOSITION
HEADING
TRUNNION DEPTH
LADDER ANGLE

RELAY OUTPUTS

FIGURE 10.10 Cutterhead position indicator. (*Courtesy, Meconaut*)

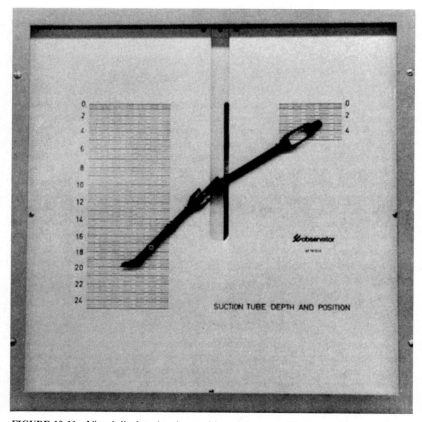

SUCTION TUBE DEPTH AND POSITION

FIGURE 10.11 Visual display showing position of dragarm. (*Courtesy, Observator*)

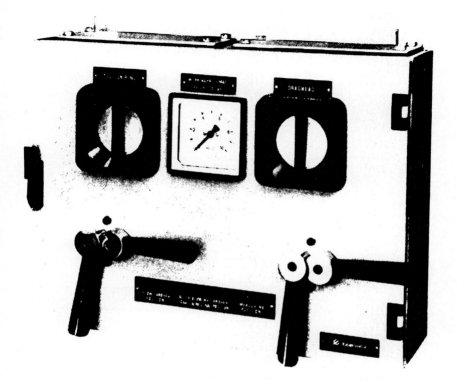

FIGURE 10.12 An air-feeding panel provides an output for feeding two to four bubbling points. (*Courtesy, Meconaut*)

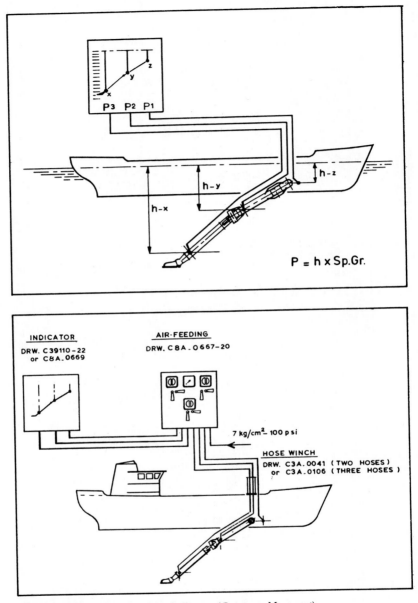

FIGURE 10.13 Dragarm position indicator. (*Courtesy, Meconaut*)

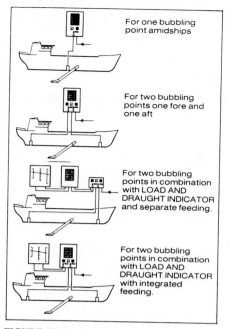

For one bubbling point amidships

For two bubbling points one fore and one aft

For two bubbling points in combination with LOAD AND DRAUGHT INDICATOR and separate feeding.

For two bubbling points in combination with LOAD AND DRAUGHT INDICATOR with integrated feeding.

FIGURE 10.14 Location of sensor for load and draft indicating system. (*Courtesy, Meconaut*)

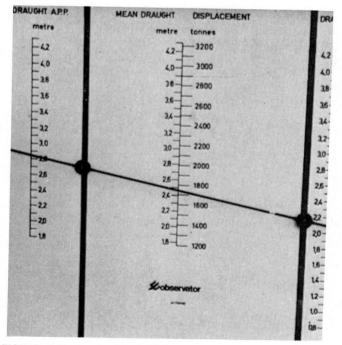
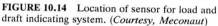

FIGURE 10.15 Load and draft display.

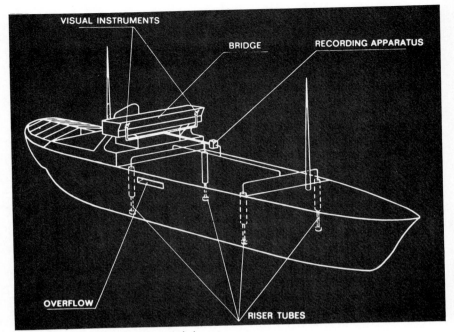

FIGURE 10.16   Loading rate system's layout.

FIGURE 10.17   Visual display of hopper loading.

hopper load at any given instant. Figure 10.16 shows the system's layout and Fig. 10.17 shows the visual display of hopper loading.

7. *Automatic light mixture overboard (ALMO):* This device increases the efficiency of the hopper loading by sensing the specific gravity of the solid-water mixture in the discharge pipe. It is set for a specified specific gravity, and a lighter mixture is discharged overboard automatically. Only a mixture above a given specific gravity, say 1.20, is allowed to discharge into the hoppers. Input to ALMO is from density and flowrate indicators and the device actuates the gate valves, diverting the mixture either to the hopper or overboard.

FIGURE 10.18 Dragtender control room on hopper dredge *Ouachita*. (*Courtesy, Gulf Coast Trailing Company*)

Several photographs show the instrumentation positioned within easy reach of the dragtender. Figure 10.18 shows the dragtender control room on hopper dredge *Ouachita*. Figures 10.19, 10.20, and 10.21 are detail displays of the dragtender control room.

FIGURE 10.19 Dragtender front display area. TV monitors, cross-pointer displays, dragarm position indicators (angle and digging depth), suction and pressure gauges. Dredge *Ouachita*. (*Courtesy, Gulf Coast Trailing Company*)

FIGURE 10.20 Dragtender control room and displays, Dredge *Ouachita*. (*Courtesy, Gulf Coast Trailing Company*)

FIGURE 10.21 Dragtender control room, Dredge *Ouachita*. (*Courtesy, Gulf Coast Trailing Company*)

AUTOMATION, SURVEY, AND PROFITABLE DREDGING*

Colin G. Weeks

President, Hydrographic Associates Inc.
Houston, TX

There are two areas in which dredging contractors could increase the profitability of their operations: dredge guidance and dredge survey. Accurate guidance ensures that the dredge removes only that material that is required to be dredged; an accurate survey ensures that the contractor will receive full payment for the work done. Achieving higher accuracies may require a prior expenditure on computers and training, but computers alone are not enough; it is equally important that a trained professional—surveyor or civil engineer—should supervise these aspects of the operation, customizing the software to the job, making the required decisions, and carrying out the necessary checks.

THE HISTORICAL RECORD

In December 1987 a *Five Year Summary of the Industry Capability Program (ICP) for Fiscal Years 1977–81* was published.[1] (The ICP program instituted competitive bidding by industry for work which had historically been undertaken by government dredges.) The most interesting statistic in the summary was that the ratio of total material excavated to the pay material excavated was 1.204—in other words, over the course of the program, 20 percent of the work performed was done without payment, because the dredge was digging in the wrong place. If anyone doubts whether this figure is typical, the summary covered 149 jobs—83 performed by industry and 66 by the government—with a total actual quantity of 130,027,013 yd^3 (99,413,400 m^3) and a total value of $164,251,785. (Industry can take some consolation from the fact that their dredges only performed 17.8 percent unpaid work whereas the figure for government dredges was 26.8 percent.) A further, hidden cost is that 26,000,000 yd^3 (19,800,000 m^3) of scarce disposal location was unnecessarily used over that 5-year period.

In dollar terms it would appear that with annual dredging contracts of the order of $500 million, some $100 million worth of work is being performed without payment each year. Looked at from the viewpoint of the individual contractor, if measures could be taken that would halve the amount of unpaid overdredge, each bid could be reduced by 10 percent, which would win more jobs, without affecting the present profit margin. Alternatively contractors could maintain their present bidding practices and increase their profit by 10 percent of the cost of each job. Of course, once all contractors have improved their efficiency, their individual margins will be back where they are now, but the bill to the taxpayer will be considerably less. In the transitional period, however, the efficient com-

---

*References for this chapter section appear at the end of the section. For convenience sake the author has used masculine terms and pronouns in this section; it is not meant to exclude females.

panies will enjoy increased profits while those who are unable to effect improvement may not survive.

A recent court case[2] has shown that there is an additional danger in overdredging. If the material outside the required prism is of a different character from that within the prism—as it often will be—the court has ruled that a claim for differing site conditions (from those advertised) is invalid, and the contractor must pay any additional cost incurred.

POSSIBLE CAUSES OF UNPAID OVERDREDGE

Hopper Dredging

It is not proposed to go too deeply into hopper dredging. Methods have changed since the ICP; shipboard computers are standard, both in the survey boats and aboard the dredge, and modern computer graphics allow the mates to see the exact location and extent of every remaining shoal as soon as that day's survey has been processed.[3] Greater use might be made of the computer in the early stages of a contract, however, to ensure that every pass is made at a different offset from the centerline—ideally a pattern of lines should be run, spaced by the width of the dragheads. On many contracts this will be unnecessary, but, if adopted as a standard practice, it would ensure that the bottom was taken down evenly and thereby avoid the development of ruts when the material was cohesive.

Hydraulic Dredging

The Five Year Summary did not distinguish between types of dredge, but it would seem probable that the nonpay ratio will be higher with a cutterhead than with a hopper dredge. Hopper dredging is a continuing process, taking a whole reach of the channel gradually down to grade. The work is checked every 2 or 3 days by a survey and, provided that the mates and dragtenders are attentive to their duties, it is unlikely that they will dredge outside the channel or go seriously overdepth. A cutterhead dredge on the other hand brings each cut down to grade before it steps forward and the survey can only be made after the dredge has moved far enough ahead for the survey boat to clear the stern. Different techniques are used for positional control; some dredges use radio positioning to fix the spud and a gyro-compass to control the swing; others place the positioning system antenna on the cutter gantry. Even in the latter case, however, it is unlikely that the cutter itself will be positioned, since the gantry rarely projects far enough forward.

In practice the civil engineer calculates the angle that the dredge must swing to reach the toe; but then he thinks of the probability that the spud will not be exactly on the centerline and that there will be errors in the positioning system and the gyro-compass, and he applies a safety margin. He probably then considers that if the dredge has to go back to clean up a toe that was not quite cleared, not only is it expensive and inefficient but it will be apparent to everyone who reads the daily report that he has miscalculated—so he adds a further safety factor. The swing angle is given to the leverman, who is rightly concerned that the dredge should not have to go back because of anything that happens on his watch, so he swings a bit further still. Now if this is what happens in a straight channel, what happens when the channel is curved or when the sideslope has to be cut without box cutting?

A COMPUTER-BASED SOLUTION

The first point to be made is that it is unlikely that the unpaid yardage can be reduced without an initial investment, certainly in equipment, possibly in personnel. The methods currently used have been developed over the course of time and are executed by crews who have, for the most part, a great deal of hard-won practical experience; it is reasonable to assume that they are doing the best that can be done with the tools with which they are supplied. Secondly, equipment and personnel alone are not enough. There must be continual management involvement and concern so that those on board are in no doubt that efficiency involves not just the number of hours worked per day or the number of cubic yards pumped per hour; it is equally important that every yard pumped should be a required yard.

It has been suggested that the primary reason for excessive overdredge is that the leverman does not know the position of the cutter relative to the channel limits and that in his natural concern to avoid undercutting he leaves an excessive safety margin. A question that might be asked is Why does he not know? since all the necessary information is available. All that is needed is a leverman who can multiply the ladder length by the cosine of the depression angle, add it to the spud to trunnion distance, and multiply the sum by the sine and cosine of the gyro heading, which, added respectively to the spud easting and northing, gives the cutter coordinates. These must then be translated to station and offset—in time to repeat the process a second later.

A more serious question is Why is a computer not used? and, as far as is known, it has not been. It is suggested that there are two principal reasons, one technical and one psychological. The psychological reason is that the present generation of dredgermen is not comfortable with computers. They are practical men, good seamen, who may not have deep theoretical knowledge but who have learned by experience the art of coaxing all manner of materials along 10 km of pipe. What they have never had occasion to use is a typewriter keyboard, still less a computer—and, like all of us, they find the unknown intimidating.

It is probably easier—and certainly preferable—to adapt the computer to the leverman than to attempt the alternative. This adaptation takes two parts; first of all, as much information as possible should be given graphically. The leverman is already subject to sensory overload and the last thing he needs is to have to pick his way through columns of figures on a computer screen in order to derive the information he requires. Not for the first time, a picture is worth a thousand words. This picture should show the channel limits, the contours of the prebid survey, any pipeline or cable crossings, wrecks or obstructions, buoys or navigation aids, etc.—all clearly differentiated by color or symbol—with a scale plan view of the dredge moving over it.

The second step has already been hinted at; it is not the computer that the leverman dislikes but the keyboard—the keyboard must go. The proposed alternative is the touchscreen, a device that might have been designed for the cutterhead dredge. When the program requires a choice to be made, the alternatives are displayed as labels on the screen and the user has only to touch the one he selects. As protection against faulty selection, the choice is normally activated when the finger is withdrawn rather than when it is pressed—if he hits the wrong word, he can slide his finger across. There are two further advantages; dredging is a 24-hr a day process and a touchscreen is as easy to use at night as it is in the daytime. Secondly, space is limited in the vicinity of the leverman and a flat surface on which to place a keyboard rarely exists. It is easier to find room for a second display, which can be at arm's length.

The technical reasons for the absence of computers are, in part, related to the psychological ones. It is only recently that high-quality color graphics and touchscreens have become available for computers that have the robustness and reliability to endure the hydraulic dredge environment, which is not entirely computer friendly. The other reason is software. The individual algorithms are not difficult but there are a lot of them, with a wide range of choices; in practice all must be operative before the system is of use. A contractor who goes to the trouble and expense of converting to such a system needs to know that he can use any positioning system, on any survey grid, in any Corps of Engineers (CE) district (each of which has its own way of defining channels and of requiring reports); he must be able to dredge off two spuds or off a sliding spud—with or without an idler barge—or off anchors—with or without a Christmas tree. The system must work as well on curved channels as on straight ones, in wideners and cut-offs, and in turning basins. When developing software one must necessarily do so a piece at a time and one would like to check each piece as completed. In this situation, however, half a loaf is not better than no bread; it is all or nothing. If the system does not give the leverman exactly what he needs—all the time—he will not fool with it.

CONSEQUENTIAL BENEFITS

Cutterhead technology has changed little over the past 30 yr and many dredges older than that are still making money for their owners. The environment in which they operate, however, has changed significantly in that time. Thirty years ago dredges were positioned visually and dredging ranges were often provided by a beneficent CE to mark the channel limits; ships relied on visual signals, their sirens, and the rule of the road to keep clear of each other and of the dredgers; dredging itself was seen as a necessity, to which the only downside was financial, not environmental—the very word *environment* did not exist as a term of everyday speech. Life for the leverman was relatively simple; his eyes told him when he was in the channel, and he had few distractions—he could concentrate on his controls and his gauges.

Contrast that with today. For the most part the dredging ranges are gone and the leverman must use some form of numeric display provided by an electronic positioning system, supplemented by a gyro compass; the positioning systems vary greatly in their clarity and ease of use, but few approach the convenience of a well-sited visual range. The standard dredging controls are little changed—on a typical cutterhead one might find 18 hydraulic controls, 14 gauges, and 8 on/off switches, all the responsibility of the leverman—but now he has three or more radio circuits, there is an internal telephone, and there may be a cellular phone as well. Most of the talk is irrelevant but when his own vessel's name is called, he must take his hand off the control to answer. This is time-wasting but, perhaps more important, the distraction can be dangerous. If he inadvertently leaves a winch clutched in or a pump running, wire cables could part or pipelines could clog or rupture.

The other big change has been the rise of the environmental lobby and its concern over the effects of dredging. Leaving on one side the question of whether the concern is well founded, the effect has been pressure on the CE, which in turn has led to increased reporting requirements from dredge contractors. A current contract calls for continuous recording of cutterhead position and depth, the latter corrected for tide. This has led to three additional electronic systems, of which the output of one has to be manually input to one of the others—by the leverman—every 10 to 15 min.

The result, from the leverman's perspective, is an ergonomic disaster—although he would doubtless find another wording—and a good case could be made for automating his task on those grounds alone. The economic argument, however, is that every distraction of the leverman reduces the efficiency of dredging which is, quite literally, in his hands. It is not proposed that a computer will ever replace the leverman; there are, however, a number of areas in which the computer could help him:

- A second graphic can show the position of the cutter superimposed on both the channel template and the appropriate predredge profile. This tells the leverman when he can swing rapidly across and when he should expect some heavy digging; it enables him to cut precise sideslopes in those situations when box-cutting is not allowed.

- Improved production reports, particularly if the load sensors are interfaced to the computer. In a hopper dredge the data for the daily report are recorded automatically; the same is possible for a cutterhead. This can be helpful to management in another way; when the leverman can see how much has been pumped in his watch, relative to his predecessor's, he is given a powerful incentive—healthy competition can spur production.

- The next step—to be taken with caution—is to control production, by controlling pumps, winches, and brakes from the computer. This is all technically feasible and, once successfully implemented, has the potential to increase the efficiency of production significantly. It is not suggested that a computer can control a dredge more effectively than a skilled leverman over short periods of time. The difference, however, is that a computer can maintain that level of efficiency for 24 hr a day, 7 days a week—and free the leverman to supervise operations, coordinate the deck crews, and talk to oncoming ships.

THE SIGNIFICANCE OF DREDGE SURVEY

The channel into Galveston is about 10 miles (18.5 km) long by 800 ft (244 m) wide; in 1984 the cost of deepening that channel by 1 ft (3 decimeters, or 3 dm) was about $4 million. It follows that a systematic error in either the pre- or postdredge survey of only 3 in (0.75 dm) would have cost either the contractor or the government $1,000,000. This is an impressive figure to one who, in 17 years of surveying for a national hydrographic office, cannot remember one survey that would have been accurate to 1 ft, let alone 3 in. This may sound strange, but the object of charting surveys is never to produce an accurate chart; it is rather to produce a safe chart, and the techniques used, the approximations made, and the consideration of errors always err on the side of caution, by showing the seabed as more shoal than in fact it is. The significance of this to the dredging industry is that national hydrographic offices tend to be regarded as the experts, and the procedures they have developed are often followed by dredging surveyors; I would suggest that this should be done with caution.

The principal cause of depth error is heave, the vertical motion of the survey vessel on the sea or swell. This is normally ignored in charting surveys, largely because until recently, the means to measure heave did not exist; the effect of ignoring heave is to make soundings shoaler than they should be—a safe error. A dredging contractor who ignores it in a postdredge survey, however, is taking money straight out of his pocket—a point that will be returned to later. Another example is the barcheck, almost universally used for calibrating depth sounders. There are theoretical grounds to suggest that use of a barcheck will lead to soundings that are 3 to 6 in (30 to 60 cm) shoal—and this has been supported by trials. It does not sound very much, but it would have meant $1 to $2 million on the Galveston project.

Much has been written about the errors of positioning systems, and a great deal of work has been done on statistical techniques to minimize them. It is well to remember, however, that statistics only apply to one category of error, the random errors, and that the other categories—systematic errors and blunders—can be of much greater practical significance. By definition, the mean of a large number of random errors is zero; we normally measure position once a second so that, if a large random error exists, the survey will be generally correct even if individual soundings are misplaced. On the other hand an uncorrected systematic error in position determination, or human error in entering shore station coordinates, for example, could move the whole dredged channel—not a result one cares to contemplate, but one that has occurred.

It is sometimes assumed that since many of the techniques of hydrographic survey are simple, particularly in the computer age, it is not necessary to employ a qualified surveyor to carry them out. In fact the opposite is true, and the more that a computer is relied on to collect and process survey data, the more important it becomes that a competent professional surveyor sets the system up at the start of a project and installs the checking procedures that are essential if accurate results are to be provided. An important part of these procedures is the *paper trail*. If an error is discovered in the computer-generated output, it must be possible to return to the raw data—a printout of time, radio ranges, and computed position, related by numbered event marks to an analog depth record. A personal preference is that this analog record should be a true analog, showing every echo in the water column independent of the digital record, rather than the modern practice of generating the analog record from whatever the depth digitizer digitized. The old-fashioned sounders showed a shoal, the seaweed growing over the shoal, and the fish feeding in the seaweed over the shoal—and the surveyor could often determine which was which. Some modern sounders show one echo only in that situation; it is a clearer record but which of the three would it be?

Besides enabling errors to be detected and corrected rapidly on site, the paper trail is almost more important after the contract is complete. The contractor is rarely paid on the results of his survey alone; a second survey will be carried out either by the client or an independent survey organization. These surveys will normally agree within a reasonable tolerance but, if they do not, it is no use comparing sheets of computer printout. The surveyor must be able to defend his survey, in court if necessary, by producing not only the raw data but also a historical record describing every calibration, bar check, and checking procedure that was carried out. In fact, however, if the surveyor is able to defend his work in this way, on a surveyor-to-surveyor level, disputes of this type will rarely go to court.

If the positioning errors have been determined and corrected and if the echo sounder has been correctly adjusted, the principal error source which remains is the changing water level; there are long-term changes—caused by astronomic tide, barometric pressure, and wind—and short-term changes—caused by heave.

Neither of these errors is easy to correct precisely. The longer-term changes are corrected by tide readings; it is important that the tide gauge be placed as close as possible to the survey area and essential that it be placed in water that is subject to the same tidal regime—in other words closeness alone is not enough. For example the CE's gauge at Fort Point, in Galveston Bay, was used to correct soundings in Galveston Entrance Channel, outside the jetties; the tidal range in the Gulf of Mexico, however, can be 1 ft (30 cm) greater than that in Galveston Bay, only 3 mi (5 km) away. The consequence is that soundings taken in the Entrance Channel at high water, and reduced by the Fort Point gauge, will be 1-ft different from soundings taken at low water in the same location—a discrepancy that caused considerable concern when first discovered.

Heave, the vertical motion of the boat on the waves, creates an error that cannot be removed in off-line processing; the sounder record taken over a flat bottom, in a sea with an amplitude of 2 ft and a wavelength of 100 ft, will appear identical to that taken in flat calm over a sand wave of similar dimensions. On navigational surveys it is normal to reduce the effect of short-period waves by a meaning process, but if this were to be done on a large-scale dredging survey, it would remove many of the features which the survey is supposed to find. There is, then, no substitute for real-time heave compensation. There are several devices available for this purpose, some of which can only be used with computer processing, some impose restrictions on the maneuvering of the survey boat, all have restrictions on the range of wave periods that can be effectively corrected, and all are costly; it is a difficult choice. For a hopper dredge, however, the cost could be recovered in a single job if swells are prevalent in the final stages. During this final cleanup the dredge is normally looking for the isolated shoals that remain above project depth; if the surveyed cross sections are degraded by heave, dredge time will be wasted either in looking for nonexistent shoals or in waiting for a calm day so that a true survey can be done. Either way the cost of the heave compensator is soon recovered.

SURVEY COMPUTER ERRORS

There is of course no limit to the range of errors which a computer can be made to produce and users must always beware. There are, however, three topics concerned with computer processing that are of particular importance; the scale correction of position data, the density of depth data, and the synchronization of depth with position.

Dredged channels are normally defined by their coordinates on the survey grid, a plane surface, while the position of the dredge and the survey boat are determined by distances measured on the earth's spheroidal surface. There is normally a difference between grid distances and true (spheroidal) distances; it is essential that the appropriate corrections are applied—but it is equally important that they are only applied once. Most computer systems used in the dredging industry will correct each range and thereafter carry out all calculations on the survey grid by plane trigonometry; this is mathematically precise provided that the positioning system is measuring "true" ranges (i.e., if it was calibrated by measuring the distance between two survey monuments, the distance calculated from difference of coordinates should first have been converted to a true distance). Alternatively, if the shipboard computer applies no correction to the measured ranges—and some of those incorporated into a positioning system do not—the corrections must be applied when the system is calibrated, either by using the distance between monuments uncorrected or by applying a scale correction to a directly measured distance. The worst-case situation is in the center (or extreme edge) of a UTM grid zone, where the difference between grid distances and true distances is 13 ft in 5.5 nautical miles (4 m in 10,000 m). If two position lines of that length cut at 90°, the positional error caused by neglecting the correction will be 18.5 ft (5.7 m), but if they cut at 30°, the error is 25 ft (7.7 m); this is a systematic error, shifting the whole channel, and if the channel is 15 nautical miles offshore, the potential error becomes 72 ft (22 m).

Turning to depth data, a sounder in shallow water will normally measure a depth 10 times every second; each of these echoes is shown on the analog record and used in manual processing. When computers were first used aboard ship, however, their slow cycle time prevented their recording depths any more often than once every 2 s. Systems of this type were widely adopted by the CE and in many Districts the shipboard computers are used primarily as data loggers, with the off-line processing performed in the District office. At first sight this is an acceptable procedure; at a sounding speed of 6 knots this is equivalent to a distance separation of 20 ft (6.1 m), about the closest that soundings can be plotted on the scale most commonly used by the CE, 1/2400. There are, however, two major disadvantages compared with more modern systems that record every depth measured by the sounder. First, if the computer only sees two depths 20 ft (6 m) apart, it is unable to discriminate between the seabed echo and the many false echoes caused by fish, ship's wakes, and floating trash. Extreme changes in depth will appear as spikes, which can be removed in a tedious editing process; the dangerous errors are those that are not obvious but which will distort the subsequent calculations.

In contrast a computer that has access to every measured depth can use a simple software gate to reject the great majority of false echoes. The second disadvantage applies particularly to hopper dredges which, in a cohesive material, can produce a highly irregular bottom contour, the detail of which just cannot be seen on a 5 percent sample but which the dredge superintendent must be aware of to avoid an unpleasant surprise on the postdredge survey (see Fig. 10.22).

A further point about data density is that this density should be regular. Several sounder manufacturers have taken advantage of a digital sounder's freedom from the constraints of a rotating stylus to maximize the quantity of depth measurements taken, by triggering a pulse as soon as the previous echo has been received. Thus in deeper water there will necessarily be fewer depths output than in shallow water—which presents the computer with an impossible problem in interpolating the position of each depth over a steep slope, an essential procedure if the channel toe is to be correctly located.

The lack of data synchronization—also known as time lag—is a problem of the pulse-type microwave and UHF systems and is caused by the time that the system takes to measure four ranges, process them, and then output over a serial interface to the computer. If this time were a constant, it could be easily corrected. In practice it is highly variable, depending on the number of remotes and the range to each, the signal propagation conditions, and the priority of the unit relative to other users; its correction can only be achieved if the positioning system manufacturer has made provision for it. A typical procedure is for a pulse to be output when a range-measuring cycle is started and for each range to be interpolated back to what it would have been had it been measured at the time of the pulse. It is then up to the computer system to use the pulse to synchronize the depth measurements. One positive feature of this error is that it is easily detected. Select a line that has a prominent depth feature (e.g., a steep sideslope)

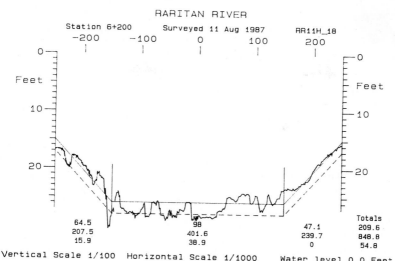

FIGURE 10.22 Example of a surveyed cross section in the wake of a hopper dredge.

and run it several times in each direction; superimposing the resulting cross sections will soon show if there is a time lag problem—this is a recommended procedure on every survey.

An example of a surveyed cross section (Fig. 10.22) illustrates the importance of using every depth measured by the sounder—typically 10 times a second, or about a foot apart over the ground—when surveying in the wake of a hopper dredge. The ruts or the high spots could be missed when depths are only read 1 or 2 s apart; they are exaggerated by the difference in scales but are typical of a cohesive material.

CONCLUSION

The introduction of automated systems into a organization is not necessarily an easy process since it will require changing practices that people have followed for all their working lives. The following are seen as prerequisites: a visible commitment by management, nomination of a computer-literate surveyor or civil engineer to be responsible for the whole system, and a corresponding commitment by the software supplier to supervise installation, train the personnel, and provide continuing support. On the positive side, acceptance should be made easier by the fact that the system as envisaged will not replace any staff on the dredge or in the survey boat; it will make their task easier and will enable them to achieve efficiency in a way that is not currently possible.

There is an element of gamble about the decision; it represents a major expenditure and must be made prior to the award of a contract if all is to be ready when work starts. A way of reducing both the risk and the up-front expenditure is to lease survey services from a contractor with experience with systems of this type. Whichever way it is achieved, a successful implementation will lead directly to increased profits and, in an increasingly competitive environment, may prove essential to the company's survival.

REFERENCES

1. U.S. Army Corps of Engineers Water Resource Support Center, *Industry Capability Program for Fiscal Years 1977–81 Five Year Summary*, Dredging Division, Fort Belvoir, VA, Dec. 1987.
2. Casey, Joseph T., Jr., "Court Ruling Points to Need for Thorough Prebid Investigations," *World Dredging*, pp. 22–25, Nov. 1987.
3. Weeks, Colin G., "Computer Guidance for Commercial Dredges...," *World Dredging*, pp. 40–42, Aug./Sept. 1988.

BIBLIOGRAPHY

Erb, T. L., "Production Meter Systems for Suction Dredges," unpublished Dredging Short Course Notes, Center for Dredging Studies, Texas A&M University, 27 pp., 1985.

Fortino, E. B., "Flow Measurement Techniques for Hydraulic Dredges," *Proc., of the ASCE, Journal of the Waterways and Hydraulic Division*, vol. 92, no. WW1, p. 109, Feb. 1966.

Pankow, V. R., "Laboratory Evaluation of Production Meter Components," *Proc., 22d Dredging Seminar*, Tacoma, Washington, Oct. 18, 1989, Center for Dredging Studies, Texas A&M University, CDS Report No. 317, pp. 77–93.

LABORATORY EVALUATION OF PRODUCTION METER COMPONENTS

Virginia R. Pankow

U.S. Army Corps of Engineers Water Resources Support Center
Navigation Data Center
Fort Belvoir, VA

BACKGROUND

A dredge production meter is a system that determines slurry velocity and slurry density and combines the two values to estimate dredge performance. This estimate of dredge production is a function of the accuracy of the individual meters. There are several types of flowmeters, for example, electromagnetic, doppler acoustic, and differential pressure, each with its own level of accuracy. The same is true of density measuring devices. Therefore, under identical conditions, a production calculation using doppler flowmeter and nuclear density gauge values may be very different from calculations made using a bend flowmeter and pressure transducer specific gravity U-loop values.

Production Meter Technology, a work unit of the Dredging Research Program, is designed to evaluate production meter performance with the aim of determining meter accuracy and ranges of best performance. This section summarizes laboratory results that can be used in the selection of production meter equipment for a variety of dredging operations.

INTRODUCTION

Georgia Iron Works, Inc. (GIW), a pump design and fabrication company, operates a research facility where their pumps are tested and evaluated. Part of this facility includes a closed test loop in which slurry flow and concentration can be monitored, controlled, and measured. Under a Dredging Research Program contract, this loop was used to evaluate the performance of several flow and density meters for a number of conditions.

Facility meters, which were used as controls, were calibrated according to procedures established at GIW, which uses an American Society for Testing and Materials standard orifice plate. Table 10.1 identifies the facility control and backup meters and the seven test meters. All test meters were installed and calibrated according to manufacturers' procedures by their factory technicians. Once the tests were started, no adjustments were made to any of the meters.

TABLE 10.1   Control and Test Instruments

Flowmeters
Control: magnetic flowmeter
Backup: bend velocity meter
Test Instruments
Doppler flowmeters
Meter 5: dual sensors
Meter 6: single sensor
Meter 7: single sensor
Magnetic flowmeters
Meter 3
Meter 4
Density Gauges
Control: specific gravity U loop with pressure transducers
Backup: densitometer
Test Instruments
Nuclear density gauge
Meter 1
Meter 2

MATERIALS AND METHODS

Two series of tests were conducted to address some of the variables that affect meter performance. The first series, Phase 1, was conducted in September 1988 and addressed meter orientation, slurry type, concentration, and velocity. Phase 1 data

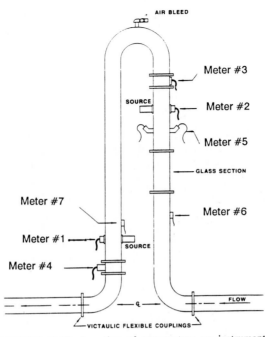

FIGURE 10.23 Location of test meters on instrument loop.

analyses resulted in several observations and the recommendation for further testing. Thus Phase 2 was conducted in June 1989 and investigated the effects of: the degree of pipe inclination, orientation of the axis of the density gauge to the axis of the pipe, air entrainment in the slurry, and two additional slurry compositions.

The Phase 1 series of tests were run with four different grain size materials each at three different concentrations (20, 30, and 40 percent concentration by weight) through a flow range from 0 to 4050 gpm (0 to 9 ft³/s), 0 to 26 ft/s (0 to 7.9 m/s). The test instruments were mounted on a U-shaped section of the pipe that could be raised 90° from a horizontal to a vertical orientation. The diameter of the pipe used for these tests was 8 in (203 mm). Data for the control and test meters were taken at 12 to 15 points along the flow range with the meters in both the vertical and the horizontal pipe position. The location of the test meters on this special section of pipe along with a glass observation section is illustrated in Fig. 10.23. Figure 10.24 shows (from top to bottom) a doppler flowmeter, nuclear density gauge, and a magnetic flowmeter located on the descending leg of the instrument loop. The control flow and density meters were located approximately 250 ft (76 m) upstream from the test instruments (Fig. 10.25). Table 10.2 lists the Phase 1 test conditions. Before each new material was added, data were collected through a range of velocities with only water in the test loop.

Phase 2 tests were run with two different types of materials of a single concentration (S.G. 1.2) through the same velocity range as Phase 1 tests. The density gauges, magnetic flowmeters, and one doppler flowmeter were in the same locations on the instrument loop: one doppler meter was recalled by the manu-

FIGURE 10.24 Doppler flowmeter, nuclear density gauge, and magnetic flowmeter on descending leg of instrument loop.

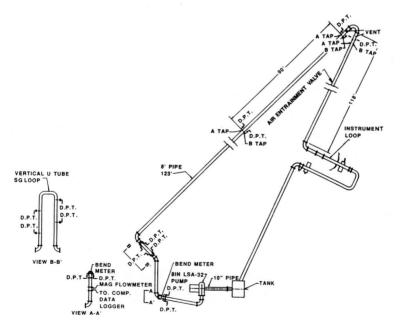

FIGURE 10.25 Schematic of test loop.

TABLE 10.2   Phase 1 Test Conditions

Material types:	Gravel	D50 16–19 mm
	Plaster sand	D50 0.70 mm
	Foundry sand	D50 0.30 mm
	Sand washings	D50 0.06 mm
Slurry velocity:	0–26 ft/s (0 to 4050 gpm)	
Concentrations:	Specific gravity (S.G.) 1.1–1.3	
	Concentration by weight (Cw) 15–40%	
Loop orientation:	Vertical (90°)	
	Horizontal (0°)	

facturer and the third was moved to a location on the test loop where it would be away from any influence from other meters, pipe bends, or flanges. The testing and data acquisition procedures for Phase 2 were the same as those used in Phase 1. Air entrainment tests were conducted in such a manner that air introduced into the system passed only through the test instruments and escaped to the atmosphere at the mixing tank and therefore did not pass through the control meters (Fig. 10.25). Table 10.3 lists the Phase 2 test conditions.

TABLE 10.3   Phase 2 Test Conditions

Material types:	Phosphate material:	D50 1 mm
		30% < 0.6 mm
		30% > 1.2 mm
	Sand/gravel mixture:	D50 0.35 mm
		60% < 0.6 mm
		20% > 12.5 mm
Slurry velocity:	0–26 ft/s (0–4050 gpm)	
Specific gravity:	1.2	
Loop orientation:	Vertical	90 deg
		60 deg
		30 deg
	Horizontal	0 deg
Air entrainment:	14 ft³/min at 50 psig	

RESULTS AND DISCUSSION

This study indicates that for determining dredged production, the most reliable instruments for measuring slurry flow and density are the magnetic flowmeter and the nuclear density gauge. Their accuracy is enhanced if they are mounted on a vertical pipe section.

The recorded flow and density values from the test instruments were compared to the control meters and each other to determine the range of readings among the instruments for a given flow and density. The relationship between the control and the test instrument values can be expressed as percentage difference using:

$$Q_d = \frac{Q_{\text{test}} - Q_{\text{cont}}}{Q_{\text{cont}}} \times 100$$

$$SM_d = \frac{SM_{\text{test}} - SM_{\text{cont}}}{SM_{\text{cont}}} \times 100$$

where
Q = flow
SM = slurry mixture specific gravity
subscripts d, test, and cont = difference, test, and control instrument values, respectively

Expressing the percentage difference of the control instrument from the test instrument will give a $+Q_d$ for test instrument values greater than the control instrument and a $-Q_d$ for test instrument values less than the control instrument.

Phase 1 Tests

With sand slurries, the nuclear density gauges had values within 1 percent of each other: for gravel this increased to almost 5 percent. The magnetic flowmeter values were within 4.7 percent of each other for sand slurries and 5.7 percent for gravel. Both performed better when mounted on the vertically oriented pipe section. The data for the doppler flowmeters show similar trends and, although each meter is fairly self-consistent, there are distinct differences among the three test doppler meter values. The doppler meters show greater differences from the control meter than did the magnetic flowmeters in all cases. The dual sensor doppler flowmeter had a larger percentage difference than either of the single sensor doppler meters. This meter had consistently higher values throughout the tests, which could indicate improper calibration. However, it was the test procedure to leave unchanged the instrument calibrations which were performed by a company representative.

Table 10.4 shows the largest percentage difference values obtained for density and flow for the range of concentrations and material types tested in both the vertical (90°) and horizontal (0°) pipe orientations.

The meters were influenced by material type in that the gravel data presented the greatest percentage difference for both the flow and density gauges. Slurry velocity had no effect on the magnetic flowmeters or the nuclear density gauges but did affect the doppler flowmeter values. The differences between the control meter values and the test doppler values increased with increasing slurry flow. Figure 10.26 illustrates that at a control meter flow of 1000 gpm (2.2 ft³/s), 6.4 ft/s (2 m/s), the doppler meter values were from 0.5 to 28 percent higher than the control meter, and at a flow of 4000 gpm (8.9 ft³/s), 26 ft/s (7.9 m/s), values were from 22 percent lower to 16 percent higher. This trend is observed with the doppler instruments in both the vertical (Fig. 10.26*a*) and the horizontal (Fig. 10.26*b*) orientation. Even without sediment, the doppler flowmeters tended to record less than the control flowmeter at high flows (Fig. 10.27). Figure 10.28 is the magnetic flowmeter data from the same 20 percent Cw foundry sand test for vertical (Fig. 10.28*a*) and horizontal (Fig. 10.28*b*) pipe orientation.

Slurry concentration appears to have a minor influence on the meter values in all but one condition. In general, the percentage difference increased slightly with the increase in slurry concentration. However, at high slurry concentration, low flow, and a horizontal pipe orientation, material settles on the bottom of the pipe

TABLE 10.4 Largest Percentage Difference for All Slurry Concentrations

	Plaster sand orientation (degrees)		Foundry sand orientation (degrees)		Sand washings orientation (degrees)		Gravel orientation (degrees)	
	90	0	90	0	90	0	90	0
Densitometers								
Meter 1	−0.1	−0.7	0.0	2.2	0.5	−1.0	4.0	5.0
Meter 2	−0.1	−2.0	−1.0	2.2	0.7	0.5	−0.7	7.5
Magnetic flowmeters								
Meter 3	−0.1	2.5	0.5	−3.7	−2.1	1.0	0.0	5.0
Meter 4	−1.0	5.0	−1.5	1.0	3.5	−1.5	−2.0	−0.7
Doppler flowmeters								
Meter 5	−7.0	−6.0	0.0	7.0	−5.0	−5.0	20.0	17.0
Meter 6	20.0	30.0	25.0	20.0	15.0	15.0	15.0	50.0
Meter 7	−15.0	−15.0	−15.0	−7.0	−10.0	−15.0	−12.0	−7.5

and moves as a sliding bed with higher velocity, less dense, fluid moving above it. This condition is conducive to dune formation and movement, which was observed. With this condition present, erroneous or erratic values could be recorded.

Phase 2 Tests

The phosphate material had a slight influence on the density gauges, their values being within 2 percent of each other as compared with the 1 percent difference for the Phase 1 sand slurries. However, density gauge performance with the sand-gravel mixture values was within 2.5 percent of each other, which was an improvement over the 5 percent difference for the Phase 1 gravel material. The magnetic flowmeters were within 1 percent of each other for both the phosphate material and the sand-gravel mixture. Of the two doppler flowmeters for Phase 2 tests, only one was on the vertical and horizontal instrument loop and therefore only data comparisons for slurry material in a horizontal pipe can be made. Apparent are the same meter trends as in the Phase 1 tests. Meter 6 performance was not improved by moving it to a long straight section of pipe, which supports the original theory of incorrect calibration. Both doppler meters performed in a similar fashion with the phosphate material and the sand-gravel mixture as with the sands and gravels of the Phase 1 tests. Table 10.5 shows the average percentage difference values obtained for the two materials and four instrument loop orientations of the Phase 2 tests.

When examining the data, care must be taken to avoid looking too closely at individual data points. Instead one must examine the general trends of meter performance as a result of a changed condition. The average percentage difference values, as shown in Table 10.5, do not clearly demonstrate the subtle improvement in meter performance as the pipeline angle of inclination increased from horizontal (0°) to vertical (90°). The improvement is in the noticeable reduction of data scatter as illustrated in Fig. 10.29. The symbols represent the percentage

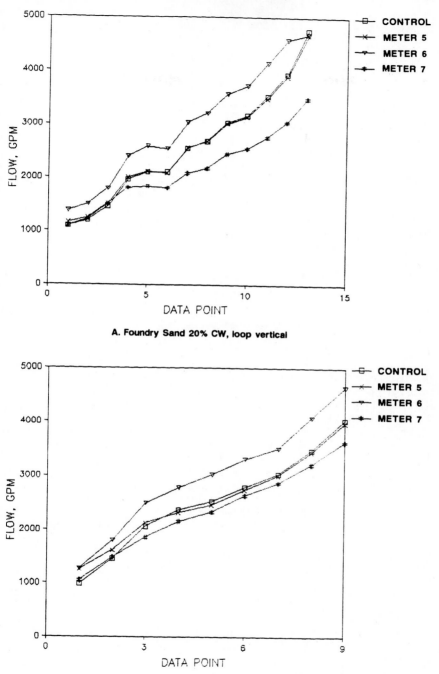

A. Foundry Sand 20% CW, loop vertical

B. Foundry Sand 20% CW, loop horizontal

FIGURE 10.26 Flow data comparison for doppler flowmeters.

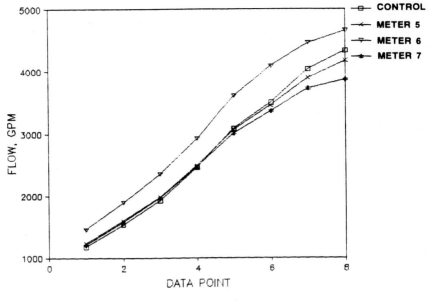

Water, loop horizontal

FIGURE 10.27 Doppler flowmeter data comparisons with water.

difference of the test instrument (values listed in Table 10.5) to the control instrument (datum) while the upper and lower bars indicate the upper and lower extremes of data for a given test and meter. Using Meter 3 as an example, as the percent difference goes from +1 percent for the 0° pipe inclination to −1.8 percent for the 90° pipe inclination, the data scatter is reduced from ±7.2 percent to ±2 percent. It appears that any inclination of the pipeline is an improvement in reducing data scatter. This trend is also recorded with the doppler flowmeter,

TABLE 10.5 Average Percentage Difference for Phase 2 Slurries

	Phosphate material pipe orientation (degrees)				Sand/gravel mixture pipe orientation (degrees)			
	0	30	60	90	0	30	60	90
	Densitometers							
Meter 1	−0.1	−0.1	−0.3	−0.6	0.8	0.05	−0.1	−0.5
Meter 2	0.4	−1.8	−1.7	−1.3	0.04	−2.4	−1.7	−1.4
	Magnetic flowmeters							
Meter 3	1.1	−1.1	−1.3	−1.3	1.3	−2.4	−1.0	0.7
Meter 4	0.8	−0.1	−0.9	−1.0	1.6	−0.8	−1.4	1.4
	Doppler flowmeters							
Meter 5	11.8	2.9	−2.6	−3.3	11.5	−4.5	−4.1	−14.7
Meter 6	60.0	(68.6)	(67.2)	(65.6)	88.7	(75.5)	(72.1)	(74.5)

( ) = values for meter on a horizontal pipe section.

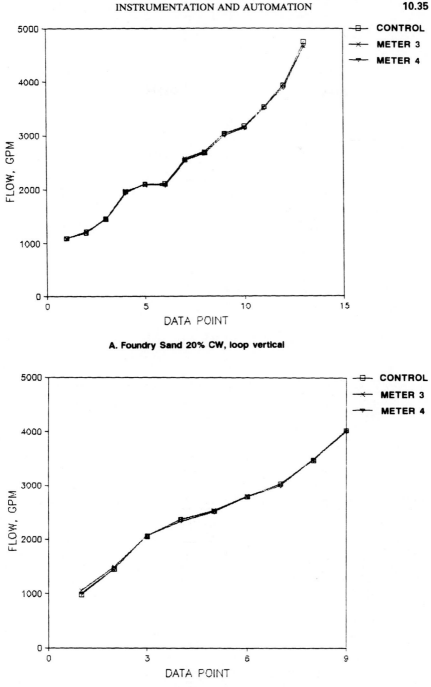

A. Foundry Sand 20% CW, loop vertical

B. Foundry Sand 20% CW, loop horizontal

FIGURE 10.28 Flow data comparisons for magnetic flowmeters.

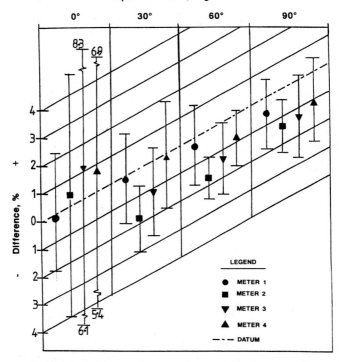

FIGURE 10.29 Influence of pipe inclination on nuclear density gauges and magnetic flowmeters.

which for the phosphate material at 0° inclination has a +11.8 percent difference with a data range of 4 to 134 percent. When the pipe was placed in the vertical position, the average percentage difference was reduced to −3.3 percent with a data range of ±55 percent. This represents a significant improvement for the meter but a wider range of data values than for the magnetic flowmeters.

The above information on doppler meter data range and percentage difference values was based on the full range of velocities used in each test (0 to 26 ft/s) (0 to 7.9 m/s). There is a noticeable reduction in the range of data values when the slurry velocity is greater than 12.8 ft/s (3.9 m/s). The average percentage difference for phosphate material, drops to −20 percent, but the range of data was reduced to ±10 percent. The change in average percentage difference is consistent with the Phase 1 doppler results, which indicated doppler values were significantly less than the control meter values at higher slurry velocities.

While the density gauges and magnetic flowmeter responses were similar for both slurry materials, it appears that the doppler flowmeter performance was better for the sand-gravel mixture than the phosphate material. The spread of data values for the sand-gravel mixture was on the order of ±35 percent, considerably less than the ±135 percent for the phosphate material. In addition, the average percentage difference for slurry velocities over 12.8 ft/s was between +1 and −10 percent.

Air Entrainment

Test conditions for air entrainment evaluation were slightly different from the standard testing procedures. For these tests, a velocity of 19 ft/s (5.8 m/s) and a density of 2.0 S.G. were maintained while air was introduced into the system at a constant rate of 14 ft³/min (0.4 m³/min) at 50 psig (3.5 kg/cm²), approximately 35 ft (10.7 m) upstream from the instrument loop. The data indicate that entrained air does affect meter values. As seen from Table 10.6, the density gauges were reading from 1 to 3 percent lower than the unaffected control meter values. However, with the introduction of air, the slurry is actually less dense and the meter values might be correct. Data spread was less than ±1.5 percent for these conditions of pipe inclination and slurry mixtures.

The magnetic flowmeters performed significantly better, for both materials, when located on vertical pipe sections. The doppler flowmeter, which is designed to perform well with aerated slurries, recorded significantly poorer values in most cases. The fact that it performed well with the sand-gravel mixture in a horizontal pipe is not clearly understood. Meter 6, mounted on a horizontal pipe section throughout the Phase 2 tests, had consistent readings for individual tests but had different percentage differences for each material.

For the Phase 1 tests, the density gauges were located on either side of the horizontal pipe (at the 3 and 9 o'clock positions). By rotating the source and sensor 45° from that position, data indicate that meter performance was improved by about 1.5 percent.

The bend meter, although not located on the instrument loop, was evaluated for performance and consistency of data. As in Phase 1, the meter performed well with a consistent 10 percent difference from the control meter. On several occasions the sensors clogged, giving poor readings until the lines could be purged. If the clogging problem could be overcome and an improved bend coefficient used, this meter has the potential to be an effective flow measuring device.

This study emphasizes that caution should be used in the selection of a production meter, especially the flowmeter components. Ideally a magnetic flowmeter should be used, giving careful consideration to such factors as accuracy,

TABLE 10.6   Average Percentage Difference for Air Entrainment Tests

	Phosphate + air pipe orientation (degrees)		Sand/gravel + air pipe orientation (degrees)	
	0	90	0	90
	Densitometers			
Meter 1	−2.50	−2.08	−0.66	−3.14
Meter 2	−2.19	−2.27	−0.90	−3.32
	Magnetic flowmeters			
Meter 3	−26.23	−4.02	−6.78	−5.97
Meter 4	−26.99	−2.83	−18.87	−9.68
	Doppler flowmeters			
Meter 5	−30.84	−30.99	1.17	−23.79
Meter 6	−11.53	(15.77)	37.90	(38.45)

( ) = values for meter on a horizontal pipe section.

initial cost, and maintenance requirements. A bend meter needs attention to the pressure taps to prevent clogging, and as the name implies, these taps must be on a pipe bend or elbow. The easiest flowmeter to install and the least expensive to purchase is the doppler-type flowmeter. However, this type of meter showed more variation in the data and differed most from the control meter. The data spread among the three doppler meters were much greater than the data spread between the two magnetic flowmeters.

The nuclear density gauge is the only readily available densitometer used on contemporary dredges. It is reliable, accurate, and safe to use. However, it does employ a radioactive source, and, therefore, a Nuclear Regulatory Commission (NRC) license is required. Many dredge operators prefer not to use this type of gauge because of the required licensing and training.

CONCLUSIONS

General conclusions that can be made from these tests are as follows:

1. The most accurate flowmeters tested were (in decreasing order) magnetic, bend, and doppler.
2. The density gauges had almost identical readings with each other and the control density meter.
3. The density gauges recorded more accurately when rotated 45° from the horizontal axis of the pipe.
4. The magnetic flowmeters had very similar readings with each other and the control flowmeter.
5. The doppler flowmeters had significantly different readings from each other and the control flowmeter.
6. The preferred pipe orientation for both the density gauge and the flowmeter is vertical or any orientation above the horizontal plane. However, a horizontal pipe is acceptable if high slurry concentrations that produce a stationary or sliding bed with dune formation are avoided. The difference between vertical and horizontal orientation is on the order of 1 percent for the density gauge, 3 percent for the magnetic flowmeters, and 5 percent for the doppler flowmeters.
7. Flow results from the sand slurries, sand-gravel mixture, and phosphate material were more consistent and accurate than those for gravel.
8. The doppler meters produced higher values than the control meter at low slurry velocities and fell off significantly, producing much lower values than the control, at higher slurry velocities.
9. As slurry velocities increased above 12 ft/s (3.6 m/s), the doppler meter experienced considerably less data scatter.

Dredge production meter values can be used as an aid in optimizing dredge operation and production. All flowmeters tested were acceptable and indeed are preferable to no meter at all. All the meters were consistent throughout the test and responded to changes in slurry concentrations, flow, and material types. The most critical element in the use of production meters is the calibrations of the individual components.

The contents of this section are not to be used for advertising, publication, or promotional purposes. Citation of trade names does not constitute an official endorsement or approval of the use of such commercial products.

ACKNOWLEDGMENTS

The work and data described are a part of the Dredging Research Program, which is sponsored by the Office of the Chief of Engineers, U.S. Army Corps of Engineers. Permission was granted by the Chief of Engineers to publish this information.

The author wishes to acknowledge the cooperation of the GIW staff especially Graeme Addie, John Maffett, and Steven Kerr, who organized and conducted the tests; Mr. R. Alan Duckworth for his data analyses and interpretations; and the manufacturers who supplied the instruments used in this study.

SUMMARY*

It has been indicated that great effort has gone into providing the dredgemaster of the modern trailing suction hopper dredge with a large degree of automatic sensing, indicating, and control equipment. Nearly every dredge system parameter has been measured and indicated. The results of this have been greater production, efficiency, and safety.

EVALUATION OF PRODUCTION METER INSTRUMENTATION AND USES

An evaluation of modern instrumentation used on hydraulic dredges has been conducted by Herbich, et al.[4] It appears that owners of modern dredges have realized the need for some type of instrumentation that would permit monitoring of the dredging process. This was concluded from the results of an extensive survey which show that 55 percent of modern dredges in this survey have flowmeters and 48 percent have density gauges. Production meters are installed on 16 percent of the dredges in the United States. However, many of the dredges that have flowmeters and density meters are capable of calculating production. Figure 10.30 presents a summary of the types of flowmeters (in this survey) in the United States and overseas. Figure 10.31 presents a summary of density meters used and Fig. 10.32 presents the summary of production meters and manufacturers of the production meters in the United States and overseas. Reliability of measuring equipment reported by the operators, rating, and frequency of maintenance are shown in Figs. 10.33, 10.34, and 10.35.

---

*This section was written by John B. Herbich.

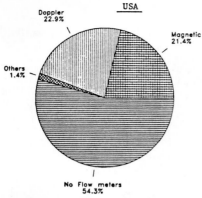
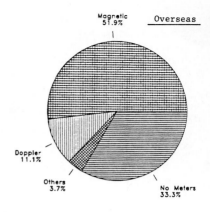

FIGURE 10.30   Flowmeters.

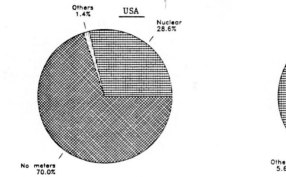
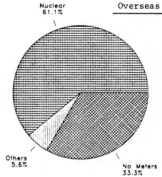

FIGURE 10.31   Density gauges.

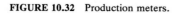
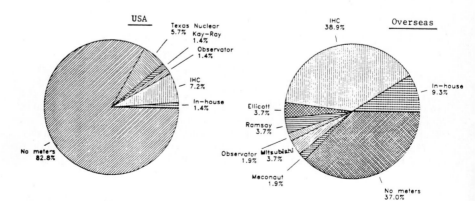

FIGURE 10.32   Production meters.

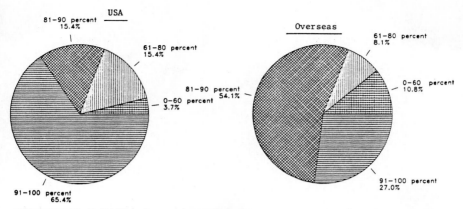

FIGURE 10.33   Reliability of measuring equipment.

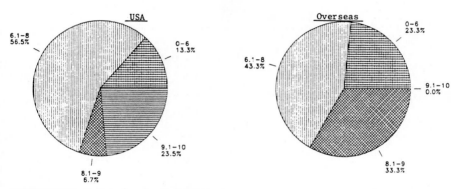

FIGURE 10.34   Rating of measuring equipment.

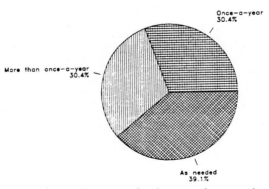

FIGURE 10.35   Frequency of maintenance for measuring equipment.

A summary of instrumentation on dredges in 1980 and in 1989 indicates a significant increase in the use of instrumentation (Table 10.7).[5] Table 10.8 shows the employment of production meters in the United States and overseas.

TABLE 10.7 Summary of Instrumentation on Dredges

1980*			1989		
United States & Canada	Overseas	Total	United States & CE	Overseas	Total
3/45 (6.7%)**	17/22 (77.3%)	20/67 (29.9%)	32/70 (46%)	40/54 (74%)	72/124 (58.1%)

*"Operating Characteristics of Cutterhead Dredges" presented by J. B. Herbich at *WODCON IX,* 1980.[5]
**Percentage indicates the ratio of total respondents having some instrumentation on their dredges versus total number of respondents.

TABLE 10.8 Summary of Production Meters on Dredges, 1989

United States & CE	Overseas	Total
12/70 (17%)	34/54 (63%)	46/124 (37%)

REFERENCES

1. Fortino, E. B., "Flow Measurement Techniques for Hydraulic Dredges," *Proc., of the ASCE, Journal of the Waterways & Harbor Division,* vol. 92, no. WW1, p. 109, Feb. 1966.

2. Erb, T. L., "Production Meter Systems for Suction Dredges," unpublished Dredging Short Course Notes, Center for Dredging Studies, Texas A&M University, 27 pp., 1985.

3. Pankow, V. R., "Laboratory Evaluation of Production Meter Components," *Proc., 22d Dredging Seminar,* Center for Dredging Studies, Texas A&M University, pp. 77–93, Oct. 1989.

4. Herbich, J. B., Lee, J. Y., Trivedi, D., Wilkinson, G. L., and De Hert, D. O., *Survey and Evaluation of Production Meter Instrumentation and Uses,* Center for Dredging Studies Report No. CDS-313, Texas A&M University, Apr. 1991.

5. Herbich, J. B., "Operating Characteristics of Cutterhead Dredges," *Ninth World Dredging Conference, WODCON IX,* Vancouver, B.C., Oct. 1980.

CHAPTER 11
PROJECT PLANNING

REQUEST FOR BIDS

Request for bids (RFB) are issued by the government or private industry. In the United States, on public projects, the RFBs are issued by the U.S. Army Corps of Engineer Districts. In this case, the projects to be completed during the next fiscal year are pretty well known and defined in advance.

Plans and specifications are provided with an RFB and should include geotechnical information based on field surveys. Dredging plans should include:

1. Location of the project.

2. Detailed cross sections of the channel and channel alignment.

3. Predredging cross sections with the required final cross sections. This would allow the contractor to estimate dredging quantities.

4. Geotechnical information including boring logs and any other information on soil including standard penetration tests (STP), shear strength, soil classification, etc. This information would allow the contractor to estimate the dredgeability of material. In many cases the geotechnical information is inadequate; either the spacing between the borings is too large or wrong sampling equipment was used or no soil testing was conducted. Samples of the cores or soil (preferably in relatively undisturbed condition) should be available for inspection by a prospective contractor. In the case of rock, unconfined compressive strength (UCS) should also be provided. The insufficient geotechnical information would usually lead to the contractor's claim of "changed conditions" and arbitration or litigation. In some cases information from previous dredging projects may be available and sometimes contractors may wish to verify the geotechnical information provided with the RFB by their own investigations. Field sampling in open water or in the ocean is always expensive and contractors may be reluctant to spend their own resources in case they are not awarded a contract.

5. Environmental information including the degree of contamination in the material to be dredged, locations of disposal of such material (usually in diked disposal areas), environmental windows, and when no dredging will be allowed because of fish spawning or for other reasons in environmentally sensitive areas. Confined land disposal areas or open-water designated areas that can be used on a given project must also be clearly described.

6. Specifications should include the starting date, completion date, liquidated damages clauses.

7. The depth of overdredging (preventive maintenance) for which the contractor will be paid should also be specified.

Huston[1] summarizes the contracts, specifications, supervision, technical provisions, and rentals in detail. Huston recommends that specifications should be studied carefully by the prospective bidder to avoid possible future controversy. One of the most common clauses in specifications is quoted from Huston:[1]

> . . .failure to acquaint himself with all the available information concerning conditions affecting the work will not relieve the contractor of the responsibility for estimating the difficulties and cost of the work, and successfully performing and completing the work as required.

BIDDING COSTS

The following are some factors that affect bidding costs:

Lump sum bids on a dredging project are not practical unless all aspects of the project are clearly defined.[2] The contractor would normally add a "contingency" cost to the bid in this case.

Cost-plus bids are followed in cases when insufficient information on the scope of work to be performed is available.

Unit price per cubic yard (or per cubic meter) of a given soil or rock are usually given on government contracts, and mobilization and demobilization costs are shown separately.

Daily rental costs are usually used in cases of emergency dredging (e.g., collision between two ships in a navigation channel, grounding, or extensive shoaling caused by a hurricane).

In some cases the contractor's dredge may have to be moved to accommodate berthing of ships; the contractor should include additional costs for maneuvering of the dredge and for the loss of production time.

OTHER REQUIREMENTS

The contractor should exercise every precaution to prevent injury or damage to property. The contractor is usually required to give warnings, signals or signs, display lights, exercise precautions against fire, and minimize the environmental impact during the dredging operations.

PAYMENTS FOR DREDGING WORK

There are several methods of payment used for dredging work:

1. A lump sum for mobilization and demobilization, and a unit price times the volume of material dredged

2. A lump sum for the entire project

3. A lump sum for mobilization, daily cost of operation of the dredging equipment and a lump sum for demobilization

The most commonly used method is the first described. Since the total payment depends on the accuracy of pre- and postdredging surveys, the surveys must be performed as accurately as possible. Cross sections of the channel should be taken at fairly close intervals, between 25 and 150 ft (7.6 and 46 m). The owner usually pays for any overdredging depth, which is agreed upon in the contract, usually 2 ft (0.6 m). However, in some heavy shoaling areas the overdredging depth may be as high as 8 ft (2.4 m).

CONTRACTING IN EUROPE

La Fédération Internationale des Ingénieurs Conseils (FIDIC; International Association of Consulting Engineers) published the fourth edition of the *Conditions of Contract for Civil Engineering Construction*[3,4] in 1987. This document defines the role of "the engineer," and discusses the risk of unforeseen circumstances, the faults of design, extension of time and claims, payments, right to suspend work, and many other matters dealing with civil engineering construction. There are standard conditions in other European countries, such as "VOB" in Germany, "UAV" in the Netherlands, the "Bestek/Cahier des Charges" in Belgium, and the "ICE" conditions in the United Kingdom.

FEDERAL PROJECTS

Descriptions of how the U.S. Army Corps of Engineers projects are conceived, authorized, and funded are described in EP-11050-2-10, which is reproduced at the end of the chapter. It should be noted that the projects must be initiated by the local, urban, and/or regional interests seeking solutions to a given problem(s). It may take 22 years from the initiation of the concept to realization of the project according to a Marine Board Study.[5]

REGULATORY PROGRAM

The U.S. Army Corps of Engineers has the regulatory authorities and responsibilities listed under Section 10 of the Rivers and Harbors Act of 1899 (33 U.S.C. 403) and Section 404 of the 1977 Clean Water Act (33 U.S.C. 1344). In addition, individual states have their own regulations. For example Texas requires that any work to be done on submerged lands in Texas requires an easement from the Texas General Land Office. Any work under Section 404 of the Clean Water Act for a Corps of Engineers dredge or fill permit requires a certification (Section 401) from the Texas Water Commission that the activity will not cause a violation of the state's water quality standards.

Jurisdiction under Section 404 covers all lakes greater than 5 acres, all streams with a rate of flow of 5 ft^3/s, and all waters subject to tides (Fig. 11.1). The ap-

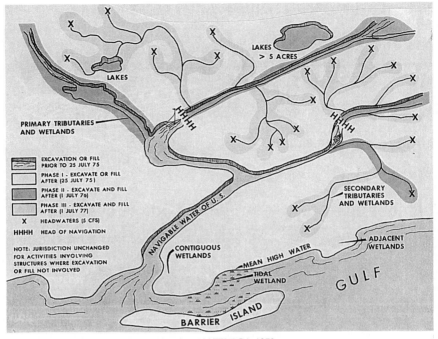

FIGURE 11.1   Jurisdiction under Section 404/FWPCA-1972.

plication information for a permit from the Corps of Engineers is reproduced from EP-1145-2-1 and is at the end of the chapter.

ENVIRONMENTAL IMPACT STATEMENTS

An environmental impact statement (EIS) must be prepared before any large project may be started. The District Engineer of the Corps of Engineers decides whether an EIS is required on smaller projects.

A description of an EIS prepared by the U.S. Army Corps of Engineers, Galveston District[6] follows.

Environmental Impact Statements—Case Study[6]

The U.S. Army Corps of Engineers is the Federal agency primarily responsible for planning Federal navigation projects and regulating activities in navigable waters of the United States and has had this responsibility even prior to this century. These civil works responsibilities began with an Act of Congress in 1824 which called for improvements to rivers and harbors for navigation. Additional legislation continued and expanded this responsibility. The River and Harbor Act of 1899 provided authority for the regulation of works and structures in navigable waters. Activities such as placement of piers, bulkheads, or pipelines and dredging, filling, or channelization in such waters are included in this act. For many years, however, such activities have taken place with little or no concern for the environment.

Today, because of increased concern for the environment, numerous laws have been enacted which have resulted in a substantial expansion of government regulatory agencies. The National Environmental Policy Act (NEPA) of 1969 is probably the single most significant law enacted in the past two decades which contribute to the expansion of regulations and regulatory agencies. It is extremely important, therefore, to be knowledgeable of and understand not only this Act, but the various laws and regulations which must be considered when evaluating a Federal navigation study or in order to obtain a permit from the Corps of Engineers for dredging and placement of dredged material.

The National Environmental Policy Act outlines the purpose and provides the basis for considering environmental aspects when evaluating various proposals. The primary purposes of the Act include:

(1) encourage productive and enjoyable harmony between man and his environment;

(2) to promote efforts which will prevent or eliminate damage to the environment and biosphere and stimulate the health and welfare of man; and

(3) to enrich the understanding of the ecological systems and natural resources important to the Nation.

Section 102 of NEPA directs that policies, regulations, and public laws will be interpreted and administered to the fullest extent possible by various agencies in accordance with the policies of the Act and impose general and specific requirements on all Federal agencies. Some of these requirements include:

(1) All agencies of the Federal government shall utilize a systematic, interdisciplinary approach which will insure the integrated use of the natural and social sciences and the environmental design arts in planning and decision making related to actions which may have an impact on man's environment;

(2) Identify methods and procedures which will insure that presently unquantified environmental amenities and values may be given appropriate consideration in decision making along with economic and technical considerations; and

(3) Include in every recommendation of report on proposals for legislation and other major Federal action significantly affecting the quality of the human environment, a detailed statement on environmental effects, or an Environmental Impact Statement (EIS).

These guidelines did not address other important provisions of the Act in areas of planning and decision making. EIS's failed to establish the link between what was learned in the environmental analysis and how that information can contribute to proper decision making. New NEPA regulations were published in final form in the Federal Register (Vol. 43, No. 230) by the Council on Environmental Quality on November 29, 1978 which emphasized three major objectives. They include:

(1) Reducing paperwork

(2) Reducing delays

(3) Better decisions

Some of the major points of the regulations which accomplish these objectives include:

- *Emphasize reasonable alternatives:* Portions of the EIS that are useful to decision makers and the public should be stressed in addition to a reduction in emphasis on background material. The most useful portions of the EIS include discussions of alternatives, affected environment, and environmental consequences.
- *Scoping process:* A process called scoping is to be used early in the planning stages to obtain input and assistance from various agencies, interested groups and proponents of an action. This action should begin as soon as possible after a decision is made to prepare an EIS. In addition to requesting participation from the various groups, several determinations are made through a cooperative effort. These include:

    (1) determine the scope of work and significant issues to be analyzed in depth;

    (2) identify and eliminate from detailed study any issues which are not significant;

    (3) identify other review and consultation requirements; and

    (4) indicate the timing of preparation of the EIS and the tentative planning and decision making schedule.

- *Format:* To encourage good analysis and clear presentation of various alternatives, the following format for EISs should be followed:

    (1) Cover Sheet

    (2) Summary

    (3) Table of Contents

    (4) Purpose of and Need for Action

    (5) Alternatives Including Proposed Action

    (6) Affected Environment

    (7) Environmental Consequences

    (8) List of Preparers

    (9) List of Agencies, Organizations, and Persons to Whom Copies of the Statement are Sent

    (10) Index

    (11) Appendices (if any)

Major sections of the EIS, previously identified as the most useful portions, are summarized as follows:

(a) *Alternatives including the proposed action:* This is one of the most important sections of the EIS. It presents in comparative form, environmental impacts of the proposal and various alternatives, thus providing the basis for choice among options by decision makers and the public.

(b) *Affected environment:* The areas to be affected by various alternatives shall be clearly and briefly described. Data and analysis should be proportionate with the importance of the impact to be described.

(c) *Environmental consequence:* This section provides the scientific and analytical basis for comparison of alternatives. The discussion will include direct and indirect impacts of all alternatives, including the proposed action, any adverse effects which cannot be avoided, any irretrievable commitments of resources which would be involved and the relationship between short-term use of man's environment and the maintenance and enhancement of long-term productivity.

- *Incorporation by reference:* Agencies are encouraged to incorporate material by reference into the EIS when the material is not of central importance and is readily available for public inspection.
- *Commenting:* Commenting on a draft EIS shall be requested of various Federal, State and local agencies, particularly those which have jurisdiction by law or special expertise with respect to any environmental impacts involved or which are authorized to develop and enforce environmental standards. Agencies with jurisdiction by law or special expertise shall comment on the EIS within the time period specified even if the agency replies that it has no comment.

  Comments on an EIS shall be as specific as possible and may address such points as the adequacy of the statement or merits of the alternatives discussed. The commenting agency shall also state that additional information is needed, if necessary, and what type is needed.

  When an agency objects to or expresses reservations about a plan with regard to environmental impacts, it shall specify mitigation measures considered necessary to balance out environmental impacts.
- *Environmental review and consultation requirements:* There are numerous other laws and regulations interrelated with NEPA which must be considered when planning a study and an EIS. These laws and regulations address all aspects of the project, from protection and enhancement of fish and wildlife resources to consistency with coastal zone management programs. They include, but are not limited to the following:

  Clean Air Act

  Clean Water Act

  Endangered Species Act

  Fish and Wildlife Coordination Act

  Marine Protection, Research and Sanctuaries Act

  National Historic Preservation Act

The following provides a brief overview of these laws and associated regulations as they relate to Corps of Engineers activities for the discharge of dredged or fill material.

(1) *Clean Air Act.* Sections 176(c) and 309 are the parts of this Act most pertinent to CE actions. Section 176(c) requires all Federal projects, licenses, permits, financial assistance, and other activities to conform to EPA approved or published state implementation plans with regard to control and abatement of air pollution. Section 309 provides that the EPA Administrator shall review and

comment on the environmental impacts of (a) legislation proposed by any Federal agency; (b) newly authorized Federal projects for construction and any other major Federal agency action requiring an EIS; and (c) proposed agency regulations. This would include for the CE, most studies for improving navigation or issuing a permit for major improvements to navigation. In order to comply with this law, an EIS would be prepared for the action and reviewed and commented on by the EPA.

(2) *Clean Water Act.* Various parts of Section 404 of this Act are applicable to CE authority over the discharge of dredged or fill material. Section 404(b) establishes guidelines by which sites for the discharge of dredged or fill material are specified. Section 404(r) establishes an exemption from provisions of this Act. A Federal project which includes the discharge of dredged or fill material specifically authorized by Congress is exempt from provisions of this Act (except toxic substances) if information on effects of the discharge, including consideration of 404(b)(1) guidelines, is included in an EIS submitted to Congress prior to the discharge in connection with authorization or appropriations. The EPA has developed guidelines pursuant to Section 404(b) of the Clean Water Act for the evaluation of sites for the discharge of dredged or fill material. These guidelines are used by the CE in evaluating Federal navigation projects. The State of Texas, however, requires a State Water Quality Certificate even though the action is Federally authorized.

(3) *Endangered Species Act.* Section 7 of this Act establishes a consultation process between Federal agencies and the Secretaries of the Interior, Commerce, or Agriculture, as appropriate, for carrying out programs for the conservation of endangered species. To accomplish this, the construction agency (CE) must request the appropriate Secretary to provide information on any listed or proposed endangered species or critical habitat in the area. Should a list be provided, the agency must prepare a biological assessment to be completed within 180 days or other specified time. Based on results of the biological assessment, the agency must initiate consultation with the Secretary if listed endangered or threatened species may be affected. The consultation shall be concluded within a 90-day period or other specified time. At the completion of consultation, the Secretary shall provide the agency an opinion on how the proposed action will affect the species or its critical habitat and if necessary, shall suggest reasonable alternatives. No agency has developed specific guidelines concerning these procedures.

(4) *Fish and Wildlife Coordination Act.* This Act provides that for any proposal for Federal work affecting any stream or other body of water, the Federal agency (CE) proposing such work must first consult with the Fish and Wildlife Service and state wildlife agency with a view to preventing losses and damages to wildlife resources and to provide for development and improvement of wildlife resources. Reports from the Secretary of the Interior and state wildlife agency shall be an integral part of any report submitted to Congress. The report submitted to Congress must give full consideration to recommendations by the Secretary of the Interior and state agency and shall include justifiable means and measures for wildlife purposes, including mitigation measures. The CE coordinates studies and the EIS with these agencies and includes their reports and comments in the EIS.

(5) *Marine Protection, Research and Sanctuaries Act.* Sections 103 and 104 of this Act establish the authority for ocean dumping of dredged material and for designating the amount, type, and location of dumping. Section 103(e) provides the Secretary of the Army with permit authority over the transportation of dredged material for the purposes of dumping in ocean waters. A permit may be issued when a determination is made that such dumping will not unreasonably

degrade or endanger human health, welfare of amities, or the marine environment, ecological systems, or economic potentialities. Section 104 requires that permits issued shall designate the amount, type, and location of the material to be dumped, and length of time for the dumping, and after consultation with the Coast Guard, provide for any special monitoring and surveillance provisions. Regulations have been published by the EPA and the CE concerning ocean dumping.

(6) *National Historic Preservation Act.* Section 106 of this Act directs the head of any Federal agency (CE) having direct or indirect jurisdiction over a proposed Federal or Federally-assisted undertaking shall, prior to the approval of expenditure of any Federal funds on the undertaking, take into account effects of the undertaking on any district, site, building, structure, or object that is included or eligible for inclusion in the National Register of Historic Places. The agency shall provide the Advisory Council on Historic Preservation and the State Historic Preservation Officer a reasonable opportunity to comment with regard to such undertaking. Regulations have been published by the Advisory Council implementing Section 106 of this Act. The CE coordinates these activities through its planning process and EIS preparation and review process.

Other Laws

Numerous other laws are somewhat applicable to Corps of Engineers dredging activities and are considered in the appropriate situation. For example, the Coastal Zone Management Act requires that any activities a Federal agency conducts or supports which directly affect the coastal zone, and any development project in the coastal zone, shall be, to the maximum extent practicable, consistent with approved State management programs.

The overall effect of these laws and regulations has been to develop better procedures for preparation of EISs and gives a complete evaluation of all potential environmental effects. The EIS constitutes an integral part of an interdisciplinary process and serves as a summation and evaluation of the effects, both beneficial and adverse, that each alternative would have on the environment. It summarizes the detailed evaluation and analyses of federal and state agencies with jurisdiction by law or special expertise with respect to environmental impacts and the concerns, views, and comments expressed by conservation and environmental action groups and the public. It is therefore extremely important to be knowledgeable of and thoroughly understand these laws and the associated regulations which must be considered in evaluating any action involving dredging and placement of dredged material.

REFERENCES

1. Huston, J., *Hydraulic Dredging, Theoretical and Applied,* Cornell Maritime Press, Inc., Centreville, MD, 332 pp., 1970.

2. Anonymous, "Dredging for Development," Report of the Dredging Task Force, The International Association of Ports and Harbors, Supplement to *Terra et Aqua,* no. 33, 24 pp., Apr. 1987.

3. Goudsmit, J. J., "Standard Contract Conditions, Characteristics, Pitfalls and Possible Amendments with Special Regard to Dredging," *Terra et Aqua,* no. 32, pp. 23–31, Sept. 1986.

4. Goudsmit, J. J., "A First Impression of the 4th Edition of the FIDIC Conditions of Contract," *Terra et Aqua,* no. 36, pp. 13–18, Apr. 1980.

5. Marine Board, National Research Council, "Dredging Coastal Ports, An Assessment of the Issues," D. E. Kash (Chairman), J. B. Herbich (Vice-Chairman), National Academy Press, Washington, DC, 212 pp., 1985.

6. U.S. Army Corps of Engineers Galveston District, "Environmental Impact Statements Case Study," unpublished Dredging Engineering Short Course Notes, Center for Dredging Studies, Texas A&M University, 8 pp., 1989.

how
U.S. Army
Corps of Engineers

projects
are

conceived,
authorized,
funded
and
implemented

BERH
- CONSIDERS VIEWS OF
  - PUBLIC
  - STATES
  - AGENCIES
- REVIEWS AND PROVIDES RECOMMENDATIONS
  - REVISED DRAFT EIS
  - FINAL FR
- TRANSMITS TO CHIEF OF ENGINEERS

13

CHIEF
- REVIEWS BOARD REPORT
- PREPARES HIS DRAFT RECOMMENDATIONS
- DISTRIBUTES FOR OUTSIDE REVIEW
  - REVISED DRAFT EIS (PUBLIC, STATES, FEDERAL DEPARTMENTS) (45-DAY REVIEW PERIOD)
  - FR (GOVERNORS, FEDERAL DEPARTMENTS) (90-DAY REVIEW PERIOD)

14

CHIEF
- REVIEWS RECEIVED COMMENTS
- MODIFIES REPORT AS APPROPRIATE
- PREPARES FINAL EIS

15

CHIEF
- FORWARDS RECOMMENDATIONS TO SECRETARY OF THE ARMY FOR CONSIDERATION
  - FINAL REPORT
  - FINAL EIS
  - SOF

16

SECRETARY OF THE ARMY
- REVIEWS
- COORDINATES WITH OMB
- PREPARES HIS RECOMMENDATIONS
- FORWARDS
  - FINAL EIS, SOF (CEQ, PUBLIC)
  - FINAL FR, FINAL EIS, SOF (CONGRESS)

17

PROJECT AUTHORIZATION
- HOLDS HEARINGS
- INCLUDES IN WATER RESOURCES DEVELOPMENT BILL OR OTHER LEGISLATION

PRESIDENT SIGNS

18

OMB
- REVIEWS CORPS BUDGET
- SUBMITS TO CONGRESS

19

PROJECT FUNDING
- CONGRESS INCLUDES IN APPROPRIATIONS BILL
- PRESIDENT SIGNS

20

LOCAL INTERESTS GUARANTEE TO FULFILL OBLIGATIONS REQUIRED BY LAW (e.g., REAL ESTATE, COST SHARING, MAINTENANCE, OPERATION, FLOOD ZONING)

21

DE
- FORMULATES PRE-CONSTRUCTION PLANNING GENERAL DESIGN MEMORANDA (GDM)
  - UPDATES EIS AS REQUIRED
  - ISSUES PUBLIC NOTICE AND CONDUCTS AT LEAST ONE PUBLIC MEETING
- OBTAINS ADDITIONAL CONGRESSIONAL AUTHORIZATION AS APPROPRIATE
- INITIATES AND COMPLETES CONSTRUCTION
- OPERATES AND MAINTAINS

22

DIVISIONS AND DISTRICTS for CIVIL WORKS ACTIVITIES

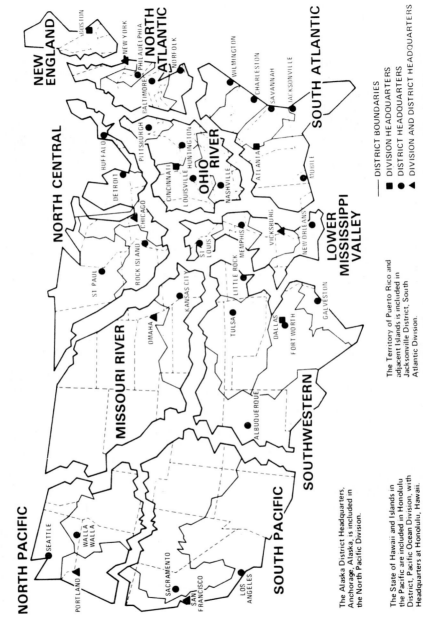

NORTH PACIFIC

NEW ENGLAND

NORTH ATLANTIC

NORTH CENTRAL

SOUTH ATLANTIC

OHIO RIVER

MISSOURI RIVER

LOWER MISSISSIPPI VALLEY

SOUTH PACIFIC

SOUTHWESTERN

SEATTLE
WALLA WALLA
PORTLAND
SACRAMENTO
SAN FRANCISCO
LOS ANGELES
ALBUQUERQUE
OMAHA
ST PAUL
ROCK ISLAND
CHICAGO
DETROIT
BUFFALO
PITTSBURGH
KANSAS CITY
TULSA
DALLAS
FORT WORTH
GALVESTON
LITTLE ROCK
ST LOUIS
MEMPHIS
VICKSBURG
NEW ORLEANS
NASHVILLE
LOUISVILLE
HUNTINGTON
CINCINNATI
ATLANTA
MOBILE
JACKSONVILLE
SAVANNAH
CHARLESTON
WILMINGTON
NORFOLK
BALTIMORE
PHILADELPHIA
NEW YORK
HOUSTON

DISTRICT BOUNDARIES

■ DIVISION HEADQUARTERS

● DISTRICT HEADQUARTERS

▲ DIVISION AND DISTRICT HEADQUARTERS

The Alaska District Headquarters, Anchorage, Alaska, is included in the North Pacific Division.

The State of Hawaii and Islands in the Pacific are included in Honolulu District, Pacific Ocean Division, with Headquarters at Honolulu, Hawaii.

The Territory of Puerto Rico and adjacent Islands is included in Jacksonville District, South Atlantic Division.

11.14

EP 1145-2-1
May 1985

United States Army Corps of Engineers

Regulatory Program

Applicant Information

This Pamphlet Supersedes EP 1145-2-1, November 1977

INTRODUCTION

This pamphlet is designed to assist you in applying for a Department of the Army permit from the Corps of Engineers. The pamphlet is not intended to be a complete description of all aspects of the permit program, but will provide general information of a non-technical nature and specific guidance on how to complete a permit application. Full explanation of the program may be found in Title 33 Code of Federal Regulations, Parts 320 through 330. These regulations are available for review at the Corps of Engineers District offices listed at the back of this pamphlet. Answers to technical questions and detailed information about special aspects of the program that pertain to your geographical area and your proposed activity may also be obtained from Corps of Engineers District offices.

John F. Wall
Major General, USA
Director of Civil Works

CONTENTS

GENERAL INFORMATION

Authority for the Regulatory Program

The U.S. Army Corps of Engineers has been regulating activities in the nation's waters since 1890. Until the 1960's the primary purpose of the regulatory program was to protect navigation. Since then, as a result of laws and court decisions, the program has been broadened so that it now considers the full public interest for both the protection and utilization of water resources.

The regulatory authorities and responsibilities of the Corps of Engineers are based on the following laws:

☐ **Section 10 of the Rivers and Harbors Act of 1899 (33 U.S.C. 403)** prohibits the obstruction or alteration of navigable waters of the United States without a permit from the Corps of Engineers.

☐ **Section 404 of the Clean Water Act (33 U.S.C. 1344).** Section 301 of this Act prohibits the discharge of dredged or fill material into waters of the United States without a permit from the Corps of Engineers.

☐ **Section 103 of the Marine Protection, Research, and Sanctuaries Act of 1972, as amended (33 U.S.C. 1413)** authorizes the Corps of Engineers to issue permits for the transportation of dredged material for the purpose of dumping it into ocean waters.

Other laws may also affect the processing of applications for Corps of Engineers permits. Among these are the National Environmental Policy Act, the Coastal Zone Management Act, the Fish and Wildlife Coordination Act, the Endangered Species Act, the National Historic Preservation Act, the Deepwater Port Act, the Federal Power Act, the Marine Mammal Protection Act, the Wild and Scenic Rivers Act, and the National Fishing Enhancement Act of 1984.

Explanation of Some Commonly Used Terms

Certain terms which are closely associated with the regulatory program are explained briefly in this section. If you need more detailed definitions, refer to the Code of Federal Regulations (33 CFR Parts 320 through 330) or contact a Corps district regulatory office.

Activity(ies) as used in this pamphlet includes structures (for example a pier, wharf, bulkhead, or jetty) and work (which includes dredging, disposal of dredged material, filling, excavation or other modification of a navigable water of the United States).

Navigable Waters of the United States are those waters of the United States that are subject to the ebb and flow of the tide shoreward to the mean high water mark and/or are presently used, or have been used in the past or may be susceptible to use to transport interstate or foreign commerce. These are waters that are navigable in the traditional sense where permits are required for certain activities pursuant to Section 10 of the Rivers and Harbors Act. This term should not be confused with the term *waters of the United States* below.

Waters of the United States is a broader term than navigable waters of the United States defined above. Included are adjacent wetlands and tributaries to navigable waters of the United States and other waters where the degradation or destruction of which could affect interstate or foreign commerce. These are the waters where permits are required for the discharge of dredged or fill material pursuant to Section 404 of the Clean Water Act.

Pre-application Consultation is one or more meetings between members of the district engineer's staff and an applicant and his agent or his consultant. A pre-application consultation is usually related to applications for major activities and may involve discussion of alternatives, environmental documents, National Environmental Policy Act procedures, and development of the scope of the data required when an environmental impact statement is required.

Public Hearings may be held to acquire information and give the public the opportunity to present views and opinions. The Corps may hold a hearing or participate in joint public hearings with other Federal or state agencies. The district engineer may specify in the public notice that a hearing will be held. In addition, any person may request in writing during the comment period that a hearing be held. Specific reasons must be given as to the need for a hearing. The district engineer may attempt to resolve the issue informally or he may set the date for a public hearing. Hearings are held at times and places that are convenient for the interested public. Very few applications involve a public hearing.

The Public Interest Review is the term which refers to the evaluation of a proposed activity to determine probable impacts. Expected benefits are balanced against reasonably foreseeable detriments. All relevant factors are weighed. Corps policy is to provide applicants with a timely and carefully weighed decision which reflects the public interest.

Public Notice is the primary method of advising interested public agencies and private parties of the proposed activity and of soliciting comments and information necessary to evaluate the probable impact on the public interest. Upon request, anyone's name will be added to the distribution list to receive public notices.

Waterbody is a river, creek, stream, lake, pool, bay, wetland, marsh, swamp, tidal flat, ocean, or other water area.

Questions That Are Frequently Asked

Various questions are often asked about the regulatory program. It is hoped that these answers will help you to understand the program better.

Q. When should I apply for a Corps permit?

A. Since two to three months is normally required to process a routine application involving a public notice, you should apply as early as possible to be sure you have all required approvals before your planned commencement date. For a large or complex activity that may take longer, it is often helpful to have a "pre-application consultation" or informal meeting with the Corps during the early planning phase of your project. You may receive helpful information at this point which could prevent delays later. When in doubt as to whether a permit may be required or what you need to do, don't hesitate to call a district regulatory office.

Q. I have obtained permits from local and state governments. Why do I have to get a permit from the Corps of Engineers?

A. It is possible you may not have to obtain an individual permit, depending on the type or location of work. The Corps has many general permits which authorize minor activities without the need for individual processing. Check with your Corps district regulatory office for information on general permits. When a general permit does not apply, you may still be required to obtain an individual permit.

Q. What will happen if I do work without getting a permit from the Corps?

A. Performing unauthorized work in waters of the United States or failure to comply with terms of a valid permit can have serious consequences. You would be in violation of Federal law and could face stiff penalties, including fines and /or requirements to restore the area.

Enforcement is an important part of the Corps regulatory program. Corps surveillance and monitoring activities are often aided by various agencies, groups, and individuals, who report suspected violations. When in doubt as to whether a planned activity needs a permit, contact the nearest district regulatory office. It could save a lot of unnecessary trouble later.

Q. How can I obtain further information about permit requirements?

A. Information about the regulatory program is available from any Corps district regulatory office. Addresses and telephone numbers of offices are listed at the back of this pamphlet. Information may also be obtained from the water resource agency in your state.

Q. Why should I waste my time and yours by applying for a permit when you probably won't let me do the work anyway?

A. Nationwide, only three percent of all requests for permits are denied. Those few applicants who have been denied permits usually have refused to change the design, timing, or location of the proposed activity. When a permit is denied, an applicant may redesign the project and submit a new application. To avoid unnecessary delays pre-application conferences, particularly for applications for major activities, are recommended. The Corps will endeavor to give you helpful information, including factors which will be considered during the public interest review, and alternatives to consider that may prove to be useful in designing a project.

Q. What is a wetland and what is its value?

A. Wetlands are areas that are periodically or permanently inundated by surface or ground water and support vegetation adapted for life in saturated soil. Wetlands include swamps, marshes, bogs and similar areas. A significant natural resource, wetlands serve important functions relating to fish and wildlife; food chain production; habitat; nesting; spawning; rearing and resting sites for aquatic and land species; protection of other areas from wave action and erosion; storage areas for storm and flood waters; natural recharge areas where ground and surface water are interconnected; and natural water filtration and purification functions.

Although individual alterations of wetlands may constitute a minor change, the cumulative effect of numerous changes often results in major damage to wetland resources. The review of applications for alteration of wetlands will include consideration of whether the proposed activity is dependent upon being located in an aquatic environment.

Q. How can I design my project to eliminate the need for a Corps permit?

A. If your activity is located in an area of tidal waters, the best way to avoid the need for a permit is to select a site that is above the high tide line and avoids wetlands or other waterbodies. In the vicinity of fresh water, stay above ordinary high water and avoid wetlands adjacent to the stream or lake. Also, it is possible that your activity is exempt and does not need a Corps permit or that it has been authorized by a nationwide or regional general permit. So, before you build, dredge or fill, contact the Corps district regulatory office in your area for specific information about location, exemptions, and regional and nationwide general permits.

THE PERMIT APPLICATION

General

The application form used to apply for a permit is Engineer Form 4345, *Application for a Department of the Army Permit*. You can obtain the application from one of the Corps of Engineers district regulatory offices listed in the back of this pamphlet. Some offices may use a slightly modified form for joint processing with a state agency; however, the required information is basically the same. It is important that you provide complete information in the requested format. If incomplete information is provided, processing of your application will be delayed. This information will be used to determine the appropriate form of authorization, and to evaluate your proposal. Some categories of activities have been previously authorized by nationwide or regional permits, and no further Corps approvals are required. Others may qualify for abbreviated permit processing, with authorizations in the form of letters of permission, in which a permit decision can usually be reached in less than 30 days. For other activities, a Public Notice may be required to notify Federal, state, and local agencies, adjacent property owners, and the general public of the proposal to allow an opportunity for review and comment or to request a public hearing. Most applications involving Public Notices are completed within four months and many are completed within 60 days.

The district engineer will begin to process your application immediately upon receipt of all required information. You will be sent an acknowledgement of its receipt and the application number assigned to your file. You should refer to this number when inquiring about your application. Your proposal will be reviewed, balancing the need and expected benefits against the probable impacts of the work, taking into consideration all comments received and other relevant factors. This process is called the *public interest review*. The Corps goal is to reach a decision regarding permit issuance or denial within 60 days of receipt of a complete application. However, some complex activities, issues, or requirements of law may prevent the district engineer from meeting this goal.

For any specific information on the evaluation process, filling out the application forms, or the status of your application, you should contact the regulatory branch of the Corps of Engineers district office which has jurisdiction over the area where you plan to do the work.

**Typical Processing Procedure for a
Standard Individual Permit**

1. Preapplication consultation (optional)
2. Applicant submits ENG Form 4345 to district regulatory office*
3. Application received and assigned identification number
4. Public notice issued (within 15 days of receiving all information)
5. 15 to 30 day comment period depending upon nature of activity
6. Proposal is reviewed** by Corps and:
   Public
   Special interest groups
   Local agencies
   State agencies
   Federal agencies
7. Corps considers all comments
8. Other federal agencies consulted, if appropriate
9. District engineer may ask applicant to provide additional information
10. Public hearing held, if needed
11. District engineer makes decision
12. Permit issued
    or
    Permit denied and applicant advised of reason

---

*A local variation, often a joint federal-state application form may be submitted.

**Review period may be extended if applicant fails to submit information or due to requirements of certain laws.

Evaluation Factors

The decision whether to grant or deny a permit is based on a public interest review of the probable impact of the proposed activity and its intended use. Benefits and detriments are balanced by considering effects on items such as:

conservation
economics
aesthetics
general environmental concerns
wetlands
cultural values
fish and wildlife values
flood hazards
floodplain values
food and fiber production
navigation
shore erosion and accretion
recreation
water supply and conservation
water quality
energy needs
safety
needs and welfare of the people
considerations of private ownership

The following general criteria will be considered in the evaluation of every application:

□ the relative extent of the public and private need for the proposed activity;

□ the practicability of using reasonable alternative locations and methods to accomplish the objective of the proposed activity; and

□ the extent and permanence of the beneficial and/or detrimental effects which the proposed activity is likely to have on the public and private uses to which the area is suited.

Section 404(b) (1) of the Clean Water Act

If your project involves the discharge of dredged or fill material, it will be necessary for the Corps to evaluate your proposed activity under the Section 404(b)(1) guidelines prepared by the Environmental Protection Agency. The guidelines restrict discharges into aquatic areas where less environmentally damaging, practicable alternatives exist.

Forms and Permits

The following forms apply to the permit process:

Application

The form that you will need to initiate the review process is ENG Form 4345 or a joint Federal-state application that may be available in your state. The appropriate form may be obtained from the district regulatory office which has jurisdiction in the area where your proposed project is located.

Individual Permits

An individual permit may be issued as either ENG Form 1721, the standard permit, or as a Letter of Permission.

☐ A standard permit is one processed through the typical review procedures, (see page 7) which include public notice, opportunity for a public hearing, and receipt of comments. It is issued following a case-by-case evaluation of a specific activity.

☐ If work is minor or routine with minimum impacts and objections are unlikely, then it may qualify for a Letter of Permission (LOP). An LOP can be issued much more quickly than a standard permit since an individual public notice is not required. The District Engineer will notify you if your proposed activity qualifies for an LOP.

General Permits

In many cases the formal processing of a permit application is not required because of general permits already issued to the public at large by the Corps of Engineers. These are issued on a regional and nationwide basis.

Separate applications may not be required for activities authorized by a general permit; nevertheless, reporting may be required. For specific information on general permits, contact a district regulatory office.

ENG Form 4336

The third form, ENG Form 4336, is used to assist with surveillance for unauthorized activities. The form, which contains a description of authorized work, should be posted at the site of an authorized activity. If the Corps decides it is appropriate for you to post this form, it will be furnished to you when you receive your permit.

Fees. Fees are required for most permits. $10.00 will be charged for a permit for a non-commercial activity; $100.00 will be charged for a permit for a commercial or industrial activity. The district engineer will make the final decision as to the amount of the fee. Do not send a fee when you submit an application. When the Corps issues a permit, you will be notified and asked to submit the required fee payable to the Treasurer of the United States. No fees are charged for transferring a permit from one property owner to another, for Letters of Permission, or for any activities authorized by a general permit or for permits to governmental agencies.

Instructions for Preparing an Application

The instructions given below, together with the sample application and drawings, should help in completing the required application form. If you have additional questions, do not hesitate to contact the district regulatory office.

Block Number 1. Application Number. Leave this block blank. When your completed application is received, it will be assigned a number for identification. You will be notified of the number in an acknowledgement letter. Please refer to this number in any correspondence or inquiry concerning your application.

Block 2. Name and address of applicant(s). Fill in name, mailing address, and telephone number(s) for all applicants. The telephone number(s) should be a number where you can be reached during business hours. If space is needed for additional names, attach a sheet of white, 8½ × 11 inch paper labeled "Block 2 Continued."

Block 3. Name, address and title of authorized agent. It is not necessary to have an agent represent you; however, if you do, fill in the agent's name, address, title and telephone number(s). If your agent is submitting and signing the application, you must fill out and sign the Statement of Authorization in Block 3.

Block 4. Detailed description of proposed activity. The written description and the drawings are the most important parts of the application. If there is not enough space in Block 4, (a), (b) or (c) attach additional sheet(s) of white, 8½ × 11 inch paper labeled "Block 4 Continued."

a. **Activity.** Describe the overall activity. Give the approximate dimensions of structures, fills, excavations (lengths, widths, heights or depths).

b. **Purpose.** Describe the purpose, need and intended use (public, private, commercial, or other use) of the proposed activity. Include a description of related facilities, if any, to be constructed on adjacent land. Give the date you plan to begin work on the activity and the date work is expected to be completed.

c. **Discharge of Dredged or Fill Material.** If the activity will involve the discharge of dredged or fill material, describe the type (rock, sand, dirt, rubble, etc.), quantity (in cubic yards), and mode of transportation to the discharge site.

Block 5. Names and addresses of adjoining property owners, lessees, etc. whose property adjoins the waterbody. List complete names, addresses and zip codes of adjacent property owners (both public and private), lessee, etc., whose property also adjoins the waterbody or wetland, in order that they may be notified of the proposed activity. This information is usually available at the local tax assessor office. If more space is needed attach a sheet of white, 8½ × 11 inch paper labeled "Block 5 Continued."

Block 6. Waterbody and location on waterbody where activity exists or is proposed. Fill in the name of the waterbody and the river mile (if known) at the location of the activity. Include easily recognizable landmarks on the shore of the waterbody to aid in locating the site of the activity.

Block 7. Location and land where activity exists or is proposed. This information is used to locate the site. Give the street address of the property where the proposed activity will take place. If the site does not have a street address, give the best descriptive location (name or waterbody), names and/or numbers of roads or highways, name of nearest community or town, name of county and state, and directions, such as 2 miles east of Brown's Store on Route 105.

Do not use your home address unless that is the location of the proposed activity. Do not use a post office box number.

Block 8. Information about completed activity. Provide information about parts of the activity which may be complete. An activity may have been authorized by a previously issued permit, may exist from a time before a Corps permit was required or may be constructed on adjacent upland.

Block 9. Information about approvals or denials by other government agencies. You may need approval or certification from other Federal, interstate, state, or local government agencies for the activity described in your application. Applications you have submitted, and approvals, certifications, or disapprovals that you have received should be recorded in Block 9. It is not necessary to obtain other Federal, state, and local permits before applying for a Corps of Engineers permit.

Block 10. Signature of applicant or agent. The application must be signed in Block 10 by the owner, lessee, or a duly authorized agent. The person named in Block 3 will be accepted as the officially designated agent of the applicant. The signature will be understood to be affirmation that the applicant possesses the requisite property interest to undertake the proposed activity.

APPLICATION FOR DEPARTMENT OF THE ARMY PERMIT (33 CFR 325)	OMB APPROVAL NO. 0702-0036 Expires 30 June 1986

The Department of the Army permit program is authorized by Section 10 of the River and Harbor Act of 1899, Section 404 of the Clean Water Act and Section 103 of the Marine, Protection, Research and Sanctuaries Act. These laws require permits authorizing activities in or affecting navigable waters of the United States, the discharge of dredged or fill material into waters of the United States, and the transportation of dredged material for the purpose of dumping it into ocean waters. Information provided on this form will be used in evaluating the application for a permit. Information in this application is made a matter of public record through issuance of a public notice. Disclosure of the information requested is voluntary; however, the data requested are necessary in order to communicate with the applicant and to evaluate the permit application. If necessary information is not provided, the permit application cannot be processed nor can a permit be issued.

One set of original drawings or good reproducible copies which show the location and character of the proposed activity must be attached to this application (see sample drawings and instructions) and be submitted to the District Engineer having jurisdiction over the location of the proposed activity. An application that is not completed in full will be returned.

1. APPLICATION NUMBER (To be assigned by Corps)

3. NAME, ADDRESS, AND TITLE OF AUTHORIZED AGENT

None

2. NAME AND ADDRESS OF APPLICANT

Fred R. Harris
852 West Branch Road
Blue Harbor, Maryland 21705

Telephone no. during business hours

A/C (301) 585-2779 (Residence)
A/C ( ) (Office)

Telephone no. during business hours

A/C ( ) _____ (Residence)
A/C ( ) _____ (Office)

Statement of Authorization   I hereby designate and authorize _____ to act in my behalf as my agent in the processing of this permit application and to furnish, upon request, supplemental information in support of the application.

SIGNATURE OF APPLICANT   DATE

4. DETAILED DESCRIPTION OF PROPOSED ACTIVITY

4a. ACTIVITY

Build timber bulkhead and pier and fill.

4b. PURPOSE

To provide boat access and prevent erosion of shoreline at my place of residence.

4c. DISCHARGE OF DREDGED OR FILL MATERIAL

Approximately 200 cubic yards of upland fill will be placed between new bulkhead and existing shoreline.

ENG FORM 4345, Apr 83   EDITION OF 1 OCT 77 IS OBSOLETE   (Proponent: DAEN CWO-N)

5. NAMES AND ADDRESSES OF ADJOINING PROPERTY OWNERS, LESSEES, ETC., WHOSE PROPERTY ALSO ADJOINS THE WATERWAY

Mary L. Clark
850 West Branch Road
Blue Harbor, Maryland 21703

(301) 585-8830

Harry N. Hampton
854 West Branch Road
Blue Harbor, Maryland 21703

(301) 585-3676

6. WATERBODY AND LOCATION ON WATERBODY WHERE ACTIVITY EXISTS OR IS PROPOSED

West Branch of the Haven River on Blue Harbor.

7. LOCATION ON LAND WHERE ACTIVITY EXISTS OR IS PROPOSED

ADDRESS:

852 West Branch Road
STREET, ROAD, ROUTE OR OTHER DESCRIPTIVE LOCATION

King Edward, Maryland 21703
COUNTY   STATE   ZIP CODE

Town of Blue Harbor
LOCAL GOVERNING BODY WITH JURISDICTION OVER SITE

8. Is any portion of the activity for which authorization is sought now complete? ☐ YES ☒ NO
If answer is "Yes" give reasons, month and year the activity was completed, and indicate the existing work on the drawings.

9. List all approvals or certifications and denials received from other federal, interstate, state or local agencies for any structures, construction, discharges or other activities described in this application.

ISSUING AGENCY	TYPE APPROVAL	IDENTIFICATION NO	DATE OF APPLICATION	DATE OF APPROVAL	DATE OF DENIAL
Town of Blue Harbor	Zoning	BH25172	6/20/82	6/30/82	
Md DNR	Certification	DNR258WQ	6/11/82	8/12/82	

10. Application is hereby made for a permit or permits to authorize the activities described herein. I certify that I am familiar with the information contained in this application, and that to the best of my knowledge and belief such information is true, complete, and accurate. I further certify that I possess the authority to undertake the proposed activities or I am acting as the duly authorized agent of the applicant.

Fred R. Harris   Oct. 15, 1982   _____   _____
SIGNATURE OF APPLICANT   DATE   SIGNATURE OF AGENT   DATE

The application must be signed by the person who desires to undertake the proposed activity (applicant) or it may be signed by a duly authorized agent if the statement in Block 3 has been filled out and signed.

18 U.S.C. Section 1001 provides that: Whoever, in any manner within the jurisdiction of any department or agency of The United States knowingly and willfully falsifies, conceals, or covers up by any trick, scheme, or device a material fact or makes any false, fictitious or fraudulent statements or representations or makes or uses any false writing or document knowing same to contain any false, fictitious or fraudulent statement or entry, shall be fined not more than $10,000 or imprisoned not more than five years, or both.

Do not send a permit processing fee with this application. The appropriate fee will be assessed when a permit is issued.

DRAWINGS

General Information

Three types of drawings—Vicinity, Plan, and Elevation—are required to accurately depict activities (See sample drawings on pages 16 and 17).

Submit one original, or good quality copy, of all drawings on 8½ × 11 inch white paper (tracing cloth or film may be used). Submit the fewest number of sheets necessary to adequately show the proposed activity. Drawings should be prepared in accordance with the general format of the samples, using block style lettering. Each page should have a title block. See check list below. Drawings do not have to be prepared by an engineer, but professional assistance may become necessary if the project is large or complex.

Leave a 1-inch margin at the top edge of each sheet for purposes of reproduction and binding.

In the title block of each sheet of drawings identify the proposed activity and include the name of the body of water; river mile (if applicable); name of county and state; name of applicant; number of the sheet and total number of sheets in set; and date the drawing was prepared.

Since drawings must be reproduced, use heavy dark lines. Color shading cannot be used; however, dot shading, hatching, or similar graphic symbols may be used to clarify line drawings.

Vicinity Map

The vicinity map you provide will be printed in any public notice that is issued and used by the Corps of Engineers and other reviewing agencies to locate the site of the proposed activity. You may use an existing road map or U.S. Geological Survey topographic map (scale 1:24,000) as the vicinity map. Please include sufficient details to simplify locating the site from both the waterbody and from land. Identify the source of the map or chart from which the vicinity map was taken and, if not already shown, add the following:

- location of activity site (draw an arrow showing the exact location of the site on the map).
- latitude, longitude, river mile, if known, and/or other information that coincides with Block 6 on the application form.
- name of waterbody and the name of the larger creek, river, bay, etc., that the waterbody is immediately tributary to.
- names, descriptions and location of landmarks.
- name of all applicable political (county, parish, borough, town, city, etc.) jurisdictions.
- name of and distance to nearest town, community, or other identifying locations.
- names or numbers of all roads in the vicinity of the site.
- north arrow.
- scale.

Plan View

The plan view shows the proposed activity as if you were looking straight down on it from above. Your plan view should clearly show the following:

- Name of waterbody (river, creek, lake, wetland, etc.) and river mile (if known) at location of activity.
- Existing shorelines.
- Mean high and mean low water lines and maximum (spring) high tide line in tidal areas.
- Ordinary high water line and ordinary low water line if the proposed activity is located on a non-tidal waterbody.

☐ Average water depths around the activity.

☐ Dimensions of the activity and distance it extends from the high water line into the water.

☐ Distances to nearby Federal projects, if applicable.

☐ Distance between proposed activity and navigation channel, where applicable.

☐ Location of structures, if any, in navigable waters immediately adjacent to the proposed activity.

☐ Location of any wetlands (marshes, swamps, tidal flats, etc.)

☐ North arrow.

☐ Scale.

☐ If dredged material is involved, you must describe the type of material, number of cubic yards, method of handling, and the location of fill and spoil disposal area. The drawing should show proposed retention levees, weirs, and/or other means for retaining hydraulically placed materials.

☐ Mark the drawing to indicate previously completed portions of the activity.

Elevation and/or Cross Section View

The elevation and/or cross section view is a scale drawing that shows the side, front, or rear of the proposed activity. If a section view is shown, it represents the proposed structure as it would appear if cut internally for display. Your elevation should clearly show the following:

☐ Water elevations as shown in the plan view.

☐ Water depth at waterward face of proposed activity or, if dredging is proposed, dredging and estimated disposal grades.

☐ Dimensions from mean high water line (in tidal waters) for proposed fill or float, or high tide line for pile supported platform. Describe any structures to be built on the platform.

☐ Cross section of excavation or fill, including approximate side slopes.

☐ Graphic or numerical scale.

☐ Principal dimensions of the activity.

Notes on Drawings*

☐ Names of adjacent property owners who may be affected. Complete names and addresses should be shown in Block 5 on ENG Form 4345.

☐ Legal property description: Number, name of subdivision, block and lot number. Section, Township and Range (if applicable) from plot, deed or tax assessment.

☐ Photographs of the site of the proposed activity are not required; however, pictures are helpful and may be submitted as part of any application.

Drawings should be as clear and simple as possible (i.e., not too "busy").

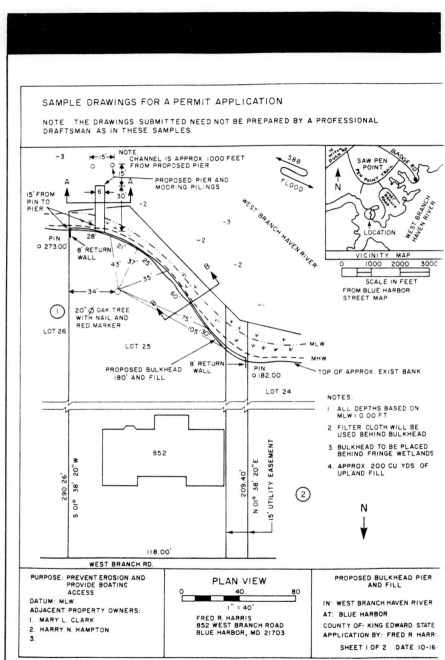

SAMPLE DRAWINGS FOR A PERMIT APPLICATION

NOTE: THE DRAWINGS SUBMITTED NEED NOT BE PREPARED BY A PROFESSIONAL DRAFTSMAN AS IN THESE SAMPLES.

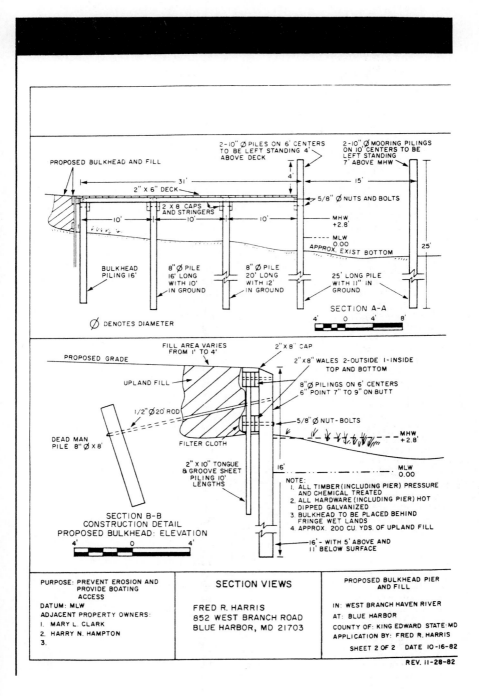

SECTION A-A

PROPOSED BULKHEAD AND FILL

2-10" Ø PILES ON 6' CENTERS TO BE LEFT STANDING 4' ABOVE DECK

2-10" Ø MOORING PILINGS ON 10' CENTERS TO BE LEFT STANDING 7' ABOVE MHW

2" X 6" DECK

2 X 8 CAPS AND STRINGERS

5/8" Ø NUTS AND BOLTS

MHW +2.8'
MLW 0.00
APPROX. EXIST BOTTOM

BULKHEAD PILING 16'

8" Ø PILE 16' LONG WITH 10' IN GROUND

8" Ø PILE 20' LONG WITH 12' IN GROUND

25' LONG PILE WITH 11" IN GROUND

Ø DENOTES DIAMETER

SECTION B-B
CONSTRUCTION DETAIL
PROPOSED BULKHEAD: ELEVATION

FILL AREA VARIES FROM 1' TO 4'

PROPOSED GRADE

UPLAND FILL

1/2" Ø 20' ROD

DEAD MAN PILE 8" Ø X 8'

FILTER CLOTH

2" X 8" CAP

2" X 8" WALES 2-OUTSIDE 1-INSIDE TOP AND BOTTOM

8" Ø PILINGS ON 6' CENTERS 6" POINT 7" TO 9" ON BUTT

5/8" Ø NUT-BOLTS

MHW +2.8'
MLW 0.00

2" X 10" TONGUE & GROOVE SHEET PILING 10' LENGTHS

NOTE:
1. ALL TIMBER (INCLUDING PIER) PRESSURE AND CHEMICAL TREATED
2. ALL HARDWARE (INCLUDING PIER) HOT DIPPED GALVANIZED
3. BULKHEAD TO BE PLACED BEHIND FRINGE WET LANDS
4. APPROX. 200 CU. YDS. OF UPLAND FILL

16' - WITH 5' ABOVE AND 11' BELOW SURFACE

PURPOSE: PREVENT EROSION AND PROVIDE BOATING ACCESS
DATUM: MLW
ADJACENT PROPERTY OWNERS:
1. MARY L. CLARK
2. HARRY N. HAMPTON
3.

SECTION VIEWS

FRED R. HARRIS
852 WEST BRANCH ROAD
BLUE HARBOR, MD 21703

PROPOSED BULKHEAD PIER AND FILL

IN: WEST BRANCH HAVEN RIVER
AT: BLUE HARBOR
COUNTY OF: KING EDWARD STATE: MD
APPLICATION BY: FRED R. HARRIS

SHEET 2 OF 2　DATE 10-16-82

REV. 11-28-82

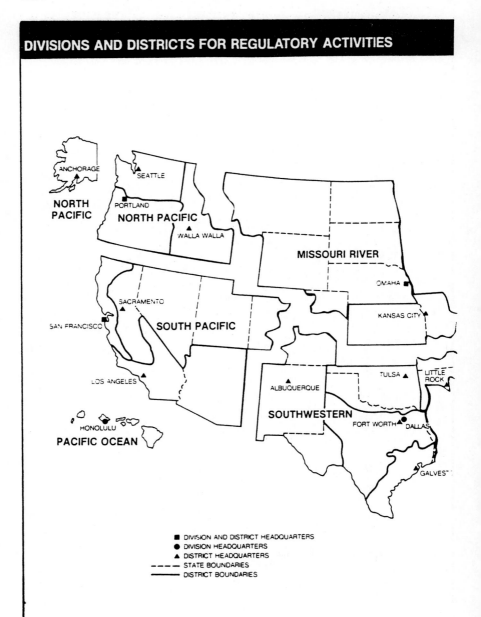

DIVISIONS AND DISTRICTS FOR REGULATORY ACTIVITIES

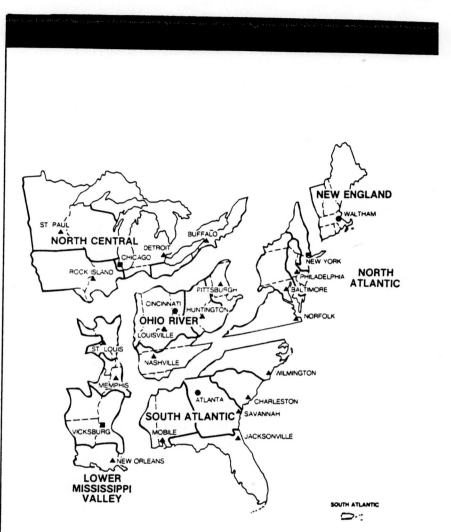

Note: In Iowa the eastern bank of the Missouri River is regulated by the Omaha office.

LOCATIONS OF REGULATORY OFFICES

Address correspondence to:

The District Engineer
U.S. Army Engineer
District
Please include attention
line in address.

ALASKA
P.O. Box 898
Anchorage, AK
99506-0898
Attention: NPACO-RF
907/753-2712

ALBUQUERQUE
P.O. Box 1580
Albuquerque, NM
87103-1580
Attention: SWACO-OR
505/766-2776

BALTIMORE
P.O. Box 1715
Baltimore, MD 21203-1715
Attention: NABOP-R
301/962-3670
Joint application with
New York, Maryland

BUFFALO
1776 Niagara Street
Buffalo, NY 14207-3199
Attention: NCBCO-S
716/876-5454 x2313
Joint application with
New York

CHARLESTON
P.O. Box 919
Charleston, SC
29402-0919
Attention: SACCO-P
803/724-4330

CHICAGO
219 S. Dearborn Street
Chicago, IL 60604-1797
Attention: NCCCO-R
312/353-6428
Joint application with
Illinois

DETROIT
P.O. Box 1027
Detroit, MI 48231-1027
Attention: NCECO-L
313/226-2218
Joint application with
Michigan

FT. WORTH
P.O. Box 17300
Ft. Worth, TX 76102-0300
Attention: SWFOD-O
817/334-2681

GALVESTON
P.O. Box 1229
Galveston, TX 77553-1229
Attention: SWGCO-R
409/766-3925

HUNTINGTON
502 8th Street
Huntington, WV 25701-2070
Attention: ORHOP-F
304/529-5487
Joint application with
West Virginia

HONOLULU
Building 230, Fort Shafter
Honolulu, HI 96858-5440
Attention: PODCO-O
808/438-9258

JACKSONVILLE
P.O. Box 4970
Jacksonville, FL 32232-0019
Attention: SAJRD
904/791-1659
Joint application with
Florida, Virgin Islands

KANSAS CITY
700 Federal Building
601 E. 12th Street
Kansas City, MO 64106-2896
Attention: MRKOD-P
816/374-3645

LITTLE ROCK
P.O. Box 867
Little Rock, AR
72203-0867
Attention: SWLCO-P
501/378-5295

LOS ANGELES
P.O. Box 2711
Los Angeles, CA 90053-232?
Attention: SPLCO-R
213/688-5606

LOUISVILLE
P.O. Box 59
Louisville, KY 40201-0059
Attention: ORLOP-F
502/582-5452
Joint application with
Illinois

MEMPHIS
Clifford Davis Federal
   Building
Room B-202
Memphis, TN 38103-1894
Attention: LMMCO-G
901/521-3471
Joint application with
Missouri, Tennessee,
Kentucky

MOBILE
P.O. Box 2288
Mobile, AL 36628-00001
Attention: SAMOP-S
205/690-2658
Joint application with
Mississippi

NASHVILLE
P.O. Box 1070
Nashville, TN 37202-1070
Attention: ORNOR-F
615/251-5181
Joint application with TVA,
Tennessee, Alabama

U.S. GOVERNMENT PRINTING OFFICE 1989-619-890/00844

NEW ORLEANS
P.O. Box 60267
New Orleans, LA
70160-0267
Attention: LMNOD-S
504/838-2255

NEW YORK
26 Federal Plaza
New York, NY 10278-0090
Attention: NANOP-R
212/264-3996

NORFOLK
803 Front Street
Norfolk, VA 23510-1096
Attention: NAOOP-P
804/446-3652
*Joint application with
Virginia*

OMAHA
P.O. Box 5
Omaha, NE 68101-0005
Attention: MROOP-N
402/221-4133

PHILADELPHIA
U.S. Custom House
2nd and Chestnut Street
Philadelphia, PA
19106-2991
Attention: NAPOP-R
215/597-2812

PITTSBURGH
Federal Building
1000 Liberty Avenue
Pittsburgh, PA 15222-4186
Attention: ORPOP-F
412/644-4204
*Joint application with
New York*

PORTLAND
P.O. Box 2946
Portland, OR 97208-2946
Attention: NPPND-RF
503/221-6995
*Joint application with
Oregon*

ROCK ISLAND
Clock Tower Building
Rock Island, IL 61201-2004
Attention: NCROD-S
309/788-6361 x6370
*Joint application with
Illinois*

SACRAMENTO
650 Capitol Mall
Sacramento, CA 95814-4794
Attention: SPKCO-O
916/440-2842

ST. LOUIS
210 Tucker Blvd., N
St. Louis, MO 63101-1986
Attention: LMSOD-F
314/263-5703
*Joint application with
Illinois, Missouri*

ST. PAUL
1135 USPO & Custom
House
St. Paul, MN 55101-1479
Attention: NCSCO-RF
612/725-5819

SAN FRANCISCO
211 Main Street
San Francisco, CA 94105-1905
Attention: SPNCO-R
415/974-0416

SAVANNAH
P.O. Box 889
Savannah, GA 31402-0889
Attention: SASOP-F
912/944-5347
*Joint application with
Georgia*

SEATTLE
P.O. Box C-3755
Seattle, WA 98124-2255
Attention: NPSOP-RF
206/764-3495
Joint application with Idaho

TULSA
P.O. Box 61
Tulsa, OK 74121-0061
Attention: SWTOD-RF
918/581-7261

VICKSBURG
P.O. Box 60
Vicksburg, MS 39180-0060
Attention: LMKOD-F
601/634-5276
*Joint application with
Mississippi*

WALLA WALLA
Building 602
City-County Airport
Walla Walla, WA
99362-9265
Attention: NPWOP-RF
509/522-6718
*Joint application with
Idaho*

WILMINGTON
P.O. Box 1890
Wilmington, NC
28402-1890
Attention: SAWCO-E
919/343-4511
*Joint application with North
Carolina*

**The Division Engineer
U.S. Army Engineer
Division**
NEW ENGLAND
424 Trapelo Road
Waltham, MA 02254-9149
Attention: NEDOD-R
617/647-8338
*Joint application with
Massachusetts, Maine*

APPENDIX 1
CONVERSION FACTORS AND QUANTITIES

TABLE A1.1 Useful Conversions

Acre feet	= 43,560 ft^3 = 325,851 gal = 1233.49 m^3	Inches of mercury	= 0.03342 atmospheres = 1.133 feet of water = 345.3 kg/m^2 = 70.73 lb/ft^2
Barrels, oil	= 42 gal, oil		= 0.4912 lb/in^2
Centimeters of mercury	= 0.01316 atmospheres = 0.4461 feet of water = 136 kg/m^3 = 27.85 lb/ft^2 = 0.1934 lb/ft^2	Inches of water	= 0.002458 atmospheres = 0.07355 inches of mercury = 25.4 kg/m^2 = 5.202 lb/ft^2 = 0.03613 lb/in^2
Ft^3/min	= 2.832 × 10^4 cm^3/min = 1723 in^3/min = 0.02832 m^3/min = 0.03704 yd^3/min = 7.48052 gal/min = 28.32 liters per minute	Kilograms (kg) Kg/cm^2 Kg/cm^2 Liters	= 2.205 lb = 14.2258 lb/in^2 = 32.86 feet of water = 10^3 cm^3
Ft^3/s	= 0.646317 million gallons per day = 448.831 gal/min		= 0.03531 ft^3 = 61.02 in^3 = 10^{-3} m^3
Feet of water	= 0.0295 atmospheres = 0.8826 inches of mercury = 304.8 kg/m^2 = 62.43 lb/ft^2	Parts per million	= 0.0584 grains per U.S. gallon = 0.07016 grains per Impe- rial gallon = 8.345 lb/million gal
Gallons	= 0.4335 lb/in^2 = 3785 cm^3 = 0.1337 ft^3 = 231 in^3 = 3.785 × 10^{-3} m^3 = 4.951 × 10^{-3} yd^3	Pound Pounds of water Pounds of water per minute	= 453.5924 g = 0.01602 ft^3 = 27.68 in^3 = 0.1198 gal = 2.670 × 10^{-4} ft^3
Gallons, Imperial Gallons, U.S.	= 1.20095 U.S. gallons = 0.83267 Imperial gallons	Lb/ft^3	= 0.01602 g/cm^3 = 16.02 kg/m^3 = 5.787 × 10^{-4} lb/in^3
Gallons, water Gallons, water per minute	= 8.3453 pounds of water = 6.0086 tons of water per 24 hours	Lb/ft^2 Lb/in^2	= 0.01602 feet of water = 0.06804 atmospheres = 2.307 feet of water = 2.036 inches of mercury
GPM GPM of dry solids	= 1.3349 ft^3/hr = tons of water per 24. hours × 0.16643 = tons per hour of dry material ÷ specific grav- ity of dry material	Tons (long)	= 703.1 kg/m^2 = 1016 kg = 2240 lb = 1.12 tons (short)

TABLE A1.1 *(Continued)*

Gal/min	= 2.228 × 10⁻³ ft³/s	Tons (metric)	= 10^3 kg
	= 0.06308 liters per second		= 2205 lb
	= 8.0208 ft³/hr		= 1.1025 tons (shore)
Grains per U.S. gallon	= 17.118 parts per million		
	= 142.86 lb/million gal.	Tons (short)	= 2000 lb
			= 907.18486 kg
Grains per Imperial gallon	= 14.254 parts per million		= 0.8927 tons (long)
Grains per liter	= 58.417 grains per gallon	Tons of water	= 240 gal
	= 8.345 lb/1000 gal		
	= 1000 parts per million	Tons of water per 24 hours	= 83.333 pounds of water per hour
			= 0.16643 gal/min
			= 1.3349 ft³/hr
m³/min	= 35.31 ft³/min		= 1.308 × 10⁻³ yd³
	= 1.308 yd³/min		= 0.2642 gallons
	= 264.2 gal/min		
		Million gallons per day	= 1.54723 ft³/s
yd³/min	= 202 gal/min		
	= 764.6 liters per minute	Miner's inches	= 1.5 ft³/min
	= 0.45 ft³/s		
	= 3.367 gal/s	Ounces (fluid)	= 1.805 in³
	= 12.74 liters per second		= 0.02957 liters

Source: Courtesy, Denver Equipment Division, Joy Manufacturing Co.

TABLE A1.2 Conversion Factors, Units of Measurement

Multiply	By	To Obtain
inches	2.54	centimeters
feet	0.3048	meters
yards	0.9144	meters
miles (U.S. statute)	1.609344	kilometers
miles (U.S. nautical)	1.852	kilometers
square inches	6.4516	square centimeters
square feet	0.092903	square meters
square yards	0.836127	square meters
square miles	2.58999	square kilometers
acres	4046.856	square meters
cubic feet	0.0283168	cubic meters
cubic feet	28.3	liters
cubic yards	0.764555	cubic meters
quarts (U.S. liquid)	946.353	cubic centimeters
tons (2000 pounds)	907.1847	kilograms
pounds per cubic foot	16.0185	kilograms per cubic meter
tons per square foot	9764.86	kilograms per square meter

Source: Courtesy, U.S. Army Engineers.

TABLE A1.3 Conversion of Feet of Water into Pressure per Square Inch and Vice Versa

Pounds per square inch	Feet head	Pounds per square inch	Feet head	Pounds per square inch	Feet head
Converting pressure per square inch into feet head of water					
1	2.31	55	126.99	180	415.61
2	4.62	60	138.54	190	438.90
3	6.93	65	150.08	200	461.78
4	9.24	70	161.63	225	519.51
5	11.54	75	173.17	250	577.24
6	13.85	80	184.72	275	643.03
7	16.16	85	196.26	300	692.69
8	18.47	90	207.81	325	730.41
9	20.78	95	219.35	350	808.13
10	23.09	100	230.90	375	865.89
15	34.63	110	253.98	400	922.58
20	46.18	120	277.07	500	1154.48
25	57.72	125	288.62	—	—
30	69.27	130	300.16	—	—
35	80.81	140	323.25	—	—
40	92.36	150	346.34	—	—
45	103.90	160	369.43	—	—
50	115.43	170	392.52	—	—

Feet head	Pounds per square inch	Feet head	Pounds per square inch	Feet head	Pounds per square inch
Converting feet head of water into pressure per square inch					
1	.43	55	23.82	190	82.29
2	.87	60	25.99	200	86.62
3	1.30	65	28.15	225	97.45
4	1.73	70	30.32	250	108.27
5	2.17	75	32.48	275	119.10
6	2.60	80	34.65	300	129.93
7	3.03	85	36.81	325	140.75
8	3.40	90	38.98	350	151.58
9	3.90	95	41.14	375	162.41
10	4.33	100	43.31	400	173.24
15	6.30	110	47.64	500	216.55
20	8.66	120	51.97	600	259.85
25	10.83	130	56.30	700	303.16
30	12.99	140	60.63	800	346.47
35	15.16	150	64.96	900	389.78
40	17.32	160	69.29	1000	433.09
45	19.49	170	73.63	—	—
50	21.63	180	77.96	—	—

Source: Courtesy, Denver Equipment Division, Joy Manufacturing Co.

TABLE A1.4   Water Quantities

Pipe diameter (in)	Ft³ of water in 1 ft of pipe	U.S. gal in 1 ft of pipe
½	0.0014	0.0102
¾	0.0031	0.0230
1	0.0055	0.0408
2	0.0218	0.1632
3	0.0491	0.3672
4	0.0873	0.6528
5	0.1364	1.020
6	0.1963	1.469
8	0.3491	2.611
10	0.5454	4.080
12	0.7854	5.875
14	1.069	7.997
16	1.396	10.44
18	1.767	13.22
20	2.182	16.32
22	2.640	19.75
24	3.142	23.50
26	3.687	27.58
28	4.276	31.99
30	4.909	36.72
32	5.585	41.78
34	6.305	47.16
36	7.069	52.88
38	7.876	58.92
40	8.727	65.28
42	9.621	71.97
44	10.559	78.99
46	11.541	86.33
48	12.566	94.00
50	13.635	102.00
52	14.748	110.3
54	15.904	119.0
56	17.104	128.0
58	18.348	137.3
60	19.635	146.9
62	20.966	156.8
64	22.340	167.1
66	23.76	177.7
68	25.22	188.7
70	26.73	200.0
72	28.27	211.5
74	29.87	223.4
76	31.50	235.6
78	33.18	248.2
80	34.91	261.1
82	36.67	274.3
84	38.48	287.9
86	40.34	301.7
88	42.24	316.0
90	44.18	330.5

TABLE A1.4 Water Quantities (*Continued*)

Pipe diameter (in)	Ft³ of water in 1 ft of pipe	U.S. gal in 1 ft of pipe
92	46.16	345.3
94	48.19	360.5
96	50.27	376.0
98	52.38	391.8
100	54.54	408.0
102	56.75	424.5
104	58.99	441.2
106	61.28	458.4
108	63.62	475.9
110	66.00	493.7
112	68.42	511.8
114	70.88	530.2
116	73.39	549.0
118	75.94	568.0
120	78.54	587.5

Source: Courtesy, Denver Equipment Division, Joy Manufacturing Co.

TABLE A1.5   Discharge Rates for Hydraulic Dredge Pipelines

Pipeline Diameter	Discharge Rate (for Flow Velocity of 12 ft/sec*)	
in.	cu ft/sec	gal/min
8	4.2	1,880
10	6.5	2,910
12	9.4	4,220
14	12.8	5,750
16	16.5	7,400
18	21.2	9,510
20	26.2	11,740
24	37.7	16,890
27	47.6	21,300
28	51.3	23,000
30	58.9	26,400
36	84.9	38,000

* To obtain discharge rates for other velocities multiply the discharge rate in this table by the velocity (ft/sec) and divide by 12.

TABLE A1.6 Conversion Factors, Non-SI to SI (Metric) Units of Measurement

Multiply	By	To obtain
Acres	4,046.873	Square meters
Cubic yards	0.7645549	Cubic meters
Degrees (angle)	0.01745329	Radians
Fahrenheit degrees	5/9	Celsius degrees or Kelvins*
Feet	0.3048	Meters
Gallons	3.785412	Cubic decimeters
Horsepower [550 foot pounds (force) per second]	745.6999	Watts
Inches	2.54	Centimeters
Micrometers	0.001	Millimeters
Miles (U.S. statute)	1.609347	Kilometers
Pounds (mass)	0.4535924	Kilograms
Square inches	6.4516	Square centimeters

*To obtain Celsius (C) temperature readings from Fahrenheit (F) readings, use the following formula: $C + (5/9)(F - 32)$. To obtain Kelvin (K) readings, use $K + (5/9)(F - 32) + 273.15$.

APPENDIX 2
FLUID PROPERTIES

TABLE A2.1 Physical Properties of Water

Temp, T (°F)	Specific weight γ (lb/ft³)	Density ρ (slugs/ft³)	Viscosity $\mu \times 10^5$ (lb-sec/ft²)	Kinematic viscosity $\nu \times 10^5$, (ft²/sec)	Surface tension* $\sigma \times 10^2$, (lb/ft)	Vapor pressure p_v/γ, (ft H₂O)	Bulk modulus of elasticity $K \times 10^{-3}$, (lb/in²)
32	62.42	1.940	3.746	1.931	0.518	0.20	293
40	62.43	1.940	3.229	1.664	0.514	0.28	294
50	62.41	1.940	2.735	1.410	0.509	0.41	305
60	62.37	1.938	2.359	1.217	0.504	0.59	311
70	62.30	1.936	2.050	1.059	0.500	0.84	320
80	62.22	1.934	1.799	0.930	0.492	1.17	322
90	62.11	1.931	1.595	0.826	0.486	1.61	323
100	62.00	1.927	1.424	0.739	0.480	2.19	327
110	61.86	1.923	1.284	0.667	0.473	2.95	331
120	61.71	1.918	1.168	0.609	0.465	3.91	333
130	61.55	1.913	1.069	0.558	0.460	5.13	334
140	61.38	1.908	0.981	0.514	0.454	6.67	330
150	61.20	1.902	0.905	0.476	0.447	8.58	328
160	61.00	1.896	0.838	0.442	0.441	10.95	326
170	60.80	1.890	0.780	0.413	0.433	13.83	322
180	60.58	1.883	0.726	0.385	0.426	17.33	318
190	60.36	1.876	0.678	0.362	0.419	21.55	313
200	60.12	1.868	0.637	0.341	0.412	26.59	308
212	59.83	1.860	0.593	0.319	0.404	33.90	300

* In contact with air
Approximate values at standard atmospheric pressure and 68°F

APPENDIX TWO

TABLE A2.2 Properties of Water

Boiling Point at Different Elevations			Weight and Heat Content at Various temperatures		
Alt. Above Sea Level in Feet	Barometer Reading in Inches	Boiling Point Degrees	temp. in Degrees	Weight per Cu. Ft. in Lbs.	B.T.U. per Lb.
15,221	16 79	184	32	62.42	0.
14,649	17 16	185	35	62.42	3.02
14,075	17 54	186	40	62.42	8.05
13,498	17 93	187	45	62.42.	13.07
12,934	18 32	188	50	62.41	18 08
12,367	18.72	189	55	62 39	23.08
11,799	19 13	190	60	62 37	28 08
11,243	19 54	191	65	62.34	33.07
10,685	19 69	192	70	62.31	38 06
10,127	20 39	193	75	62 28	43.05
9,579	20 82	194	80	62.23	48 03
9,031	21 26	195	85	62.18	53 02
8,481	21.71	196	90	62.13	58 00
7,932	22.17	197	95	62.08	62.99
7,381	22.64	198	100	62.02	67.97
6,843	23 11	199	105	61.96	72 95
6,304	23.59	200	110	61.89	77.94
5,764	24.08	201	115	61 82	82.92
5,225	24.58	202	120	61 74	87.91
4,697	25.08	203	125	61 65	92.90
4,169	25.59	204	130	61 56	97.89
3,642	26.11	205	135	61 47	102 88
3,115	26.64	206	140	61 37	107.87
2,589	27 18	207	145	61.28	112 86
2,063	27.73	208	150	61.18	117.86
1,809	28.00	208.5	155	61.08	122.86
1,539	28.29	209	160	60.98	127.86
1,290	28.56	209.5	165	60.87	132.86
1,025	28.85	210	170	60.77	137.87
754	29.15	210.5	175	60.66	142.87
512	29.42	211	180	60.55	147.88
255	29.71	211.5	185	60.44	152.89
0—Sea Level	30.00	212	190	60.32	157.91
−261 Below S.L.	30.30	212.5	195	60.20	162.92
−511 Below S.L.	30.59	213	200	60.07	167.94
			205	59.95	172 96
			210	59.82	177.99
			212	59.76	180.00

Source: Courtesy, Denver Equipment Division, Joy Manufacturing Co.

TABLE A2.3 Physical Properties of Common Liquids

Liquid	Specific gravity	Dynamic viscosity μ (lb-sec/ft^2)	Surface tension* σ (lb/ft)	Vapor pressure (lb/in^2 abs.]	Volume modulus (lb/in^2)
Benzene	0.90	1.4×10^{-5}	2.0×10^{-3}	1.50	150,00
Gasoline	0.68	6.2×10^{-6}			
Glycerin	1.26	3.1×10^{-2}	4.3×10^{-3}	2×10^{-6}	630,00
Kerosene	0.81	4.0×10^{-5}	1.7×10^{-3}		
Mercury	13.55	3.3×10^{-5}	3.2×10^{-2}	2.33×10^{-5}	3,800,00
SAE 10 oil	0.92	1.7×10^{-3}	2.5×10^{-3}		
SAE 30 oil	0.92	9.2×10^{-3}	2.4×10^{-3}		
Water	1.00	2.1×10^{-5}	5.0×10^{-3}	0.34	300,00

* In contact with air
  Approximate values at standard atmospheric pressure and 68°F

APPENDIX 3

CONCENTRATION OF MIXTURES, FALL VELOCITY OF PARTICLES, AND APPARENT KINEMATIC VISCOSITY

TABLE A3.1 Concentrations by Volume (C_v) and by Weight (C_w) for Different Materials and Different Specific Gravities of Mixture

Specific Gravity of Mixture	Clay (%)		Sand (%)		Coal (%)		Flyash (%)	
SG_M	C_v	C_w	C_v	C_w	C_v	C_w	C_v	C_w
1.00	−1.45	−3.99	−1.54	−4.08	−5.26	−7.89	−2.46	−5.02
1.10	4.35	10.87	4.62	11.12	15.79	21.53	7.39	13.70
1.20	10.14	23.25	10.77	23.78	36.84	46.05	17.24	29.31
1.30	15.94	33.72	16.92	34.50	57.89	66.80	27.09	42.52
1.40	21.74	42.70	23.08	43.68	78.95	84.59	36.95	53.84
1.50	27.54	50.48	29.23	51.64	100.00	100.00	46.80	63.65
1.60	33.33	57.29	35.38	58.61	121.05	112.49	56.65	72.25
1.70	39.13	63.30	41.54	64.75	141.11	125.39	66.50	79.80
1.80	44.93	68.64	47.69	70.21	163.16	135.96	76.35	86.54
1.90	50.72	73.42	53.85	75.10	184.21	145.43	86.21	92.56
2.00	56.52	77.72	60.00	79.50	205.26	153.95	96.06	97.98

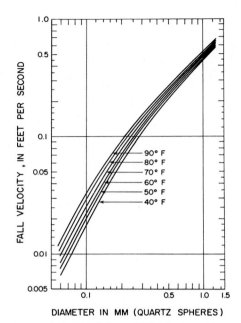

FIGURE A3.1 Variation of fall velocity with temperature. (*Rubey, 1933*)

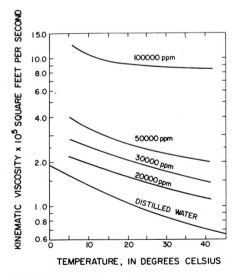

FIGURE A3.2 ''Apparent'' kinematic viscosity of aqueous dispersions of bentonite as a function of temperature. (*Guy, 1966*)

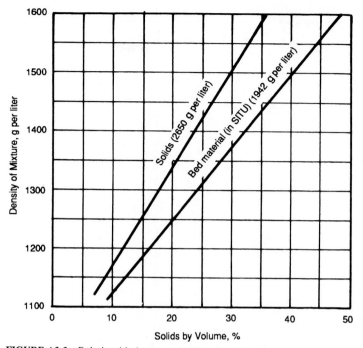

FIGURE A3.3 Relationship between density of mixture pumped and percentage of solids by volume. (*Franco, 1967*)

REFERENCES

1. Rubey, W. W., "Settling Velocities of Gravel, Sand and Silt Particles," *American Journal of Science,* vol. 25, no. 148, Apr. 1933.

2. Guy, H. P., Simons, D. B., and Richardson, E. V., "Summary of Alluvial Channel Data from Flume Experiments, 1959–61," U.S.G.S. Professional Paper no. 462-I, 1966.

3. Franco, J. J., "Model Study of Hopper Dredge Dragheads," U.S. Army Engineers, Waterways Experiment Station, TR No. 2-755, Jan. 1967.

APPENDIX 4
GOVERNMENT REGULATIONS

POLICY AND LEGISLATION PERTINENT TO DREDGING

(a)   Public laws.
   (1)   American Folklife Preservation Act, Pub. L. 94-201; 20
         U.S.C. 2101, et seq.
   (2)   Anadromous Fish Conservation Act, Pub. L. 89-304; 16 U.S.C.
         757, et seq.
   (3)   Antiquities Act of 1906, Pub. L. 59-209; 16 U.S.C. 431, et
         seq.
   (4)   Archeological and Historic Preservation Act, Pub. L.
         93-291; 16 U.S.C. 469, et seq. (Also known as the Reservoir
         Salvage Act of 1960, as amended; Public Law 93-291, as
         amended; the Moss-Bennett Act; and the Preservation of
         Historic and Archeological Data Act of 1974.)
   (5)   Bald Eagle Act; 16 U.S.C. 666.
   (6)   Clean Air Act, as amended, Pub. L. 91-604; 42 U.S.C.
         1857h-7, et seq.
   (7)   Clean Water Act, Pub. L. 92-500; 33 U.S.C. 1251, et seq.
         (Also known as the Federal Water Pollution Control Act; and
         Public Law 92-500, as amended.)
   (8)   Coastal Zone Management Act of 1972, as amended, Pub. L.
         92-583; 16 U.S.C. 1451, et seq.
   (9)   Endangered Species Act of 1973, as amended, Pub. L. 93-205;
         16 U.S.C. 1531, et seq.
   (10) Estuary Protection Act, Pub. L. 90-454; 16 U.S.C. 1221, et
         seq.
   (11) Federal Environmental Pesticide Control Act, Pub. L.
         92-516; 7 U.S.C. 136.
   (12) Federal Water Project Recreation Act, as amended, Pub. L.
         89-72; 16 U.S.C. 460-1(12), et seq.
   (13) Fish and Wildlife Coordination Act of 1958, as amended,
         Pub. L. 85-624; 16 U.S.C. 661, et seq. (Also known as the
         Coordination Act.)
   (14) Historic Sites of 1935, as amended, Pub. L. 74-292; 16
         U.S.C. 461, et seq.
   (15) Land and Water Conservation Fund Act, Pub. L. 88-578; 16
         U.S.C. 4601-4601-11, et seq.
   (16) Marine Mammal Protection Act of 1972, Pub. L. 92-522; 16
         U.S.C. 1361, et seq.
   (17) Marine Protection, Research and Sanctuaries Act of 1972,
         Pub. L. 92-532; 33 U.S.C. 1401, et seq.
   (18) Migratory Bird Conservation Act of 1928; 16 U.S.C. 715.
   (19) Migatory Bird Treaty Act of 1918; 16 U.S.C. 703, et seq.
   (20) National Environmental Policy Act of 1969, as amended, Pub.
         L. 91-190; 42 U.S.C. 4321, et seq. (Also known as NEPA;
         often incorrectly cited as the National Environmental
         Protection Act.)
   (21) National Historic Preservation Act of 1966, as amended,
         Pub. L. 89-655; 16 U.S.C. 470a, et seq.
   (22) Native American Religious Freedom Act, Pub. L. 95-341; 42
         U.S.C. 1996, et seq.

(23)   Resource Conservation and Recovery Act of 1976; Pub. L. 94-580; 7 U.S.C. 1010, et seq.
(24)   River and Harbor Act of 1899, 33 U.S.C. 403, et seq. (Also known as the Refuse Act of 1899.)
(25)   Submerged Lands Act of 1953, Pub. L. 82-3167; 43 U.S.C. 1301, et seq.
(26)   Surface Mining Control and Reclamation Act of 1977, Pub. L. 95-89; 30 U.S.C. 1201, et seq.
(27)   Toxic Substances Control Act, Pub. L. 94-469; 15 U.S.C. 2601, et seq.
(28)   Watershed Protection and Flood Prevention Act, as amended, Pub. L. 83-566; 16 U.S.C. 1001, et seq.
(29)   Wild and Scenic Rivers Act, as amended, Pub. L. 90-542; 16 U.S.C. 1271, et seq.

(b)   Executive orders.
(1)   Executive Order, 11593, Protection and Enhancement of the Cultural Environment, May 13, 1979 (36 FR 8921; May 15, 1971).
(2)   Executive Order, 11988, Floodplain Management, May 24, 1977 (42 FR 26951; May 25, 1977).
(3)   Executive Order, 11990, Protection of Wetlands, May 24, 1977 (42 FR 26961; May 25, 1977).
(4)   Executive Order, 11514, Protection and Enhancement of Environmental Quality, March 5, 1970, as amended by Executive Order, 11991, May 24, 1977.
(5)   Executive Order, 12088, Federal Compliance with Pollution Control Standards, October 13, 1978.

(c)   Other Federal policies.
(1)   Council on Environmental Quality Memorandum of August 11, 1980: Analysis of Impacts on Prime or Unique Agricultural Lands in Implementing the National Environmental Policy Act.
(2)   Council on Environmental Quality Memorandum of August 10, 1980: Interagency Consultation to Avoid or Mitigate Adverse Effects on Rivers in the Nationwide Inventory.
(3)   Migratory Bird Treaties and other international agreements listed in the Endangered Species Act of 1973, as amended, Section 2(a)(4).

(d)   Selected state legislation, lead agencies, and concerns
(1) Maine: Department of Environmental Protection for coastal and great ponds projects (Tidal Wetlands Act 38 Maine Revised Statutes Annotated Sections 471-478 and Great Ponds Act 38 MRSA Sections 386-396, respectively). Department of Inland Fish and Wildlife for fill projects on rivers and streams (Alteration of Rivers, Streams and Brooks Act 12 MRSA Sections 7776-7780). The Board of Environmental Protection may establish any reasonable requirement to ensure that the applicant does not contravene environmental quality.

(2) New Hampshire: Water Supply and Pollution Control Commission (Resource Statutes Annotated, Subsection 149.8A) and the Wetlands Board (RSA, Subsection 483A). The Water Supply and

Pollution Control Commission requires that there be no
degradation of water quality.

(3) Massachusetts: Conservation Commission of locality directly
affected by the project (State Wetlands Protection Law,
Chapter 131, Section 40). A local Conservation Commission
may attach special conditions to an application to ensure
proper response to its concerns when discharge to a wetlands
is proposed.

(4) Rhode Island: Coastal Resources Management Council.
(General Laws, Chapter 279, Section 1). The Coastal
Resources Management Council is concerned with state coastal
plan consistency and permitting activities in territorial
waters and saltwater wetlands.

(5) Connecticut: Commissioner of the Department of Environmental
Protection, (Marine Mining Statute, Section 25-7d for new
dredging work and structures and Dredging Statute, Section
25-11 for regulating building of marina structures). The
Department of Environmental Protection requires containment
of materials disposed of on upland sites. In-water disposal
permits may require special conditions to protect fish and
wildlife recommended by the U.S. Fish and Wildlife Service
and/or the National Marine Fisheries Service.

(6) New York Department of Environmental Conservation
(Environmental Conservation Law, articles). The Department
of Environmental Conservation may specify seasonal
restrictions to protect spawning. It may also specify
certain types of dredging and containment procedures to
alleviate environmental impact.

(7) California: California Coastal Commission (Proposition 20,
1972; California Coastal Zone Act, 1976). Requires port
master plan; lead agency for review of port projects. Water
Control Board (California Resources Code). Permit authority
over effects of dredging/filling on water quality.
Department of Fish and Game (California Resources Code).
Review and comment authority over effects of proposed
projects on fish and wildlife. Air Resources Board
(California Resources Code). Permit review authority over
sources of stationary (point-source) air pollution has been
applied to dredging equipment and port facilities.

(8) Oregon: Department of Land Conservation and Development.
Statewide goals and guidelines for coastal resources.
Department of Fish and Wildlife (Oregon Administrative
Rule). Classifies estuaries.

(9) Washington: Shoreline Hearings Board (Shoreline Management
Act). Permit appeal authority. Department of Ecology
(Washington Resources Code). Water and air quality permit
authority; review of proposed projects for effects on fish
and wildlife.

PERTINENT SECTIONS FROM THE FEDERAL REGISTER

PART 323—PERMITS FOR DIS-CHARGES OF DREDGED OR FILL MATERIAL INTO WATERS OF THE UNITED STATES

Sec.
323.1 General.
323.2 Definitions.
323.3 Discharges requiring permits.
323.4 Discharges not requiring permits.
323.5 Program transfer to states.
323.6 Special policies and procedures.

AUTHORITY: 33 U.S.C. 1344.

SOURCE: 51 FR 41232, Nov. 13, 1986, unless otherwise noted.

§ 323.1 General.

This regulation prescribes, in addition to the general policies of 33 CFR Part 320 and procedures of 33 CFR Part 325, those special policies, practices, and procedures to be followed by the Corps of Engineers in connection with the review of applications for DA permits to authorize the discharge of dredged or fill material into waters of the United States pursuant to section 404 of the Clean Water Act (CWA) (33 U.S.C. 1344) (hereinafter referred to as section 404). (See 33 CFR 320.2(g).) Certain discharges of dredged or fill material into waters of the United States are also regulated under other authorities of the Department of the Army. These include dams and dikes in navigable waters of the United States pursuant to section 9 of the Rivers and Harbors Act of 1899 (33 U.S.C. 401; see 33 CFR Part 321) and certain structures or work in or affecting navigable waters of the United States pursuant to section 10 of the Rivers and Harbors Act of 1899 (33 U.S.C. 403; see 33 CFR Part 322). A DA permit will also be required under these additional authorities if they are applicable to activities involving discharges of dredged or fill material into waters of the United States. Applicants for DA permits under this part should refer to the other cited authorities and implementing regulations for these additional permit requirements to determine whether they also are applicable to their proposed activities.

§ 323.2 Definitions.

For the purpose of this part, the following terms are defined:

(a) The term "waters of the United States" and all other terms relating to the geographic scope of jurisdiction are defined at 33 CFR Part 328.

(b) The term "lake" means a standing body of open water that occurs in a natural depression fed by one or more streams from which a stream may flow, that occurs due to the widening or natural blockage or cutoff of a river or stream, or that occurs in an isolated natural depression that is not a part of a surface river or stream. The term also includes a standing body of open water created by artificially blocking or restricting the flow of a river, stream, or tidal area. As used in this regulation, the term does not include artificial lakes or ponds created by excavating and/or diking dry land to collect and retain water for such purposes as stock watering, irrigation, settling basins, cooling, or rice growing.

(c) The term "dredged material" means material that is excavated or dredged from waters of the United States.

(d) The term "discharge of dredged material" means any addition of dredged material into the waters of the United States. The term includes, without limitation, the addition of dredged material to a specified discharge site located in waters of the United States and the runoff or overflow from a contained land or water disposal area. Discharges of pollutants into waters of the United States resulting from the onshore subsequent processing of dredged material that is extracted for any commercial use (other than fill) are not included within this term and are subject to section 402 of the Clean Water Act even though the extraction and deposit of such material may require a permit from the Corps of Engineers. The term does not include plowing, cultivating, seeding and harvesting for the production of food, fiber, and forest products (See § 323.4 for the definition of these terms). The term does not include *de minimis*, incidental soil movement occurring during normal dredging operations.

(e) The term "fill material" means any material used for the primary purpose of replacing an aquatic area with dry land or of changing the bottom elevation of an waterbody. The term does not include any pollutant discharged into the water primarily to dispose of waste, as that activity is regulated under section 402 of the Clean Water Act.

(f) The term "discharge of fill material" means the addition of fill material into waters of the United States. The term generally includes, without limitation, the following activities: Placement of fill that is necessary for the construction of any structure in a water of the United States; the building of any structure or impoundment requiring rock, sand, dirt, or other material for its construction; site-development fills for recreational, industrial, commercial, residential, and other uses; causeways or road fills; dams and dikes; artificial islands; property protection and/or reclamation devices such as riprap, groins, seawalls, breakwaters, and revetments; beach nourishment; levees; fill for structures such as sewage treatment facilities, intake and outfall pipes associated with power plants and subaqueous utility lines; and artificial reefs. The term does not include plowing, cultivating, seeding and harvesting for the production of food, fiber, and forest products (See § 323.4 for the definition of these terms).

(g) The term "individual permit" means a Department of the Army authorization that is issued following a case-by-case evaluation of a specific project involving the proposed discharge(s) in accordance with the procedures of this part and 33 CFR Part 325 and a determination that the proposed discharge is in the public interest pursuant to 33 CFR Part 320.

(h) The term "general permit" means a Department of the Army authorization that is issued on a nationwide or regional basis for a category or categories of activities when:

(1) Those activities are substantially similar in nature and cause only minimal individual and cumulative environmental impacts; or

(2) The general permit would result in avoiding unnecessary duplication of regulatory control exercised by another Federal, state, or local agency provided it has been determined that the environmental consequences of the action are individually and cumulatively minimal. (See 33 CFR 325.2(e) and 33 CFR Part 330.)

§ 323.3   Discharges requiring permits.

(a) *General.* Except as provided in § 323.4 of this Part, DA permits will be required for the discharge of dredged or fill material into waters of the United States. Certain discharges specified in 33 CFR Part 330 are permitted by that regulation ("nationwide per-

Corps of Engineers, Dept. of the Army, DoD § 323.4

mits"). Other discharges may be authorized by district or division engineers on a regional basis ("regional permits"). If a discharge of dredged or fill material is not exempted by § 323.4 of this Part or permitted by 33 CFR Part 330, an individual or regional section 404 permit will be required for the discharge of dredged or fill material into waters of the United States.

(b) *Activities of Federal agencies.* Discharges of dredged or fill material into waters of the United States done by or on behalf of any Federal agency, other than the Corps of Engineers (see 33 CFR 209.145), are subject to the authorization procedures of these regulations. Agreement for construction or engineering services performed for other agencies by the Corps of Engineers does not constitute authorization under the regulations. Division and district engineers will therefore advise Federal agencies and instrumentalities accordingly and cooperate to the fullest extent in expediting the processing of their applications.

§ 323.4 **Discharges not requiring permits.**

(a) *General.* Except as specified in paragraphs (b) and (c) of this section, any discharge of dredged or fill material that may result from any of the following activities is not prohibited by or otherwise subject to regulation under section 404:

(1)(i) Normal farming, silviculture and ranching activities such as plowing, seeding, cultivating, minor drainage, and harvesting for the production of food, fiber, and forest products, or upland soil and water conservation practices, as defined in paragraph (a)(1)(iii) of this section.

(ii) To fall under this exemption, the activities specified in paragraph (a)(1)(i) of this section must be part of an established (i.e., on-going) farming, silviculture, or ranching operation and must be in accordance with definitions in § 323.4(a)(1)(iii). Activities on areas lying fallow as part of a conventional rotational cycle are part of an established operation. Activities which bring an area into farming, silviculture, or ranching use are not part of an established operation. An operation ceases to be established when the area on which it was conducted has been

coverted to another use or has lain idle so long that modifications to the hydrological regime are necessary to resume operations. If an activity takes place outside the waters of the United States, or if it does not involve a discharge, it does not need a section 404 permit, whether or not it is part of an established farming, silviculture, or ranching operation.

(iii) (A) Cultivating means physical methods of soil treatment employed within established farming, ranching and silviculture lands on farm, ranch, or forest crops to aid and improve their growth, quality or yield.

(B) Harvesting means physical measures employed directly upon farm, forest, or ranch crops within established agricultural and silvicultural lands to bring about their removal from farm, forest, or ranch land, but does not include the construction of farm, forest, or ranch roads.

(C)(*1*) Minor Drainage means:

(*i*) The discharge of dredged or fill material incidental to connecting upland drainage facilities to waters of the United States, adequate to effect the removal of excess soil moisture from upland croplands. (Construction and maintenance of upland (dryland) facilities, such as ditching and tiling, incidental to the planting, cultivating, protecting, or harvesting of crops, involve no discharge of dredged or fill material into waters of the United States, and as such never require a section 404 permit.);

(*ii*) The discharge of dredged or fill material for the purpose of installing ditching or other such water control facilities incidental to planting, cultivating, protecting, or harvesting of rice, cranberries or other wetland crop species, where these activities and the discharge occur in waters of the United States which are in established use for such agricultural and silvicultural wetland crop production;

(*iii*) The discharge of dredged or fill material for the purpose of manipulating the water levels of, or regulating the flow or distribution of water within, existing impoundments which have been constructed in accordance with applicable requirements of CWA, and which are in established use for the production of rice, cranberries, or other wetland crop species. (The provisions of paragraphs (a)(1)(iii)(C)(*1*) (*ii*) and (*iii*) of this section apply to areas that are in established use exclusively for wetland crop production as well as areas in established use for conventional wetland/non-wetland crop rotation (e.g., the rotations of rice and soybeans) where such rotation results in the cyclical or intermittent temporary dewatering of such areas.)

(*iv*) The discharges of dredged or fill material incidental to the emergency removal of sandbars, gravel bars, or other similar blockages which are formed during flood flows or other events, where such blockages close or constrict previously existing drainageways and, if not promptly removed, would result in damage to or loss of existing crops or would impair or prevent the plowing, seeding, harvesting or cultivating of crops on land in established use for crop production. Such removal does not include enlarging or extending the dimensions of, or changing the bottom elevations of, the affected drainageway as it existed prior to the formation of the blockage. Removal must be accomplished within one year of discovery of such blockages in order to be eligible for exemption.

(*2*) Minor drainage in waters of the U.S. is limited to drainage within areas that are part of an established farming or silviculture operation. It does not include drainage associated with the immediate or gradual conversion of a wetland to a non-wetland (e.g., wetland species to upland species not typically adapted to life in saturated soil conditions), or conversion from one wetland use to another (for example, silviculture to farming). In addition, minor drainage does not include the construction of any canal, ditch, dike or other waterway or structure which drains or otherwise significantly modifies a stream, lake, swamp, bog or any other wetland or aquatic area constituting waters of the United States. Any discharge of dredged or fill material into the waters of the United States incidental to the construction of any such structure or waterway requires a permit.

(D) Plowing means all forms of primary tillage, including moldboard, chisel, or wide-blade plowing, discing, harrowing and similar physical means utilized on farm, forest or ranch land for the breaking up, cutting, turning over, or stirring of soil to prepare it for the planting of crops. The term does not include the redistribution of soil, rock, sand, or other surficial materials in a manner which changes any area of the waters of the United States to dry land. For example, the redistribution of surface materials by blading, grading, or other means to fill in wetland areas is not plowing. Rock crushing activities which result in the loss of natural drainage characteristics, the reduction of water storage and recharge capabilities, or the overburden of natural water filtration capacities do not constitute plowing. Plowing as described above will never involve a discharge of dredged or fill material.

(E) Seeding means the sowing of seed and placement of seedlings to produce farm, ranch, or forest crops and includes the placement of soil beds for seeds or seedlings on established farm and forest lands.

(2) Maintenance, including emergency reconstruction of recently damaged parts, of currently serviceable structures such as dikes, dams, levees, groins, riprap, breakwaters, causeways, bridge abutments or approaches, and transportation structures. Maintenance does not include any modification that changes the character, scope, or size of the original fill design. Emergency reconstruction must occur within a reasonable period of time after damage occurs in order to qualify for this exemption.

(3) Construction or maintenance of farm or stock ponds or irrigation ditches, or the maintenance (but not construction) of drainage ditches. Discharges associated with siphons, pumps, headgates, wingwalls, weirs, diversion structures, and such other facilities as are appurtenant and functionally related to irrigation ditches are included in this exemption.

(4) Construction of temporary sedimentation basins on a construction site which does not include placement of fill material into waters of the U.S.

The term "construction site" refers to any site involving the erection of buildings, roads, and other discrete structures and the installation of support facilities necessary for construction and utilization of such structures. The term also includes any other land areas which involve land-disturbing excavation activities, including quarrying or other mining activities, where an increase in the runoff of sediment is controlled through the use of temporary sedimentation basins.

(5) Any activity with respect to which a state has an approved program under section 208(b)(4) of the CWA which meets the requirements of sections 208(b)(4) (B) and (C).

(6) Construction or maintenance of farm roads, forest roads, or temporary roads for moving mining equipment, where such roads are constructed and maintained in accordance with best management practices (BMPs) to assure that flow and circulation patterns and chemical and biological characteristics of waters of the United States are not impaired, that the reach of the waters of the United States is not reduced, and that any adverse effect on the aquatic environment will be otherwise minimized. These BMPs which must be applied to satisfy this provision shall include those detailed BMPs described in the state's approved program description pursuant to the requirements of 40 CFR 233.22(i), and shall also include the following baseline provisions:

(i) Permanent roads (for farming or forestry activities), temporary access roads (for mining, forestry, or farm purposes) and skid trails (for logging) in waters of the U.S. shall be held to the minimum feasible number, width, and total length consistent with the purpose of specific farming, silvicultural or mining operations, and local topographic and climatic conditions;

(ii) All roads, temporary or permanent, shall be located sufficiently far from streams or other water bodies (except for portions of such roads which must cross water bodies) to minimize discharges of dredged or fill material into waters of the U.S.;

(iii) The road fill shall be bridged, culverted, or otherwise designed to prevent the restriction of expected flood flows;

(iv) The fill shall be properly stabilized and maintained during and following construction to prevent erosion;

(v) Discharges of dredged or fill material into waters of the United States to construct a road fill shall be made in a manner that minimizes the encroachment of trucks, tractors, bulldozers, or other heavy equipment within waters of the United States (including adjacent wetlands) that lie outside the lateral boundaries of the fill itself;

(vi) In designing, constructing, and maintaining roads, vegetative disturbance in the waters of the U.S. shall be kept to a minimum;

(vii) The design, construction and maintenance of the road crossing shall not disrupt the migration or other movement of those species of aquatic life inhabiting the water body;

(viii) Borrow material shall be taken from upland sources whenever feasible;

(ix) The discharge shall not take, or jeopardize the continued existence of, a threatened or endangered species as defined under the Endangered Species Act, or adversely modify or destroy the critical habitat of such species;

(x) Discharges into breeding and nesting areas for migratory waterfowl, spawning areas, and wetlands shall be avoided if practical alternatives exist;

(xi) The discharge shall not be located in the proximity of a public water supply intake;

(xii) The discharge shall not occur in areas of concentrated shellfish production;

(xiii) The discharge shall not occur in a component of the National Wild and Scenic River System;

(xiv) The discharge of material shall consist of suitable material free from toxic pollutants in toxic amounts; and

(xv) All temporary fills shall be removed in their entirety and the area restored to its original elevation.

(b) If any discharge of dredged or fill material resulting from the activities listed in paragraphs (a) (1) through (6) of this section contains any toxic pollutant listed under section 307 of the CWA such discharge shall be subject to any applicable toxic effluent standard or prohibition, and shall require a Section 404 permit.

(c) Any discharge of dredged or fill material into waters of the United States incidental to any of the activities identified in paragraphs (a) (1) through (6) of this section must have a permit if it is part of an activity whose purpose is to convert an area of the waters of the United States into a use to which it was not previously subject, where the flow or circulation of waters of the United States nay be impaired or the reach of such waters reduced. Where the proposed discharge will result in significant discernible alterations to flow or circulation, the presumption is that flow or circulation may be impaired by such alteration. For example, a permit will be required for the conversion of a cypress swamp to some other use or the conversion of a wetland from silvicultural to agricultuial use when there is a discharge of dredged or fill material into waters of the United States in conjunction with construction of dikes, drainage ditches or other works or structures used to effect such conversion. A conversion of a Section 404 wetland to a non-wetland is a change in use of an area of waters of the United States. A discharge which elevates the bottom of waters of the United States without converting it to dry land does not thereby reduce the reach of, but may alter the flow or circulation of, waters of the United States.

(d) Federal projects which qualify under the criteria contained in section 404(r) of the CWA are exempt from section 404 permit requirements, but may be subject to other state or Federal requirements.

§ 323.5 Program transfer to states.

Section 404(h) of the CWA allows the Administrator of the Environmental Protection Agency (EPA) to transfer administration of the section 404 permit program for discharges into certain waters of the United States to qualified states. (The program cannot be transferred for those waters which are presently used, or are susceptible to use in their natural condition or by reasonable improvement as a means to transport interstate or foreign commerce shoreward to their ordinary high water mark, including all waters which are subject to the ebb and flow of the tide shoreward to the high tide linc, including wetlands adjacent thereto). See 40 CFR Parts 233 and 124 for procedural regulations for transferring Section 404 programs to states. Once a state's 404 program is approved and in effect, the Corps of Engineers will suspend processing of section 404 applications in the applicable waters and will transfer pending applications to the state agency responsible for administering the program. District engineers will assist EPA and the states in any way practicable to effect transfer and will develop appropriate procedures to ensure orderly and expeditious transfer.

§ 323.6 Special policies and procedures.

(a) The Secretary of the Army has delegated to the Chief of Engineers the authority to issue or deny section 404 permits. The district engineer will review applications for permits for the discharge of dredged or fill material into waters of the United States in accordance with guidelines promulgated by the Administrator, EPA, under authority of section 404(b)(1) of the CWA. (see 40 CFR Part 230.) Subject to consideration of any economic impact on navigation and anchorage pursuant to section 404(b)(2), a permit will be denied if the discharge that would be authorized by such a permit would not comply with the 404(b)(1) guidelines. If the district engineer determines that the proposed discharge would comply with the 404(b)(1) guidelines, he will grant the permit unless issuance would be contrary to the public interest.

(b) The Corps will not issue a permit where the regional administrator of EPA has notified the district engineer and applicant in writing pursuant to 40 CFR 231.3(a)(1) that he intends to issue a public notice of a proposed determination to prohibit or withdraw the specification, or to deny, restrict or withdraw the use for specification, of any defined area as a disposal site in accordance with section 404(c) of the Clean Water Act. However the Corps will continue to complete the

Corps of Engineers, Dept. of the Army, DoD

§ 324.3

administrative processing of the application while the section 404(c) procedures are underway including completion of final coordination with EPA under 33 CFR Part 325.

PART 324—PERMITS FOR OCEAN DUMPING OF DREDGED MATERIAL

Sec.
324.1 General.
324.2 Definitions.
324.3 Activities requiring permits.
324.4 Special procedures.

AUTHORITY: 33 U.S.C. 1413.

SOURCE: 51 FR 41235, Nov. 13, 1986, unless otherwise noted.

§ 324.1 General.

This regulation prescribes in addition to the general policies of 33 CFR Part 320 and procedures of 33 CFR Part 325, those special policies, practices and procedures to be followed by the Corps of Engineers in connection with the review of applications for Department of the Army (DA) permits to authorize the transportation of dredged material by vessel or other vehicle for the purpose of dumping it in ocean waters at dumping sites designated under 40 CFR Part 228 pursuant to section 103 of the Marine Protection, Research and Sanctuaries Act of 1972, as amended (33 U.S.C. 1413) (hereinafter referred to as section 103). See 33 CFR 320.2(h). Activities involving the transportation of dredged material for the purpose of dumping in the ocean waters also require DA permits under Section 10 of the Rivers and Harbors Act of 1899 (33 U.S.C. 403) for the dredging in navigable waters of the United States. Applicants for DA permits under this Part should also refer to 33 CFR Part 322 to satisfy the requirements of Section 10.

§ 324.2 Definitions.

For the purpose of this regulation, the following terms are defined:

(a) The term "ocean waters" means those waters of the open seas lying seaward of the base line from which the territorial sea is measured, as provided for in the Convention on the Territorial Sea and the Contiguous Zone (15 UST 1606: TIAS 5639).

(b) The term "dredged material" means any material excavated or dredged from navigable waters of the United States.

(c) The term "transport" or "transportation" refers to the conveyance and related handling of dredged material by a vessel or other vehicle.

§ 324.3 Activities requiring permits.

(a) *General.* DA permits are required for the transportation of dredged material for the purpose of dumping it in ocean waters.

(b) *Activities of Federal agencies.* (1) The transportation of dredged material for the purpose of disposal in ocean waters done by or on behalf of any Federal agency other than the activities of the Corps of Engineers is subject to the procedures of this regulation. Agreement for construction or engineering services performed for other agencies by the Corps of Engineers does not constitute authorization under these regulations. Division and district engineers will therefore advise Federal agencies accordingly and cooperate to the fullest extent in the expeditious processing of their applications. The activities of the Corps of Engineers that involve the transportation of dredged material for disposal in ocean waters are regulated by 33 CFR 209.145.

(2) The policy provisions set out in 33 CFR 320.4(j) relating to state or local authorizations do not apply to work or structures undertaken by Federal agencies, except where compliance with non-Federal authorization is required by Federal law or Executive policy. Federal agencies are responsible for conformance with such laws and policies. (See EO 12088, October 18, 1978.) Federal agencies are not required to obtain and provide certification of compliance with effluent limitations and water quality standards from state or interstate water pollution control agencies in connection with activities involving the transport of dredged material for dumping into ocean waters beyond the territorial sea.

§ 324.4 Special procedures.

The Secretary of the Army has delegated to the Chief of Engineers the authority to issue or deny section 103 permits. The following additional procedures shall also be applicable under this regulation.

(a) *Public notice.* For all applications for section 103 permits, the district engineer will issue a public notice which shall contain the information specified in 33 CFR 325.3.

(b) *Evaluation.* Applications for permits for the transportation of dredged material for the purpose of dumping it in ocean waters will be evaluated to determine whether the proposed dumping will unreasonably degrade or endanger human health, welfare, amenities, or the marine environment, ecological systems or economic potentialities. District engineers will apply the criteria established by the Administrator of EPA pursuant to section 102 of the Marine Protection, Research and Sanctuaries Act of 1972 in making this evaluation. (See 40 CFR Parts 220–229) Where ocean dumping is determined to be necessary, the district engineer will, to the extent feasible, specify disposal sites using the recommendations of the Administrator pursuant to section 102(c) of the Act.

(c) *EPA review.* When the Regional Administrator, EPA, in accordance with 40 CFR 225.2(b), advises the district engineer, in writing, that the proposed dumping will comply with the criteria, the district engineer will complete his evaluation of the application under this part and 33 CFR Parts 320 and 325. If, however, the Regional Administrator advises the district engineer, in writing, that the proposed dumping does not comply with the criteria, the district engineer will proceed as follows:

(1) The district engineer will determine whether there is an economically feasible alternative method or site available other than the proposed ocean disposal site. If there are other feasible alternative methods or sites available, the district engineer will evaluate them in accordance with 33 CFR Parts 320, 322, 323, and 325 and this Part, as appropriate.

(2) If the district engineer determines that there is no economically feasible alternative method or site available, and the proposed project is otherwise found to be not contrary to the public interest, he will so advise the Regional Administrator setting forth his reasons for such determination. If the Regional Administrator has not removed his objection within 15 days, the district engineer will submit a report of his determination to the Chief of Engineers for further coordination with the Administrator, EPA, and decision. The report forwarding the case will contain the analysis of whether there are other economically feasible methods or sites available to dispose of the dredged material.

(d) *Chief of Engineers review.* The Chief of Engineers shall evaluate the permit application and make a decision to deny the permit or recommend its issuance. If the decision of the Chief of Engineers is that ocean dumping at the proposed disposal site is required because of the unavailability of economically feasible alternatives, he shall so certify and request that the Secretary of the Army seek a waiver from the Administrator, EPA, of the criteria or of the critical site designation in accordance with 40 CFR 225.4.

PART 325—PROCESSING OF DEPARTMENT OF THE ARMY PERMITS

Sec.
325.1   Applications for permits.
325.2   Processing of applications.
325.3   Public notice.
325.4   Conditioning of permits.
325.5   Forms of permits.
325.6   Duration of permits.
325.7   Modification, suspension, or revocation of permits.
325.8   Authority to issue or deny permits.
325.9   Authority to determine jurisdiction.
325.10   Publicity.

APPENDIX A—PERMIT FORM AND SPECIAL CONDITIONS
APPENDIX B—NEPA IMPLEMENTATION PROCEDURES FOR THE REGULATORY PROGRAM
APPENDIX C—PROCEDURES FOR THE PROTECTION OF HISTORIC PROPERTIES

AUTHORITY: 33 U.S.C. 401 et seq.; 33 U.S.C. 1344; 33 U.S.C. 1413.

SOURCE: 51 FR 41236, Nov. 13, 1986, unless otherwise noted.

§ 325.1 Applications for permits.

(a) *General.* The processing procedures of this Part apply to any Department of the Army (DA) permit. Special procedures and additional information are contained in 33 CFR Parts 320 through 324, 327 and Part 330. This Part is arranged in the basic timing sequence used by the Corps of Engineers in processing applications for DA permits.

(b) *Pre-application consultation for major applications.* The district staff element having responsibility for administering, processing, and enforcing federal laws and regulations relating to the Corps of Engineers regulatory program shall be available to advise potential applicants of studies or other information foreseeably required for later federal action. The district engineer will establish local procedures and policies including appropriate publicity, programs which will allow potential applicants to contact the district engineer or the regulatory staff element to request pre-application consultation. Upon receipt of such request, the district engineer will assure the conduct of an orderly process which may involve other staff elements and affected agencies (Federal, state, or local) and the public. This early process should be brief but thorough so that the potential applicant may begin to assess the viability of some of the more obvious potential alternatives in the application. The district engineer will endeavor, at this stage, to provide the potential applicant with all helpful information necessary in pursuing the application, including factors which the Corps must consider in its permit decision making process. Whenever the district engineer becomes aware of planning for work which may require a DA permit and which may involve the preparation of an environmental document, he shall contact the principals involved to advise them of the requirement for the permit(s) and the attendant public interest review including the development of an environmental document. Whenever a potential applicant indicates the intent to submit an application for work which may require the preparation of an environmental document, a single point of

contact shall be designated within the district's regulatory staff to effectively coordinate the regulatory process, including the National Environmental Policy Act (NEPA) procedures and all attendant reviews, meetings, hearings, and other actions, including the scoping process if appropriate, leading to a decision by the district engineer. Effort devoted to this process should be commensurate with the likelihood of a permit application actually being submitted to the Corps. The regulatory staff coordinator shall maintain an open relationship with each potential applicant or his consultants so as to assure that the potential applicant is fully aware of the substance (both quantitative and qualitative) of the data required by the district engineer for use in preparing an environmental assessment or an environmental impact statement (EIS) in accordance with 33 CFR Part 230, Appendix B.

(c) *Application form.* Applicants for all individual DA permits must use the standard application form (ENG Form 4345, OMB Approval No. OMB 49–R0420). Local variations of the application form for purposes of facilitating coordination with federal, state and local agencies may be used. The appropriate form may be obtained from the district office having jurisdiction over the waters in which the activity is proposed to be located. Certain activities have been authorized by general permits and do not require submission of an application form but may require a separate notification.

(d) *Content of application.* (1) The application must include a complete description of the proposed activity including necessary drawings, sketches, or plans sufficient for public notice (detailed engineering plans and specifications are not required); the location, purpose and need for the proposed activity; scheduling of the activity; the names and addresses of adjoining property owners; the location and dimensions of adjacent structures; and a list of authorizations required by other federal, interstate, state, or local agencies for the work, including all approvals received or denials already made. See § 325.3 for information required to be in public notices. District and division engineers are not authorized to develop additional information forms but may request specific information on a case-by-case basis. (See § 325.1(e)).

(2) All activities which the applicant plans to undertake which are reasonably related to the same project and for which a DA permit would be required should be included in the same permit application. District engineers should reject, as incomplete, any permit application which fails to comply with this requirement. For example, a permit application for a marina will include dredging required for access as well as any fill associated with construction of the marina.

(3) If the activity would involve dredging in navigable waters of the United States, the application must include a description of the type, composition and quantity of the material to be dredged, the method of dredging, and the site and plans for disposal of the dredged material.

(4) If the activity would include the discharge of dredged or fill material into the waters of the United States or the transportation of dredged material for the purpose of disposing of it in ocean waters the application must include the source of the material; the purpose of the discharge, a description of the type, composition and quantity of the material; the method of transportation and disposal of the material; and the location of the disposal site. Certification under section 401 of the Clean Water Act is required for such discharges into waters of the United States.

(5) If the activity would include the construction of a filled area or pile or float-supported platform the project description must include the use of, and specific structures to be erected on, the fill or platform.

(6) If the activity would involve the construction of an impoundment structure, the applicant may be required to demonstrate that the structure complies with established state dam safety criteria or that the structure has been designed by qualified persons and, in appropriate cases, independently reviewed (and modified as the review would indicate) by similiarly qualified persons. No specific design criteria are to be prescribed nor is an independent detailed engineering review to be made by the district engineer.

(7) *Signature on application.* The application must be signed by the person who desires to undertake the proposed activity (i.e. the applicant) or by a duly authorized agent. When the applicant is represented by an agent, that information will be included in the space provided on the application or by a separate written statement. The signature of the applicant or the agent will be an affirmation that the applicant possesses or will possess the requisite property interest to undertake the activity proposed in the application, except where the lands are under the control of the Corps of Engineers, in which cases the district engineer will coordinate the transfer of the real estate and the permit action. An application may include the activity of more than one owner provided the character of the activity of each owner is similar and in the same general area and each owner submits a statement designating the same agent.

(8) If the activity would involve the construction or placement of an artificial reef, as defined in 33 CFR 322.2(g), in the navigable waters of the United States or in the waters overlying the outer continental shelf, the application must include provisions for siting, constructing, monitoring, and managing the artificial reef.

(9) *Complete application.* An application will be determined to be complete when sufficient information is received to issue a public notice (See 33 CFR 325.1(d) and 325.3(a).) The issuance of a public notice will not be delayed to obtain information necessary to evaluate an application.

(e) *Additional information.* In addition to the information indicated in paragraph (d) of this section, the applicant will be required to furnish only such additional information as the district engineer deems essential to make a public interest determination including, where applicable, a determination of compliance with the section 404(b)(1) guidelines or ocean dumping criteria. Such additional information may include environmental data and information on alternate methods and sites as may be necessary for the preparation of the required environmental documentation.

(f) *Fees.* Fees are required for permits under section 404 of the Clean Water Act, section 103 of the Marine Protection, Research and Sanctuaries Act of 1972, as amended, and sections 9 and 10 of the Rivers and Harbors Act of 1899. A fee of $100.00 will be charged when the planned or ultimate purpose of the project is commercial or industrial in nature and is in support of operations that charge for the production, distribution or sale of goods or services. A $10.00 fee will be charged for permit applications when the proposed work is non-commercial in nature and would provide personal benefits that have no connection with a commercial enterprise. The final decision as to the basis for a fee (commercial vs. non-commercial) shall be solely the responsibility of the district engineer. No fee will be charged if the applicant withdraws the application at any time prior to issuance of the permit or if the permit is denied. Collection of the fee will be deferred until the proposed activity has been determined to be not contrary to the public interest. Multiple fees are not to be charged if more than one law is applicable. Any modification significant enough to require publication of a public notice will also require a fee. No fee will be assessed when a permit is transferred from one property owner to another. No fees will be charged for time extensions, general permits or letters of permission. Agencies or instrumentalities of federal, state or local governments will not be required to pay any fee in connection with permits.

§ 325.2 **Processing of applications.**

(a) *Standard procedures.* (1) When an application for a permit is received the district engineer shall immediately assign it a number for identification, acknowledge receipt thereof, and advise the applicant of the number assigned to it. He shall review the application for completeness, and if the application is incomplete, request from the applicant within 15 days of receipt of the application any additional information necessary for further processing.

(2) Within 15 days of receipt of an application the district engineer will either determine that the application is complete (see 33 CFR 325.1(d)(9) and issue a public notice as described in § 325.3 of this Part, unless specifically exempted by other provisions of this regulation or that it is incomplete and notify the applicant of the information necessary for a complete application. The district engineer will issue a supplemental, revised, or corrected public notice if in his view there is a change in the application data that would affect the public's review of the proposal.

(3) The district engineer will consider all comments received in response to the public notice in his subsequent actions on the permit application. Receipt of the comments will be acknowledged, if appropriate, and they will be made a part of the administrative record of the application. Comments received as form letters or petitions may be acknowledged as a group to the person or organization responsible for the form letter or petition. If comments relate to matters within the special expertise of another federal agency, the district engineer may seek the advice of that agency. If the district engineer determines, based on comments received, that he must have the views of the applicant on a particular issue to make a public interest determination, the applicant will be given the opportunity to furnish his views on such issue to the district engineer (see § 325.2(d)(5)). At the earliest practicable time other substantive comments will be furnished to the applicant for his information and any views he may wish to offer. A summary of the comments, the actual letters or portions thereof, or representative comment letters may be furnished to the applicant. The applicant may voluntarily elect to contact objectors in an attempt to resolve objections but will not be required to do so. District engineers will ensure that all parties are informed that the Corps alone is responsible for reaching a decision on the merits of any application. The district engineer may also offer Corps regulatory staff to be present at meetings between applicants and objectors, where appropriate, to provide informa-

tion on the process, to mediate differences, or to gather information to aid in the decision process. The district engineer should not delay processing of the application unless the applicant requests a reasonable delay, normally not to exceed 30 days, to provide additional information or comments.

(4) The district engineer will follow Appendix B of 33 CFR Part 230 for environmental procedures and documentation required by the National Environmental Policy Act of 1969. A decision on a permit application will require either an environmental assessment or an environmental impact statement unless it is included within a categorical exclusion.

(5) The district engineer will also evaluate the application to determine the need for a public hearing pursuant to 33 CFR Part 327.

(6) After all above actions have been completed, the district engineer will determine in accordance with the record and applicable regulations whether or not the permit should be issued. He shall prepare a statement of findings (SOF) or, where an EIS has been prepared, a record of decision (ROD), on all permit decisions. The SOF or ROD shall include the district engineer's views on the probable effect of the proposed work on the public interest including conformity with the guidelines published for the discharge of dredged or fill material into waters of the United States (40 CFR Part 230) or with the criteria for dumping of dredged material in ocean waters (40 CFR Parts 220 to 229), if applicable, and the conclusions of the district engineer. The SOF or ROD shall be dated, signed, and included in the record prior to final action on the application. Where the district engineer has delegated authority to sign permits for and in his behalf, he may similarly delegate the signing of the SOF or ROD. If a district engineer makes a decision on a permit application which is contrary to state or local decisions (33 CFR 320.4(j) (2) & (4)), the district engineer will include in the decision document the significant national issues and explain how they are overriding in importance. If a permit is warranted, the district engineer will determine the special conditions, if

any, and duration which should be incorporated into the permit. In accordance with the authorities specified in § 325.8 of this Part, the district engineer will take final action or forward the application with all pertinent comments, records, and studies, including the final EIS or environmental assessment, through channels to the official authorized to make the final decision. The report forwarding the application for decision will be in a format prescribed by the Chief of Engineers. District and division engineers will notify the applicant and interested federal and state agencies that the application has been forwarded to higher headquarters. The district or division engineer may, at his option, disclose his recommendation to the news media and other interested parties, with the caution that it is only a recommendation and not a final decision. Such disclosure is encouraged in permit cases which have become controversial and have been the subject of stories in the media or have generated strong public interest. In those cases where the application is forwarded for decision in the format prescribed by the Chief of Engineers, the report will serve as the SOF or ROD. District engineers will generally combine the SOF, environmental assessment, and findings of no significant impact (FONSI), 404(b)(1) guideline analysis, and/or the criteria for dumping of dredged material in ocean waters into a single document.

(7) If the final decision is to deny the permit, the applicant will be advised in writing of the reason(s) for denial. If the final decision is to issue the permit and a standard individual permit form will be used, the issuing official will forward the permit to the applicant for signature accepting the conditions of the permit. The permit is not valid until signed by the issuing official. Letters of permission require only the signature of the issuing official. Final action on the permit application is the signature on the letter notifying the applicant of the denial of the permit or signature of the issuing official on the authorizing document.

(8) The district engineer will publish monthly a list of permits issued or denied during the previous month.

The list will identify each action by public notice number, name of applicant, and brief description of activity involved. It will also note that relevant environmental documents and the SOF's or ROD's are available upon written request and, where applicable, upon the payment of administrative fees. This list will be distributed to all persons who may have an interest in any of the public notices listed.

(9) Copies of permits will be furnished to other agencies in appropriate cases as follows:

(i) If the activity involves the construction of artificial islands, installations or other devices on the outer continental shelf, to the Director, Defense Mapping Agency, Hydrographic Center, Washington, DC 20390 Attention, Code NS12, and to the Charting and Geodetic Services, N/CG222, National Ocean Service NOAA, Rockville, Maryland 20852.

(ii) If the activity involves the construction of structures to enhance fish propagation (e.g., fishing reefs) along the coasts of the United States, to the Defense Mapping Agency, Hydrographic Center and National Ocean Service as in paragraph (a)(9)(i) of this section and to the Director, Office of Marine Recreational Fisheries, National Marine Fisheries Service, Washington, DC 20235.

(iii) If the activity involves the erection of an aerial transmission line, submerged cable, or submerged pipeline across a navigable water of the United States, to the Charting and Geodetic Services N/CG222, National Ocean Service NOAA, Rockville, Maryland 20852.

(iv) If the activity is listed in paragraphs (a)(9) (i), (ii), or (iii) of this section, or involves the transportation of dredged material for the purpose of dumping it in ocean waters, to the appropriate District Commander, U.S. Coast Guard.

(b) *Procedures for particular types of permit situations.*—(1) *Section 401 Water Quality Certification.* If the district engineer determines that water quality certification for the proposed activity is necessary under the provisions of section 401 of the Clean Water Act, he shall so notify the applicant and obtain from him or the certifying agency a copy of such certification.

(i) The public notice for such activity, which will contain a statement on certification requirements (see § 325.3(a)(8)), will serve as the notification to the Administrator of the Environmental Protection Agency (EPA) pursuant to section 401(a)(2) of the Clean Water Act. If EPA determines that the proposed discharge may affect the quality of the waters of any state other than the state in which the discharge will originate, it will so notify such other state, the district engineer, and the applicant. If such notice or a request for supplemental information is not received within 30 days of issuance of the public notice, the district engineer will assume EPA has made a negative determination with respect to section 401(a)(2). If EPA determines another state's waters may be affected, such state has 60 days from receipt of EPA's notice to determine if the proposed discharge will affect the quality of its waters so as to violate any water quality requirement in such state, to notify EPA and the district engineer in writing of its objection to permit issuance, and to request a public hearing. If such occurs, the district engineer will hold a public hearing in the objecting state. Except as stated below, the hearing will be conducted in accordance with 33 CFR Part 327. The issues to be considered at the public hearing will be limited to water quality impacts. EPA will submit its evaluation and recommendations at the hearing with respect to the state's objection to permit issuance. Based upon the recommendations of the objecting state, EPA, and any additional evidence presented at the hearing, the district engineer will condition the permit, if issued, in such a manner as may be necessary to insure compliance with applicable water quality requirements. If the imposition of conditions cannot, in the district engineer's opinion, insure such compliance, he will deny the permit.

(ii) No permit will be granted until required certification has been obtained or has been waived. A waiver may be explicit, or will be deemed to occur if the certifying agency fails or refuses to act on a request for certification within sixty days after receipt of such a request unless the district engineer determines a shorter or longer period is reasonable for the state to act. In determining whether or not a waiver period has commenced or waiver has occurred, the district engineer will verify that the certifying agency has received a valid request for certification. If, however, special circumstances identified by the district engineer require that action on an application be taken within a more limited period of time, the district engineer shall determine a reasonable lesser period of time, advise the certifying agency of the need for action by a particular date, and that, if certification is not received by that date, it will be considered that the requirement for certification has been waived. Similarly, if it appears that circumstances may reasonably require a period of time longer than sixty days, the district engineer, based on information provided by the certifying agency, will determine a longer reasonable period of time, not to exceed one year, at which time a waiver will be deemed to occur.

(2) *Coastal Zone Management Consistency.* If the proposed activity is to be undertaken in a state operating under a coastal zone management program approved by the Secretary of Commerce pursuant to the Coastal Zone Management (CZM) Act (see 33 CFR 320.3(b)), the district engineer shall proceed as follows:

(i) If the applicant is a federal agency, and the application involves a federal activity in or affecting the coastal zone, the district engineer shall forward a copy of the public notice to the agency of the state responsible for reviewing the consistency of federal activities. The federal agency applicant shall be responsible for complying with the CZM Act's directive for ensuring that federal agency activities are undertaken in a manner which is consistent, to the maximum extent practicable, with approved CZM Programs. (See 15 CFR Part 930.) If the state coastal zone agency objects to the proposed federal activity on the basis of its inconsistency with the state's approved CZM Program, the district engineer shall not make a final decision on the application until the disagreeing parties have had an opportunity to utilize the procedures specified by the CZM Act for resolving such disagreements.

(ii) If the applicant is not a federal agency and the application involves an activity affecting the coastal zone, the district engineer shall obtain from the applicant a certification that his proposed activity complies with and will be conducted in a manner that is consistent with the approved state CZM Program. Upon receipt of the certification, the district engineer will forward a copy of the public notice (which will include the applicant's certification statement) to the state coastal zone agency and request its concurrence or objection. If the state agency objects to the certification or issues a decision indicating that the proposed activity requires further review, the district engineer shall not issue the permit until the state concurs with the certification statement or the Secretary of Commerce determines that the proposed activity is consistent with the purposes of the CZM Act or is necessary in the interest of national security. If the state agency fails to concur or object to a certification statement within six months of the state agency's receipt of the certification statement, state agency concurrence with the certification statement shall be conclusively presumed. District engineers will seek agreements with state CZM agencies that the agency's failure to provide comments during the public notice comment period will be considered as a concurrence with the certification or waiver of the right to concur or non-concur.

(iii) If the applicant is requesting a permit for work on Indian reservation lands which are in the coastal zone, the district engineer shall treat the application in the same manner as prescribed for a Federal applicant in paragraph (b)(2)(i) of this section. However, if the applicant is requesting a permit on non-trust Indian lands, and the state CZM agency has decided to assert jurisdiction over such lands, the district engineer shall treat the application in the same manner as prescribed for a non-Federal applicant in paragraph (b)(2)(ii) of this section.

(3) *Historic Properties.* If the proposed activity would involve any property listed or eligible for listing in the National Register of Historic Places, the district engineer will proceed in accordance with Corps National Historic Preservation Act implementing regulations.

(4) *Activities Associated with Federal Projects.* If the proposed activity would consist of the dredging of an access channel and/or berthing facility associated with an authorized federal navigation project, the activity will be included in the planning and coordination of the construction or maintenance of the federal project to the maximum extent feasible. Separate notice, hearing, and environmental documentation will not be required for activities so included and coordinated, and the public notice issued by the district engineer for these federal and associated non-federal activities will be the notice of intent to issue permits for those included non-federal dredging activities. The decision whether to issue or deny such a permit will be consistent with the decision on the federal project unless special considerations applicable to the proposed activity are identified. (See § 322.5(c).)

(5) *Endangered Species.* Applications will be reviewed for the potential impact on threatened or endangered species pursuant to section 7 of the Endangered Species Act as amended. The district engineer will include a statement in the public notice of his current knowledge of endangered species based on his initial review of the application (see 33 CFR 325.2(a)(2)). If the district engineer determines that the proposed activity would not affect listed species or their critical habitat, he will include a statement to this effect in the public notice. If he finds the proposed activity may affect an endangered or threatened species or their critical habitat, he will initiate formal consultation procedures with the U.S. Fish and Wildlife Service or National Marine Fisheries Service. Public notices forwarded to the U.S. Fish and Wildlife Service or National Marine Fisheries Service will serve as the request for information on whether any listed or proposed to be listed endangered or threatened species may be present in the area which would be affected by the proposed activity, pursuant to section 7(c) of the Act. References, definitions, and consultation procedures are found in 50 CFR Part 402.

(c) [Reserved]

(d) *Timing of processing of applications.* The district engineer will be guided by the following time limits for the indicated steps in the evaluation process:

(1) The public notice will be issued within 15 days of receipt of all information required to be submitted by the applicant in accordance with paragraph 325.1.(d) of this Part.

(2) The comment period on the public notice should be for a reasonable period of time within which interested parties may express their views concerning the permit. The comment period should not be more than 30 days nor less than 15 days from the date of the notice. Before designating comment periods less than 30 days, the district engineer will consider: (i) Whether the proposal is routine or noncontroversial,

(ii) Mail time and need for comments from remote areas,

(iii) Comments from similar proposals, and

(iv) The need for a site visit. After considering the length of the original comment period, paragraphs (a)(2) (i) through (iv) of this section, and other pertinent factors, the district engineer may extend the comment period up to an additional 30 days if warranted.

(3) District engineers will decide on all applications not later than 60 days after receipt of a complete application, unless (i) precluded as a matter of law or procedures required by law (see below),

(ii) The case must be referred to higher authority (see § 325.8 of this Part),

(iii) The comment period is extended,

(iv) A timely submittal of information or comments is not received from the applicant,

(v) The processing is suspended at the request of the applicant, or

(vi) Information needed by the district engineer for a decision on the application cannot reasonably be obtained within the 60-day period. Once the cause for preventing the decision from being made within the normal 60-day period has been satisfied or eliminated, the 60-day clock will start running again from where it was suspended. For example, if the comment period is extended by 30 days, the district engineer will, absent other restraints, decide on the application within 90 days of receipt of a complete application. Certain laws (e.g., the Clean Water Act, the CZM Act, the National Environmental Policy Act, the National Historic Preservation Act, the Preservation of Historical and Archeological Data Act, the Endangered Species Act, the Wild and Scenic Rivers Act, and the Marine Protection, Research and Sanctuaries Act) require procedures such as state or other federal agency certifications, public hearings, environmental impact statements, consultation, special studies, and testing which may prevent district engineers from being able to decide certain applications within 60 days.

(4) Once the district engineer has sufficient information to make his public interest determination, he should decide the permit application even though other agencies which may have regulatory jurisdiction have not yet granted their authorizations, except where such authorizations are, by federal law, a prerequisite to making a decision on the DA permit application. Permits granted prior to other (non-prerequisite) authorizations by other agencies should, where appropriate, be conditioned in such manner as to give those other authorities an opportunity to undertake their review without the applicant biasing such review by making substantial resource commitments on the basis of the DA permit. In unusual cases the district engineer may decide that due to the nature or scope of a specific proposal, it would be prudent to defer taking final action until another agency has acted on its authorization. In such cases, he may advise the other agency of his position on the DA permit while deferring his final decision.

(5) The applicant will be given a reasonable time, not to exceed 30 days, to respond to requests of the district en-

gineer. The district engineer may make such requests by certified letter and clearly inform the applicant that if he does not respond with the requested information or a justification why additional time is necessary, then his application will be considered withdrawn or a final decision will be made, whichever is appropriate. If additional time is requested, the district engineer will either grant the time, make a final decision, or consider the application as withdrawn.

(6) The time requirements in these regulations are in terms of calendar days rather than in terms of working days.

(e) *Alternative procedures.* Division and district engineers are authorized to use alternative procedures as follows:

(1) *Letters of permission.* Letters of permission are a type of permit issued through an abbreviated processing procedure which includes coordination with Federal and state fish and wildlife agencies, as required by the Fish and Wildlife Coordination Act, and a public interest evaluation, but without the publishing of an individual public notice. The letter of permission will not be used to authorize the transportation of dredged material for the purpose of dumping it in ocean waters. Letters of permission may be used:

(i) In those cases subject to section 10 of the Rivers and Harbors Act of 1899 when, in the opinion of the district engineer, the proposed work would be minor, would not have significant individual or cumulative impacts on environmental values, and should encounter no appreciable opposition.

(ii) In those cases subject to section 404 of the Clean Water Act after:

(A) The district engineer, through consultation with Federal and state fish and wildlife agencies, the Regional Administrator, Environmental Protection Agency, the state water quality certifying agency, and, if appropriate, the state Coastal Zone Management Agency, develops a list of categories of activities proposed for authorization under LOP procedures;

(B) The district engineer issues a public notice advertising the proposed list and the LOP procedures, requesting comments and offering an opportunity for public hearing; and

(C) A 401 certification has been issued or waived and, if appropriate, CZM consistency concurrence obtained or presumed either on a generic or individual basis.

(2) *Regional permits.* Regional permits are a type of general permit as defined in 33 CFR 322.2(f) and 33 CFR 323.2(n). They may be issued by a division or district engineer after compliance with the other procedures of this regulation. After a regional permit has been issued, individual activities falling within those categories that are authorized by such regional permits do not have to be further authorized by the procedures of this regulation. The issuing authority will determine and add appropriate conditions to protect the public interest. When the issuing authority determines on a case-by-case basis that the concerns for the aquatic environment so indicate, he may exercise discretionary authority to override the regional permit and require an individual application and review. A regional permit may be revoked by the issuing authority if it is determined that it is contrary to the public interest provided the procedures of § 325.7 of this Part are followed. Following revocation, applications for future activities in areas covered by the regional permit shall be processed as applications for individual permits. No regional permit shall be issued for a period of more than five years.

(3) *Joint procedures.* Division and district engineers are authorized and encouraged to develop joint procedures with states and other Federal agencies with ongoing permit programs for activities also regulated by the Department of the Army. Such procedures may be substituted for the procedures in paragraphs (a)(1) through (a)(5) of this section provided that the substantive requirements of those sections are maintained. Division and district engineers are also encouraged to develop management techniques such as joint agency review meetings to expedite the decision-making process. However, in doing so, the applicant's rights to a full public interest review and independent deci-

sion by the district or division engineer must be strictly observed.

(4) *Emergency procedures.* Division engineers are authorized to approve special processing procedures in emergency situations. An "emergency" is a situation which would result in an unacceptable hazard to life, a significant loss of property, or an immediate, unforeseen, and significant economic hardship if corrective action requiring a permit is not undertaken within a time period less than the normal time needed to process the application under standard procedures. In emergency situations, the district engineer will explain the circumstances and recommend special procedures to the division engineer who will instruct the district engineer as to further processing of the application. Even in an emergency situation, reasonable efforts will be made to receive comments from interested Federal, state, and local agencies and the affected public. Also, notice of any special procedures authorized and their rationale is to be appropriately published as soon as practicable.

§ 325.3 Public notice.

(a) *General.* The public notice is the primary method of advising all interested parties of the proposed activity for which a permit is sought and of soliciting comments and information necessary to evaluate the probable impact on the public interest. The notice must, therefore, include sufficient information to give a clear understanding of the nature and magnitude of the activity to generate meaningful comment. The notice should include the following items of information:

(1) Applicable statutory authority or authorities;

(2) The name and address of the applicant;

(3) The name or title, address and telephone number of the Corps employee from whom additional information concerning the application may be obtained;

(4) The location of the proposed activity;

(5) A brief description of the proposed activity, its purpose and intend-

447

ed use, so as to provide sufficient information concerning the nature of the activity to generate meaningful comments, including a description of the type of structures, if any, to be erected on fills or pile or float-supported platforms, and a description of the type, composition, and quantity of materials to be discharged or disposed of in the ocean;

(6) A plan and elevation drawing showing the general and specific site location and character of all proposed activities, including the size relationship of the proposed structures to the size of the impacted waterway and depth of water in the area;

(7) If the proposed activity would occur in the territorial seas or ocean waters, a description of the activity's relationship to the baseline from which the territorial sea is measured;

(8) A list of other government authorizations obtained or requested by the applicant, including required certifications relative to water quality, coastal zone management, or marine sanctuaries;

(9) If appropriate, a statement that the activity is a categorical exclusion for purposes of NEPA (see paragraph 7 of Appendix B to 33 CFR Part 230);

(10) A statement of the district engineer's current knowledge on historic properties;

(11) A statement of the district engineer's current knowledge on endangered species (see § 325.2(b)(5));

(12) A statement(s) on evaluation factors (see § 325.3(c));

(13) Any other available information which may assist interested parties in evaluating the likely impact of the proposed activity, if any, on factors affecting the public interest;

(14) The comment period based on § 325.2(d)(2);

(15) A statement that any person may request, in writing, within the comment period specified in the notice, that a public hearing be held to consider the application. Requests for public hearings shall state, with particularity, the reasons for holding a public hearing;

(16) For non-federal applications in states with an approved CZM Plan, a statement on compliance with the approved Plan; and

(17) In addition, for section 103 (ocean dumping) activities:

(i) The specific location of the proposed disposal site and its physical boundaries;

(ii) A statement as to whether the proposed disposal site has been designated for use by the Administrator, EPA, pursuant to section 102(c) of the Act;

(iii) If the proposed disposal site has not been designated by the Administrator, EPA, a description of the characteristics of the proposed disposal site and an explanation as to why no previously designated disposal site is feasible;

(iv) A brief description of known dredged material discharges at the proposed disposal site;

(v) Existence and documented effects of other authorized disposals that have been made in the disposal area (e.g., heavy metal background reading and organic carbon content);

(vi) An estimate of the length of time during which disposal would continue at the proposed site; and

(vii) Information on the characteristics and composition of the dredged material.

(b) *Public notice for general permits.* District engineers will publish a public notice for all proposed regional general permits and for significant modifications to, or reissuance of, existing regional permits within their area of jurisdiction. Public notices for statewide regional permits may be issued jointly by the affected Corps districts. The notice will include all applicable information necessary to provide a clear understanding of the proposal. In addition, the notice will state the availability of information at the district office which reveals the Corps' provisional determination that the proposed activities comply with the requirements for issuance of general permits. District engineers will publish a public notice for nationwide permits in accordance with 33 CFR 330.4.

(c) *Evaluation factors.* A paragraph describing the various evaluation factors on which decisions are based shall be included in every public notice.

(1) Except as provided in paragraph (c)(3) of this section, the following will be included:

Corps of Engineers, Dept. of the Army, DoD   **§ 325.3**

"The decision whether to issue a permit will be based on an evaluation of the probable impact including cumulative impacts of the proposed activity on the public interest. That decision will reflect the national concern for both protection and utilization of important resources. The benefit which reasonably may be expected to accrue from the proposal must be balanced against its reasonably foreseeable detriments. All factors which may be relevant to the proposal will be considered including the cumulative effects thereof; among those are conservation, economics, aesthetics, general environmental concerns, wetlands, historic properties, fish and wildlife values, flood hazards, floodplain values, land use, navigation, shoreline erosion and accretion, recreation, water supply and conservation, water quality, energy needs, safety, food and fiber production, mineral needs, considerations of property ownership and, in general, the needs and welfare of the people."

(2) If the activity would involve the discharge of dredged or fill material into the waters of the United States or the transportation of dredged material for the purpose of disposing of it in ocean waters, the public notice shall also indicate that the evaluation of the impact of the activity on the public interest will include application of the guidelines promulgated by the Administrator, EPA, (40 CFR Part 230) or of the criteria established under authority of section 102(a) of the Marine Protection, Research and Sanctuaries Act of 1972, as amended (40 CFR Parts 220 to 229), as appropriate. (See 33 CFR Parts 323 and 324).

(3) In cases involving construction of artificial islands, installations and other devices on outer continental shelf lands which are under mineral lease from the Department of the Interior, the notice will contain the following statement: "The decision as to whether a permit will be issued will be based on an evaluation of the impact of the proposed work on navigation and national security."

(d) *Distribution of public notices.* (1) Public notices will be distributed for posting in post offices or other appropriate public places in the vicinity of the site of the proposed work and will be sent to the applicant, to appropriate city and county officials, to adjoining property owners, to appropriate state agencies, to appropriate Indian Tribes or tribal representatives, to concerned Federal agencies, to local, regional and national shipping and other concerned business and conservation organizations, to appropriate River Basin Commissions, to appropriate state and areawide clearing houses as prescribed by OMB Circular A-95, to local news media and to any other interested party. Copies of public notices will be sent to all parties who have specifically requested copies of public notices, to the U.S. Senators and Representatives for the area where the work is to be performed, the field representative of the Secretary of the Interior, the Regional Director of the Fish and Wildlife Service, the Regional Director of the National Park Service, the Regional Administrator of the Environmental Protection Agency (EPA), the Regional Director of the National Marine Fisheries Service of the National Oceanic and Atmospheric Administration (NOAA), the head of the state agency responsible for fish and wildlife resources, the State Historic Preservation Officer, and the District Commander, U.S. Coast Guard.

(2) In addition to the general distribution of public notices cited above, notices will be sent to other addressees in appropriate cases as follows:

(i) If the activity would involve structures or dredging along the shores of the seas or Great Lakes, to the Coastal Engineering Research Center, Washington, DC 20016.

(ii) If the activity would involve construction of fixed structures or artificial islands on the outer continental shelf or in the territorial seas, to the Assistant Secretary of Defense (Manpower, Installations, and Logistics (ASD(MI&L)), Washington, DC 20310; the Director, Defense Mapping Agency (Hydrographic Center) Washington, DC 20390, Attention, Code NS12; and the Charting and Geodetic Services, N/CG222, National Ocean Service NOAA, Rockville, Maryland 20852, and to affected military installations and activities.

(iii) If the activity involves the construction of structures to enhance fish propagation (e.g., fishing reefs) along the coasts of the United States, to the Director, Office of Marine Recreational Fisheries, National Marine Fisheries Service, Washington, DC 20235.

(iv) If the activity involves the construction of structures which may affect aircraft operations or for purposes associated with seaplane operations, to the Regional Director of the Federal Aviation Administration.

(v) If the activity would be in connection with a foreign-trade zone, to the Executive Secretary, Foreign-Trade Zones Board, Department of Commerce, Washington, DC 20230 and to the appropriate District Director of Customs as Resident Representative, Foreign-Trade Zones Board.

(3) It is presumed that all interested parties and agencies will wish to respond to public notices; therefore, a lack of response will be interpreted as meaning that there is no objection to the proposed project. A copy of the public notice with the list of the addresses to whom the notice was sent will be included in the record. If a question develops with respect to an activity for which another agency has responsibility and that other agency has not responded to the public notice, the district engineer may request its comments. Whenever a response to a public notice has been received from a member of Congress, either in behalf of a constitutent or himself, the district engineer will inform the member of Congress of the final decision.

(4) District engineers will update public notice mailing lists at least once every two years.

§ 325.4 Conditioning of permits.

(a) District engineers will add special conditions to Department of the Army permits when such conditions are necessary to satisfy legal requirements or to otherwise satisfy the public interest requirement. Permit conditions will be directly related to the impacts of the proposal, appropriate to the scope and degree of those impacts, and reasonably enforceable.

(1) Legal requirements which may be satisfied by means of Corps permit conditions include compliance with the 404(b)(1) guidelines, the EPA ocean dumping criteria, the Endangered Species Act, and requirements imposed by conditions on state section 401 water quality certifications.

(2) Where appropriate, the district engineer may take into account the existence of controls imposed under other federal, state, or local programs which would achieve the objective of the desired condition, or the existence of an enforceable agreement between the applicant and another party concerned with the resource in question, in determining whether a proposal complies with the 404(b)(1) guidelines, ocean dumping criteria, and other applicable statutes, and is not contrary to the public interest. In such cases, the Department of the Army permit will be conditioned to state that material changes in, or a failure to implement and enforce such program or agreement, will be grounds for modifying, suspending, or revoking the permit.

(3) Such conditions may be accomplished on-site, or may be accomplished off-site for mitigation of significant losses which are specifically identifiable, reasonably likely to occur, and of importance to the human or aquatic environment.

(b) District engineers are authorized to add special conditions, exclusive of paragraph (a) of this section, at the applicant's request or to clarify the permit application.

(c) If the district engineer determines that special conditions are necessary to insure the proposal will not be contrary to the public interest, but those conditions would not be reasonably implementable or enforceable, he will deny the permit.

(d) *Bonds.* If the district engineer has reason to consider that the permittee might be prevented from completing work which is necessary to protect the public interest, he may require the permittee to post a bond of sufficient amount to indemnify the government against any loss as a result of corrective action it might take.

§ 325.5 Forms of permits.

(a) *General discussion.* (1) DA permits under this regulation will be in the form of individual permits or general permits. The basic format shall be ENG Form 1721, DA Permit (Appendix A).

Corps of Engineers, Dept. of the Army, DoD § 325.6

(2) The general conditions included in ENG Form 1721 are normally applicable to all permits; however, some conditions may not apply to certain permits and may be deleted by the issuing officer. Special conditions applicable to the specific activity will be included in the permit as necessary to protect the public interest in accordance with § 325.4 of this Part.

(b) *Individual permits—*(1) *Standard permits.* A standard permit is one which has been processed through the public interest review procedures, including public notice and receipt of comments, described throughout this Part. The standard individual permit shall be issued using ENG Form 1721.

(2) *Letters of permission.* A letter of permission will be issued where procedures of § 325.2(e)(1) have been followed. It will be in letter form and will identify the permittee, the authorized work and location of the work, the statutory authority, any limitations on the work, a construction time limit and a requirement for a report of completed work. A copy of the relevant general conditions from ENG Form 1721 will be attached and will be incorporated by reference into the letter of permission.

(c) *General permits—*(1) *Regional permits.* Regional permits are a type of general permit. They may be issued by a division or district engineer after compliance with the other procedures of this regulation. If the public interest so requires, the issuing authority may condition the regional permit to require a case-by-case reporting and acknowledgment system. However, no separate applications or other authorization documents will be required.

(2) *Nationwide permits.* Nationwide permits are a type of general permit and represent DA authorizations that have been issued by the regulation (33 CFR Part 330) for certain specified activities nationwide. If certain conditions are met, the specified activities can take place without the need for an individual or regional permit.

(3) *Programmatic permits.* Programmatic permits are a type of general permit founded on an existing state, local or other Federal agency program and designed to avoid duplication with that program.

(d) *Section 9 permits.* Permits for structures in interstate navigable waters of the United States under section 9 of the Rivers and Harbors Act of 1899 will be drafted at DA level.

§ 325.6 **Duration of permits.**

(a) *General.* DA permits may authorize both the work and the resulting use. Permits continue in effect until they automatically expire or are modified, suspended, or revoked.

(b) *Structures.* Permits for the existence of a structure or other activity of a permanent nature are usually for an indefinite duration with no expiration date cited. However, where a temporary structure is authorized, or where restoration of a waterway is contemplated, the permit will be of limited duration with a definite expiration date.

(c) *Works.* Permits for construction work, discharge of dredged or fill material, or other activity and any construction period for a structure with a permit of indefinite duration under paragraph (b) of this section will specify time limits for completing the work or activity. The permit may also specify a date by which the work must be started, normally within one year from the date of issuance. The date will be established by the issuing official and will provide reasonable times based on the scope and nature of the work involved. Permits issued for the transport of dredged material for the purpose of disposing of it in ocean waters will specify a completion date for the disposal not to exceed three years from the date of permit issuance.

(d) *Extensions of time.* An authorization or construction period will automatically expire if the permittee fails to request and receive an extension of time. Extensions of time may be granted by the district engineer. The permittee must request the extension and explain the basis of the request, which will be granted unless the district engineer determines that an extension would be contrary to the public interest. Requests for extensions will be processed in accordance with the regular procedures of § 325.2 of this Part, including issuance of a public notice, except that such processing is not required where the district engineer determines that there have been no significant changes in the attendant circumstances since the authorization was issued.

(e) *Maintenance dredging.* If the authorized work includes periodic maintenance dredging, an expiration date for the authorization of that maintenance dredging will be included in the permit. The expiration date, which in no event is to exceed ten years from the date of issuance of the permit, will be established by the issuing official after evaluation of the proposed method of dredging and disposal of the dredged material in accordance with the requirements of 33 CFR Parts 320 to 325. In such cases, the district engineer shall require notification of the maintenance dredging prior to actual performance to insure continued compliance with the requirements of this regulation and 33 CFR Parts 320 to 324. If the permittee desires to continue maintenance dredging beyond the expiration date, he must request a new permit. The permittee should be advised to apply for the new permit six months prior to the time he wishes to do the maintenance work.

§ 325.7 Modification, suspension, or revocation of permits.

(a) *General.* The district engineer may reevaluate the circumstances and conditions of any permit, including regional permits, either on his own motion, at the request of the permittee, or a third party, or as the result of periodic progress inspections, and initiate action to modify, suspend, or revoke a permit as may be made necessary by considerations of the public interest. In the case of regional permits, this reevaluation may cover individual activities, categories of activities, or geographic areas. Among the factors to be considered are the extent of the permittee's compliance with the terms and conditions of the permit; whether or not circumstances relating to the authorized activity have changed since the permit was issued or extended, and the continuing adequacy of or need for the permit conditions; any significant objections to the authorized activity

which were not earlier considered; revisions to applicable statutory and/or regulatory authorities; and the extent to which modification, suspension, or other action would adversely affect plans, investments and actions the permittee has reasonably made or taken in reliance on the permit. Significant increases in scope of a permitted activity will be processed as new applications for permits in accordance with § 325.2 of this Part, and not as modifications under this section.

(b) *Modification.* Upon request by the permittee or, as a result of reevaluation of the circumstances and conditions of a permit, the district engineer may determine that the public interest requires a modification of the terms or conditions of the permit. In such cases, the district engineer will hold informal consultations with the permittee to ascertain whether the terms and conditions can be modified by mutual agreement. If a mutual agreement is reached on modification of the terms and conditions of the permit, the district engineer will give the permittee written notice of the modification, which will then become effective on such date as the district engineer may establish. In the event a mutual agreement cannot be reached by the district engineer and the permittee, the district engineer will proceed in accordance with paragraph (c) of this section if immediate suspension is warranted. In cases where immediate suspension is not warranted but the district engineer determines that the permit should be modified, he will notify the permittee of the proposed modification and reasons therefor, and that he may request a meeting with the district engineer and/or a public hearing. The modification will become effective on the date set by the district engineer which shall be at least ten days after receipt of the notice by the permittee unless a hearing or meeting is requested within that period. If the permittee fails or refuses to comply with the modification, the district engineer will proceed in accordance with 33 CFR Part 326. The district engineer shall consult with resource agencies before modifying any permit terms or conditions, that would result in greater impacts, for a project about which

452

that agency expressed a significant interest in the term, condition, or feature being modified prior to permit issuance.

(c) *Suspension.* The district engineer may suspend a permit after preparing a written determination and finding that immediate suspension would be in the public interest. The district engineer will notify the permittee in writing by the most expeditious means available that the permit has been suspended with the reasons therefor, and order the permittee to stop those activities previously authorized by the suspended permit. The permittee will also be advised that following this suspension a decision will be made to either reinstate, modify, or revoke the permit, and that he may within 10 days of receipt of notice of the suspension, request a meeting with the district engineer and/or a public hearing to present information in this matter. If a hearing is requested, the procedures prescribed in 33 CFR Part 327 will be followed. After the completion of the meeting or hearing (or within a reasonable period of time after issuance of the notice to the permittee that the permit has been suspended if no hearing or meeting is requested), the district engineer will take action to reinstate, modify, or revoke the permit.

(d) *Revocation.* Following completion of the suspension procedures in paragraph (c) of this section, if revocation of the permit is found to be in the public interest, the authority who made the decision on the original permit may revoke it. The permittee will be advised in writing of the final decision.

(e) *Regional permits.* The issuing official may, by following the procedures of this section, revoke regional permits for individual activities, categories of activities, or geographic areas. Where groups of permittees are involved, such as for categories of activities or geographic areas, the informal discussions provided in paragraph (b) of this section may be waived and any written notification nay be made through the general public notice procedures of this regulation. If a regional permit is revoked, any permittee may then apply for an individual permit which shall be processed in accordance with these regulations.

§ 325.8 **Authority to issue or deny permits.**

(a) *General.* Except as otherwise provided in this regulation, the Secretary of the Army, subject to such conditions as he or his authorized representative may from time to time impose, has authorized the Chief of Engineers and his authorized representatives to issue or deny permits for dams or dikes in intrastate waters of the United States pursuant to section 9 of the Rivers and Harbors Act of 1899; for construction or other work in or affecting navigable waters of the United States pursuant to section 10 of the Rivers and Harbors Act of 1899; for the discharge of dredged or fill material into waters of the United States pursuant to section 404 of the Clean Water Act; or for the transportation of dredged material for the purpose of disposing of it into ocean waters pursuant to section 103 of the Marine Protection, Research and Sanctuaries Act of 1972, as amended. The authority to issue or deny permits in interstate navigable waters of the United States pursuant to section 9 of the Rivers and Harbors Act of March 3, 1899 has not been delegated to the Chief of Engineers or his authorized representatives.

(b) *District engineer's authority.* District engineers are authorized to issue or deny permits in accordance with these regulations pursuant to sections 9 and 10 of the Rivers and Harbors Act of 1899; section 404 of the Clean Water Act; and section 103 of the Marine Protection, Research and Sanctuaries Act of 1972, as amended, in all cases not required to be referred to higher authority (see below). It is essential to the legality of a permit that it contain the name of the district engineer as the issuing officer. However, the permit need not be signed by the district engineer in person but may be signed for and in behalf of him by whomever he designates. In cases where permits are denied for reasons other than navigation or failure to obtain required local, state, or other federal approvals or certifications, the Statement of Findings must conclusively justify a denial decision. District engineers are authorized to deny permits without issuing a public notice or taking other procedural steps where required local, state, or other federal permits for the proposed activity have been denied or where he determines that the activity will clearly interfere with navigation except in all cases required to be referred to higher authority (see below). District engineers are also authorized to add, modify, or delete special conditions in permits in accordance with § 325.4 of this Part, except for those conditions which may have been imposed by higher authority, and to modify, suspend and revoke permits according to the procedures of § 325.7 of this Part. District engineers will refer the following applications to the division engineer for resolution:

(1) When a referral is required by a written agreement between the head of a Federal agency and the Secretary of the Army;

(2) When the recommended decision is contrary to the written position of the Governor of the state in which the work would be performed;

(3) When there is substantial doubt as to authority, law, regulations, or policies applicable to the proposed activity;

(4) When higher authority requests the application be forwarded for decision; or

(5) When the district engineer is precluded by law or procedures required by law from taking final action on the application (e.g. section 9 of the Rivers and Harbors Act of 1899, or territorial sea baseline changes).

(c) *Division engineer's authority.* Division engineers will review and evaluate all permit applications referred by district engineers. Division engineers may authorize the issuance or denial of permits pursuant to section 10 of the Rivers and Harbors Act of 1899; section 404 of the Clean Water Act; and section 103 of the Marine Protection, Research and Sanctuaries Act of 1972, as amended; and the inclusion of conditions in accordance with § 325.4 of this Part in all cases not required to be referred to the Chief of Engineers. Division engineers will refer the following applications to the Chief of Engineers for resolution:

(1) When a referral is required by a written agreement between the head of a Federal agency and the Secretary of the Army;

(2) When there is substantial doubt as to authority, law, regulations, or policies applicable to the proposed activity;

(3) When higher authority requests the application be forwarded for decision; or

(4) When the division engineer is precluded by law or procedures required by law from taking final action on the application.

§ 325.9　**Authority to determine jurisdiction.**

District engineers are authorized to determine the area defined by the terms "navigable waters of the United States" and "waters of the United States" except:

(a) When a determination of navigability is made pursuant to 33 CFR 329.14 (division engineers have this authority); or

(b) When EPA makes a section 404 jurisdiction determination under its authority.

§ 325.10　**Publicity.**

The district engineer will establish and maintain a program to assure that potential applicants for permits are informed of the requirements of this regulation and of the steps required to obtain permits for activities in waters of the United States or ocean waters. Whenever the district engineer becomes aware of plans being developed by either private or public entities which might require permits for implementation, he should advise the potential applicant in writing of the statutory requirements and the provisions of this regulation. Whenever the district engineer is aware of changes in Corps of Engineers regulatory jurisdiction, he will issue appropriate public notices.

PART 335—OPERATION AND MAINTENANCE OF ARMY CORPS OF ENGINEERS CIVIL WORKS PROJECTS INVOLVING THE DISCHARGE OF DREDGED OR FILL MATERIAL INTO WATERS OF THE U.S. OR OCEAN WATERS

AUTHORITY: 33 U.S.C. 1344; 33 U.S.C. 1413.

SOURCE: 53 FR 14911, Apr. 26, 1988, unless otherwise noted.

§ 335.1 Purpose.

This regulation prescribes the practices and procedures to be followed by the Corps of Engineers to ensure compliance with the specific statutes governing Army Civil Works operations and maintenance projects involving the discharge of dredged or fill material into waters of the U.S. or the transportation of dredged material for the purpose of disposal into ocean waters. These practices and procedures should be employed throughout the decision/management process concerning methodologies and alternatives to be used to ensure prudent operation and maintenance activities.

§ 335.2 Authority.

Under authority delegated from the Secretary of the Army and in accordance with section 404 of the Clean Water Act of 1977 (CWA) and section 103 of the Marine Protection, Research, and Sanctuaries Act of 1972, hereinafter referred to as the Ocean Dumping Act (ODA), the Corps of Engineers regulates the discharge of dredged or fill material into waters of the United States and the transportation of dredged material for the purpose of disposal into ocean waters. Section 404 of the CWA requires public notice with opportunity for public hearing for discharges of dredged or fill material into waters of the U.S. and that discharge sites can be specified through the application of guidelines developed by the Administrator of the Environmental Protection Agency (EPA) in conjunction with the Secretary of the Army. Section 103 of the ODA requires public notice with opportunity for public hearing for the transportation for disposal of dredged material for disposal in ocean waters. Ocean disposal of dredged material must be evaluated using the criteria developed by the Administrator of EPA in consultation with the Secretary of the Army. Section 103(e) of the ODA provides that the Secretary of the Army may, in lieu of permit procedures, issue regulations for Federal projects involving the transportation of dredged material for ocean disposal which require the application of the same criteria, procedures, and requirements which apply to the issuance of permits. Similarly, the Corps does not issue itself a CWA permit to authorize Corps discharges of dredged material or fill material into U.S. waters, but does apply the 404(b)(1) guidelines and other substantive requirements of the CWA and other environmental laws.

§ 335.3 Applicability.

This regulation (33 CFR Parts 335 through 338) is applicable to the Corps of Engineers when undertaking operation and maintenance activities at Army Civil Works projects.

§ 335.4 Policy.

The Corps of Engineers undertakes operations and maintenance activities where appropriate and environmentally acceptable. All practicable and reasonable alternatives are fully considered on an equal basis. This includes the discharge of dredged or fill material into waters of the U.S. or ocean waters in the least costly manner, at the least costly and most practicable location, and consistent with engineering and environmental requirements.

§ 335.5 Applicable laws.

(a) The Clean Water Act (33 U.S.C. 1251 et seq.) (also known as the Federal Water Pollution Control Act Amendments of 1972, 1977, and 1987).

(b) The Marine Protection, Research, and Sanctuaries Act of 1972 (33 U.S.C. 1401 et seq.) (commonly referred to as the Ocean Dumping Act (ODA)).

§ 335.6 Related laws and Executive Orders.

(a) The National Historic Preservation Act of 1966 (16 U.S.C. 470a et seq.), as amended.

(b) The Reservoir Salvage Act of 1960 (16 U.S.C. 469), as amended.

(c) The Endangered Species Act (16 U.S.C. 1531 et seq.), as amended.

(d) The Estuary Protection Act (16 U.S.C. 1221).

(e) The Fish and Wildlife Coordination Act (16 U.S.C. 661 et seq.), as amended.

(f) The National Environmental Policy Act (42 U.S.C. 4341 et seq.), as amended.

(g) The Wild and Scenic Rivers Act (16 U.S.C. 1271 et seq.) as amended.

(h) Section 307(c) of the Coastal Zone Management Act of 1976 (16 U.S.C. 1456 (c)), as amended.

(i) The Water Resources Development Act of 1976 (Pub. L. 94-587).

(j) Executive Order 11593, *Protection and Enhancement of the Cultural Environment*, May 13, 1971, (36 FR 8921, May 15, 1971).

(k) Executive Order 11988, *Floodplain Management*, May 24, 1977, (42 FR 26951, May 25, 1977).

(l) Executive Order 11990, *Protection of Wetlands*, May 24, 1977, (42 FR 26961, May 25, 1977).

(m) Executive Order 12372, *Intergovernmental Review of Federal Programs*, July 14, 1982, (47 FR 3959, July 16, 1982).

(n) Executive Order 12114, *Environmental Effects Abroad of Major Federal Actions*, January 4, 1979.

§ 335.7 Definitions.

The definitions of 33 CFR Parts 323, 324, 327, and 329 are hereby incorporated. The following terms are defined or interpreted from Parts 320 through 330 for purposes of 33 CFR Parts 335 through 338.

"Beach nourishment" means the discharge of dredged or fill material for the purpose of replenishing an eroded beach or placing sediments in the littoral transport process.

"Emergency" means a situation which would result in an unacceptable hazard to life or navigation, a significant loss of property, or an immediate and unforeseen significant economic hardship if corrective action is not taken within a time period less than the normal time needed under standard procedures.

"Federal standard" means the dredged material disposal alternative or alternatives identified by the Corps which represent the least costly alternatives consistent with sound engineering practices and meeting the environmental standards established by the 404(b)(1) evaluation process or ocean dumping criteria.

"Navigable waters of the U.S." means those waters of the U.S. that are subject to the ebb and flow of the tide shoreward to the mean high water mark, and/or are presently used, have been used in the past, or may be susceptible to use with or without reasonable improvement to transport interstate or foreign commerce. A more complete definition is provided in 33 CFR Part 329. For the purpose of this regulation, the term also includes the confines of Federal navigation approach channels extending into ocean waters beyond the territorial sea which are used for interstate or foreign commerce.

Part 336

"Practicable" means available and capable of being done after taking into consideration cost, existing technology, and logistics in light of overall project purposes.

"Statement of Findings (SOF)" means a comprehensive summary compliance document signed by the district engineer after completion of appropriate environmental documentation and public involvement.

"Territorial sea" means the belt of the seas measured from the line of ordinary low water along that portion of the coast which is in direct contact with the open sea and the line marking the seaward limit of inland waters, extending seaward a distance of three miles as described in the convention on the territorial sea and contiguous zone, 15 U.S.T. 1606.

PART 336—FACTORS TO BE CONSIDERED IN THE EVALUATION OF ARMY CORPS OF ENGINEERS DREDGING PROJECTS INVOLVING THE DISCHARGE OF DREDGED MATERIAL INTO WATERS OF THE U.S. AND OCEAN WATERS

Sec.
336.0 General.
336.1 Discharges of dredged or fill material into waters of the U.S.
336.2 Transportation of dredged material for the purpose of disposal into ocean waters.

AUTHORITY: 33 U.S.C. 1344; 33 U.S.C. 1413.

SOURCE: 53 FR 14912, Apr. 26, 1988, unless otherwise noted.

§ 336.0 General.

Since the jurisdiction of the CWA extends to all waters of the U.S., including the territorial sea, and the jurisdiction of the ODA extends over ocean waters including the territorial sea, the following rules are established to assure appropriate regulation of discharges of dredged or fill material into waters of the U.S. and ocean waters.

(a) The disposal into ocean waters, including the territorial sea, of dredged material excavated or dredged from navigable waters of the U.S. will be evaluated by the Corps in accordance with the ODA.

(b) In those cases where the district engineer determines that the discharge of dredged material into the territorial sea would be for the primary purpose of fill, such as the use of dredged material for beach nourishment, island creation, or construction of underwater berms, the discharge will be evaluated under section 404 of the CWA.

(c) For those cases where the district engineer determines that the materials proposed for discharge in the territorial sea would not be adequately evaluated under the section 404(b)(1) guidelines of the CWA, he may evaluate that material under the ODA.

§ 336.1 Discharges of dredged or fill material into waters of the U.S.

(a) *Applicable laws.* Section 404 of the CWA governs the discharge of dredged or fill material into waters of the U.S. Although the Corps does not process and issue permits for its own activities, the Corps authorizes its own discharges of dredged or fill material by applying all applicable substantive legal requirements, including public notice, opportunity for public hearing, and application of the section 404(b)(1) guidelines.

(1) The CWA requires the Corps to seek state water quality certification for discharges of dredged or fill material into waters of the U.S.

(2) Section 307 of the Coastal Zone Management Act (CZMA) requires that certain activities that a Federal agency conducts or supports be consistent with the Federally-approved state management plan to the maximum extent practicable.

(b) *Procedures.* If changes in a previously approved disposal plan for a Corps navigation project warrant reevaluation under the CWA, the following procedures should be followed by district enginers prior to discharging dredged material into waters of the U.S. except where emergency action as described in § 337.7 of this chapter is required.

(1) A public notice providing opportunity for a public hearing should be issued at the earliest practicable time. The public notification procedures of

Corps of Engineers, Dept. of the Army, DoD   **§ 336.1**

§ 337.1 of this chapter should be followed.

(2) The public hearing procedures of 33 CFR Part 327 should be followed.

(3) As soon as practicable, the district engineer will request from the state a 401 water quality certification and, if applicable, provide a coastal zone consistency determination for the Corps activity using the procedures of § 336.1(b) (8) and (9), respectively, of this part.

(4) Discharges of dredged material will be evaluated using the guidelines authorized under section 404(b)(1) of the CWA, or using the ODA regulations, where appropriate. If the guidelines alone would prohibit the designation of a proposed discharge site, the economic impact on navigation and anchorage of the failure to use the proposed discharge site will also be considered in evaluating whether the proposed discharge is to be authorized under CWA section 404(b)(2).

(5) The EPA Administrator can prohibit or restrict the use of any defined area as a discharge site under 404(c) whenever he determines, after notice and opportunity for public hearing and after consultation with the Secretary of the Army, that the discharge of such materials into such areas will have an unacceptable adverse effect on municipal water supplies, shellfish beds and fishery areas, wildlife, or recreation areas. Upon notification of the prohibition of a discharge site by the Administrator the district engineer will complete the administrative processing of the proposed project up to the point of signing the Statement of Findings (SOF) or Record of Decision (ROD). The unsigned SOF or ROD along with a report described in § 337.8 of this chapter will be forwarded through the appropriate Division office to the Dredging Division, Office of the Chief of Engineers.

(6) In accordance with the National Environmental Policy Act (NEPA), and the regulations of the Council on Environmental Quality (40 CFR Parts 1500-1508), an Environmental Impact Statement (EIS) or Environmental Assessment (EA) will be prepared for all Corps of Engineers projects involving the discharge of dredged or fill material, unless such projects are included

within a categorical exclusion found at 33 CFR Part 230 or addressed within an existing EA or EIS. If a proposed maintenance activity will result in a deviation in the operation and maintenance plan as described in the EA or EIS, the district engineer will determine the need to prepare a new EA, EIS, or supplement. If a new EA, EIS, or supplement is required, the procedures of 33 CFR Part 230 will be followed.

(7) If it can be anticipated that related work by other Federal or non-Federal interests will occur in the same area as Corps projects, the district engineer should use all reasonable means to include it in the planning, processing, and review of Corps projects. Related work normally includes, but is not necessarily limited to, maintenance dredging of approach channels and berthing areas connected to Federal navigation channels. The district engineer should coordinate the related work with interested Federal, state, regional, and local agencies and the general public at the same time he does so for the Corps project. The district engineer should ensure that related work meets all substantive and procedural requirements of 33 CFR Parts 320 through 330. Documents covering Corps maintenance activities normally should also include an appropriate discussion of ancillary maintenance work. District engineers should assist local interests to obtain from the state any necessary section 401 water quality certification and, if required, the section 307 coastal zone consistency concurrence. The absence of such certification or concurrence by the state or the denial of a Corps permit for related work shall not be cause for delay of the Federal project. Local sponsors will be responsible for funding any related work. If permitting of the related work complies with all legal requirements and is not contrary to the public interest, section 10, 404, and 103 permits normally will be issued by the district engineer in a separate SOF or ROD. Authorization by nationwide or regional general permit may be appropriate. If the related work does not receive a necessary state water quality certification and/or CZMA consistency concurrence, or are determined to

be contrary to the public interest the district engineer should re-examine the project viability to ensure that continued maintenance is warranted.

(8) *State water quality certification:* Section 401 of the CWA requires the Corps to seek state water quality certification for dredged material disposal into waters of the U.S. The state certification request must be processed to a conclusion by the state within a reasonable period of time. Otherwise, the certification requirements of section 401 are deemed waived. The district engineer will request water quality certification from the state at the earliest practicable time using the following procedures:

(i) In addition to the Corps section 404 public notice, information and data demonstrating compliance with state water quality standards will be provided to the state water quality certifying agency along with the request for water quality certification. The information and data may be included within the 404(b)(1) evaluation. The district engineer will request water quality certification to be consistent with the maintenance dredging schedule for the project. Submission of the public notice, including information and data demonstrating compliance with the state water quality standards, will constitute a valid water quality certification request pursuant to section 401 of the CWA.

(ii) If the proposed disposal activity may violate state water quality standards, after consideration of disposal site dilution and dispersion, the district engineer will work with the state to acquire data to satisfy compliance with the state water quality standards. The district engineer will use the technical manual "Management Strategy for Disposal of Dredged Material: Contaminant Testing and Controls" or its appropriate updated version as a guide for developing the appropriate tests to be conducted on such dredged material.

(iii) If the state does not take final action on a request for water quality certification within two months from the date of the initial request, the district engineer will notify the state of his intention to presume a waiver as provided by section 401 of the CWA. If the state agency, within the two-month period, requests an extension of time, the district engineer may approve one 30-day extension unless, in his opinion, the magnitude and complexity of the information contained in the request warrants a longer or additional extension period. The total period of time in which the state must act should not exceed six months from the date of the initial request. Waiver of water quality certification can be conclusively presumed after six months from the date of the initial request.

(iv) The procedures of § 337.2 will be followed if the district engineer determines that the state data acquisition requirements exceed those necessary in establishment of the Federal standard.

(9) *State coastal zone consistency:* Section 307 of the CZMA requires that activities subject to the CZMA which a Federal agency conducts or supports be consistent with the Federally approved state management program to the maximum extent practicable. The state is provided a reasonable period of time as defined in § 336.1(b)(9)(iv) to take final action on Federal consistency determinations; otherwise state concurrence can be presumed. The district engineer will provide the state a consistency determination at the earliest practicable time using the following procedures:

(i) The Corps section 404 public notice and any additional information that the district engineer determines to be appropriate will be provided the state coastal zone management agency along with the consistency determination. The consistency determination will consider the maintenance dredging schedule for the project. Submission of the public notice and, as appropriate, any additional information as determined by the district engineer will constitute a valid coastal zone consistency determination pursuant to section 307 of the CZMA.

(ii) If the district engineer decides that a consistency determination is not required for a Corps activity, he may provide the state agency a written determination that the CZMA does not apply.

(iii) The district engineer may provide the state agency a general consistency determination for routine or repetitive activities.

(iv) If the state fails to provide a response within 45 days from receipt of the initial consistency determination, the district engineer will presume state agency concurrence. If the state agency, within the 45-day period, requests an extension of time, the district engineer will approve one 15-day extension unless, in his opinion, the magnitude and complexity of the information contained in the consistency determination warrants a longer or additional extension period. The longer or additional extension period shall not exceed six months from the date of the initial consistency determination.

(v) If the district engineer determines that the state recommendations to achieve consistency to the maximum degree practicable exceed either his authority or funding for a proposed dredging or disposal activity, he will so notify the state coastal zone management agency indicating that the Corps has complied to the maximum extent practicable with the state's coastal zone management program. If the district engineer determines that state recommendations to achieve consistency to the maximum degree practicable do not exceed his authority or funding but, nonetheless, are excessive, he will follow the procedures of § 337.2.

(c) *Evaluation factors.* The following factors will be used, as appropriate, to evaluate the discharge of dredged material into waters of the U.S. Other relevant factors may also be evaluated, as needed.

(1) *Navigation and Federal standard.* The maintenance of a reliable Federal navigation system is essential to the economic well-being and national defense of the country. The district engineer will give full consideration to the impact of the failure to maintain navigation channels on the national and, as appropriate, regional economy. It is the Corps' policy to regulate the discharge of dredged material from its projects to assure that dredged material disposal occurs in the least costly, environmentally acceptable manner,

consistent with engineering requirements established for the project. The environmental assessment or environmental impact statement, in conjunction with the section 404(b)(1) guidelines and public notice coordination process, can be used as a guide in formulating environmentally acceptable alternatives. The least costly alternative, consistent with sound engineering practices and selected through the 404(b)(1) guidelines or ocean disposal criteria, will be designated the Federal standard for the proposed project.

(2) *Water quality.* The 404(b)(1) guidelines at 40 CFR Part 230 and ocean dumping criteria at 40 CFR Part 220 implement the environmental protection provisions of the CWA and ODA, respectively. These guidelines and criteria provide general regulatory guidance and objectives, but not a specific technical framework for evaluating or managing contaminated sediment that must be dredged. Through the section 404(b)(1) evaluation process (or ocean disposal criteria for the territorial sea), the district engineer will evaluate the water quality impacts of the proposed project. The evaluation will include consideration of state water quality standards. If the district engineer determines the dredged material to be contaminated, he will follow the guidance provided in the most current published version of the technical manual for contaminant testing and controls. This manual is currently cited as: Francingues, N.R., Jr., et al. 1985. "Management Strategy for Disposal of Dredged Material: Contaminant Testing and Controls," Miscellaneous Paper D-85-1, U.S. Army Waterways Experiment Station, Vicksburg, Mississippi. The procedures of § 336.1(b)(8) will be followed for state water quality certification requests.

(3) *Coastal zone consistency.* As appropriate, the district engineer will determine whether the proposed project is consistent with the state coastal zone management program to the maximum extent practicable. The procedures of § 336.1(b)(9) will be followed for coastal zone consistency determinations.

(4) *Wetlands.* Most wetland areas constitute a productive and valuable public resource, the unnecessary alter-

ation or destruction of which should be discouraged as contrary to the public interest. The district engineer will, therefore, follow the guidance in 33 CFR 320.4(b) and EO 11990, dated May 24, 1977, when evaluating Corps operations and maintenance activities in wetlands.

(5) *Endangered species.* All Corps operations and maintenance activities will be reviewed for the potential impact on threatened or endangered species, pursuant to the Endangered Species Act of 1973. If the district engineer determines that the proposed activity will not affect listed species or their critical habitat, a statement to this effect should be included in the public notice. If the proposed activity may affect listed species or their critical habitat, appropriate discussions will be initiated with the U.S. Fish and Wildlife Service or National Marine Fisheries Service, and a statement to this effect should be included in the public notice. (See 50 CFR Part 402).

(6) *Historic resources.* Archeological, historical, or architectural resource surveys may be required to locate and identify previously unrecorded historic properties in navigation channels and at dredged or fill material disposal sites. If properties that may be historic are known or found to exist within the navigation channel or proposed disposal area, field testing and analysis may sometimes be necessary in order to evaluate the properties against the criteria of the National Register of Historic Places. Such testing should be limited to the amount and kind needed to determine eligibility for the National Register; more detailed and extensive work on a property may be prescribed later, as the outcome of review under section 106 of the National Historic Preservation Act. Historic properties are not normally found in previously constructed navigation channels or previously used disposal areas. Therefore, surveys to identify historic properties should not be conducted for maintenance dredging and disposal activities proposed within the boundaries of previously constructed navigation channels or previously used disposal areas unless there is good reason to believe that historic properties exist there.

(i) The district engineer will establish whether historic properties located in navigation channels or at disposal sites are eligible for inclusion in the National Register of Historic Places in accordance with applicable regulations of the Advisory Council on Historic Preservation and the Department of the Interior.

(ii) The district engineer will take into account the effects of any proposed actions on properties included in or eligible for inclusion in the National Register of Historic Places, and will request the comments of the Advisory Council on Historic Preservation, in accordance with applicable regulations of the Advisory Council on Historic Preservation.

(7) *Scenic and recreational values.* (i) Maintenance dredging and disposal activities may involve areas which possess recognized scenic, recreational, or similar values. Full evaluation requires that due consideration be given to the effect which dredging and disposal of the dredged or fill material may have on the enhancement, preservation, or development of such values. Recognition of these values is often reflected by state, regional, or local land use classification or by similar Federal controls or policies. Operations and maintenance activities should, insofar as possible, be consistent with and avoid adverse effects on the values or purposes for which such resources have been recognized or set aside, and for which those classifications, controls, or policies were established. Special consideration must be given to rivers named in section 3 of the Wild and Scenic Rivers Act and those proposed for inclusion as provided by section 4 and 5 of the Act, or by later legislation.

(ii) Any other areas named in Acts of Congress or Presidential Proclamations, such as National Rivers, National Wilderness Areas, National Seashores, National Parks, and National Monuments, should be given full consideration when evaluating Corps operations and maintenance activities.

(8) *Fish and wildlife.* (i) In those cases where the Fish and Wildlife Coordination Act (FWCA) applies, district engineers will consult, through the public notification process, with

the Regional Directors of the U.S. Fish and Wildlife Service and the National Marine Fisheries Service and the head of the agency responsible for fish and wildlife for the state in which the work is to be performed, with a view to the conservation of fish and wildlife resources by considering ways to prevent their direct and indirect loss and damage due to the proposed operation and maintenance activity. The district engineer will give full consideration to these views on fish and wildlife conservation in evaluating the activity. The proposed operations may be modified in order to lessen the damage to such resources. The district engineer should include such justifiable means and measures for fish and wildlife resources that are found to be appropriate. Corps funding of Fish and Wildlife Service activities under the Transfer of Funds Agreement between the Fish and Wildlife Service and the Corps is not applicable for Corps operation and maintenance projects.

(ii) District engineers should consider ways of reducing unavoidable adverse environmental impacts of dredging and disposal activities. The determination as to the extent of implementation of such measures will be done by the district engineer after weighing the benefits and detriments of the maintenance work and considering applicable environmental laws, regulations, and other relevant factors.

(9) *Marine sanctuaries.* Operations and maintenance activities involving the discharge of dredged or fill material in a marine sanctuary established by the Secretary of Commerce under authority of section 302 of the ODA should be evaluated for the impact on the marine sanctuary. In such a case, certification should be obtained from the Secretary of Commerce that the proposed project is consistent with the purposes of Title III of the ODA and can be carried out within the regulations promulgated by the Secretary of Commerce to control activities within the marine sanctuary.

(10) *Other state requirements.* District engineers will make all reasonable efforts to comply with state water quality standards and Federally approved coastal zone programs using the procedures of §§ 336.1(b) (8), (9), and 337.2. District engineers should not seek state permits or licenses unless authorized to do so by a clear, explicit, and unambiguous Congressional waiver of Federal sovereign immunity, giving the state authority to impose that requirement on Federal activities (e.g., CWA sections 401 and 404(t), and CZMA section 307 (c)(1) and (c)(2)).

(11) *Additional factors.* In addition to the factors described in paragraphs (c)(1) through (9) of this section, the following factors should also be considered.

(i) The evaluation of Corps operations and maintenance activities involving the discharge of dredged or fill material into waters of the U.S. is a continuing process and should proceed concurrently with the processing of state water quality certification and, if required, the provision of a coastal zone consistency determination to the state. If a local agency having jurisdiction over or concern with the particular activity comments on the project through the public notice coordination, due consideration should be given to those official views as a reflection of local factors.

(ii) Where officially adopted state, regional, or local land use classifications, determinations, or policies are applicable, they normally will be presumed to reflect local views and will be considered in addition to other national factors.

§ 336.2　**Transportation of dredged material for the purpose of disposal into ocean waters.**

(a) *Applicable law.* Section 103(a) of the ODA provides that the Corps of Engineers may issue permits, after notice and opportunity for public hearing, for the transportation of dredged material for disposal into ocean waters.

(b) *Procedures.* The following procedures will be followed by district engineers for dredged material disposal into ocean waters except where emergency action as described in § 337.7 of this chapter is required.

(1) In accordance with the provisions of section 103 of the ODA, the district engineer should issue a public notice giving opportunity for public hearing, following the procedures described in § 337.1 of this chapter for Corps operation and maintenance activities involving disposal of dredged material in ocean waters, as well as dredged material transported through the territorial sea for ocean disposal.

(2) The public hearing procedures of 33 CFR Part 327 should be followed.

(c) *State permits and licenses.* The terms and legislative history of the ODA leave some doubt regarding whether a state has legal authority to exert control over ocean dumping activities of the Corps in the territorial sea covered under the Act (see section 106(d)). Notwithstanding this legal question, the Corps will voluntarily as a matter of comity apply for state section 401 water quality certification and determine consistency with a Federally-approved coastal zone management plan for Corps ocean disposal of dredged material within the three-mile extent of the territorial sea. Moreover, the Corps will attempt to comply with any reasonable requirement imposed by a state in the course of the 401 certification process or the CZMA consistency determination process. Nevertheless, the Corps reserves its legal rights regarding any case where a state unreasonably denies or conditions a 401 water quality certification for proposed Corps ocean disposal of dredged material within the limits of the territorial sea, or asserts that such disposal would not be consistent with an approved state CZMA plan. If such a circumstance arises, the district engineer shall so notify the division engineer who then decides on consultation with CECW-D, CECW-Z, and CECC-E for purposes of determining the Corps of Engineers' appropriate response and course of action.

(d) *Evaluation factors.* (1) In addition to the appropriate evaluation factors of § 336.1(c), activities involving the transportation of dredged material for the purpose of disposal in ocean waters will be evaluated by the Corps to determine whether the proposed disposal will unreasonably degrade or endanger human health, welfare, or amenities, or the marine environment, ecological systems or economic potentialities. In making this evaluation, the district engineer, in addition to considering the criteria developed by EPA on the effects of the dumping, will also consider navigation, economic and industrial development, and foreign and domestic commerce, as well as the availability of alternatives to ocean disposal, in determining the need for ocean disposal of dredged material. Where ocean disposal is determined to be appropriate, the district engineer will, to the extent feasible, specify disposal sites which have been designated by the Administrator pursuant to section 102(c) of the ODA.

(2) As provided by the EPA regulations at 40 CFR 225.2(b-e) for implementing the procedures of section 102 of the ODA, the regional administrator of EPA may make an independent evaluation of dredged material disposal activities regulated under section 103 of the ODA related to the effects of dumping. The EPA regulations provide that the regional administrator make said evaluation within 15 days after receipt of all requested information. The regional administrator may request from the district engineer an additional 15-day period for a total of to 30 days. The EPA regulations provide that the regional administrator notify the district engineer of non-compliance with the environmental impact criteria or with any restriction relating to critical areas on the use of an EPA recommended disposal site designated pursuant to section 102(c) of the ODA. In cases where the regional administrator has notified the district engineer in writing that the proposed disposal will not comply with the criteria related to the effects of dumping or related to critical area restriction, no dredged material disposal may occur unless and until the provisions of 40 CFR 225.3 are followed and the Administrator grants a waiver of the criteria pursuant to section 103(d) of the ODA.

(3) If the regional administrator advises the district engineer that the proposed disposal will comply with the criteria, the district engineer will com-

Corps of Engineers, Dept. of the Army, DoD   § 337.1

plete the administrative record and sign the SOF.

(4) In situations where an EPA-designated site is not feasible for use or where no site has been designated by the EPA, the district engineer, in accordance with the ODA and in consultation with EPA, may select a site pursuant to section 103. Appropriate NEPA documentation should be used to support site selections. District engineers should address site selection factors in the NEPA document. District engineers will consider the criteria of 40 CFR Parts 227 and 228 when selecting ocean disposal sites, as well as other technical and economic considerations. Emphasis will be placed on evaluation to determine the need for ocean disposal and other available alternatives. Each alternative should be fully considered on an equal basis, including the no dredging option.

(5) If the regional administrator advises the district engineer that a proposed ocean disposal site or activity will not comply with the criteria, the district engineer should proceed as follows.

(i) The district engineer should determine whether there is an economically feasible alternative method or site available other than the proposed ocean disposal site. If there are other feasible alternative methods or sites available, the district engineer will evaluate the engineering and economic feasibility and environmental acceptability of the alternative sites.

(ii) If the district engineer makes a determination that there is no economically feasible alternative method or site available, he will so advise the regional administrator of his intent to proceed with the proposed action setting forth his reasons for such determination.

(iii) If the regional administrator advises, within 15 days of the notice of the intent to issue, that he will commence procedures specified by section 103(c) of the ODA to prohibit use of a proposed disposal site, the case will be forwarded through the respective Division office and CECW-D to the Secretary of the Army or his designee for further coordination with the Administrator of EPA and final resolution. The report forwarding the case should be in the format described in § 337.8 of this chapter.

(iv) The Secretary of the Army or his designee will evaluate the proposed project and make a final determination on the proposed disposal. If the decision of the Secretary of the Army or his designee is that ocean disposal at the proposed site is required because of the unavailability of economically feasible alternatives, he will seek a waiver from the Administrator, EPA, of the criteria or of the critical site designation in accordance with section 103(d) of the ODA.

589

PART 220—GENERAL

Sec.
220.1 Purpose and scope.
220.2 Definitions.
220.3 Categories of permits.
220.4 Authorities to issue permits.

AUTHORITY: 33 U.S.C. 1412 and 1418.

SOURCE: 42 FR 2468, Jan. 11, 1977, unless otherwise noted.

PART 225—CORPS OF ENGINEERS DREDGED MATERIAL PERMITS

Sec.
225.1 General.
225.2 Review of Dredged Material Permits.
225.3 Procedure for invoking economic impact.
225.4 Waiver by Administrator.

AUTHORITY: 33 U.S.C. 1412 and 1418.

SOURCE: 42 FR 2475, Jan. 11, 1977, unless otherwise noted.

§ 225.1 General.

Applications and authorizations for Dredged Material Permits under section 103 of the Act for the transportation of dredged material for the purpose of dumping it in ocean waters will be evaluated by the U.S. Army Corps of Engineers in accordance with the criteria set forth in Part 227 and processed in accordance with 33 CFR 209.120 with special attention to § 209.120(g)(17) and 33 CFR 209.145.

§ 225.2 Review of Dredged Material Permits.

(a) The District Engineer shall send a copy of the public notice to the appropriate Regional Administrator, and set forth in writing all of the following information:

(1) The location of the proposed disposal site and its physical boundaries;

(2) A statement as to whether the site has been designated for use by the Administrator pursuant to section 102(c) of the Act;

(3) If the proposed disposal site has not been designated by the Administrator, a statement of the basis for the proposed determination why no previously designated site is feasible and a description of the characteristics of the proposed disposal site necessary for its designation pursuant to Part 228 of this Subchapter H;

(4) The known historical uses of the proposed disposal site;

(5) Existence and documented effects of other authorized dumpings that have been made in the dumping

Environmental Protection Agency

area (e.g., heavy metal background reading and organic carbon content);

(6) An estimate of the length of time during which disposal will continue at the proposed site;

(7) Characteristics and composition of the dredged material; and

(8) A statement concerning a preliminary determination of the need for and/or availability of an environmental impact statement.

(b) The Regional Administrator will within 15 days of the date the public notice and other information required to be submitted by paragraph (a) of § 225.2 are received by him, review the information submitted and request from the District Engineer any additional information he deems necessary or appropriate to evaluate the proposed dumping.

(c) Using the information submitted by the District Engineer, and any other information available to him, the Regional Administrator will within 15 days after receipt of all requested information, make an independent evaluation of the proposed dumping in accordance with the criteria and respond to the District Engineer pursuant to paragraph (d) or (e) of this section. The Regional Administrator may request an extension of this 15 day period to 30 days from the District Engineer.

(d) When the Regional Administrator determines that the proposed dumping will comply with the criteria, he will so inform the District Engineer in writing.

(e) When the Regional Administrator determines that the proposed dumping will not comply with the criteria he shall so inform the District Engineer in writing. In such cases, no Dredged Material Permit for such dumping shall be issued unless and until the provisions of § 225.3 are followed and the Administrator grants a waiver of the criteria pursuant to § 225.4.

§ 225.3 Procedure for invoking economic impact.

(a) When a District Engineer's determination to issue a Dredged Material Permit for the dumping of dredged material into ocean waters has been

161

rejected by a Regional Administrator upon application of the Criteria, the District Engineer may determine whether, under section 103(d) of the Act, there is an economically feasible alternative method or site available other than the proposed dumping in ocean waters. If the District Engineer makes any such preliminary determination that there is no economically feasible alternative method or site available, he shall so advise the Regional Administrator setting forth his reasons for such determination and shall submit a report of such determination to the Chief of Engineers in accordance with 33 CFR 209.120 and 209.145.

(b) If the decision of the Chief of Engineers is that ocean dumping at the designated site is required because of the unavailability of feasible alternatives, he shall so certify and request that the Secretary of the Army seek a waiver from the Administrator of the Criteria or of the critical site designation in accordance with § 225.4.

§ 225.4　Waiver by Administrator.

The Administrator shall grant the requested waiver unless within 30 days of his receipt of the notice, certificate and request in accordance with paragraph (b) of § 225.3 he determines in accordance with this section that the proposed dumping will have an unacceptable adverse effect on municipal water supplies, shellfish beds and fishery areas (including spawning and breeding areas), wildlife, or recreational areas. Notice of the Administrator's final determination under this section shall be given to the Secretary of the Army.

PART 227—CRITERIA FOR THE EVALUATION OF PERMIT APPLICATIONS FOR OCEAN DUMPING OF MATERIALS

Subpart A—General

Sec.
227.1　Applicability.
227.2　Materials which satisfy the environmental impact criteria of Subpart B.

Sec.
227.3　Materials which do not satisfy the environmental impact criteria set forth in Subpart B.

Subpart B—Environmental Impact

227.4　Criteria for evaluating environmental impact.
227.5　Prohibited materials.
227.6　Constituents prohibited as other than trace contaminants.
227.7　Limits established for specific wastes or waste constituents.
227.8　Limitations on the disposal rates of toxic wastes.
227.9　Limitations on quantities of waste materials.
227.10　Hazards to fishing, navigation, shorelines or beaches.
227.11　Containerized wastes.
227.12　Insoluble wastes.
227.13　Dredged materials.

Subpart C—Need for Ocean Dumping

227.14　Criteria for evaluating the need for ocean dumping and alternatives to ocean dumping.
227.15　Factors considered.
227.16　Basis for determination of need for ocean dumping.

Subpart D—Impact of the Proposed Dumping on Esthetic, Recreational and Economic Values

227.17　Basis for determination.
227.18　Factors considered.
227.19　Assessment of impact.

Subpart E—Impact of the Proposed Dumping on Other Uses of the Ocean

227.20　Basis for determination.
227.21　Uses considered.
227.22　Assessment of impact.

Subpart F—Special Requirements for Interim Permits Under Section 102 of the Act

227.23　General requirement.
227.24　Contents of environmental assessment.
227.25　Contents of plans.
227.26　Implementation of plans.

Subpart G—Definitions

227.27　Limiting permissible concentration (LPC).
227.28　Release zone.
227.29　Initial mixing.
227.30　High-level radioactive waste.
227.31　Applicable marine water quality criteria.

Environmental Protection Agency   § 227.2

Sec.
227.32 Liquid, suspended particulate, and solid phases of a material.

AUTHORITY: 33 U.S.C. 1412 and 1418.

SOURCE: 42 FR 2476, Jan. 11, 1977, unless otherwise noted.

Subpart A—General

§ 227.1 Applicability.

(a) Section 102 of the Act requires that criteria for the issuance of ocean disposal permits be promulgated after consideration of the environmental effect of the proposed dumping operation, the need for ocean dumping, alternatives to ocean dumping, and the effect of the proposed action on esthetic, recreational and economic values and on other uses of the ocean. This Parts 227 and 228 of this Subchapter H together constitute the criteria established pursuant to section 102 of the Act. The decision of the Administrator, Regional Administrator or the District Engineer, as the case may be, to issue or deny a permit and to impose specific conditions on any permit issued will be based on an evaluation of the permit application pursuant to the criteria set forth in this Part 227 and upon the requirements for disposal site management pursuant to the criteria set forth in Part 228 of this Subchapter H.

(b) With respect to the criteria to be used in evaluating disposal of dredged materials, this section and Subparts C, D, E, and G apply in their entirety. To determine whether the proposed dumping of dredged material complies with Subpart B, only §§ 227.4, 227.5, 227.6, 227.9, 227.10 and 227.13 apply. An applicant for a permit to dump dredged material must comply with all of Subparts C, D, E, G and applicable sections of B, to be deemed to have met the EPA criteria for dredged material dumping promulgated pursuant to section 102(a) of the Act. If, in any case, the Chief of Engineers finds that, in the disposition of dredged material, there is no economically feasible method or site available other than a dumping site, the utilization of which would result in noncompliance with the criteria established pursuant to Subpart B relating to the effects of dumping or with the restrictions es-

tablished pursuant to section 102(c) of the Act relating to critical areas, he shall so certify and request that the Secretary of the Army seek a waiver from the Administrator pursuant to Part 225.

(c) The Criteria of this Part 227 are established pursuant to section 102 of the Act and apply to the evaluation of proposed dumping of materials under Title I of the Act. The Criteria of this Part 227 deal with the evaluation of proposed dumping of materials on a case-by-case basis from information supplied by the applicant or otherwise available to EPA or the Corps of Engineers concerning the characteristics of the waste and other considerations relating to the proposed dumping.

(d) After consideration of the provisions of §§ 227.28 and 227.29, no permit will be issued when the dumping would result in a violation of applicable water quality standards.

§ 227.2 Materials which satisfy the environmental impact criteria of Subpart B.

(a) If the applicant satisfactorily demonstrates that the material proposed for ocean dumping satisfies the environmental impact criteria set forth in Subpart B, a permit for ocean dumping will be issued unless:

(1) There is no need for the dumping, and alternative means of disposal are available, as determined in accordance with the criteria set forth in Subpart C; or

(2) There are unacceptable adverse effects on esthetic, recreational or economic values as determined in accordance with the criteria set forth in Subpart D; or

(3) There are unacceptable adverse effects on other uses of the ocean as determined in accordance with the criteria set forth in Subpart E.

(b) If the material proposed for ocean dumping satisfies the environmental impact criteria set forth in Subpart B, but the Administrator or the Regional Administrator, as the case may be, determines that any one of the considerations set forth in paragraph (a)(1), (2) or (3) of this section applies, he will deny the permit application; provided however, that he may

163

issue an interim permit for ocean dumping pursuant to paragraph (d) of § 220.3 and Subpart F of this Part 227 when he determines that:

(1) The material proposed for ocean dumping does not contain any of the materials listed in § 227.5 or listed in § 227.6, except as trace contaminants; and

(2) In accordance with Subpart C there is a need to ocean dump the material and no alternatives are available to such dumping; and

(3) The need for the dumping and the unavailability of alternatives, as determined in accordance with Subpart C, are of greater significance to the public interest than the potential for adverse effect on esthetic, recreational or economic values, or on other uses of the ocean, as determined in accordance with Subparts D and E, respectively.

§ 227.3 Materials which do not satisfy the environmental impact criteria set forth in Subpart B.

If the material proposed for ocean dumping does not satisfy the environmental impact criteria of Subpart B, the Administrator or the Regional Administrator, as the case may be, will deny the permit application; provided however, that he may issue an interim permit pursuant to paragraph (d) of § 220.3 and Subpart F of this Part 227 when he determines that:

(a) The material proposed for dumping does not contain any of the materials listed in § 227.6 except as trace contaminants, or any of the materials listed in § 227.5;

(b) In accordance with Subpart C there is a need to ocean dump the material; and

(c) Any one of the following factors is of greater significance to the public interest than the potential for adverse impact on the marine environment, as determined in accordance with Subpart B:

(1) The need for the dumping, as determined in accordance with Subpart C; or

(2) The adverse effects of denial of the permit on recreational or economic values as determined in accordance with Subpart D; or

(3) The adverse effects of denial of the permit on other uses of the ocean, as determined in accordance with Subpart E.

Subpart B—Environmental Impact

§ 227.4 Criteria for evaluating environmental impact.

This Subpart B sets specific environmental impact prohibitions, limits, and conditions for the dumping of materials into ocean waters. If the applicable prohibitions, limits, and conditions are satisfied, it is the determination of EPA that the proposed disposal will not unduly degrade or endanger the marine environment and that the disposal will present:

(a) No unacceptable adverse effects on human health and no significant damage to the resources of the marine environment;

(b) No unacceptable adverse effect on the marine ecosystem;

(c) No unacceptable adverse persistent or permanent effects due to the dumping of the particular volumes or concentrations of these materials; and

(d) No unacceptable adverse effect on the ocean for other uses as a result of direct environmental impact.

§ 227.5 Prohibited materials.

The ocean dumping of the following materials will not be approved by EPA or the Corps of Engineers under any circumstances:

(a) High-level radioactive wastes as defined in § 227.30;

(b) Materials in whatever form (including without limitation, solids, liquids, semi-liquids, gases or organisms) produced or used for radiological, chemical or biological warfare;

(c) Materials insufficiently described by the applicant in terms of their compositions and properties to permit application of the environmental impact criteria of this Subpart B;

(d) Persistent inert synthetic or natural materials which may float or remain in suspension in the ocean in such a manner that they may interfere materially with fishing, navigation, or other legitimate uses of the ocean.

§ 227.6 Constituents prohibited as other than trace contaminants.

(a) Subject to the exclusions of paragraphs (f), (g) and (h) of this section, the ocean dumping, or transportation for dumping, of materials containing the following constituents as other than trace contaminants will not be approved on other than an emergency basis:

(1) Organohalogen compounds;

(2) Mercury and mercury compounds;

(3) Cadmium and cadmium compounds;

(4) Oil of any kind or in any form, including but not limited to petroleum, oil sludge, oil refuse, crude oil, fuel oil, heavy diesel oil, lubricating oils, hydraulic fluids, and any mixtures containing these, transported for the purpose of dumping insofar as these are not regulated under the FWPCA;

(5) Known carcinogens, mutagens, or teratogens or materials suspected to be carcinogens, mutagens, or teratogens by responsible scientific opinion.

(b) These constituents will be considered to be present as trace contaminants only when they are present in materials otherwise acceptable for ocean dumping in such forms and amounts in liquid, suspended particulate, and solid phases that the dumping of the materials will not cause significant undesirable effects, including the possibility of danger associated with their bioaccumulation in marine organisms.

(c) The potential for significant undesirable effects due to the presence of these constituents shall be determined by application of results of bioassays on liquid, suspended particulate, and solid phases of wastes according to procedures acceptable to EPA, and for dredged material, acceptable to EPA and the Corps of Engineers. Materials shall be deemed environmentally acceptable for ocean dumping only when the following conditions are met:

(1) The liquid phase does not contain any of these constituents in concentrations which will exceed applicable marine water quality criteria after allowance for initial mixing; provided that mercury concentrations in the disposal site, after allowance for initial mixing, may exceed the average normal ambient concentrations of mercury in ocean waters at or near the dumping site which would be present in the absence of dumping, by not more than 50 percent; and

(2) Bioassay results on the suspended particulate phase of the waste do not indicate occurrence of significant mortality or significant adverse sublethal effects including bioaccumulation due to the dumping of wastes containing the constituents listed in paragraph (a) of this section. These bioassays shall be conducted with appropriate sensitive marine organisms as defined in § 227.27(c) using procedures for suspended particulate phase bioassays approved by EPA, or, for dredged material, approved by EPA and the Corps of Engineers. Procedures approved for bioassays under this section will require exposure of organisms for a sufficient period of time and under appropriate conditions to provide reasonable assurance, based on consideration of the statistical significance of effects at the 95 percent confidence level, that, when the materials are dumped, no significant undesirable effects will occur due either to chronic toxicity or to bioaccumulation of the constituents listed in paragraph (a) of this section; and

(3) Bioassay results on the solid phase of the wastes do not indicate occurrence of significant mortality or significant adverse sublethal effects due to the dumping of wastes containing the constituents listed in paragraph (a) of this section. These bioassays shall be conducted with appropriate sensitive benthic marine organisms using benthic bioassay procedures approved by EPA, or, for dredged material, approved by EPA and the Corps of Engineers. Procedures approved for bioassays under this section will require exposure of organisms for a sufficient period of time to provide reasonable assurance, based on considerations of statistical significance of effects at the 95 percent confidence level, that, when the materials are dumped, no significant undesirable effects will occur due either to chronic toxicity or to bioaccumulation of the

constituents listed in paragraph (a) of this section; and

(4) For persistent organohalogens not included in the applicable marine water quality criteria, bioassay results on the liquid phase of the waste show that such compounds are not present in concentrations large enough to cause significant undesirable effects due either to chronic toxicity or to bioaccumulation in marine organisms after allowance for initial mixing.

(d) When the Administrator, Regional Administrator or District Engineer, as the case may be, has reasonable cause to believe that a material proposed for ocean dumping contains compounds identified as carcinogens, mutagens, or teratogens for which criteria have not been included in the applicable marine water quality criteria, he may require special studies to be done prior to issuance of a permit to determine the impact of disposal on human health and/or marine ecosystems. Such studies must provide information comparable to that required under paragraph (c)(3) of this section.

(e) The criteria stated in paragraphs (c)(2) and (3) of this section will become mandatory as soon as announcement of the availability of acceptable procedures is made in the FEDERAL REGISTER. At that time the interim criteria contained in paragraph (e) of this section shall no longer be applicable. As interim measures the criteria of paragraphs (c)(2) and (3) of this section may be applied on a case-by-case basis where interim guidance on acceptable bioassay procedures is provided by the Regional Administrator or, in the case of dredged material, by the District Engineer; or, in the absence of such guidance, permits may be issued for the dumping of any material only when the following conditions are met, except under an emergency permit:

(1) Mercury and its compounds are present in any solid phase of a material in concentrations less than 0.75 mg/kg, or less than 50 percent greater than the average total mercury content of natural sediments of similar lithologic characteristics as those at the disposal site; and

(2) Cadmium and its compounds are present in any solid phase of a materi-al in concentrations less than 0.6 mg/kg, or less than 50 percent greater than the average total cadmium content of natural sediments of similar lithologic characteristics as those at the disposal site; and

(3) The total concentration of organohalogen constituents in the waste as transported for dumping is less than a concentration of such constituents known to be toxic to marine organisms. In calculating the concentration of organohalogens, the applicant shall consider that these constituents are all biologically available. The determination of the toxicity value will be based on existing scientific data or developed by the use of bioassays conducted in accordance with approved EPA procedures; and

(4) The total amounts of oils and greases as identified in paragraph (a)(4) of this section do not produce a visible surface sheen in an undisturbed water sample when added at a ratio of one part waste material to 100 parts of water.

(f) The prohibitions and limitations of this section do not apply to the constituents identified in paragraph (a) of this section when the applicant can demonstrate that such constituents are (1) present in the material only as chemical compounds or forms (e.g., inert insoluble solid materials) non-toxic to marine life and non-bioaccumulative in the marine environment upon disposal and thereafter, or (2) present in the material only as chemical compounds or forms which, at the time of dumping and thereafter, will be rapidly rendered non-toxic to marine life and non-bioaccumulative in the marine environment by chemical or biological degradation in the sea; provided they will not make edible marine organisms unpalatable; or will not endanger human health or that of domestic animals, fish, shellfish, or wildlife.

(g) The prohibitions and limitations of this section do not apply to the constituents identified in paragraph (a) of this section for the granting of research permits if the substances are rapidly rendered harmless by physical, chemical or biological processes in the sea; provided they will not make edible marine organisms unpalatable and will not endanger human health or that of domestic animals.

(h) The prohibitions and limitations of this section do not apply to the constituents identified in paragraph (a) of this section for the granting of permits for the transport of these substances for the purpose of incineration at sea if the applicant can demonstrate that the stack emissions consist of substances which are rapidly rendered harmless by physical, chemical or biological processes in the sea. Incinerator operations shall comply with requirements which will be established on a case-by-case basis.

[42 FR 2476, Jan. 11, 1977; 43 FR 1071, Jan. 6, 1978]

§ 227.7 Limits established for specific wastes or waste constituents.

Materials containing the following constituents must meet the additional limitations specified in this section to be deemed acceptable for ocean dumping:

(a) Liquid waste constituents immiscible with or slightly soluble in seawater, such as benzene, xylene, carbon disulfide and toluene, may be dumped only when they are present in the waste in concentrations below their solubility limits in seawater. This provision does not apply to materials which may interact with ocean water to form insoluble materials;

(b) Radioactive materials, other than those prohibited by § 227.5, must be contained in accordance with the provisions of § 227.11 to prevent their direct dispersion or dilution in ocean waters;

(c) Wastes containing living organisms may not be dumped if the organisms present would endanger human health or that of domestic animals, fish, shellfish and wildlife by:

(1) Extending the range of biological pests, viruses, pathogenic microorganisms or other agents capable of infesting, infecting or extensively and permanently altering the normal populations of organisms;

(2) Degrading uninfected areas; or

(3) Introducing viable species not indigenous to an area.

(d) In the dumping of wastes of highly acidic or alkaline nature into the ocean, consideration shall be given to:

(1) The effects of any change in acidity or alkalinity of the water at the disposal site; and

(2) The potential for synergistic effects or for the formation of toxic compounds at or near the disposal site. Allowance may be made in the permit conditions for the capability of ocean waters to neutralize acid or alkaline wastes; provided, however, that dumping conditions must be such that the average total alkalinity or total acidity of the ocean water after allowance for initial mixing, as defined in § 227.29, may be changed, based on stoichiometric calculations, by no more than 10 percent during all dumping operations at a site to neutralize acid or alkaline wastes.

(e) Wastes containing biodegradable constituents, or constituents which consume oxygen in any fashion, may be dumped in the ocean only under conditions in which the dissolved oxygen after allowance for initial mixing, as defined in § 227.29, will not be depressed by more than 25 percent below the normally anticipated ambient conditions in the disposal area at the time of dumping.

§ 227.8 Limitations on the disposal rates of toxic wastes.

No wastes will be deemed acceptable for ocean dumping unless such wastes can be dumped so as not to exceed the limiting permissible concentration as defined in § 227.27; *Provided*, That this § 227.8 does not apply to those wastes for which specific criteria are established in § 227.11 or § 227.12. Total quantities of wastes dumped at a site may be limited as described in § 228.8.

§ 227.9 Limitations on quantities of waste materials.

Substances which may damage the ocean environment due to the quantities in which they are dumped, or which may seriously reduce amenities, may be dumped only when the quantities to be dumped at a single time and place are controlled to prevent long-term damage to the environment or to amenities.

§ 227.10 Hazards to fishing, navigation, shorelines or beaches.

(a) Wastes which may present a serious obstacle to fishing or navigation may be dumped only at disposal sites and under conditions which will insure no unacceptable interference with fishing or navigation.

(b) Wastes which may present a hazard to shorelines or beaches may be dumped only at sites and under conditions which will insure no unacceptable danger to shorelines or beaches.

§ 227.11 Containerized wastes.

(a) Wastes containerized solely for transport to the dumping site and expected to rupture or leak on impact or shortly thereafter must meet the appropriate requirements of §§ 227.6, 227.7, 227.8, 227.9, and 227.10.

(b) Other containerized wastes will be approved for dumping only under the following conditions:

(1) The materials to be disposed of decay, decompose or radiodecay to environmentally innocuous materials within the life expectancy of the containers and/or their inert matrix; and

(2) Materials to be dumped are present in such quantities and are of such nature that only short-term localized adverse effects will occur should the containers rupture at any time; and

(3) Containers are dumped at depths and locations where they will cause no threat to navigation, fishing, shorelines, or beaches.

§ 227.12 Insoluble wastes.

(a) Solid wastes consisting of inert natural minerals or materials compatible with the ocean environment may be generally approved for ocean dumping provided they are insoluble above the applicable trace or limiting permissible concentrations and are rapidly and completely settleable, and they are of a particle size and density that they would be deposited or rapidly dispersed without damage to benthic, demersal, or pelagic biota.

(b) Persistent inert synthetic or natural materials which may float or remain in suspension in the ocean as prohibited in paragraph (d) of § 227.5 may be dumped in the ocean only when they have been processed in such a fashion that they will sink to the bottom and remain in place.

§ 227.13 Dredged materials.

(a) Dredged materials are bottom sediments or materials that have been dredged or excavated from the navigable waters of the United States, and their disposal into ocean waters is regulated by the U.S. Army Corps of Engineers using the criteria of applicable sections of Parts 227 and 228. Dredged material consists primarily of natural sediments or materials which may be contaminated by municipal or industrial wastes or by runoff from terrestrial sources such as agricultural lands.

(b) Dredged material which meets the criteria set forth in the following paragraphs (b)(1), (2), or (3) of this section is environmentally acceptable for ocean dumping without further testing under this section:

(1) Dredged material is composed predominantly of sand, gravel, rock, or any other naturally occurring bottom material with particle sizes larger than silt, and the material is found in areas of high current or wave energy such as streams with large bed loads or coastal areas with shifting bars and channels; or

(2) Dredged material is for beach nourishment or restoration and is composed predominantly of sand, gravel or shell with particle sizes compatible with material on the receiving beaches; or

(3) *When:* (i) The material proposed for dumping is substantially the same as the substrate at the proposed disposal site; and

(ii) The site from which the material proposed for dumping is to be taken is far removed from known existing and historical sources of pollution so as to provide reasonable assurance that such material has not been contaminated by such pollution.

(c) When dredged material proposed for ocean dumping does not meet the criteria of paragraph (b) of this section, further testing of the liquid, suspended particulate, and solid phases, as defined in § 227.32, is required. Based on the results of such testing,

Environmental Protection Agency

dredged material can be considered to be environmentally acceptable for ocean dumping only under the following conditions:

(1) The material is in compliance with the requirements of § 227.6; and

(2)(i) All major constituents of the liquid phase are in compliance with the applicable marine water quality criteria after allowance for initial mixing; or

(ii) When the liquid phase contains major constituents not included in the applicable marine water quality criteria, or there is reason to suspect synergistic effects of certain contaminants, bioassays on the liquid phase of the dredged material show that it can be discharged so as not to exceed the limiting permissible concentration as defined in paragraph (a) of § 227.27; and

(3) Bioassays on the suspended particulate and solid phases show that it can be discharged so as not to exceed the limiting permissible concentration as defined in paragraph (b) of § 227.27.

(d) For the purposes of paragraph (c)(2) of this section, major constituents to be analyzed in the liquid phase are those deemed critical by the District Engineer, after evaluating and considering any comments received from the Regional Administrator, and considering known sources of discharges in the area.

PART 228—CRITERIA FOR THE MANAGEMENT OF DISPOSAL SITES FOR OCEAN DUMPING

AUTHORITY: 33 U.S.C. 1412 and 1418.

SOURCE: 42 FR 2482, Jan. 11, 1977, unless otherwise noted.

PART 230—SECTION 404(b)(1) GUIDELINES FOR SPECIFICATION OF DISPOSAL SITES FOR DREDGED OR FILL MATERIAL

169

A4.48 APPENDIX FOUR

PART 233–404 STATE PROGRAM REGULATIONS

59110   Federal Register / Vol. 56, No. 226 / Friday, November 22, 1991 / Rules and Regulations

DEPARTMENT OF DEFENSE

Corps of Engineers, Department of the Army

33 CFR Part 330

Final Rule for Nationwide Permit Program Regulations and Issue, Reissue, and Modify Nationwide Permits

AGENCY: U.S. Army Corps of Engineers, DOD.

ACTION: Final rule.

SUMMARY: The Corps of Engineers is hereby amending its nationwide permit program regulations at 33 CFR part 330. The amendments will simplify and clarify the nationwide permit program and reduce the effort expended in regulating activities with minimal impacts.

The Corps is also reissuing the existing nationwide permits, some with modifications, issuing 10 new nationwide permits, and adding new conditions to all of the nationwide permits.

EFFECTIVE DATE: January 21, 1992.

ADDRESSES: Information can be obtained by writing to: The Chief of Engineers, U.S. Army Corps of Engineers, ATTN: CECW-OR, Washington, DC 20314–1000.

FOR FURTHER INFORMATION CONTACT: Mr. Sam Collinson or Mr. John Studt at (202) 272–1782.

SUPPLEMENTARY INFORMATION: On April 10, 1991, the Corps published its proposed revision to the Nationwide Permit Program regulations and its proposal to issue, reissue, and modify the nationwide permits (56 FR 14598). The changes were proposed with the intent to simplify and clarify the nationwide permit program and to reduce the effort expended in regulating activities with minimal impacts. In addition, we proposed to reissue the existing 26 nationwide permits, some with modifications, to issue 13 new nationwide permits, to add new

conditions to all of the nationwide permits. A public hearing on the proposed rule and nationwide permits was held on May 10, 1991, in Washington, DC. We received over 700 comments in response to the proposed regulations and there were 17 speakers at the public hearing. In response to these comments, we made a number of revisions to the nationwide permit program regulations and to the nationwide permits.

The Corps is restructuring the regulations governing the nationwide permit (NWP) program. In addition, the Corps is adopting changes that will allow the district engineer (DE) to assert a discretionary authority to modify, suspend, or revoke NWPs for individual activities; broaden the basis for asserting discretionary authority to include all public interest factors; provide that the DE require an individual permit whenever he determines that an activity would have more than minimal adverse environmental effects, either individually or cumulatively, or would be contrary to the public interest; and, modify the predischarge notification (PDN) process required by some NWPs.

The Corps is also reissuing the existing NWPs; issuing 10 new NWPs; modifying some of the existing NWPs; converting the best management practices (BMPs) to permit conditions to increase their enforceability; and, clarifying recurring questions about the applicability of some of the NWPs to certain situations.

Upon the expiration of the NWPs in five years from their effective date, we will remove appendix A from the CFR and issue the NWPs separately from the regulations governing their use. In this way, issuance of the NWPs will follow procedures similar to those for individual permits and regional general permits. Until the NWPs in appendix A are removed from the CFR, the proposed issuance, reissuance, modification, and revocation of NWPs would be published in the **Federal Register** concurrent with regional public notices issued by district engineers, to solicit comments and to provide the opportunity to request a public hearing. All comments would be included in the administrative record, and substantive comments addressed in a decision document for each NWP. The final decisions on the NWPs will be announced by publication in the **Federal Register** concurrent with regional public notices issued by district engineers.

All the changes taken together should result in an overall increase in protection of the aquatic environment and an overall decrease in workload. Any workload savings will be devoted to more efficient individual permit evaluation and increased enforcement and compliance activities.

APPENDIX 5

ENGINEERING MANUALS PERTAINING TO DREDGING ACTIVITIES

EEDP-06-3
March 1988

Environmental
Effects of Dredging
Technical Notes

ENGINEER MANUAL SERIES ON DREDGING AND DREDGED MATERIAL DISPOSAL

PURPOSE: This technical note describes a series of Engineer Manuals (EMs) on dredging and dredged material disposal being published by the Office, Chief of Engineers, US Army. The note describes the purpose of the manual series, intended audience, major topics covered, availability of published manuals, and the status of future manuals.

BACKGROUND: Manuals already published in the series include EM 1110-2-5025, "Dredging and Dredged Material Disposal"; EM 1110-2-5026, "Dredged Material Beneficial Uses"; and EM 1110-2-5027, "Confined Disposal of Dredged Material." This manual series is the first comprehensive guidance on dredging and disposal developed for routine Corps use.

The guidance contained in the manuals was developed based on experience of the Corps Districts and Divisions and research conducted under the Dredged Material Research Program (DMRP) and subsequent research programs managed under the Environmental Effects of Dredging Programs (EEDP). As additional information becomes available, the EM series will be updated with published changes or new manuals.

ADDITIONAL INFORMATION OR QUESTIONS: Contact the EEDP program manager, Dr. Robert M. Engler, commercial or FTS: (601) 634-3624. Questions on the content of respective manuals and suggestions for changes or additions can be directed to the author, Dr. Michael R. Palermo, (601) 634-3753, for EM 1110-2-5025 and EM 1110-2-5027, or Dr. Mary C. Landin, (601) 634-2942, for EM 1110-2-5026.

Description of Engineer Manual Series

Technical guidance for planning, design, operation, and management of Corps of Engineers projects is normally published in Engineer Manuals (EMs). The information and procedures contained in EMs are not considered policy or regulation, but rather official guidance. Use of alternate procedures should be justified on a technical basis.

US Army Engineer Waterways Experiment Station, Environmental Laboratory
PO Box 631, Vicksburg, Mississippi 39180-0631

The Office, Chief of Engineers, US Army is publishing a series of EMs on dredging and dredged material disposal. Manuals published thus far include the following:

EM 1110-2-5025, "Dredging and Dredged Material Disposal"

EM 1110-2-5026, "Dredged Material Beneficial Uses"

EM 1110-2-5027, "Confined Disposal of Dredged Material"

These manuals have been developed for routine use by engineers and scientists in Corps Districts and Divisions involved in all aspects of dredging projects. The information contained in the manuals is applicable to all functional areas (i.e. planning, design, construction, operations, and maintenance). Descriptions of the purpose and scope of each manual are given in the following paragraphs.

EM 1110-2-5025, "Dredging and Dredged Material Disposal"

EM 1110-2-5025 is the "umbrella" manual of the series. The manual includes a description of dredging equipment and disposal techniques used in the United States and provides guidance for activities associated with both new work and maintenance projects. The manual also provides basic guidance on evaluating and selecting dredging equipment. A descriptive overview of disposal alternatives is provided, since more detailed guidance on disposal alternatives is available in other manuals in the series.

The major topic areas contained in EM 1110-2-5025 are as follows:

Design considerations for dredging projects

Dredging equipment and techniques

Factors in equipment selection

Dredge operating characteristics

Advances in dredging technology

Environmental considerations for dredging

Sediment resuspension due to dredging

Evaluation of dredged material pollution potential

Influence of disposal conditions on impacts

Overview of open water disposal

Overview of confined disposal

Habitat development as a disposal alternative

EEDP-06-3
March 1988

EM 1110-2-5026, "Dredged Material Beneficial Uses"

EM 1110-2-5026 provides guidance for planning, designing, developing, and managing dredged material for beneficial uses. The manual incorporates ecological concepts and engineering designs with biological, economical, and social feasibility.

The major topic areas contained in EM 1110-2-5026 are as follows:

Dredged material as a resource
Logistical considerations for beneficial use
Habitat development case studies
Habitat development selection process
Wetland habitat development with dredged material
Upland habitat development with dredged material
Island habitat development
Aquatic habitat development
Beaches and beach nourishment
Aquaculture
Parks and recreation uses
Agricultural and related uses
Strip mine reclamation and landfill cover use
Multipurpose and other land use
Construction and industrial/commercial uses
Baseline data collection and monitoring techniques
Site valuation

EM 1110-2-5027, "Confined Disposal of Dredged Material"

EM 1110-2-5027 provides guidance for planning, designing, constructing, operating, and managing confined dredged material disposal areas. Site design to retain suspended solids during disposal operations and to provide adequate short- and long-term storage capacity is included.

Major topic areas contained in EM 1110-2-5027 are as follows:

Field investigations and sampling
Site selection to avoid groundwater impacts
Settling tests for evaluation of solids retention

3

APPENDIX 6
EXAMPLE DESIGN CALCULATIONS FOR CONTAINMENT AREAS

This appendix presents example calculations for containment area designs.* The examples are developed to illustrate use of field and laboratory data and include designs for sedimentation, weir design, and requirements for storage capacity. Only those calculations necessary to illustrate the procedure are included in the examples.

EXAMPLE 1: CONTAINMENT AREA DESIGN METHOD FOR FRESHWATER SEDIMENTS

Project Information

Each year an average of 300,000 yd^3 of fine-grained channel sediment is dredged from a harbor on Lake Michigan. A new in-water containment area is being constructed to accommodate the long-term dredged material disposal needs in this harbor. However, the new containment area will not be ready for approximately 2 yr. One containment area in the harbor has some remaining storage capacity, but it is not known whether the remaining capacity is sufficient to accommodate the immediate disposal requirements. Design procedures must be followed to determine the detection time needed to meet effluent requirements of 4 grams per liter and the storage volume required for the 300,000 yd^3 of channel sediment. These data will be used to determine if the existing containment area storage capacity is sufficient for the planned dredged material disposal activity. The existing containment area is about 3 miles from the dredging activity.

Records indicate that for the last three dredging operations, an 18-in pipeline dredge was contracted to do the work. The average working time was 17 hr per day, and the dredging rate was 600 yd^3/hr of in situ channel sediment. The project depth in the harbor is 50 ft.

---

*Taken almost in its entirety from "Guidelines for Designing, Operating and Managing Dredged Material Containment Areas," EM 1110-2-5006, Department of the Army, Corps of Engineers, Office of Chief of Engineers, Sept. 30, 1980.

A6.1

TABLE A6.1 Observed Flocculent Settling Concentrations with Depth, in Grams per Liter

Time (min)	Depth from top of settling column (ft)						
	1	2	3	4	5	6	7
0	132.00	132.00	132.00	132.00	132.00	132	132
30	46.00	99.00	115.00	125.00	128.00	135	146
60	25.00	49.00	72.00	96.00	115.00	128	186
120	14.00	20.00	22.00	55.00	78.00	122	227
180	11.00	14.00	16.00	29.00	75.00	119	
240	6.80	10.20	12.00	18.00	65.00	117	
360	3.60	5.80	7.50	10.00	37.00	115	
600	2.80	2.90	3.90	4.40	14.00	114	
720	1.01	1.60	1.90	3.10	4.50	110	
1020	0.90	1.40	1.70	2.40	3.26	106	
1260	0.83	1.14	1.20	1.40	1.70	105	
1500	0.74	0.96	0.99	1.10	1.20	92	
1740	0.63	0.73	0.81	0.85	0.94	90	

Data from actual test on freshwater sediments. Although a 6-ft test depth is recommended, an 8-ft depth was used in this test.
Source: Montgomery, 1978.

Results of Containment Area Survey

The existing containment area has the following dimensions:

1. Size: 96 acres
2. Shape: length-to-width ratio of about 3
3. Volume: 1,548,800 yd^3 (average depth, from surveys, is 10 ft)
4. Weir length: 24 ft (rectangular weir)

Results of Laboratory Tests and Analysis of Data

The following data were obtained from laboratory tests:

1. Salinity: <1 ppt.
2. Channel sediment in situ water content w: 85 percent.
3. Specific gravity G_s: 2.69.
4. Observed flocculent settling concentrations as a function of depth (see Table A6.1).
5. Percent of initial concentration with time (see Table A6.2). This is determined as follows: Column concentration at beginning of tests is 132 grams per liter. Concentration at 1-ft level at time = 30 min is 46 grams per liter (Table A6.1). Percentage of initial concentration = 46 ÷ 132 = 0.35 = 35 percent. These calculations are repeated for each time and depth to develop Table A6.2.
6. Plot the percent of initial concentration versus depth profile for each time interval from data given in Table A6.2 (see Fig. A6.1).

TABLE A6.2 Percentage of Initial Concentration with Time

Time (min)	Depth from top of settling column (ft)		
	1	2	3
0	100	100	100
30	35	75	87
60	19	37	55
120	11	15	17
180	8	11	12
240	5	8	9
360	3	4	6
600	2.0	2.2	3.0
720	1.0	1.2	1.4

Note: Initial suspended solids concentration = 132 grams per liter.
Source: Montgomery, 1978.

7. Determine concentration as a function of time (15-day settling column data) (see Table A6.3).

8. Plot time versus concentration from data in Table A6.3 as shown in Fig. A6.2.

9. Laboratory tests indicate that 20 percent of the sediment is coarse-grained material (> no. 40 sieve); therefore, the volume of coarse-grained material V_{sd} is

$$V_{sd} = 300,000(0.20) = 60,000 \text{ yd}^3$$

and the volume of fine-grained material V_i is:

$$V_i = 300,000 - 60,000 = 240,000 \text{ yd}^3$$

Compute Detention Time Required for Sedimentation

The design detention time is computed as follows:

1. Calculate removal percentages for depths of 1, 2, and 3 ft. An example calculation is

$$t = 30 \text{ min}$$
$$d = 1 \text{ ft}$$
$$C_i = 132 \text{ grams per liter}$$
$$H_{pd} = 2 \text{ ft}$$
$$C_e = 4 \text{ grams per liter}$$

Calculating the total area down to a depth of 1 ft from Fig. A6.1 gives an area of 100 (scale units). Calculating the area to the right of the 30-min time line down to a depth of 1 ft gives 82.5 (scale units). These areas could also have been determined by planimetering the plot. Compute removal percentages as follows:

$$R = \frac{82.5}{100}(100) = 82.5$$

For a settling time of 30 min, 82.5 percent of the suspended solids are removed from the water column above the 1-ft depth.

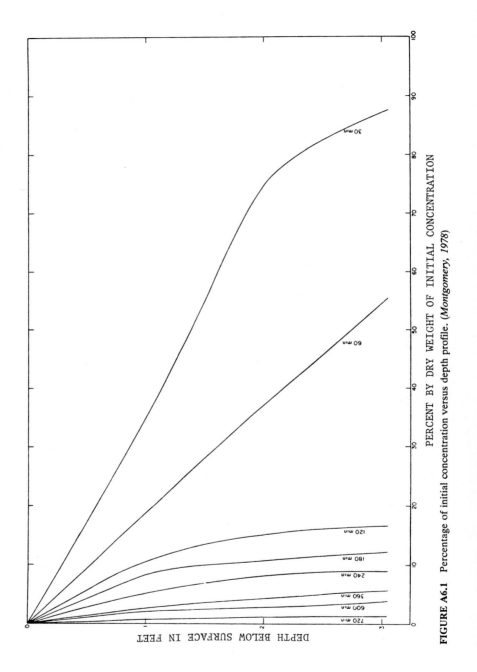

FIGURE A6.1 Percentage of initial concentration versus depth profile. (*Montgomery, 1978*)

TABLE A6.3 Concentration as a Function of Time

Time (days)	Concentration (grams per liter)
1	190
2	217
3	230
4	237
5	240
6	242
7	244
9	249
10	247
15	256

Source: Montgomery, 1978.

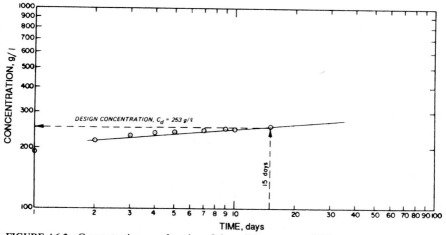

FIGURE A6.2 Concentration as a function of time. (*Montgomery, 1978*)

TABLE A6.4 Removal Percentages as a Function of Settling Time

Time (min)	Depth from top of settling column (ft)		
	1	2	3
30	82.5	62.0	47.0
60	91.0	81.0	73.0
120	93.7	90.2	88.1
180	95.8	93.1	91.5
240	97.4	95.5	94.2
360	98.0	97.0	96.2
600	98.9	98.4	98.1
720	99.6	99.3	99.1

Source: Montgomery, 1978.

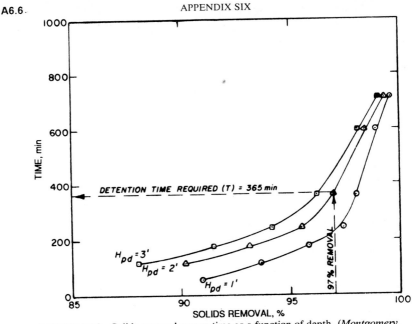

FIGURE A6.3 Solids removal versus time as a function of depth. (*Montgomery, 1978*).

2. The calculations illustrated in step 1 are repeated for each depth as a function of time and the results are tabulated in Table A6.4.

3. Plot the data in Table A6.4 as shown in Fig. A6.3.

4. Since the average ponding depth H_{pd} is 2 ft, use the 2-ft depth curve shown in Fig. A6.3 and determine the theoretical detention time required to meet the 4-grams per liter effluent suspended solids requirement:

$$\text{Required solids removal} = \frac{C_i - C_e}{C_i}$$

$$= \frac{132 - 4}{132} = 0.97 \text{ or } 97 \text{ percent}$$

5. From Fig. A6.3, $T = 365$ min.

6. Increase the theoretical detention time T by a factor of 2.25:

$$T_d = 2.25T$$

$$T_d = 2.25(365)$$

The design detention time T_d equals 822 min.

Compute Volume Required for Sedimentation

Compute the volume required for sedimentation as follows:

$$V_B = Q_i T_d$$

$$Q_i = \frac{(18 \text{ in}^2/12)\pi}{4}(15 \text{ ft/s})$$

$$= 26.5 \text{ ft}^3/\text{s}$$

$$= 1590.4 \text{ ft}^3/\text{min}$$

$$V_B = 1590.4(82) \approx 1,300,000 \text{ ft}^3$$

Compute Design Concentration

Compute the design concentration as follows:

1. Project information:
   a. Dredge size: 18 in
   b. Volume to be dredged: 300,000 yd³
   c. Average operating time: 17 hr per day
   d. Production: 600 yd³/hr
2. Estimate time of dredging activity:

$$\frac{300,000 \text{ yd}^3}{600 \text{ yd}^3/\text{hr}} = 500 \text{ hr}$$

$$\frac{500 \text{ hr}}{17 \text{ hr/day}} = 29.4 \approx 30 \text{ days}$$

3. Average time for dredged material consolidation:

$$\frac{30 \text{ days}}{2} = 15 \text{ days}$$

4. Design solids concentration C_d is the concentration shown in Fig. A6.2 at 15 days:

$$C_d = 253 \text{ grams per liter}$$

Estimate Volume Required for Dredged Material

Estimate the volume required for dredged material as follows:

1. Compute average void ratio e_o:

$$e_o = \frac{G_s \gamma_w}{\gamma_d} - 1$$

$$G_s = 2.69$$

$$\gamma_w \approx 1000 \text{ grams per liter}$$

$$\gamma_d = 253 \text{ grams per liter}$$

$$e_o = \frac{2.69(1000)}{253} - 1$$

$$e_o = 9.63$$

2. Compute the change in volume of fine-grained channel sediments after disposal in containment area:

$$\Delta V = V_i \frac{e_o - e_i}{1 + e_i}$$

$$e_i = \frac{wG_s}{S_D}$$

$$= \frac{(85/100)(2.69)}{1.00}$$

$$e_i = 2.29$$

$$V_i = 240{,}000 \text{ yd}^3$$

$$\Delta V = \frac{9.63 - 2.29}{1 + 2.29}(240{,}000)$$

$$\Delta V = 535{,}440 \text{ yd}^3$$

Estimate Volume Required by Dredged Material in Containment Area

$$V = V_i + \Delta V + V_{sd}$$

$$V_i = 240{,}000 \text{ yd}^3$$

$$\Delta V = 535{,}440 \text{ yd}^3$$

$$\underline{V_{sd} = 60{,}000 \text{ yd}^3}$$

$$V = \qquad 835{,}440 \text{ yd}^3$$

Determine Maximum Dike Height

Foundation conditions limit dike heights to 10 ft.

Determine Design Area

Design area is equal to existing surface area:

$$A_d = 96 \text{ acres } (43{,}560 \text{ ft}^2 \text{ per acre})$$

$$A_d = 4{,}181{,}760 \text{ ft}^2$$

Evaluate Volume Available for Sedimentation Near the End of the Disposal Operation

Determine this value from:

$$V^* = H_{pd}A_d$$

$$V^* = 2 \text{ ft } (4{,}181{,}760 \text{ ft}^2)$$
$$V^* = 8{,}363{,}520 \text{ ft}^3$$

Compare V^* and V_B

Since $V^* > V_B$, a 96-acre containment area will meet the suspended solids effluent requirement of 4 grams per liter for the entire disposal operation.

Estimate Thickness of Dredged Material Layer

Determine this from:

$$H_{dm} = \frac{V}{A_d}$$
$$= \frac{835{,}440 \text{ yd}^3 \ (27)}{4{,}181{,}760 \text{ ft}^2}$$
$$H_{dm} = 5.4 \text{ ft}$$

Determine Required Containment Area Depth

This depth is determined from:

$$D = H_{dm} + H_{pd} + H_{fb}$$
$$= 5.4 \text{ ft} + 2 \text{ ft} + 2 \text{ ft}$$
$$D = 9.4 \text{ ft}$$

Since $D = 9.4$ ft is less than the average basin depth of 10 ft, sufficient volume is available for the project.

Check Weir Length

The existing effective weir length L_e equals the weir crest length L for rectangular weirs:

$$L_e = 48 \text{ ft}$$
$$C_e = 4 \text{ grams per liter}$$
$$Q_i = 26.5 \text{ ft}^3/\text{s}$$
$$H_{pd} = 2 \text{ ft}$$

With an average ponding depth within the containment area H_{pd} of 2 ft, the ponding depth at the weir D_p is estimated to be in excess of 3 ft, accounting for a dredged material surface which slopes toward the weir. Using Fig. A6.4 from the main text, a 3-ft ponding depth at the weir requires an effective weir length of

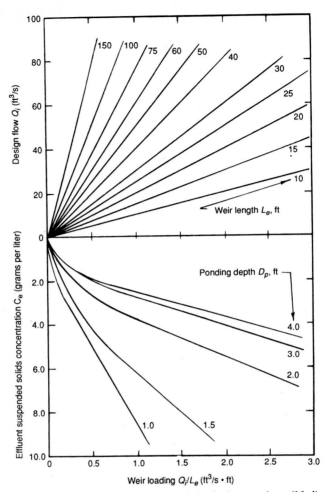

FIGURE A6.4 Weir design nomograph for freshwater clays. (*Modified from Walski and Schroeder, 1978*)

approximately 13 ft. The existing 24-ft weir length should therefore be adequate, but effluent suspended solids should be monitored periodically.

The remaining volume of 1,548,800 yd³ in the existing containment area is sufficient to accommodate disposal of the 300,000 yd³ of maintenance channel sediment into the basin under a continuous disposal operation. Since the required basin depth is less than the existing depth, no upgrading will be necessary to accommodate the first dredging operation. See the following example for determination of storage requirements for a second annual dredging.

EXAMPLE 2: ESTIMATION OF ADDITIONAL STORAGE CAPACITY REQUIREMENTS FOR AN EXISTING CONTAINMENT AREA

Project Information

General project data are identical with that used for the design method for fresh-water sediments. Since the new disposal facility will not be available for 2 yr, the available storage capacity must be determined for a second dredging. Estimates must therefore be made of the total settlement which will occur following placement of the first dredging. The following data were determined in the previous example:

1. V_i = 240,000 yd³ = volume of fine-grained channel sediment (annual requirement).
2. V_{sd} = 60,000 yd³ = volume of sand (annual requirement).
3. ΔV = 535,440 yd³ = change in volume of fine-grained sediments after disposal in the containment area.
4. e_o = 9.63 = average void ratio of dredged material at end of dredging.
5. G_s = 2.69 = specific gravity of solids.
6. D = 10 ft = average depth presently available as determined by survey.
7. A = 96 acres = surface area available for disposal or containment surface area requirement.
8. H_{dm} = 5.4 ft = thickness of dredged material layer at end of the dredging operation (for the first dredging).

Results of Containment Area Field Investigations

Field investigations were conducted at the containment area to define foundation conditions and to obtain samples for laboratory tests. Simplified foundation conditions as defined by the investigations are shown in Fig. A6.5.

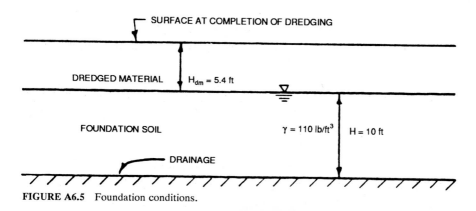

FIGURE A6.5 Foundation conditions.

Results of Laboratory Tests

Consolidation tests were performed on samples of channel sediment and samples of the compressible foundation soils. Representative void ratio-log pressure and coefficient of consolidation-log pressure relationships were selected and are presented in Figs. A6.6 through A6.9.

Settlement Due to Consolidation (Dredged Material)

Compute the effective stress acting at midheight of the dredged material layer:

$$\bar{p}_f = 1/2(H_{dm}\gamma_w) \frac{G_s - 1}{1 + e_o}$$

$$= 1/2 \ (5.4 \ \text{ft}) \ (62.4 \ \text{lb/ft}^3) \ \frac{2.69 - 1}{1 + 9.63}$$

$$\bar{p}_f = 26.78 \approx 27 \ \text{lb/ft}^2 = 0.0135 \ \text{ton/ft}^2$$

The void ratio e_f corresponding to an effective stress $\bar{p}_f$ of 0.0135 ton/ft² is found using the e-log p curve as shown in Fig. A6.6 ($e_f = 4.15$).

The final settlement of the dredged material layer at completion of primary consolidation ΔH is computed as follows:

$$\Delta H = H_{dm} \frac{e_o - e_f}{1 + e_o}$$

$$= 5.4 \ \text{ft} \ \frac{9.63 - 4.15}{1 + 9.63}$$

$$\Delta H = 2.78 \ \text{ft}$$

Settlement Due to Consolidation (Foundation Soil)

The initial consolidation pressure acting at midheight of the foundation soil layer may be found using the effective unit weight as follows:

$$p_1 = 1/2(10.0 \ \text{ft})(110 \ \text{lb/ft}^3 - 62.4 \ \text{lb/ft}^3)$$

$$p_1 = 238 \ \text{lb/ft}^2 = 0.119 \ \text{ton/ft}^2$$

Assuming that the groundwater tables remain in a perched condition, the increase in consolidation pressure because of placement of the dredged material is equal to the total load applied. The total unit weight of dredged material is computed as:

$$\gamma = \frac{\gamma_w(e_o + G_s)}{1 + e_o}$$

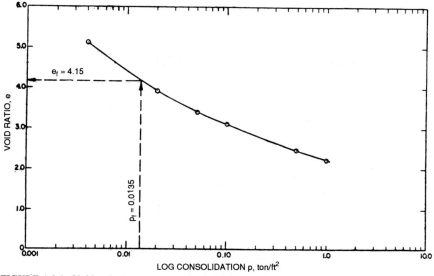

FIGURE A6.6 Void ratio-log pressure relationship for dredged material.

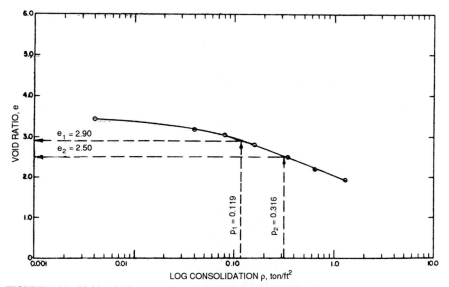

FIGURE A6.7 Void ratio-log pressure relationship for foundation soil.

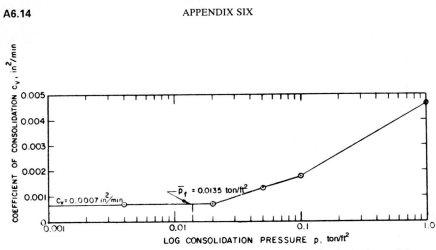

FIGURE A6.8 Coefficient of consolidation-log pressure relationship for dredged material.

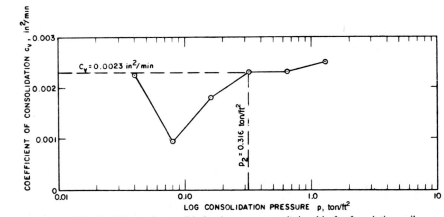

FIGURE A6.9 Coefficient of consolidation-log pressure relationship for foundation soil.

$$\gamma = \frac{62.4(9.63 + 2.69)}{1 + 9.63}$$

$$\gamma = 72.32 \approx 73 \text{ lb/ft}^3$$

The increase in load Δp is then computed and is added to the initial consolidation pressure p_1 to obtain the total pressure p_2 acting at midheight of the foundation soil layer:

$$\Delta p = 5.4 \text{ ft}(73 \text{ lb/ft}^3) = 394 \text{ lb/ft}^2 = 0.197 \text{ ton/ft}^2$$

$$p_2 = p_1 + \Delta p = 0.119 + 0.197 = 0.316 \text{ ton/ft}^2$$

The void ratios e_1 and e_2 corresponding to pressures p_1 and p_2 are found using the e-log p curve in Fig. A6.7 (e_1 = 2.90 and e_2 = 2.50).

The settlement of the foundation layer caused by placement of dredged material ΔH is computed as follows:

$$\Delta H = \frac{e_1 - e_2}{1 + e_1} (H)$$

$$= \frac{2.90 - 2.50}{1 + 2.90} (10 \text{ ft})$$

$$\Delta H = 1.03 \text{ ft}$$

Time-Rate of Consolidation

The coefficient of consolidation c_{vf} corresponding to the average effective stress $\overline{p}_f$ acting at midheight of the dredged material layer is found using Fig. A6.8 to be c_{vf} = 0.007 in²/min. Times required for the dredged material layer to reach various percentages of total consolidation are computed. Settlements for various times are also computed. The resulting settlement versus time relationship is plotted in Fig. A6.10. The time-rate of consolidation for the foundation soil was determined in a similar manner and the resulting settlement versus time relationship is also plotted in Fig. A6.10.

Determination of Storage Capacity Requirement

The site will be required for the second dredging project 11 months following placement of the first dredging. The first dredging will be placed to an initial layer

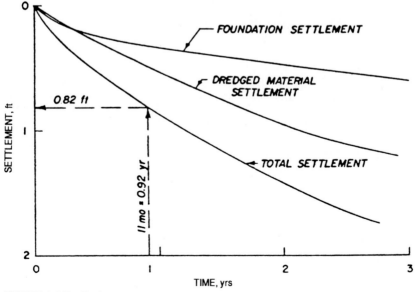

FIGURE A6.10 Settlement versus time relationships.

thickness of 5.4 ft during the 30-day dredging phase. Subsequent total settlement of the dredged material and foundation soil in the following 11 months (0.92 year) can be estimated using the settlement versus time relationship developed in Fig. A6.10. Total settlement equals 0.82 ft.

The remaining height available for storage of the second dredging following settlement of the first dredging is determined as follows:

$$H_{dm} \text{ (available for second dredging)}$$
$$= D - H_{pd} - H_{fb} - H_{dm} \text{ (first dredging)}$$
$$+ \text{ total settlement first dredging}$$
$$= 10.0 - 2.0 - 2.0 - 5.4 + 0.82$$
$$= 1.42 \text{ ft}$$

The storage capacity available for the second dredging may then be obtained as follows:

$$V = A_d H_{dm} \text{ (available for second dredging)}$$
$$= 96 \text{ acres } (43,560 \text{ ft}^3/\text{acre}) \, 1.42 \text{ ft}$$
$$V = 5,938,099 \text{ ft}^3 = 219,930 \text{ yd}^3$$

In order to determine if the dikes must be raised for the second dredging, the available storage capacity must be compared with the storage capacity required to accommodate the total second dredging operation. This is most easily done by using the equivalent thicknesses of dredged material layers since it will result in a direct determination of the amount of dike raising necessary. Because the volume required for the total second dredging is equivalent to a thickness of dredged material H_{dm} of 5.4 ft and because the remaining height available is 1.42 ft, the dikes must be raised approximately 4 ft ($5.4 - 1.42 = 3.98$ ft).

EXAMPLE 3: CONTAINMENT AREA DESIGN METHOD FOR SALTWATER SEDIMENTS

Project Information

Fine-grained maintenance dredged material is scheduled to be dredged from a harbor maintained to a project depth of 50 ft. Channel surveys indicated that 500,000 yd³ of channel sediment must be dredged. All available disposal areas are filled near the dredging activity, but land is available for a new site 2 miles from the dredging project. Since this harbor has to be dredged once every 2 yr, the containment area must be designed to accommodate long-term disposal needs while meeting effluent suspended solids levels of 4 grams per liter. The largest dredge contracted for the maintenance dredging in previous years has been a 24-in pipeline dredge. This is the largest size dredge located in the area.

Results of Laboratory Tests

The following data were obtained from laboratory tests:

1. Salinity = 15 ppt.

TABLE A6.5 Depth to Solids Interface (Feet) as a Function of Settling Time (Hours)

Time	\multicolumn{8}{c}{Initial suspended solids concentration (grams per liter)}
	55	73	120	143	163	215	243	310
0	0	0	0	0	0	0	0	0
0.25	0.230	0.145	0.065	0.050	0.065	0.026	0.010	—
0.50	0.390	0.290	0.165	0.090	0.138	0.050	0.020	0.005
0.75	0.530	0.435	0.270	0.170	0.210	0.075	0.030	—
1.00	0.620	0.535	0.360	0.230	0.276	0.100	0.040	0.009
2.00	0.690	0.635	0.490	0.420	0.430	0.225	0.080	0.020
3.00	0.740	0.680	0.535	0.475	0.467	0.340	0.100	0.025
4.00	0.770	0.700	0.555	0.505	0.495	0.365	0.122	0.035
5.00	0.805	0.710	0.580	0.530	0.510	0.390	0.140	0.050
6.00	0.820	0.730	0.585	0.553	0.515	0.410	0.160	0.070
7.00	0.830	—	—	0.565	—	0.440	0.188	—
8.00	0.840	—	—	0.575	—	0.440	0.188	—
10.00	—	—	—	0.595	—	0.459	0.212	—
20.00	—	—	—	0.655	—	0.522	0.259	0.190
30.00	—	—	—	0.690	—	0.564	0.292	0.250

Source: Montgomery, 1978.

2. Channel sediment in situ water content w = 92.3 percent.

3. Specific gravity G_s = 2.71.

4. Depth to solids interface as a function of time (settling column data; see Table A6.5).

5. Zone settling velocity as a function of concentration (see Table A6.6).

6. Zone settling velocity versus concentration curve (see Fig. A6.11).

7. Calculations of solids loading values (use data given in Fig. A6.11 to develop Table A6.7).

8. Solids loading versus solids concentration (use data in Table A6.7 to develop Fig. A6.12).

TABLE A6.6 Zone Settling Velocity as a Function of Suspended Solids Concentration

\multicolumn{3}{c}{Concentration}	Zone settling velocity (ft/hr)
grams per liter	%	lb/ft³	
55	5.2	3.4	1.238
73	6.8	4.5	0.571
120	10.8	7.5	0.410
143	12.7	9.0	0.245
163	14.3	10.2	0.282
215	18.5	13.5	0.133
243	20.7	15.2	0.041
310	25.8	19.5	0.015

Source: Montgomery, 1978.

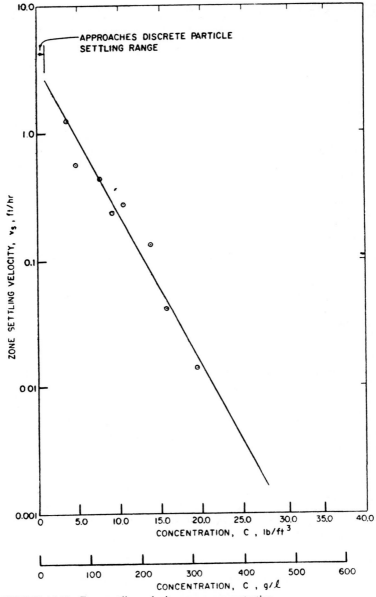

FIGURE A6.11   Zone settling velocity versus concentration.

TABLE A6.7 Calculations of Solids Loading Values*

Suspended solids concentration C			Zone settling velocity, v_s (ft/hr)	Solids loading $S = v_s C$ (lb/hr-ft^2)
%	grams per liter	lb/ft^3		
6.1	65	4	1.150	4.60
7.4	80	5	0.880	4.40
14.2	160	10	0.230	2.30
20.4	240	15	0.060	0.87
26.0	320	20	0.020	0.29
31.2	400	25	0.004	0.09

*Developed from curve shown in Figure A6.12.
Source: Montgomery, 1978.

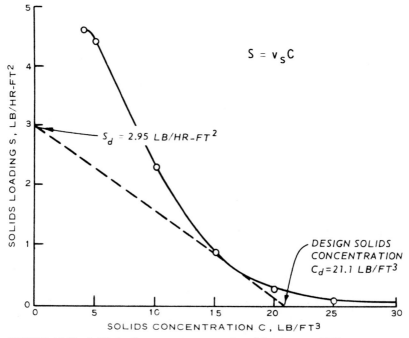

FIGURE A6.12 Solids loading versus concentration. (*Montgomery, 1978*)

9. Concentration as a function of time data (15-day settling column data; see Table A6.8).
10. Concentration versus time curve (see Fig. A6.13).
11. Representative samples of channel sediments tested in the laboratory indicate that 15 percent of the sediment is coarse-grained material (> no. 40 sieve).

$$V_{sd} = 500,000(0.15) = 75,000 \text{ yd}^3$$

$$V_i = 500,000 - 75,000 = 425,000 \text{ yd}^3$$

TABLE A6.8   Concentration as a Function of
Time

Time (days)	Concentration (grams per liter)
1	192
2	215
3	219
4	240
5	251
6	272
8	280
10	290
15	320

Source: Montgomery, 1978.

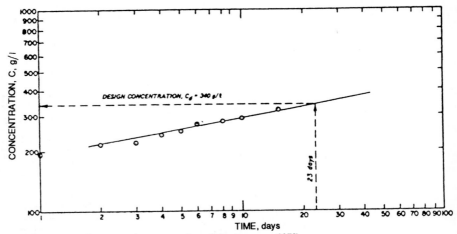

FIGURE A6.13   Concentration versus time. (*Montgomery, 1978*)

Compute Design Concentration

Compute the design concentration as follows:

1. Project information:
   Dredge size: 24 in
   Volume to be dredged: 500,000 yd^3
2. Good records are available from past years of maintenance dredging in this harbor. They show that each time a 24-in dredge was used, the dredge averaged operating 12 hr per day and dredged an average of 900 yd^3/hr.
3. Estimate time of dredging activity:

$$\frac{500,000 \text{ yd}^3}{900 \text{ yd}^3/\text{hr}} = 556 \text{ hr}$$

   Operating time per day = 12 hr

$$\frac{556 \text{ hr}}{12 \text{ hr}} \approx 46 \text{ days}$$

4. Average time for dredged material consolidation:

$$\frac{46 \text{ days}}{2} = 23 \text{ days}$$

5. Design solids concentration is the concentration shown in Fig. A6.13 at 23 days:

$$C_d = 340 \text{ grams per liter, or } 21.1 \text{ lb/ft}^3$$

Compute Area Required for Sedimentation

This value is computed as follows:

1. Construct operating line from design concentration (21.1 lb/ft^3) tangent to the loading curve (Fig. A6.12).

2. Design solids loading S_d = 2.95 lb/hr-ft^2.

3. Compute area requirement:

$$A = \frac{Q_i C_i}{S_d}$$

$$Q_i = A_p V_d$$

$$V_d = 15 \text{ ft/s}$$

$$C_i = 9.2 \text{ lb/ft}^3$$

$$S_d = 2.95 \text{ lb/hr-ft}^2$$

$$Q_i = \frac{(24 \text{ in}/12)^2 \pi}{4} (15 \text{ ft/s})$$

$$= 47.12 \text{ ft}^3/\text{s}$$

$$Q_i = 169{,}632 \text{ ft}^3/\text{hr}$$

$$A = \frac{169{,}632(9.2)}{2.95}$$

$$= 529{,}022 \text{ ft}^2$$

$$A = \frac{529{,}022}{43{,}560} = 12.14 \text{ acres}$$

4. Increase the area by a factor of 2.25 (assumes containment area can be constructed with a length-to-width ratio of approximately 3):

$$A_d = 2.25(12.14 \text{ acres})$$

$$A_d = 27.3 \text{ acres}$$

Thus, the area required for sedimentation is 27.3, or 27 acres.

Estimate Volume Required for Dredged Material

This volume is estimated as follows:

1. Compute average void ratio:

$$e_o = \frac{G_s \gamma_w}{\gamma_d} - 1$$

$$G_s = 2.71$$

$$\gamma_w \approx 1000 \text{ grams per liter}$$

$$\gamma_d = 340 \text{ grams per liter} = \text{design concentration } C_d$$
(Fig. A6.13)

$$e_o = \frac{2.71(1000)}{340} - 1$$

$$e_o = 6.97$$

2. Compute the change in volume of fine-grained channel sediments after disposal in containment area:

$$\Delta V = V_i \frac{e_o - e_i}{1 - e_i}$$

$$e_i = \frac{wG_s}{S_D}$$

$$e_i = \frac{(92.3/100)(2.71)}{1.00}$$

$$e_i = 2.5$$

$$V_i = 425,000 \text{ yd}^3$$

$$\Delta V = \frac{6.97 - 2.50}{1 + 2.50}(425,000)$$

$$= 542,785 \text{ yd}^3$$

3. Estimate the volume required by dredged material in the containment area:

$$V = V_i + \Delta V + V_{sd}$$

$$V_i = 425,000 \text{ yd}^3$$

$$\Delta V = 542,785 \text{ yd}^3$$

$$V_{sd} = 75,000 \text{ yd}^3$$

$$V = 425,000 + 542,785 + 75,000$$

$$= 1,042,785 \text{ yd}^3$$

Estimate Thickness of Dredged Material at End of Disposal Operation

$$H_{dm} = \frac{V}{A_d}$$

$$= \frac{1,042,785 \text{ yd}^3(27)}{27 \text{ acres}(43,560)}$$

$$H_{dm} = 23.4 \text{ ft}$$

Because of foundation problems, dike heights are limited to 15 ft. Therefore, the area of the disposal area must be increased to accommodate the storage requirements. Allowable dredged material height is determined as follows:

$$D = H_{dm} + H_{pd} + H_{fb}$$

$$D = 15 \text{ ft}$$

$$H_{pd} = 2 \text{ ft}$$

$$H_{fb} = 2 \text{ ft}$$

$$H_{dm} = D - H_{pd} - H_{fb}$$

$$H_{dm} = 15 - 2 - 2$$

$$H_{dm} = 11 \text{ ft}$$

Compute New Area Requirement

$$H_{dm} = \frac{V}{A_d}$$

$$A_d = \frac{1,042,785 \text{ yd}^3(27)}{11}$$

$$= 2,559,563 \text{ ft}^2$$

$$A_d = 59 \text{ acres}$$

Design for Weir

The design parameters are:

$$Q_i = 47.12 \text{ ft}^3/\text{s}$$

$$C_e = 4 \text{ grams per liter}$$

Using Fig. A6.14, operating lines constructed at $Q_i = 47.12$ ft^3/s and $C_e = 4$ grams per liter indicate possible combinations of ponding depth and effective weir length required. Assuming that a 1-ft ponding depth at the weir is the minimum that could be allowed, a weir length of 35 ft is required. However, a ponding depth of 2 ft is recommended during the operation to provide a margin of safety. It should be noted that 59 acres is the *minimum* area required for storage of one dredging of 500,000 yd^3 and will not meet the long-term storage capacity requirement. See the following example for determination of the area required to meet this requirement.

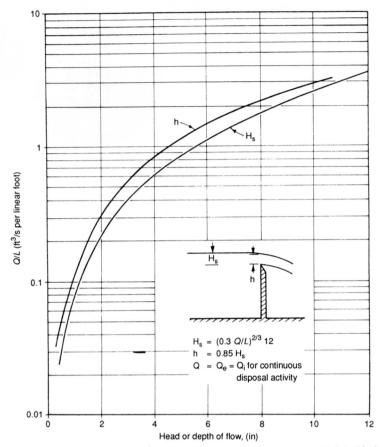

FIGURE A6.14 Relationship between flow rate, weir length, and head. (*Modified from B. J. Gallagher and Co., 1978*)

EXAMPLE 4: ESTIMATION FOR LONG-TERM CAPACITY REQUIREMENTS

Project Information

General project data are identical with that for the design method for saltwater sediments. It is required that the containment area be designed for a service life of 10 yr, accommodating a biannual dredging of 500,000 yd³ of in situ channel sediment. The following data were determined in the previous example:

1. V_i = 425,000 yd³ = volume of fine-grained channel sediments (biannual requirement)
2. V_{sd} = 75,000 yd³ = volume of sand (biannual requirement)
3. ΔV = 542,785 yd³ = change in volume of fine-grained sediments after disposal in the containment area

4. G_s = 2.71 = specific gravity of solids
5. H_{dm} (maximum) = 11 ft = thickness of the dredged material layer at the end of the dredging operation (maximum allowable thickness at end of last dredging operation due to foundation conditions)
6. Time required for each dredging = 46 days

Results of Containment Area Field Investigations

Field investigations were conducted at the containment area to define foundation conditions and to obtain samples for laboratory tests. Simplified foundation conditions as defined by the investigations are shown in Fig. A6.15.

Results of Laboratory Tests

Consolidation tests were performed on samples of channel sediment and samples of the compressible foundation soil. Representative void ratio-log pressure and coefficient of consolidation-log pressure relationships were selected and are presented in Figs. A6.6 through A6.9.

Determination of Surface Area Required

Since the total dredging requirement equals five dredging operations (10-yr service life, biannual dredging), the minimum required surface area for one dredging will not meet the long-term requirement. Increasing the surface area in use will result in decreased initial dredged material layer thicknesses, allowing for a greater degree of consolidation between dredging operations. The optimum surface area for the long-term storage required cannot be directly determined since the magnitude of consolidation is dependent on layer thickness and loading, which is also a function of surface area. The solution must therefore be deter-

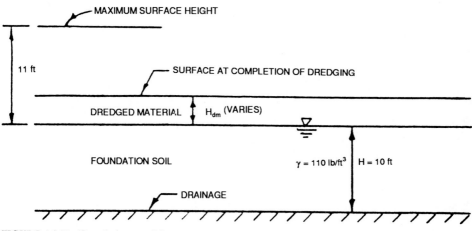

FIGURE A6.15 Foundation conditions.

mined by trial.* A convenient method for selection of trials is to first establish upper and lower bounds on the required surface area. The lower bound may be selected using the minimum required surface area for storage of a single dredging. The upper bound on surface area may be determined assuming no consolidation takes place. The area required for storage of one dredging was determined earlier to be 59 acres. The upper bound for trial surface area may be established by multiplying the total number of dredging operations times the minimum surface area required for storage of a single dredging as follows:

$$A_{d\ max} = \text{number of dredging operations} \times A_{d\ min}$$

$$A_{d\ max} = 5 \times 59 \text{ acres} = 295 \text{ acres}$$

Use of Mathematical Model

The multiple dredging involved would require unduly complex computations; therefore, the use of a mathematical model is desirable to estimate optimum surface area needed to meet the 10-yr storage requirement. Trial runs of the model will allow selection of an optimum surface area. After examining the upper and lower bounds as described in the previous paragraph, the following trials were selected:

Trial no.	Trial surface area (acres)	Corresponding lift thickness per individual dredging (ft)
1	319	2
2	159	4
3	106	6

Johnson's model[4] was selected for use. Input data were coded for each trial in accordance with Johnson[4] using the project information stated earlier and laboratory test results shown in Figs. A6.6 through A6.9. Relationships for the coefficient of consolidation, coefficient of permeability, and coefficient of volume change versus consolidation pressure for dredged material required for the model were developed from the laboratory data and are shown plotted in Figs. A6.16 through A6.18.

Results of the trial model runs are interpreted in Figs. A6.19 through A6.21, which show projected surface heights versus time. The service life of the containment area for each trial run is also indicated in Figs. A6.19 through A6.21. The optimum surface area for a 10-yr service life may then be estimated by plotting the surface area versus service life for all trials, as shown in Fig. A6.22. For this example, the design surface area is 255 acres. The containment area should therefore be constructed with dike heights of 15 ft, total enclosed area of 255 acres, and a length-to-width ratio of approximately 3:1.

---

*If the surface area was predetermined such as for an existing site or within available right-of-way limits, trial runs would not be required and the service life could be determined directly.

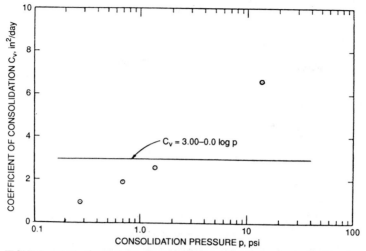

FIGURE A6.16 Coefficient of consolidation-log pressure relationship for dredged material.

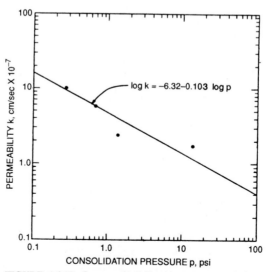

FIGURE A6.17 Log permeability-log pressure relationship for dredged material.

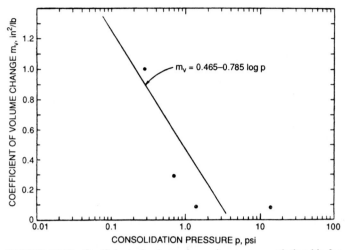

FIGURE A6.18 Coefficient of volume change-log pressure relationship for dredged material.

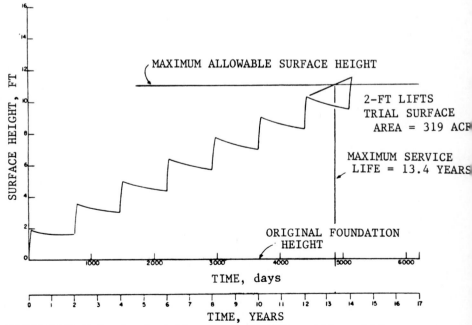

FIGURE A6.19 Projected surface height versus time for trial layer thickness of 2 ft.

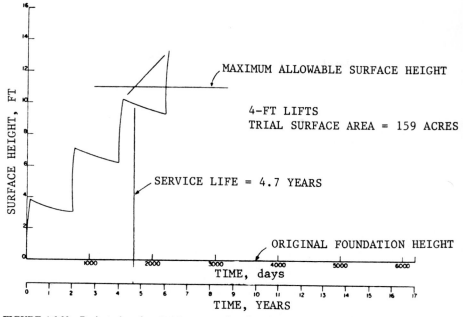

FIGURE A6.20   Projected surface height versus time for trial layer thickness of 4 ft.

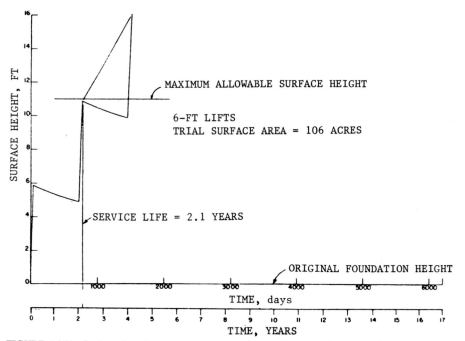

FIGURE A6.21   Projected surface height versus time for trial layer thickness of 6 ft.

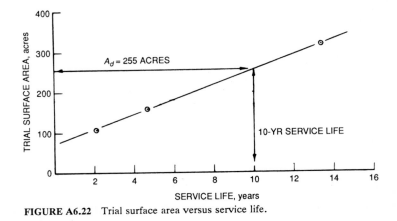

FIGURE A6.22 Trial surface area versus service life.

REFERENCES

1. Montgomery, R. L., *Methodology for Design of Fine-Grained Dredged Material Containment Areas for Solids Retention,* T.R. D-78-56, U.S. Army Engineer Waterways Experiment Station, Vicksburg, MS, Dec. 1978.

2. Walski, T. M., and Schroeder, P. R., *Weir Design to Maintain Effluent Quality from Dredged Material Containment Areas,* T.R. D-78-18, U.S. Army Engineer Waterways Experiment Station, Vicksburg, MS, May 1978.

3. Gallagher, B. J., and Co., *Investigation of Containment Area Design to Maximize Hydraulic Efficiency,* T.R. D-78-12, U.S. Army Engineer Waterways Experiment Station, Vicksburg, MS, May 1978.

4. Johnson, L. D., *Mathematical Model for Predicting the Consolidation of Dredged Material in Confined Disposal Areas,* T.R. D-76-1, U.S. Army Engineer Waterways Experiment Station, Vicksburg, MS, Jan. 1976.

APPENDIX 7
DREDGING RESEARCH PROGRAM—U.S. ARMY ENGINEERS

US Army Corps
of Engineers
Waterways Experiment
Station

Dredging Research Program

Improving technology
related to the Corps'
dredging mission

Dredging

History of the Corps' Involvement

Early attempts at dredging in what was to become the continental United States were made by the French in the 1720's at the mouth of the Mississippi River. Congressional approval in 1824 to remove sandbars and snags from major navigable rivers led to Federal involvement in dredging and a mission for the Army Engineers to implement this Congressional Act.

Numerous dredging projects were undertaken using a variety of dredges, most modelled on European dredges. However, the first seagoing hydraulic hopper dredge, *General Moultrie*, was built in the United States in 1855 and operated in this country under a contract with the Corps of Engineers.

The present dredging mission of the Corps includes maintenance and improvement of 25,000 miles of commercially navigable channels and service to 400 ports.

The Corps' Present and Future Dredging Mission

The dredging mission of the Corps now includes--

- Annual dredging of 250 to 300 million cubic yards of sediment.
- Expenditures of $400 million yearly in performing this mission.

In addition, authorization of improvements to US waterways and harbors calls for an average expenditure of $200 million annually over 10 years for new work alone.

With the increase in the scope and extent of the Corps' dredging mission, research needs developed in the areas of the fate of dredged material, dredging equipment, and dredging systems.

Program Development

The extent and nature of the Corps' involvement in dredging has changed dramatically over the years. The National Environmental Policy Act of 1969 and subsequent legislation drastically changed the Corps' role in dredging and emphasized environmental concerns associated with dredging. Also a once-large government fleet has given way to contracted dredging services. The emphasis of the Corps' dredging program changed again with the passage of the Water Resources Development Act of 1986. The Corps would now be expected not only to maintain existing waterways, but also to undertake major new improvements to existing navigation projects.

The Dredging Research Program, or DRP, is the Corps' response to its newest dredging challenge. The work units of the five problem areas of the DRP were designed to provide new or improved technologies to the field, based on the actual needs of the field offices. Such a plan would ensure practical, usable technology.

Problem Area 1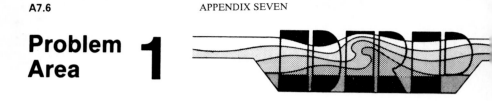

Analysis of Dredged Material Placed in Open Waters

Open-water placement through bottom-dump scow operations

Objectives:

- Calculate boundary layer fluid properties and sediment motion for analyzing behavior of open-water disposal areas.
- Acquire field data sets for improving calculation of fluid and sediment motion.
- Improve and develop computational techniques to predict short- and long-term fate of dredged material.
- Collect field data to improve simulation methods and site-monitoring techniques.

Problem Area 2

Material Properties Related to Navigation and Dredging

Bottom and subbottom definition and sampling by geotechnical and geophysical methods

Objectives:

- Develop intruments and operating procedures for rapid surveys of fluid mud properties
- Define navigable depth in fine-grain sediment.
- Develop intruments for analyzing properties of consolidated sediments.
- Establish dredging-related soil and rock descriptors.

Problem Area 3

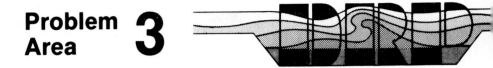

Dredge Plant Equipment and System Processes

Draghead design improvements

Objectives:

- Improve draghead design for dredging compacted fine sands and cohesive muds.
- Improve eductor (jet pump) designs for sand bypassing operations.
- Develop systems to monitor and equipment to increase dredge payloads for fine-grain sediments.
- Design portable single-point mooring buoy for hopper dredge direct pumpout.

Problem Area 4

Vessel Positioning, Survey Controls, and Dredge Monitoring Systems

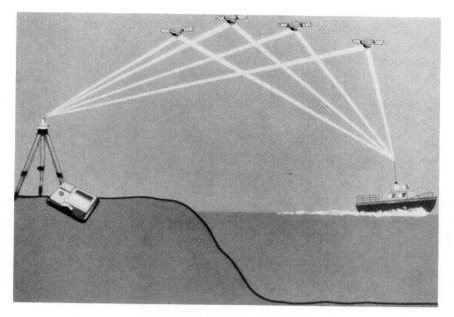

Global Positioning System (GPS)

Objectives:

- Develop real-time system for measuring project site tide and wave conditions in offshore open waters.

- Develop three-dimensional positioning system for dredging and hydrographic surveying operations using GPS satellite constellation.

- Evaluate productions meters used in various dredging situations.

- Develop automated inspection monitoring and reporting system for use on any type of dredge.

Problem Area 5

Management of Dredging Projects

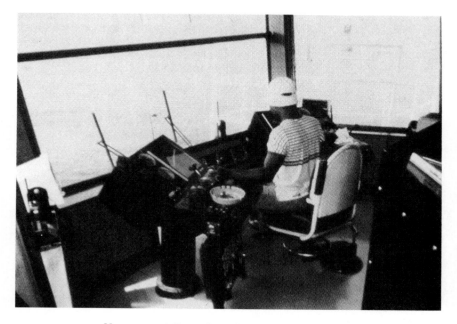

Management through optimum dredge operation

Objectives:

- Evaluate the effects of dredging decisions and project changes.
- Optimize use of open-water disposal sites.
- Analyze dredging cost-estimating techniques.
- Prepare dredging manuals incorporating state-of-the-art technology.

Technology Transfer

Technology transfer for the Dredging Research Program began with the technological need. Work units to develop or modify technology were designed after consultation with field personnel. This approach acknowledges that technology transfer is a complete circle, starting with a technological need and ending with a technology in use.

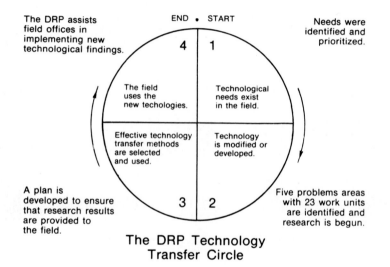

The DRP assists field offices in implementing new technological findings.

END • START

Needs were identified and prioritized.

4 | 1

The field uses the new techologies.

Technological needs exist in the field.

Effective technology transfer methods are selected and used.

Technology is modified or developed.

A plan is developed to ensure that research results are provided to the field.

3 | 2

Five problems areas with 23 work units are identified and research is begun.

The DRP Technology
Transfer Circle

Technology transfer methods to be used by the DRP include:

- Technical reports
- Technical manuals
- Engineer manuals
- Engineer technical letters
- Technical notes
- Executive notes
- Information exchange bulletin
- Specifications
- Video reports
- Computer software & user's manuals
- Training courses
- Meetings
- Workshops
- Symposiums
- One-stop consultations
- Field demonstrations

Dredging Research Program

Dredging Research

Dredging Research is an information exchange bulletin published by the Dredging Research Program. To obtain a copy, write:

Commander and Director
USAE Waterways Experiment Station
ATTN: CEWES-CP-D
3909 Halls Ferry Road
Vicksburg, MS 39180-6199

US Army Corps of Engineers

*D*redging *R*esearch

VOL DRP-89-2   INFORMATION EXCHANGE BULLETIN   NOV 1989

Aerial view of beach north of inlet

Jet Pump Sand Bypassing, Indian River Inlet, Delaware

Gus Rambo
US Army Engineer District, Philadelphia
and
James E. Clausner
US Army Engineer Waterways Experiment Station

Indian River Inlet is located on the Atlantic Coast of Delaware approximately 10 miles north of Ocean City, MD (Figure 1). The 500-foot-wide inlet is stabilized by parallel, rubble-mound jetties. Since the inlet's construction in 1938-40, erosion related to the jetties has occurred. Net northerly transport has resulted in a sizeable accretion fillet adjacent to the south jetty that has remained stable since 1954 and corresponding erosion of the beach north of the inlet. Material that bypasses the south fillet is trapped in flood and ebb tidal shoals.

The major erosion problem along the ocean shoreline north of Indian River Inlet is the threat of breaching the Route 1 roadway. Since 1957, beach erosion has been mitigated by periodic placement of beach fill along a zone extending up to 5,000 feet north of the north jetty. Required nourishment for

AUTHOR INDEX

This is an index of authors mentioned in the text of this handbook.

SUBJECT INDEX

ABOUT THE AUTHOR

John B. Herbich is the W.H. Bauer Professor of Dredging Engineering and Director of the Center for Dredging Studies at Texas A&M University, College Station, Texas. One of the leading experts in coastal and ocean engineering education and research, Dr. Herbich is a Fellow and life member of the American Society of Civil Engineers and has served on the panels and committees of the Marine Board, the National Research Council, and Academy of Sciences, and is on the Board of Directors of the Western Dredging Association. He was on the faculty of Lehigh University prior to joining Texas A&M University and served as a Project Manager for UNDP in India. He also taught and conducted research at WES Graduate Institute and Coastal Engineering Research Center, Waterways Experiment Station, Vicksburg, MS. He has written more than 200 professional reports and papers and is also the author or editor of *Handbook of Coastal and Ocean Engineering, Seafloor Scour, Design Guidelines for Ocean-Founded Structures, Offshore Pipelines: Design Elements*, and *Coastal and Deep Ocean Dredging*.